Cyber-Physical and Gentelligent Systems in Manufacturing and Life Cycle

Cyber-Physical and Gentelligent Systems in Manufacturing and Life Cycle

Genetics and Intelligence – Keys to Industry 4.0

Edited By

Berend Denkena
Tobias Mörke

ELSEVIER

ACADEMIC PRESS
An imprint of Elsevier

Academic Press is an imprint of Elsevier
125 London Wall, London EC2Y 5AS, United Kingdom
525 B Street, Suite 1800, San Diego, CA 92101-4495, United States
50 Hampshire Street, 5th Floor, Cambridge, MA 02139, United States
The Boulevard, Langford Lane, Kidlington, Oxford OX5 1GB, United Kingdom

Library of Congress Cataloging-in-Publication Data
A catalog record for this book is available from the Library of Congress

British Library Cataloguing-in-Publication Data
A catalogue record for this book is available from the British Library

ISBN: 978-0-12-811939-6

For information on all Academic Press publications
visit our website at https://www.elsevier.com/books-and-journals

Publisher: Matthew Deans
Acquisition Editor: Brian Guerin
Editorial Project Manager: Jennifer Pierce
Production Project Manager: Sruthi Satheesh
Cover Designer: Christian Bilbow

Typeset by SPi Global, India

Contents

List of contributors

Sebastian Barton Institute of Material Science, Leibniz Universität Hannover, Garbsen, Germany

Bernd-Arno Behrens Institute of Forming Technology and Machines, Leibniz Universität Hannover, Garbsen, Germany

Berend Denkena Institute of Production Engineering and Machine Tools, Leibniz Universität Hannover, Garbsen, Germany

Martin Bonhage Institute of Forming Technology and Machines, Leibniz Universität Hannover, Garbsen, Germany

Christian Bonk Institute of Forming Technology and Machines, Leibniz Universität Hannover, Garbsen, Germany

Haythem Boujnah Institute of Production Engineering and Machine Tools, Leibniz Universität Hannover, Garbsen, Germany

Bernd Breidenstein Institute of Production Engineering and Machine Tools, Leibniz Universität Hannover, Garbsen, Germany

Dominik Dahlmann Institute of Production Engineering and Machine Tools, Leibniz Universität Hannover, Garbsen, Germany

Quang Huy Dao Institute of Microwave and Wireless Systems, Leibniz Universität Hannover, Hannover, Germany

Christian Demminger Institute of Material Science, Leibniz Universität Hannover, Garbsen, Germany

Marc-André Dittrich Institute of Production Engineering and Machine Tools, Leibniz Universität Hannover, Garbsen, Germany

Jan Friedrich Düsing Laser Zentrum Hannover e.V., Hannover, Germany

Bernd Geck Institute of Microwave and Wireless Systems, Leibniz Universität Hannover, Hannover, Germany

Philipp Gottwald Institute of Product Development, Leibniz Universität Hannover, Hannover, Germany

Thilo Grove Institute of Production Engineering and Machine Tools, Leibniz Universität Hannover, Garbsen, Germany

Bechir Hachicha Institute of Transport- and Automation Technology, Leibniz Universität Hannover, Garbsen, Germany

Rolf Hockauf Institute of Production Engineering and Machine Tools, Leibniz Universität Hannover, Garbsen, Germany

Lisa Jogschies Institute of Micro Production Technology, Leibniz Universität Hannover, Garbsen, Germany

Johann Kiesner Institute of Production Engineering and Machine Tools, Leibniz Universität Hannover, Garbsen, Germany

Daniel Klaas Institute of Micro Production Technology, Leibniz Universität Hannover, Garbsen, Germany

Christian Klose Institute of Material Science, Leibniz Universität Hannover, Garbsen, Germany

Jürgen Koch Laser Zentrum Hannover e.V., Hannover, Germany

Roland Lachmayer Institute of Product Development, Leibniz Universität Hannover, Hannover, Germany

Hans Jürgen Maier Institute of Material Science, Leibniz Universität Hannover, Garbsen, Germany

Irfan Yousaf Malik Institute of Forming Technology and Machines, Leibniz Universität Hannover, Garbsen, Germany

Tobias Mörke Institute of Production Engineering and Machine Tools, Leibniz Universität Hannover, Garbsen, Germany

Iryna Mozgova Institute of Product Development, Leibniz Universität Hannover, Hannover, Germany

Gregor Mroz Institute of Material Science, Leibniz Universität Hannover, Garbsen, Germany

Markus Mücke Institute of Production Engineering and Machine Tools, Leibniz Universität Hannover, Garbsen, Germany

Thomas Neff Institute of Production Engineering and Machine Tools, Leibniz Universität Hannover, Garbsen, Germany

Peter Nyhuis Institute of Production Systems and Logistics, Leibniz Universität Hannover, Garbsen, Germany

Jörn Ostermann Institute of Information Processing, Leibniz Universität Hannover, Hannover, Germany

Ludger Overmeyer Institute of Transport and Automation Technology, Leibniz Universität Hannover, Garbsen; Laser Zentrum Hannover e.V., Hannover, Germany

Melissa Quirico Institute of Production Systems and Logistics, Leibniz Universität Hannover, Garbsen, Germany

Wilfried Reimche Institute of Material Science, Leibniz Universität Hannover, Garbsen, Germany

Bastian Sauthoff Institute of Product Development, Leibniz Universität Hannover, Hannover, Germany

Wieben Scheidel Institute of Product Development, Leibniz Universität Hannover, Hannover, Germany

Christopher Schmidt Institute of Production Engineering and Machine Tools, Leibniz Universität Hannover, Garbsen, Germany

Alexander Seibel Institute of Production Engineering and Machine Tools, Leibniz Universität Hannover, Garbsen, Germany

Benjamin Spitschan Institute of Information Processing, Leibniz Universität Hannover, Hannover, Germany

Oliver Suttmann Laser Zentrum Hannover e.V., Hannover, Germany

Piriya Taptimthong Institute of Micro Production Technology, Leibniz Universität Hannover, Garbsen, Germany

Florian Uhlich Institute of Production Engineering and Machine Tools, Leibniz Universität Hannover, Garbsen, Germany

Christoph von der Ahe Institute of Transport and Automation Technology, Leibniz Universität Hannover, Garbsen, Germany

Philipp von Witzendorff Laser Zentrum Hannover e.V., Hannover, Germany

Maximilian Winkens Institute of Production Systems and Logistics, Leibniz Universität Hannover, Garbsen, Germany

Florian Winter Institute of Production Engineering and Machine Tools, Leibniz Universität Hannover, Garbsen, Germany

Marc Christopher Wurz Institute of Micro Production Technology, Leibniz Universität Hannover, Garbsen, Germany

Preface

Production technology on its way to the 4th industrial revolution

Gentelligent Components in their Lifecycle — Utilization of inheritable component information in product engineering: This is the title of the Collaborative Research Centre 653 (CRC 653). With its research, the CRC 653 has created a substantial basis for an "Industry 4.0" compliant cross-linked production. In 17 subprojects, 40 scientists of the faculties of mechanical engineering, electrical engineering, and computer science of the Leibniz Universität Hannover worked interdisciplinarily on the realization of tomorrow's production. With ten research institutes involved, the CRC 653 depicted the entire process chain of manufacturing.

This book summarizes the results of twelve years of research and addresses not only the scientific community, but also industry professionals. In this book, the benefits of the developed technologies are illustrated by examples of application. For industry professionals, concrete suggestions for realizable implementations of the developed methods are offered. For the scientific community, contributions to the individual technologies provide technical bases for further development and additional research potential is pointed out.

Turning the CRC's vision into reality required fundamental research in the fields of *Sensors, Data Storage and Communication Technologies*. This first group of themes presents developments utilizing electric and magnetic effects as well as changes of a component's microstructure and geometric features to merge component, sensor, and data storage to an inseparable entity. The benefits of advanced sensors along with components, carrying a summary of all relevant information on their production is described in the second group of themes: *Applications in Manufacturing*. Two concrete applications are presented pointing out the use of gentelligent components for process planning and control in a networked production by means of a Virtual Planner as well as the potential of intelligent manufacturing systems. The third group of themes, *Applications in the Product Life Cycle*, demonstrates the benefits of technologies able to sense mechanical loads of components for a condition-based maintenance as well as a design optimization on the basis of real load data from the component's life cycle. Finally, five *Examples of Application* along the value chain of a product are presented to illustrate the potential of gentelligent components and to verify the practical benefits due to the developed technologies.

The support of the German Research Foundation over a period of twelve years has made possible the results of the research that you find in this book. The conveyance of this supported fundamental research for industrial usage is indicatory for the cross-linked production of "Industry 4.0." As the speaker of CRC 653, I would like to thank the DFG very much for its support — also in the name of all those who participated in these projects.

Prof. Dr.-Ing. Berend Denkena
(Speaker CRC 653)

Introduction

1

How has digitalization changed your life over the last 15–20 years? The internet, mobile communication, and social media are ubiquitous. Today, one almost takes for granted that information is available on smartphones and can be exchanged at any time. Furthermore, sensors and control systems make our daily life easier. Embedded assistance systems in automobiles, for example, are taking on more and more tasks. Proximity control systems and automatic parking, which are almost the norm in new vehicles, are just two examples. The self-driving car is no longer merely a vision. The old fashion computer itself is increasingly disappearing and being replaced by intelligent devices. Digitalization fundamentally determines the way we live today, but how has digitalization changed production engineering at the same time? Most of the key advancements regarding our daily life are still visions of tomorrow's production. Smart factories, smart products, and the fourth industrial revolution stand for a world in which components and machine tools are able to determine and communicate their condition and make decisions. So far, it is a vision that detailed information on the production process is available and can be exchanged at any time. Assistance systems which autonomously change a component's production route due to current events are not yet available.

"Industry 4.0" is characterized by three main aspects. First, all relevant information on a product is available by cross-linking all elements of its value chain. Second, the entire value chain of a product from its production throughout its life cycle is organized and controlled a novel way. And third, changeable and self-optimizing value chains are built by the interaction of people, product, and technology. Therefore the key issue is: What are the key features of components and machines that enable the realization of smart components, smart factories, and an "Industry 4.0"?

Since 2005, the vision of inherently connecting components and information and enabling these components to make decisions and communicate has been driving scientists at the Collaborative Research Centre 653—Gentelligent Components in their Lifecycle. Promoted by the German Research Foundation, the CRC 653 is striving for components that will monitor their conditions independently, know their life expectancy and, when necessary, autonomously initiate an inspection. Accidents caused by untimely fatigue failure as well as expensive product recalls, like in the automobile industry, could be avoided. Its realization will influence tomorrow's production in a sustainable way. With biology as its basis, the term "gentelligent" was created. CRC 653 identified the combination of genetic and intelligent properties as key features of a technical component and machines. Gentelligent components will thereby be enabled to collect, store, and communicate information during their life cycle. With that, the basis for transferring the principles of biology will be created in terms of passing on information to subsequent component generations and lifelong learning. The genetic information of a component is fundamental for its identification or reproduction, e.g., its geometry and workpiece material. This information is stored

Cyber-Physical and Gentelligent Systems in Manufacturing and Life Cycle. http://dx.doi.org/10.1016/B978-0-12-811939-6.00001-7

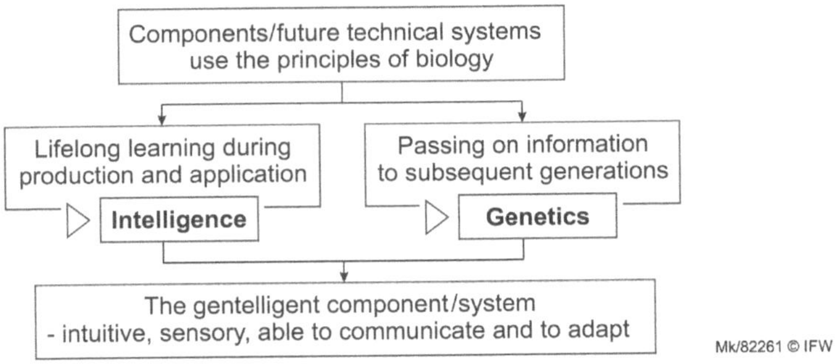

Fig. 1.1 Gentelligent components follow the principles of biology.

as static and unalterable data in the component. Besides that, the component is enabled to contain information about its own production. The intelligence of gentelligent components is created by their technical ability to autonomously and inherently store and process information such as the effects of mechanical and thermal load during their production and life cycle. This is achieved by using the appropriate materials and sensor technologies, which are integrated into the component. The entirety of information is inherently connected to the gentelligent component and always retrievable. Thus a gentelligent component is identified by its inherent sensorial properties and its ability to inherently store data and communication (Fig. 1.1).

Gentelligent technologies cancel the separation of components and information. They make it possible to gather, store, read, combine, and use information throughout the production as well as the life cycle of a product. New possibilities to improve the production, usage, maintenance, and design of future components result from these new features.

1.1 Development of sensors, data storage and communication technologies

In order to implement the approach of gentelligent components and machines, the CRC 653 has developed advanced sensor, data storage, and communication technologies. Components are supposed to gather and store information during their manufacturing as well as their application. The CRC demands the provision of these advanced properties by means of the manufacturing process rather than by attaching additional sensors, data storage components, or communication devices. Necessary features are inseparably merged with the component itself. For the first time, deterministic and stochastic geometric features of a component's surface have been made useable for data storage and component identification (Chapter 2.1). Furthermore, the selective manipulation and monitoring of the components subsurface bears the potential of storing data and detecting load (Chapter 2.2). To enable the employment of magnetic effects for lightweight materials a novel magnetic magnesium alloy has been developed (Chapter 2.3). By integrating microsensors into the component, electrical effects are used to realize the CRC's vision of "feeling" components (Chapter 2.4). Concluding, integrated communication systems that enable the data exchange within gentelligent systems and components have been developed (Chapter 2.5).

Therefore geometrical and subsurface properties as well as magnetic and electrical effects are utilized to store data within a component. Furthermore, detailed information on a component's manufacturing and life cycle is available. Since a component is capable of identifying itself and bearing information on its production, new possibilities have arisen with respect to the planning of its manufacturing process. Moreover, the direct feedback of detailed process information and the condition of available machine tools provide vital information for an advanced controlling of manufacturing processes.

1.2 Applications in manufacturing

The efficient planning and controlling of manufacturing processes require—with an increasing emphasis on economics—more information about the production stage. Especially when manufacturing lot size one, the planning of manufacturing processes is costly and requires lots of data due to the constantly changing product portfolio. In order to satisfy this hunger for information, all the data that had accrued in companies anyway but not been used so far was processed at CRC in the first instance. The data from the control system of a modern machine tool, for example, could be used to make BDE terminals superfluous. The current number of orders can be recorded to the split second and reported back to the production planning by using a combination of the NC program and the program's progress. Where the available data is insufficient, this information gap can be closed by introducing additional sensors. The availability of a significantly increased amount of data requires appropriate process algorithms to unfold its potential. For example, the quality of process planning and the production control will be significantly increased by recording and processing the data of past work processes to the split second. High productivity and product quality are crucial in order to survive in international competition. A challenge for some companies is the increasing individualization of production. In the single-item and small-lot production, it leads to a very large variety of versions and to a high number of possible processes and process chains. Retracing current status information from the tool machines into the monitoring of the process, for example, enables a teachless observation process when manufacturing single components.

Focusing on the feedback of process information, it is demonstrated how intelligent manufacturing systems can be used as a key to optimize production processes. Machine tools and components are being equipped with integrated sensors and communication systems distributed in the machine in a way which is similar to human nerves and nerve tracts. Inspired by a human who holds a tool in one hand and a component in the other the machine "feels" the conditions of the process, the component, and the tool. The information can be retraced, which makes it possible for the production to react autonomously to events which require the individual rescheduling and adaptation of procedures and work processes. Retracing information has an enormous potential to increase the quality and safety of production processes while increasing productivity at the same time. The "feeling" machines developed in CRC provide detailed information about the condition of the machine and the process. With that, the "feeling" machines enable an independent adaption of production parameters. The sensors developed by the CRC show a higher sensitivity and are therefore suitable for usage in the stiff structures of tool machines. Data exchange between machines, tools, and components are made possible by the CRC's communication technologies. The information gained from the manufacturing process of the "feeling" machine is processed directly. These data flow into a self-learning regression model aimed at minimizing the rejection rate at consistent maximum productivity.

1.3 Applications in the product life cycle

Information on loads of a component during its life cycle can be made useful for the optimization of the maintenance plan and component design. Components with sensitive properties, which can store information while being used, offer an enormous potential. Real data from the utilization phase of a component replace models and simulations: Manufacturing costs can be minimized, a component optimization, a maintenance plan, and a life expectancy prognosis based on actual mechanical load can be carried out. Compared to conventional components, the introduced components are equipped with additional functions. Among others, they record and store mechanical load during operation. As a demonstration, for example, a load-sensitive wheel carrier was produced by using a newly developed magnetic magnesium alloy. It records load-induced stress and stores the information in the material itself. Additional measuring and storage technology is not necessary. Furthermore, the integration of an eddy current sensor enables a dynamic recording of the loads. With that, the obtained data is available for subsequent analysis routines. They are, for example, used for a component status-driven maintenance: Optimal times for the maintenance of components can be derived from that. On the one hand, this enables a high exploitation of the rest potential and, on the other hand, it minimizes the risk of production cancelations. In this way, cost-intensive, spontaneous cancellations can be prevented. Furthermore, the mechanical load information can be used when designing future component generations. With a genetic algorithm, they are assessed for the optimal adaptation of the component design, adapted according to their real load stresses experience and developed under consideration of various target criteria, such as lightweight construction and stiffness.

1.4 Examples of application

An exemplary value chain has been realized to illustrate the potential of gentelligent components and verify the practical benefits due to the developed technologies. The wheel carrier of a race car was selected as an exemplary component since it represents a small-lot production component, which experiences high loads during its application and whose production involves several manufacturing processes. The scenario

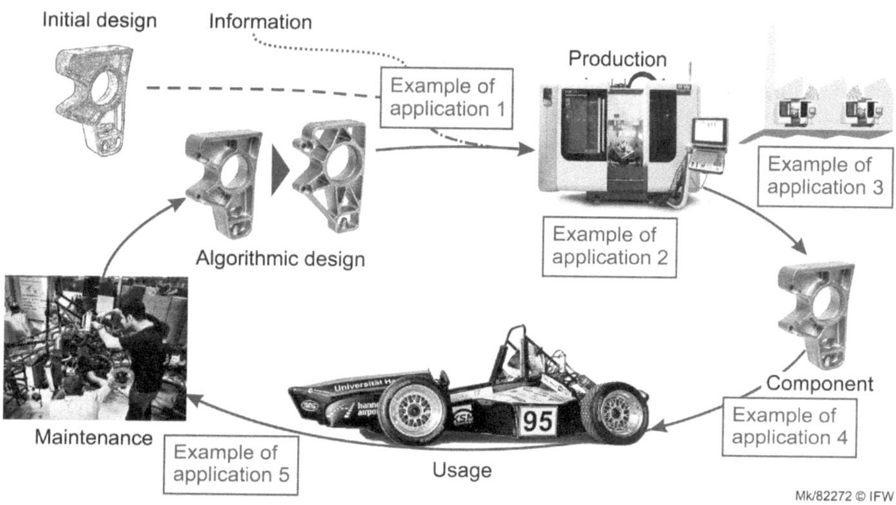

Fig. 1.2 Placement of the application examples in the production cycle.

illustrated in Fig. 1.2 covers the design, production, usage, and maintenance. Five examples of application demonstrate the possible deployment of developments by the CRC. Starting with an initial design, the unmachined part enters the production cycle. It carries the summary of information on its production using one of the CRC data storage or communication technologies (example of application 1).

The following examples of application cover the production cycle including the optimization of machining processes (example of application 2), the work and process planning (example of application 3), as well as the life cycle of the product. During application, the data storage technologies of the CRC enable the identification of the component at any time (example of application 4). Combined with the capability of sensing mechanical loads, future product design and a condition-based maintenance planning are put into practice (example of application 5 is given in Chapters 4.3). Algorithmic Design Evolution is based on an analysis of product life cycle information and consists of multiple steps, which are descripted in detail in Chapter 4.2. All examples mentioned in the book have been realized within the framework of CRC.

The research work and the results of CRC 653 bound in this book represent a considerable contribution to future production. In 2005 the CRC 653 created a vision of inherently connecting components and information and the enabling of these components to make decisions and communicate. Even now, this vision goes beyond today's demands of "Industry 4.0." Gentelligent technologies physically merge components and information. They make it possible to gather, store, read, combine, and use information, not only during the production but also throughout the product's life cycle. Furthermore, passing on information from one generation to another enables feedback into product design and production.

In the meantime, "Industry 4.0" has become a key word worldwide for the economic development and the increasing integration of digital information. Its corresponding physical component has found its way into industry. The work described in this book shows the benefits of transferring the CRC 653 research results into practice. At the same time, the collective contributions highlight the enormous potential of research for digital and cross-linked production as well as for the remaining gaps still to be overcome by research.

Sensors, data storage and communication technologies

2

Gentelligent components are able to gather, store, read, combine, and use information throughout the production as well as the life cycle of a product. The CRC 653 demands to provide these advanced properties by the manufacturing process rather than to attach additional sensors, data storage components, or communication devices. Necessary features are inseparably merged with the component itself. Thereby, gentelligent technologies cancel the separation of components and information.

In the following, innovative sensor technologies developed by the CRC 653 are described, which overcome the physical separation of sensor and component by using, for instance, electrical and magnetic properties as well as properties of the component's microstructure (see Table 2.1).

Novel approaches of manufacturing microsensors have led to a significant increase of sensitivity and accuracy compared to commercial sensors. One development deposits sensor layers on thin and flexible polymer foils. The rigid carrier substrate lapses and the sensor is positioned closer to the stressed surface, whose strain is to be determined (Table 2.1, a). For depositing sensors directly on large components, the CRC has developed a mobile coating system. It allows the direct deposition of sensors on components of arbitrary size (Table 2.1, b). Another approach can be taken if the coating of the component with a layer system is

Table 2.1 Sensor technologies of the CRC 653

	Sensor technology	Used effect	Type	Chapter
a	Sensor deposition on thin flexible polymer films	Electrical properties	Reversible	2.4.1
b	Direct deposition of sensors on technical surfaces	Electrical properties	Reversible	2.4.2
c	3D laser pattering of layer systems	Electrical properties	Reversible	2.4.3
d	Inverse magnetostriction	Magnetic properties	Reversible	2.3.2
e	Local heat treatment of metastable austenitic materials	Microstructural properties	Irreversible	2.2.2
f	Residual stress relaxation	Microstructural properties	Irreversible	2.2.1

Cyber-Physical and Gentelligent Systems in Manufacturing and Life Cycle. http://dx.doi.org/10.1016/B978-0-12-811939-6.00002-9

practicable. If that is the case, a laser removal process can be applied to directly manufacture sensors onto the component's surface (Table 2.1, c). Aside from electrical effects, the technologies developed within the CRC exploit magnetic effects as well as changes of the material's microstructure to merge component and sensor. Lightweight components should be enabled to "feel" mechanical load stress in order to gather information on a component's loading during its life cycle. Therefore a magnetic magnesium alloy has been developed. It possesses this capability by means of inverse magnetostriction (Table 2.1, d). Additionally, the CRC's research goes deeper into the material's microstructure. Load stress-induced phase transformations (Table 2.1, e) as well as the relaxation of residual stresses (Table 2.1, f) are made accessible to gather information on the load stresses gentelligent components experienced. By using these effects, sensory properties are given to different metallic alloys. Table 2.1 summarizes the advanced sensor technologies developed.

To make the second key feature of gentelligent components possible—the ability to remember—data storage technologies were developed. By using geometrical and magnetic properties as well as properties of a component's microstructure, data storage and component are physically merged (Table 2.2).

The CRC uses magnetic effects to realize a random accessible memory. A heat-assisted magnetic recording procedure following the state-of-the-art hard drives enables the data storage within magnetic materials. Structures are magnetized in patches and, by doing so, the coding of information in a binary system is performed (Table 2.2, a). Additionally, three extremely robust methods were implemented, using geometrical properties as well as properties of the material's microstructure to merge components and their corresponding data as a read-only memory. Piezo-actuated turning and milling tools are applied to cut micro patterns within the component's surface representing complex graphics

Table 2.2 Data storage technologies of the CRC 653

	Data storage by	Used effect	Type	Chapter
a	Heat-assisted magnetic recording	Magnetic properties	Random access memory	2.3.3
b	Milling of micro patterns	Geometric properties	Read-only memory	2.1.2
c	Sintering using foreign particles	Microstructural properties	Read-only memory	2.2.4
d	3D data pattern by local heat treatment	Microstructural properties	Read-only memory	2.2.3

or coded data (Table 2.2, b). Storing information into the component's volume by means of a sintering process is carried out by introducing foreign powder during the primary shaping process (Table 2.2, c). Three-dimensional data matrix codes are created within the component by the use of local heat treatment (Table 2.2, d). These technical procedures are extremely robust against mechanical and chemical alteration. Therefore they are perfectly suited for a component identification in a harsh production environment and meet the requirements of a heavy duty use during a component's life cycle. Table 2.2 summarizes the developed approaches.

Besides having a central data storage on server systems, the CRC strives for the decentralized storage of data on the component itself. The potential of the developed sensor and data storage technologies in a production environment unfolds by merging advanced communication technologies and machine tools or components. The CRC has developed the following optical and radio frequency communication (Table 2.3).

High speed energy and data transmission within machine tools has been built up by optical signal coupling. For this, optical waveguides are induced into the surface of metallic structures (Table 2.3, a). As an alternative, highly flexible optoelectronically integrated radio frequency communication modules have been developed. They are solar powered and capable for operation framed by metallic structures (Table 2.3, b).

The following subchapters give detailed information on the sensor, data storage, and communication technologies developed by the CRC 653. They are structured following the underlying effects. Chapters 2.1 and 2.2 include technologies that employ geometrical and subsurface properties. The employment of magnetic effects for lightweight materials (Chapter 2.3) required the development of a novel magnetic magnesium alloy. Detailed information on sensor technologies based on electrical effects is given in Chapter 2.4. In conclusion, Chapter 2.5 presents integrated communication systems enabling data exchange within gentelligent systems and components.

Table 2.3 Communication technologies of the CRC 653

	Communication by	Used effect	Chapter
a	Energy and data transmission by optical signal coupling	Optical transmission	2.5.1
b	Optoelectronically integrated radio frequency communication	Wireless transmission	2.5.2

2.1 Employment of geometrical properties

In this chapter it will be demonstrated how geometrical properties of a component's surface can be used for its distinct identification, and how a component's surface can be used as a data storage unit.

The surface of a component is defined as the interface between its material structure and the surrounding area. It determines many component quality indicators such as optical, tribological, or haptical properties. Most of the technically used components obtain their surface by machining processes. Either a surface is generated with a geometrically defined or with a geometrically undefined cutting edge, the created surface can be described by deterministical and random properties. On the one hand, deterministical effects are mainly caused by the macro and micro geometry of the inserted cutting tool and the kinematic of the machining process (see Fig. 2.1.1, top). On the other hand, random effects are caused by the ductility of the machined material, the wear behavior of the cutting tool, or irregularities during the machining process like vibrations. In general, the proportion of random effects increases with the geometrical coefficient of determination of the tool. A grinding tool with a ceramic bonding changes its topography after each revolution during the cutting process by microscopic wear. Thus the generated surface has a high proportion of random effects (see Fig. 2.1.1, bottom). Galvanically bonded grinding tools or inserts for turning and grinding applications, in turn, generate surface topographies which are geometrically defined to a high proportion.

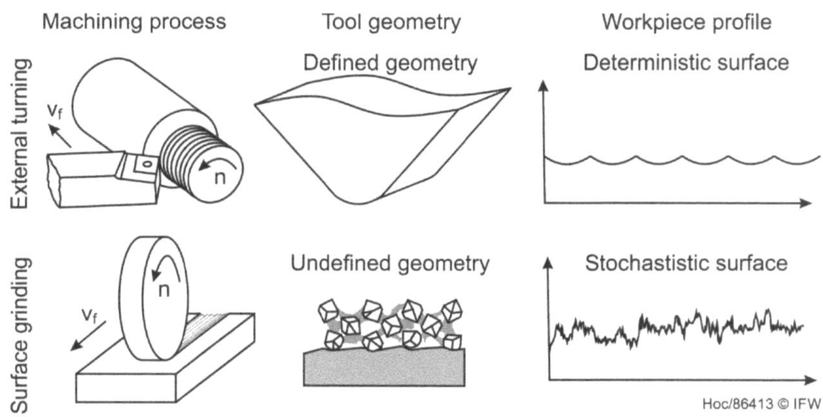

Fig. 2.1.1 Surface generation in grinding and turning.

The CRC 653's research shows that cutting processes innately provide all prerequisites to inseparably merge a component and corresponding data. Within this chapter, two approaches are presented demonstrating how a component's surface topography is enabled to "store" relevant information. This chapter presents a method which utilizes the random effect of a component's surface as a plagiarism security or self-authentication system. This technology is able to read and identify the random effect of a component. Unique features gained from single roughness profiles of the surface are used to generate an individual fingerprint. Without any additional marking (such as bar code, RFID, or hologram) or further manufacturing processes a plagiarism security and self-authentication is enabled (see Chapter 2.1.1). After the introduction of passive usage of surface effects an active technology for surface structuring is presented using the deterministic surface parts. A new generation of piezo-actuated turning and milling tools is developed to induce information in form of micro patterns into the component's surface representing binary codes. Random data can therefore be cut into the surface of the component. In addition to the machining technology, a reading method to select and decode the turned and milled structures is presented (see Chapter 2.1.2).

See the following chapters for further information

2.1.1 Component identification by means of unique topography features
2.1.2 Data storage within a component's surface by cutting of micro patterns

2.1.1 Component identification by means of unique topography features

Bernd Breidenstein, Tobias Mörke*, Rolf Hockauf*, Jörn Ostermann†, Benjamin Spitschan†*

*Institute of Production Engineering and Machine Tools, Leibniz Universität Hannover, Garbsen, Germany, †Institute of Information Processing, Leibniz Universität Hannover, Hannover, Germany

2.1.1.1 Generation of generically machined component surfaces

In this chapter, a classification of deterministically/periodically and stochastically/randomly occurring surface effects will be made. These effects can be used for a future identification scheme of components for plagiarism detection or for identification (Auto-ID). In general, surface deviations between geometrical surface (drawing requirement) and measured surface occur which can be divided into 6 orders. In Fig. 2.1.1.1 the first four deviation orders are depicted. The combination of these 4 orders leads to the surface roughness of a real component topography.

The first order describes deviations due to wrong clamping of the workpiece, a tilting of the guidance system, or a run out of the machine spindle. The second

Surface diviations	Turning	Plane grinding
Order 1: 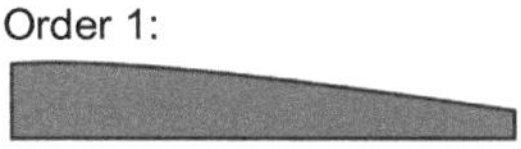	Skewness, curvature, form, shape	
Order 2: 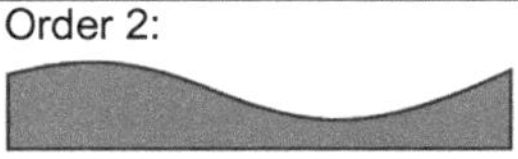	Waviness, chatter marks	
Order 3: 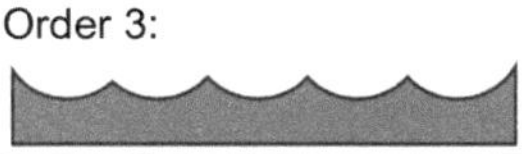	Feed marks	Grain grooves
Order 4: 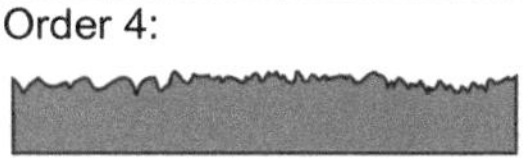	Scale, side flow, burr, groove, dimple, slip steps	Grain grooves

Hoc/86401 © IFW

Fig. 2.1.1.1 Surface deviation according to DIN 4760 [1].

order of surface deviation is caused by vibrations in most cases. The vibrations generate a variable depth of cut, resulting into an undulating surface. In case of a self-created vibration the wavelength on the surface changes due to the natural frequency. In the following two sections the generation and the created effects on turned and plane ground surfaces will be discussed. The deviations of the third, fourth, and fifth order describe different gradations of roughness. The gradation of roughness is defined as the relation between groove distance and its depth. The relation in orders three to five has to be between 100:1 and 5:1. The fifth and the sixth order describe the microstructure of a material and therefore it is difficult to depict them in a figure.

2.1.1.1.1 Generation of turned surfaces

Turning is a machining process in which a geometrically defined cutting tool, typically a nonrotary tool insert, describes a helical tool path by moving in feed direction while the workpiece rotates. The generation of turned surfaces and surface effects are influenced by the machined material as well as the tool and process parameters. For a classification of surface effects caused by geometrically defined cutting edges the effects are divided into periodically and randomly occurring ones.

2.1.1.1.1.1 Periodical effects on turned surfaces

Periodical effects are surface deviations of 3rd and 4th order, which can be specifically introduced by using the same initial parameters for different workpieces. The characteristic deviations for a turned surface are helical feed marks. Due to the feed *f* and the corner radius r_ε the feed marks get a theoretical shape as it is described in Fig. 2.1.1.2A.

These feed marks belong to the third order of surface deviation. If the tool wears over time, outbreaks, scars, and cracks can be found on the flank face of the tool which creates a negative imprint in the feed marks (see Fig. 2.1.1.2B). These 4th-order

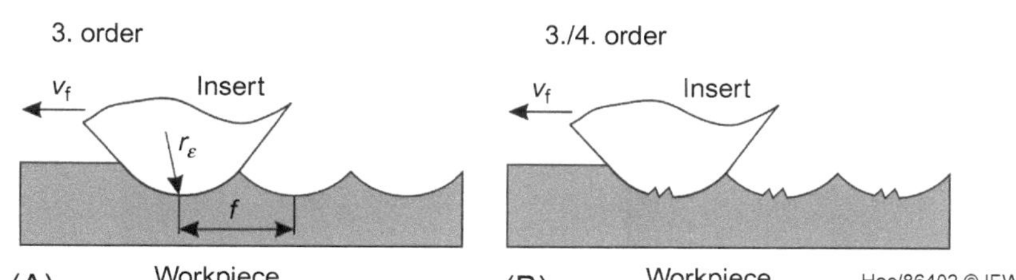

Fig. 2.1.1.2 Generation of feed marks by turning [2,3].

effects can be considered as periodical over a period of time. Beside the wear behavior, scars in the feed mark can be caused by chipping in the coating [2,3].

2.1.1.1.1.2 Random effects on turned surfaces

Random effects are surface deviations of 4th and higher order which cannot be repeatedly introduced in a workpiece's surface and occur irregularly. Their generation mainly depends on the machined workpiece material and its heat treatment. For this article all investigations were performed with AISI 1060 steel in delivery state. A typical effect for ductile materials is called side flow. During the chip formation a plastic deformation of workpiece material opposite to the feed direction takes place at the secondary cutting edge. With increasing feed and depth of cut the ridges of the feed marks become higher because of the plastic deformation (Fig. 2.1.1.3A) [4,6,7]. By increasing the cross-section of an undeformed chip again, the ridge cracks and builds burrs (Fig. 2.1.1.3B) [5,8]. These burrs are material leftovers of the chip formation. From a cross-section of an undeformed chip of 1.5 mm^2 the ridge begins to tear off partially (Fig. 2.1.1.3C). The reason is the cohesion in the chip which tears off the thin ridge. In investigations with cutting speeds of 50, 200, and 300 m/min the decisive factor is the cross-section of the undeformed chip.

Another surface deviation of the 4th order is the irregular creation of scales. Scales are structures which occur in the feed marks ground at low cutting speed because of strain hardening of chip particles. The accumulation of hardened particles creates a built-up edge (BUE) which acts as the extension of the cutting edge. As shown in Fig. 2.1.1.4 the depth of cut is increased to the required cutting process and the flank face has no contact to the workpiece material. Due to instabilities small particles of the BUE are detached from the cutting edge and crushed under the flank face. The crushed material is welded with the workpiece and builds a

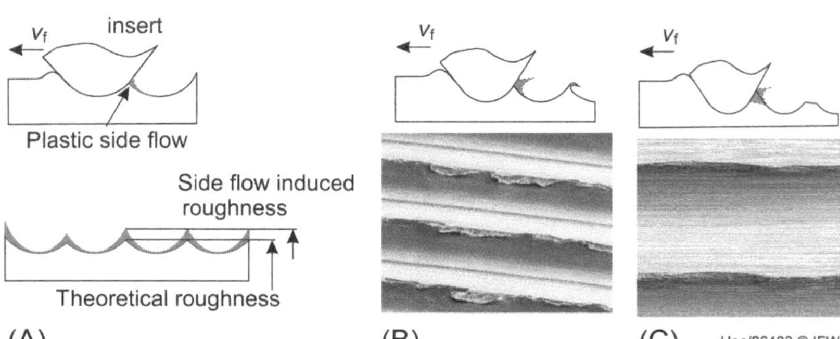

Fig. 2.1.1.3 Random effects by side flow [4–8].

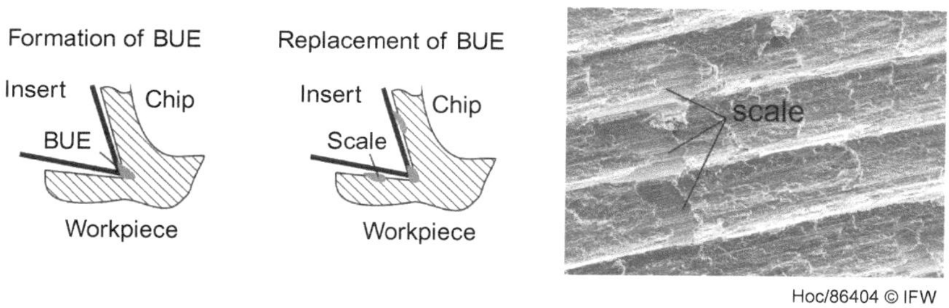

Fig. 2.1.1.4 Formation of scales by BUE [9].

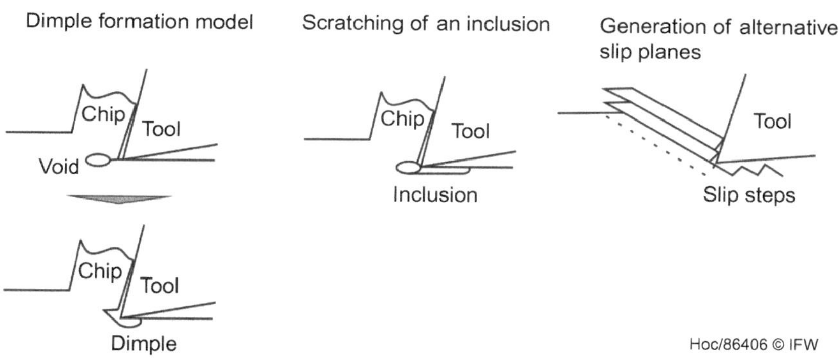

Fig. 2.1.1.5 Micro deviation by turning [2,10,11].

scale [9]. The main factors for the generation of a BUE are the cutting speed and the temperature in the contact zone. Shaw et al. describe the "no built-up edge boundary" at 750°C, 150 m/min at feeds of 0.05, 0.1, and 0.2 mm. Above these values the generation of BUE is implausible. Negative rake angles cause BUE more likely than positive ones [10].

Material inhomogeneity can also be a reason for random deviations. Inhomogeneity manifests itself in the form of voids and inclusions in the microstructure of the material. These may cause the generation of dimples or inclusion scratches and also alternative slip planes (Fig. 2.1.1.5). In the case of slip steps these are produced because deformation is energetically more favorable in certain glide planes. The shape of these steps depends on the orientation of the crystalline structure. Slip steps, inclusion scratches, and dimples are nontypical surface marks that produce undesirable effects in application [11,12].

2.1.1.1.2 Generation of flat ground surfaces

Flat grinding is a machining process with geometrically undefined cutting edge in which a rotationally symmetric tool consisting of grains, bonding, and pores performs a linear feed motion over a flat workpiece while spinning around its rotatory axis. The generation of ground surfaces largely depends on the type of grinding tool that is used. The most commonly used abrasives are diamond, cubic boron nitride, and corundum. The combination with the three main bonding types ceramic, metallic, and synthetic resin results in different amounts and sizes of pore and grain protrusion. In contrast to machining with geometrically defined cutting edges the wear behavior and the resulting change of tool topography are main factors for the description of surface generation in grinding. In addition, the hardness of the machined material as well as process parameters also influence the surface. A classification of surface effects into periodical and random is difficult without knowing the grinding wheel topography. Therefore investigations with galvanically bonded tools were performed. Laser triangulation is used to obtain the entire tool topography by scanning the complete circumference. By summing up the maximum points of the topography in tool width direction an envelope profile results. Due to the high rotational speed of tools in grinding the envelope profile shows a high conformance to the real workpiece profile (Fig. 2.1.1.6)

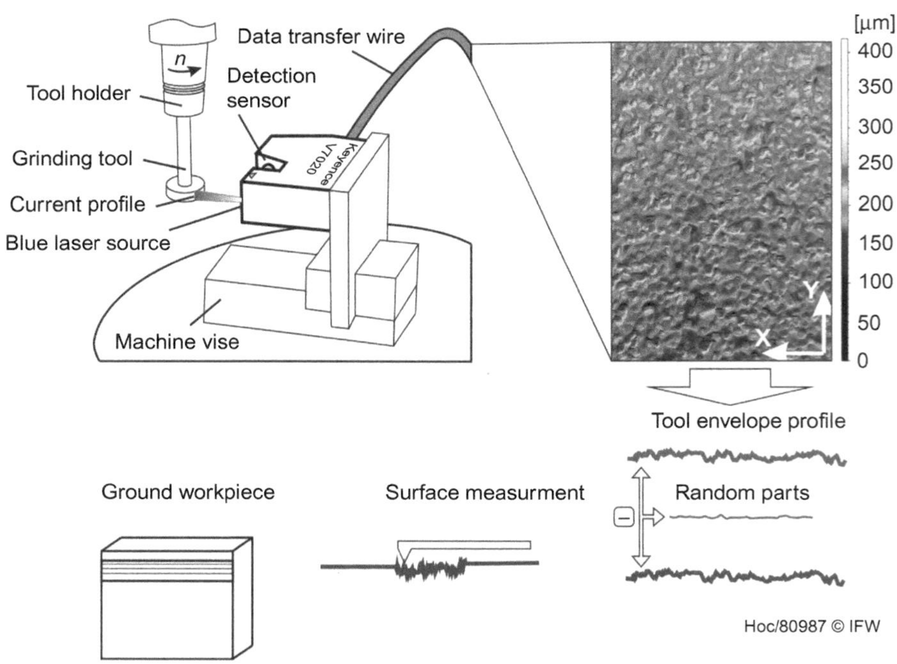

Fig. 2.1.1.6 Setup for measuring grinding wheel topography.

[13]. The remaining difference between envelope and workpiece profile is caused by random wear effects of the grinding wheel or a brittle separating behavior of the workpiece material.

In the contact zone between workpiece and grinding tool, high temperatures and pressures on grains and bonding occur. Due to the extreme conditions in the grain surface, oxidation and diffusion processes reduce the resistance against abrasion. The first signs of wear are initial cracks which cause the outbreak of grain particles. In other areas the grain gets dull by rounding or deposits of workpiece material. Those grains have a bad cutting behavior and increase pressure and temperature in the contact zone. Subsequently, the surrounding grains and their bonding are damaged which can cause bonding cracks and outbreaks of grains as well as entire grain groups. These wear effects, depicted in Fig. 2.1.1.7, have to be considered in process design [14–17]. Due to the tool specification the wear behavior increases the random character of the generated surface. Especially for galvanically bonded tools the description of wear effects becomes possible with the approach given earlier. This is due to the high grain protrusion and retention forces as well as a low grain wear. As an example the outbreak of a B151 grain is shown in Fig. 2.1.1.8.

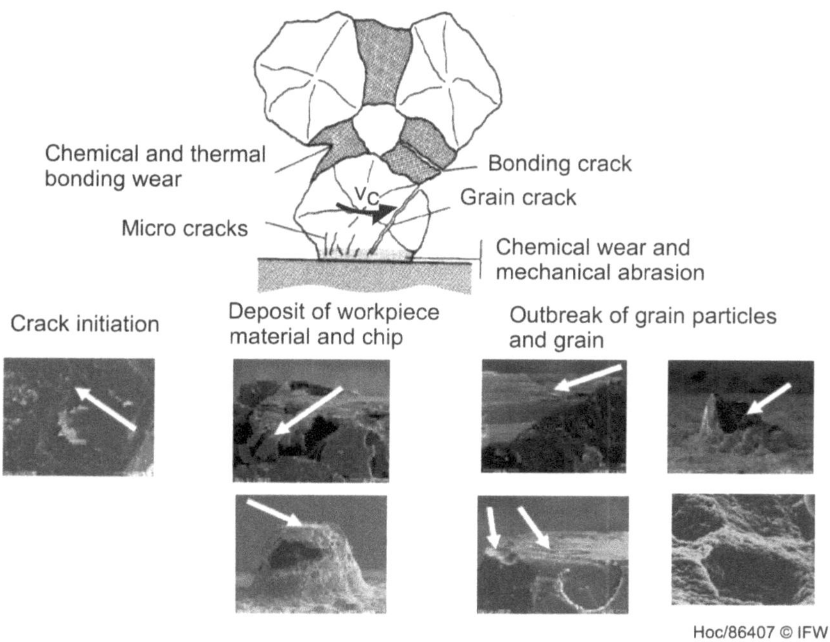

Fig. 2.1.1.7 Wear effects of grinding tools [14,15].

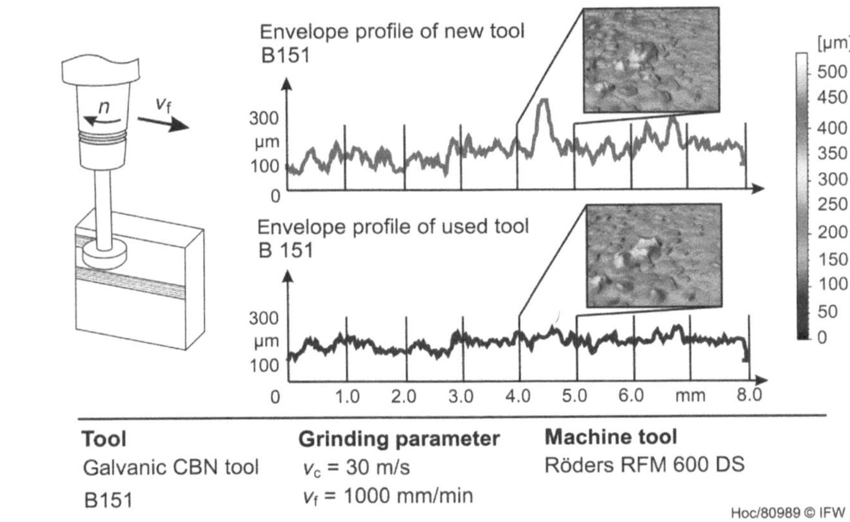

Fig. 2.1.1.8 Detection of grain outbreaks of a B151 grinding tool [13].

2.1.1.2 Fingerprinting of machined surfaces

As seen in Chapter 1, machined surfaces generally exhibit a certain amount of randomness which manifests itself as spatial fluctuation of the surface topography. The main idea of the method depicted in this article relies on the fact that this random content is unique to every individual component, i.e., to every single instance of a component. In this case, the topography is the unique realization of a random process and may serve as a physical unclonable function (PUF).

2.1.1.2.1 Physical unclonable functions (PUFs)

Physical unclonable functions (PUFs) address the general problem of any ex-post plagiarism protection scheme. Ex-post protection is the retrospective detection of plagiarism in contrast to constructive means of prevention. Such a scheme commonly employs markers that the plagiarist is either not aware of, i.e., they need to be sufficiently hidden or concealed, or that he is technologically not able to reproduce. Previous approaches include the use of particular isotopes in copper cables, ceramic microparticles that are cast into metals or plastics, or pigments with distinct spectral reflectances [18]. As soon as, however, these requirements are not fulfilled anymore, the protection scheme is void. The solution is to use such markers that are first distinct for each individual part, and second can in no way be reproduced. PUFs can act as such markers. They usually consist of a stochastic random source. Unlike true (physical) hardware random number generators based upon quantum phenomena, PUFs yield the same unpredictable—but repeatable—signal/sequence in each read-out round.

In order to analyze surface profiles with high stochastic content toward their suitability, the autocorrelation function (ACF)

$$\mathrm{ACF}_{h(x)}(u) = \int_{-\infty}^{\infty} h(x+u)h(x)dx \tag{2.1.1.1}$$

of a profile $h(x)$ was used, with u denoting the lag, i.e., the displacement between the profile and its shifted version. Commonly, the ACF is normalized in order to eliminate the influence of signal length and different measurement units. Figs. 2.1.1.9 and 2.1.1.10 show an example of a profile signal and its normalized ACF, respectively.

Apart from some low-frequent waviness, the energy is well compacted around the zero-lag coefficient ($u = 0$ mm) of the ACF. Experiments showed that in the vast amount of cases, carefully conducted surface scans exhibit this energy concentration, indicating a high degree of randomness of the signal. They may therefore serve as PUFs for practical purposes.

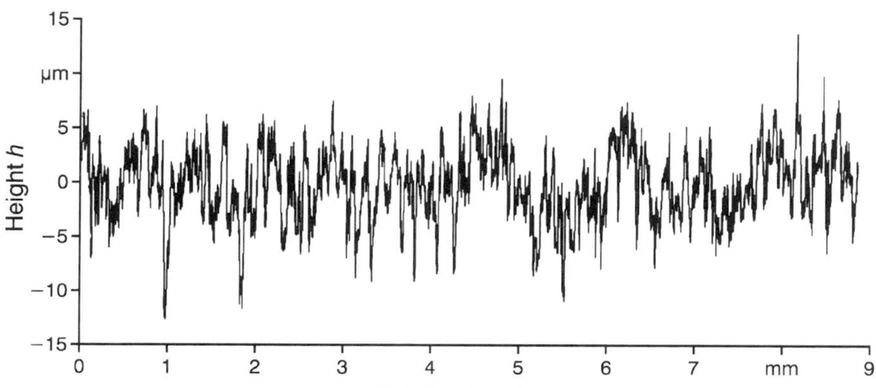

Fig. 2.1.1.9 One-dimensional signal of surface profile $h(x)$.

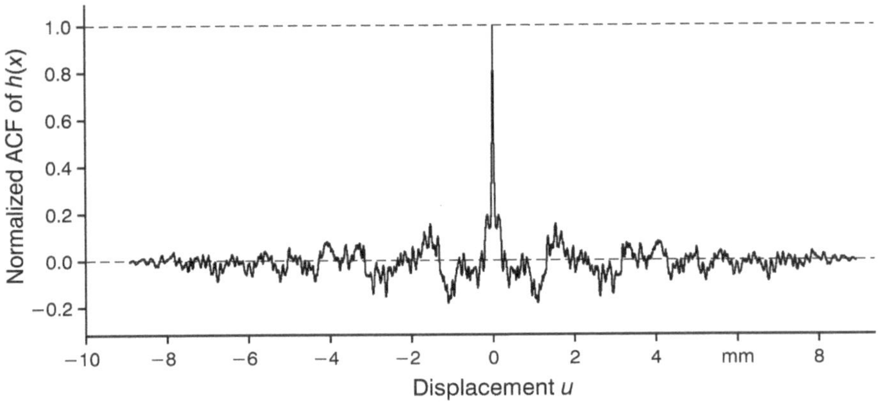

Fig. 2.1.1.10 Normalized autocorrelation function $\mathrm{ACF}_{h(x)}(u)$ of profile $h(x)$.

2.1.1.2.2 The location-dependent spectrum

In order to match a particular roughness profile against another one or even against an entire database, a conformance metric must be used. The naïve approach is to use the normalized cross-correlation (NCC), i.e., the cross-correlation of two profile signals $h_1(x)$ and $h_2(x)$ centered around their means $\bar{h}_1$, $\bar{h}_2$ and normalized to their standard deviations σ_{h_1}, σ_{h_2} according to

$$\mathrm{NCC}_{h_1(x),h_2(x)} = \int_{-\infty}^{\infty} \frac{(h_1(x) - \bar{h}_1)(h_2(x) - \bar{h}_2)}{\sigma_{h_1}\sigma_{h_2}} dx \tag{2.1.1.2}$$

An exemplary comparison using the NCC of two surface profiles recorded by confocal microscopy is depicted in Fig. 2.1.1.11. First, a profile h_1 is matched against a second, entirely different profile h_2 (top). The maximum of the NCC is at a low level of 0.12. Second, the profile h_1 is matched against another recording h'_1 of that profile that has been measured directly after the first scan (bottom). A high degree of conformance can be determined by calculating the NCC. Differences in the two profile scans result from measurement imperfections such as sensing noise.

However, this only works well under strict geometrical and mechanical constraints. The NCC is highly susceptible to missing data, i.e., surface damages caused by scratches, corrosion, sensing errors, or differences in profile length, and exhibits invariance neither to scaling nor to translation. Therefore a more robust scheme for profile comparison must be devised.

A frequent solution to pattern recognition tasks is to draw analogies to approaches made by humans. In the case of matching two profile scans, a common human

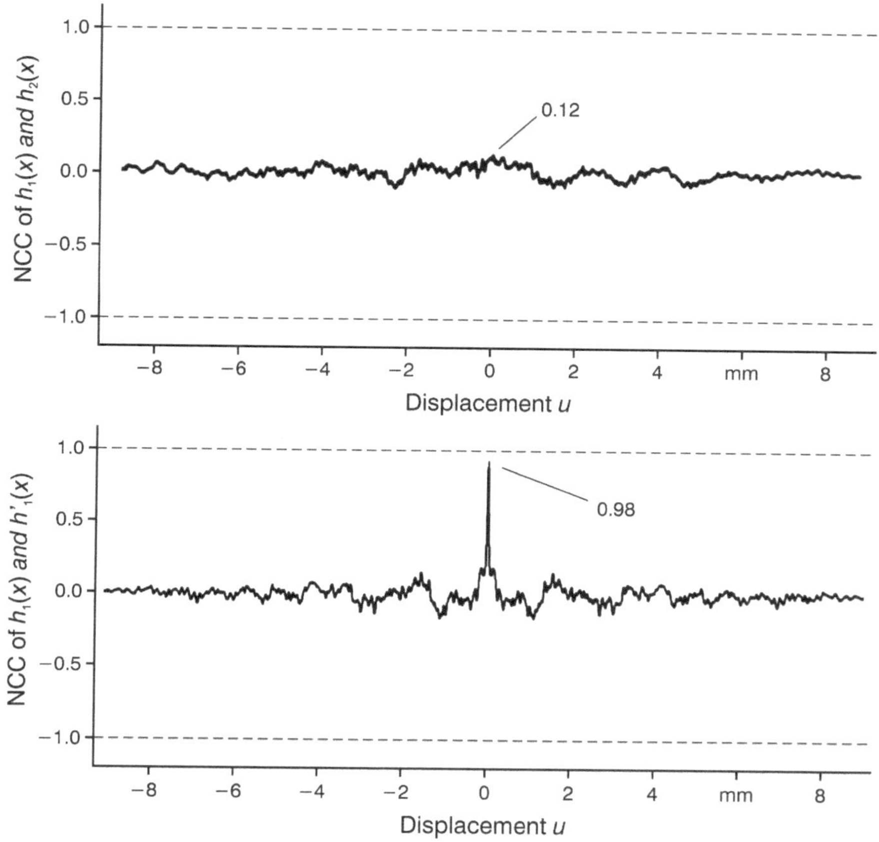

Fig. 2.1.1.11 Normalized cross-correlations (NCC) of a profile h_1 with an entirely different profile h_2 (top) and with a different recording h_1' of the same profile (bottom).

observer would probably look out for distinct, characteristic signal sections with prominent shape such as spikes or half-waves, as marked in Fig. 2.1.1.12. By determining their location and shape, one can extract an individual data set describing the unique characteristic of the surface, analogous to a human fingerprint.

Under the assumption that the shape of a distinctive signal section corresponds to its spectrum, time-frequency distributions were chosen as a mean to describe the shape. They have the property that when used as a mathematical analysis tool, they are able to denote spectral content at each signal location separately.

Since in the case of 1D surface profiles we deal with signals over location, time-frequency distributions should rather be called location-spatial frequency distributions.

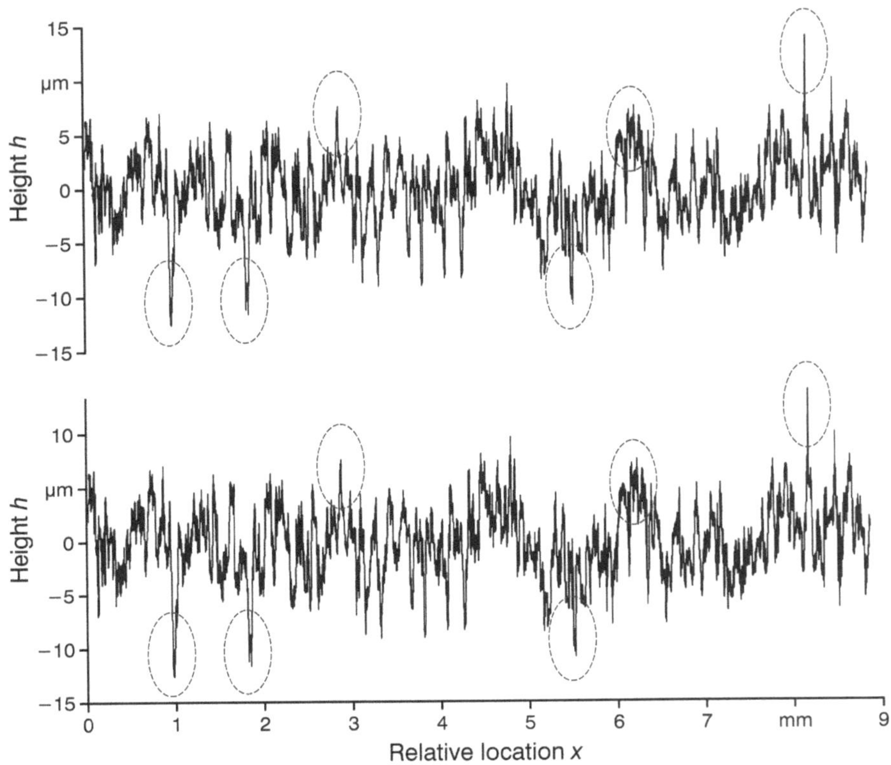

Fig. 2.1.1.12 Matching two profiles by checking for common distinctive signal sections.

2.1.1.2.2.1 The Continuous Wavelet Transform

Among others such as the Wigner distribution or the short-time Fourier transform (STFT) along with its special case, the Gabor Transform, the Continuous Wavelet Transform (CWT) is an integral transform that constructs a suitable time-frequency distribution. For the intended purposes it is particularly well suited due to its computational efficiency if appropriate analysis wavelets are used.

The CWT of a one-dimensional height signal $h(x)$ is defined as follows [19]

$$\mathrm{CWT}\{h(x)\} = S_{\mathrm{CWT}}(x, \omega) = \int_{-\infty}^{\infty} h(\xi)\psi^*((\xi - x)\omega)\, d\xi \tag{2.1.1.3}$$

and yields the location-dependent spectral decomposition, i.e., the spectral response S_{CWT} with respect to both spatial frequency ω and spatial location x. ψ is the so-called

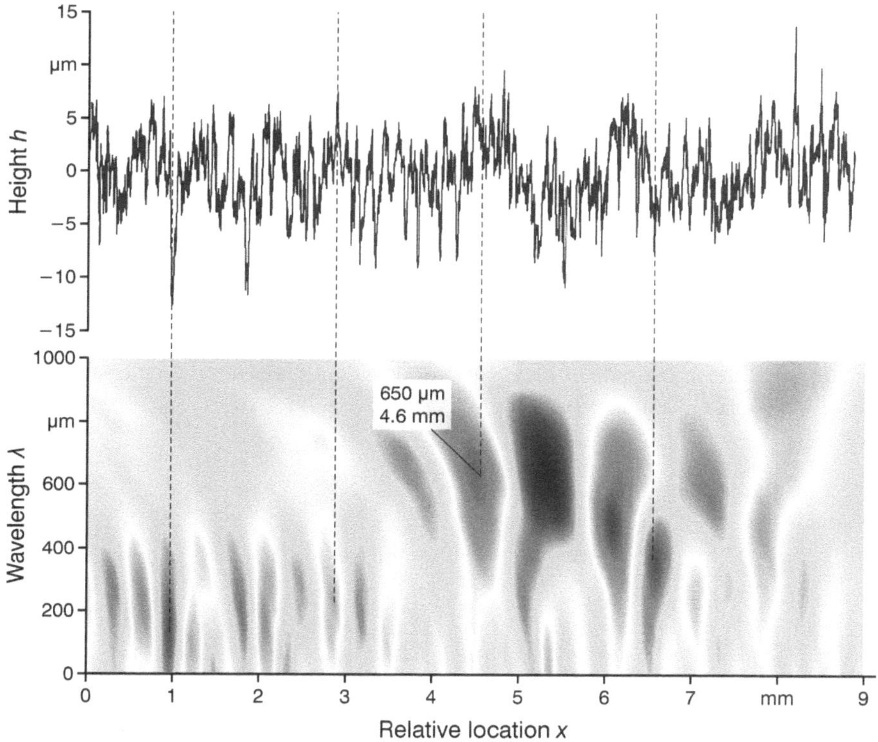

Fig. 2.1.1.13 1D height profile (top) and corresponding Continuous Wavelet Transform (CWT) spectrum (normalized modulus, bottom).

mother wavelet, i. e., the analysis kernel. For the purpose of profile analysis, a suitable mother wavelet was empirically determined. Both Daubechies and Morlet wavelets were considered due to their frequent and well-studied use in a plentitude of signal processing applications. Exploratory experiments on a data set of 104 typical grinding and turning surfaces showed that the *db2* wavelet from the Daubechies family of wavelets serves best the requirements of selectivity in the space-spatial frequency domain and reasonable computational expenses.

An example CWT decomposition is given in Fig. 2.1.1.13. Both plots are arranged in a way that they have the relative location in common. In the lower plot, the modulus of the CWT, i. e., its absolute magnitude, is depicted with dark red indicating large positive values and dark blue indicating large negative values. Accordances between the waveforms in the profile signal and its CWT spectrum can easily be spotted, e. g., the relatively wide positive half-wave at 4.6 mm, which finds its accordance at a strong peak in the spectrum at a wavelength of approximately 650 μm, is neighbored by the corresponding negative half-wave.

By transforming the profile signal into space-spatial frequency domain, distinctive sections can now be described by the 2D location in the location-dependent spatial-frequency spectrum instead of their location and shape in the spatial domain.

2.1.1.2.3 The "fingerprint"

2.1.1.2.3.1 Feature extraction

Local extrema (minima and maxima, or "peaks" and "valleys" when interpreted as a height map) of the magnitude of the location-dependent spectrum indicate salient sections of the profile, as marked in Fig. 2.1.1.14.

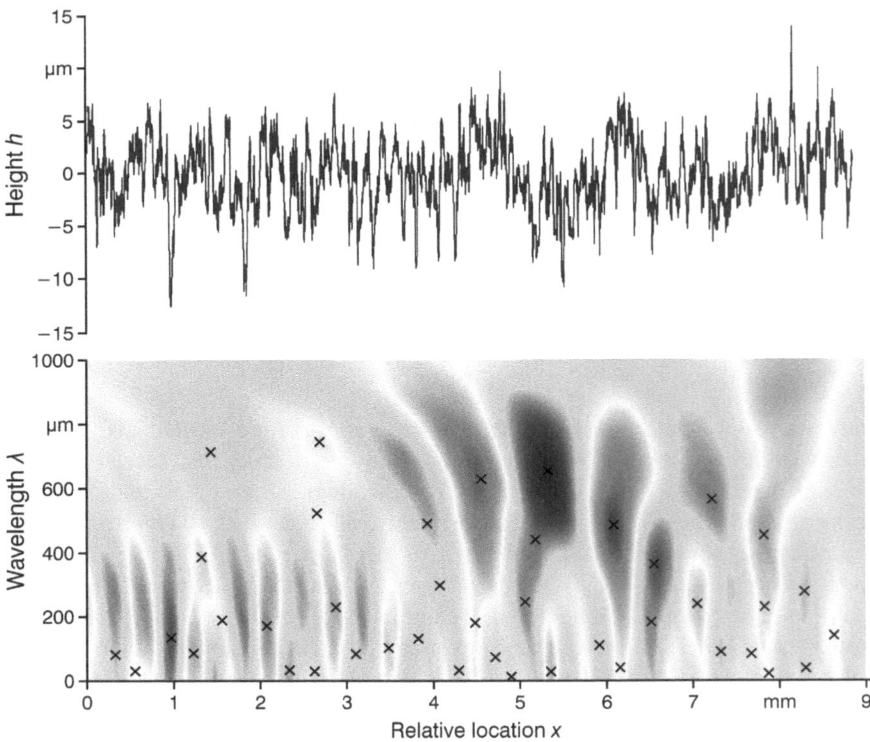

Fig. 2.1.1.14 1D height profile (top) and local extrema of CWT spectrum (normalized modulus, bottom).

Several non-maximum suppression algorithms exist, which fully automatically determine the location of those local extrema and avoid clustered and noisy detections.

In a second step, a feature descriptor is now constructed by evaluating the direct surrounding of the found minima and maxima. The neighborhood itself in the magnitude spectrum, i.e., in wavelet space, is utilized. Its size depends on the current spatial and spatial-frequency resolution. The resulting feature descriptors are subsequently normalized in order to accommodate for level offset or different height scaling or resolution.

The advantage of using a reduced compact representation of the original signal comprising information about all striking signal sections is a high robustness toward local perturbation, e.g., surface damages. Furthermore, invariance toward translation (i.e., a slightly different profile location) and scaling (e.g., a slightly different resolution) is achieved.

The set of feature descriptors together with their location in space-spatial frequency space now form a "fingerprint" that is unique and distinct for a particular surface. Depending on the use case, such a data set is subsequently saved to a database together with a serial identification number for each individual component.

2.1.1.2.3.2 Matching

In order to find the corresponding identifier to a given surface profile at any later time after the initial fingerprinting, a matching stage needs to be implemented.

As a first step, the individual comparison of two given fingerprints is comparatively simple both algorithmically and computationally. The local correspondence candidates among the descriptors are determined by calculating the pairwise Euclidean distance. Since the descriptors are normalized, this is effectively the cosine similarity. A threshold of 0.8...0.95 that is empirically adapted needs to be applied then. Subsequently, the global geometric constellation is verified using RANSAC [20] on the precalculated candidates applying a simple 2D translatory model in the space-spatial-frequency domain. Simulations show that the false-positive rate, i.e., the probability of a fake component being incorrectly identified as original, or the probability of a component being identified with a wrong ID, is less than 10^{-20}. The confusion matrix with the number of correspondences between the fingerprints i, j of 45 different profiles of ground surface samples is depicted in Fig. 2.1.1.15. A relatively large displacement of ± 0.5 mm was assumed so that the robustness toward positioning errors can be verified. Hence, the fingerprints i and j are constructed from profiles measured in a distance of 1 mm on each sample.

The computation time of performing one comparison is around 30...45 ms. When used in a production environment, a fingerprint, however, needs to be matched within less than a second against huge databases of several ten thousands to

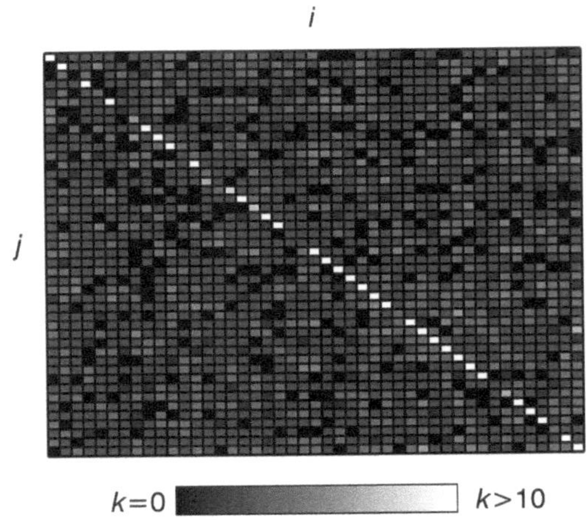

Fig. 2.1.1.15 Confusion matrix with the number of correspondences k for fingerprint matching of 45 ground surfaces [21].

millions of entries. A brute-force approach that simply iterates through the entire database and returns that match with the smallest distance is not feasible for more than a dozen parts. Therefore a novel matching and searching technique was developed. It adopts tree-based search methods that were initially devised for image search [22]. Using the new approach, a search time of 50 ms for a database of 12,000 data sets was accomplished.

2.1.1.3 Application scenarios

The presented fingerprinting scheme is highly flexible with respect to the surface measurement technique. Both standardized tactile or optical profilometry as well as noncalibrated surface imaging methods were verified for profile signal generation. In high-speed, high-volume applications, e. g., in cost-sensitive production environments, industrial off-the-shelf cameras in conjunction with directed, inclined illumination provide real-time surface images that are well suited for fingerprinting.

In Chapter 5.1 it is shown that these fingerprints can serve as passive, intrinsic identifiers for Auto-ID scenarios, enabling a component to control its own further manufacturing steps autonomously when traveling through the production.

In Chapter 5.4 an example of application is given in which the fingerprinting method is used for plagiarism detection. An example workflow as well as a prototype implementation is presented, depicting how it can be employed in the production chain.

2.1.1.4 Conclusion

A novel method for Auto-ID and ex post plagiarism detection has been presented. It utilizes the stochastic content of surface fluctuations that unintentionally occur during manufacturing and has been experimentally verified with grinding and turning processes. Due to the fact that only information that is already intrinsically connected to even very simple components, very low-cost solutions become viable. The technique has high potential to be a significant step toward "Industry 4.0" applications for machined high-volume components.

References

[1] DIN 4760:1982-06, Gestaltabweichungen, Begriffe, Ordnungssystem, Beuth, Berlin, 1982.

[2] D.J. Whitehouse, Handbook of surface and nanometrology, CRC Press, Boca Raton, 2011. http://www.crcnetbase.com/isbn/9781420082029.

[3] H.H. Shahabi, M.M. Ratnam, Simulation and measurement of surface roughness via grey scale image of tool in finish turning, Precis. Eng. 43 (2016) 146–153.

[4] K. Liu, S.N. Melkote, Effect of plastic side flow on surface roughness in micro-turning process, Int. J. Mach. Tools Manuf. 46 (2006) 1778–1785.

[5] M.S. Selvam, K. Balakrishnanthe, Study of machined surfaces roughness by random analysis, Wear 41 (1977) 287–293.

[6] C.L. He, W.J. Zong, T. Sun, Origins for the size effect of surface roughness in diamond turning, Int. J. Mach. Tools Manuf. 106 (2016) 22–42.

[7] T. Zhang, Z. Liu, Z. Shi, C. Xu, Size effect on surface roughness in micro turning, Int. J. Precis. Eng. Manuf. 14 (3) (2013) 335–349.

[8] P.H. Brammertz, Die Entstehung der Oberflächenrauheit beim Feindrehen, Industrie-Anzeiger 2 (1961) 25–32.

[9] H. Opitz, M. Gappisch, Die Aufbauschneidenbildung bei der spanabhebenden Bearbeitung, Forschungsbericht des Landes Nordrhein-Westfalen Nr. 1405, Westdeutscher Verlag, Köln und Opladen, 1964.

[10] M.C. Shaw, Metal Cutting Principles, Clarendon Press, Oxford, 1984.

[11] D. Gillibrand, Micro-Defects on Machined Carbon Steel Surfaces, Tribology International, IPC Business Press, 1979. pp. 165–168.

[12] A. Simoneau, E. Ng, M.A. Elbestawi, Surface defects during microcutting, Int. J. Mach. Tools Manuf. 46 (2006) 1378–1387.

[13] B. Breidenstein, B. Denkena, T. Grove, R. Hockauf, Enabling of component identification by high speed measuring of grinding wheel topography, Procedia Technol. 26 (2016) 252–258.

[14] F. Klocke, W. König, Fertigungsverfahren 2 – Schleifen, Honen, Läppen, Springer-Verlag, Berlin Heidelberg, 2005.
[15] U.R. Zitt, Modellierung und Simulation von Hochleistungsschleifprozessen (PhD thesis), Universität Kaiserslautern, 1999.
[16] R.J. Stokes, ABN-Kornverschleiß im Schleifprozeß, Industrie Diamanten Rundschau (IDR) 4 (1983) 197–209.
[17] J. Fromlowitz, Entwicklung mathematischer Ansätze zur rechnergestützten Simulation von Verschleißvorgängen, Final report, RWTH Aachen, 1993.
[18] J. Gausemeier, R. Glatz, U. Lindemann, Präventiver Produktschutz – Leitfaden und Anwendungsbeispiele, Carl Hanser Verlag, München, 2012.
[19] S. Mallat, A Wavelet Tour of Signal Processing, second ed., Academic Press, Cambridge, 1999.
[20] M.A. Fischler, R.C. Bolles, Random sample consensus: a paradigm for model fitting with applications to image analysis and automated cartography, Commun. ACM 24 (6) (1981) 381–395.
[21] R. Dragon, T. Mörke, B. Rosenhahn, J. Ostermann, Fingerprints for machines—characterization and optical identification of grinding imprints, in: Proceedings of the 33rd Annual Symposium of the German Association for Pattern Recognition (DAGM 2011), Frankfurt/Main, August 30th to September 2nd, 2011, 2011.
[22] D. Nistér, H. Stewenius, Scalable recognition with a vocabulary tree, in: Proceedings of the Conference on Computer Vision and Pattern Recognition, 2006.

2.1.2 Data storage within the surface of a component by cutting micro patterns

Bernd Breidenstein, Tobias Mörke*, Rolf Hockauf*,*
Christopher Schmidt, Jörn Ostermann†, Benjamin Spitschan†*
*Institute of Production Engineering and Machine Tools, Leibniz Universität Hannover, Garbsen, Germany, †Institute of Information Processing, Leibniz Universität Hannover, Hannover, Germany

2.1.2.1 Introduction of information into the component's surface

Intrinsically incorporating component-specific information enables the communication of workpiece data without the use of additional aids, such as Kanban cards, printed QR codes, or bar codes. By storing product and process-related information together it is possible to identify a component throughout its life cycle.

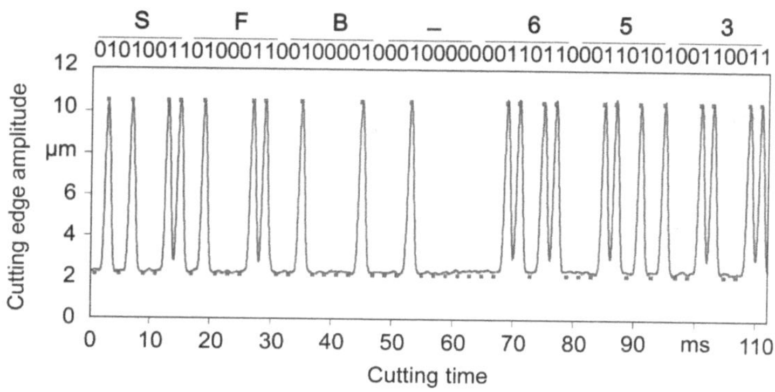

Fig. 2.1.2.1 Progression of actuator position (graph), demodulated symbols (dots), and decoded ASCII characters (top) using binary digital amplitude modulation as the chosen method.

The information is inserted by means of piezo-electrically actuated turning and milling tools which create micro patterns during the process of machining. Those micro patterns are able to encode data by means of basic binary encoding with the presence or absence of a structure representing a binary "1" or "0", respectively. The method was developed for the turning process and subsequently transferred to the face-milling process. By the analysis of the cutting tool's transmission behavior it was found that modulation and demodulation conditions apply, similar to data transmission by electromagnetic waves. The tool can be modeled as a linear time-invariant (LTI) system as long as it does not resonate.

Initially, binary nonreturn to zero (NRZ) base-band modulation was investigated as the simplest form of channel coding. After the identification of the transmission function (Fig. 2.1.2.1), it turned out that common requirements of distortion-free transmission were not satisfied. The first criterion is adequate signal-noise separation, which can be achieved by a minimum amplitude existing at 0 Hz. The second criterion, i.e., a constant signal group delay, was not satisfied, however. The group delay, which is equivalent to a linear phase response, was found to be nonconstant near 0 Hz in the base band because the actuator is controlled by an integration element. That means that the modulated signal must be free of direct components. The phase response is also nonlinear in the resonance range around 6 kHz where, however, the linear signal model cannot be applied. The next type of modulation chosen was a sine wave for the bit "1" and a constant zero progression for the "0" bit, which equates to binary digital amplitude modulation. The constant zero progression means reduced mean power and/or the mean amplitude can be doubled.

The generated signal sequence employing this type of modulation was cut into the component surface. For the rapid analysis of the signal and for the simulation of

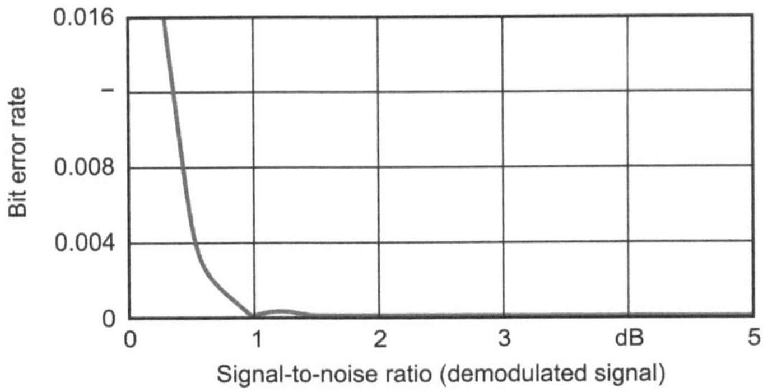

Fig. 2.1.2.2 Dependency of signal-to-noise ratio of demodulated signal and bit error rate of decoded data.

modulation and decoding, the actual actuator progression was recorded during the writing process (Fig. 2.1.2.1).

For encoding the data, standard methods of message transmission were examined together with the general parameters existing in this case, such as the signal delay due to the mass of the writing actuator. For channel coding, a turbo code was chosen, which is well suited for the kind of application, i.e., a low signal-to-noise ratio combined with high data volumes. The results were verified by simulations (Fig. 2.1.2.2). The bit error rates occurring in real conditions were determined by investigations using a channel model [1].

The encoded information needs be inserted directly into the workpiece in order to turn the workpiece into an information carrier. For this purpose, a new cutting tool was developed.

2.1.2.2 Piezo-electric actuator-driven turning tool

The piezo-electric actuator-driven turning tool was developed at the *Institute of Production Engineering and Machine Tools*. Referred to as a fast tool servo (FTS), the tool (Fig. 2.1.2.3) overlays the amplitude of a radially oscillating piezo-electric actuator with the cutting and feed rate in an external longitudinal turning process. Thus a pattern distribution can be specified by the variation of amplitude, amplitude progression, and frequency.

By driving the piezo-electric stack actuators braced by the base of the housing, the expansion of the solid-state actuator is transmitted directly to the motion of the indexable insert. The tool amplitude can be adjusted by connecting the actuators in series and in parallel.

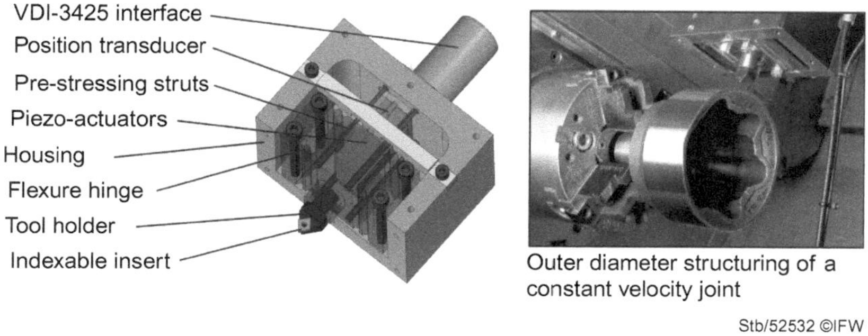

Fig. 2.1.2.3 Piezo-electric actuator-driven turning tool.

In order to be able to determine the deployment behavior of the structuring tool and to limit its range of use, a modal analysis is carried out. The determined first-order natural frequency of the tool is approximately 6500 Hz. The natural frequencies of the machine tool are identified in the range of 70 Hz and 7000 Hz, indicating that no superimposition of the various natural frequencies is to be expected. In order to avoid nonlinear effects arising from the natural frequencies, all investigations for the determination of suitable process parameters for the FTS to cut micro patterns are carried out in the range of 2000–5500 Hz. When the tool cutting edge is in contact with the workpiece, the amplitude progression shows virtually constant attenuation without a shift in the natural frequency. As a result, a constant reduction of the tool tip deflection at different frequencies in the working range and with varying amplification factors can be assumed [2,3].

Once the working range of the actuated turning tool has been determined, it can be used to introduce surface micro patterns. Fig. 2.1.2.4 shows micro patterns within a surface. In order to achieve clearer track depths and widths, the signal is substantially amplified so that on real components the pattern depths are within the range of 4–6 μm. Before storing the information, the alphanumeric sequence "SFB_653" is converted into a signal sequence using the modulation methods described earlier. The information can be read out in cutting direction. Thus a track profile contains the signals for the digit "1", observable as dark areas, and for "0", observable as light areas. The signals have a defined separation from each other, which results from the proper combination of excitation frequency f_s and cutting speed v_c. The data density along a track cannot be infinitely increased as the natural frequency of the tool must not be exceeded and the cutting speed should not be less than $v_c = 200$ m/min for the technological reasons set out earlier. The feed rate f also affects the data density, at least in component axis direction. For technological reasons, the data density cannot be infinitely reduced either. Therefore the maximum data density is restricted to

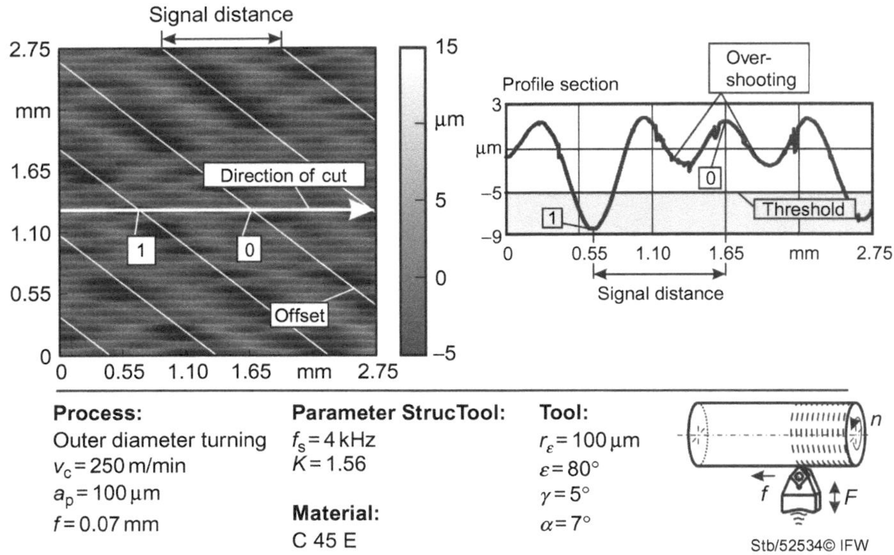

Fig. 2.1.2.4 Code consisting of micro patterns.

approximately 4.2 kbit/cm^2. Another process variable is the amplification factor K, which affects the depth of the micro patterns [4].

2.1.2.3 Piezo-electric actuator-driven milling tool

Another main focus of the research is the development of a spindle system to produce micro patterns on flat surfaces and the description of the interrelationships between the manipulated and system variables of the new process and the produced surface. Based on the knowledge gained from the development of the activated turning tool, a piezo-electrically actuated tool for the production of micro patterns in a milling process is developed. Fig. 2.1.2.5 shows the design of the milling tool [5,6].

To produce axial deflection of the milling cutter tip, a ring actuator with a maximum force of F_{max} = 20 kN and a deflection of a_{max} = 60 µm at a maximum permissible voltage of A_{max} = 1.0 V was used. The system is mounted by two diaphragms made of spring steel acting as solid-state joints. The use of a ring actuator increases the performance capacity compared with cylinder actuators due to the larger surface area and the associated better thermal properties. Furthermore, the ring design allows rotationally symmetrical, central spring loading by disc springs. For use in machine tools, the milling tool has a standard HSK-A 63 shank. Due to the high power to be transmitted, a slip ring is employed. In addition to the voltage, this slip ring also

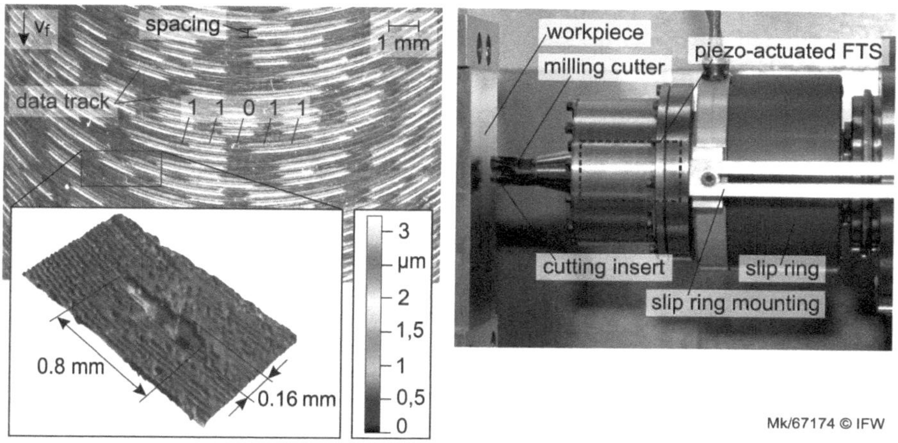

Fig. 2.1.2.5 Left: milled, information-bearing micro patterns; right: actuated tool holder.

transmits a temperature signal and a strain gauge signal to determine the expansion of the actuator. The use of titanium for the tool body reduces the rotating mass including a conventional screw-in milling head ($D = 20$ mm) to 320 g. This enables a highly dynamic controlled excitation (natural frequency $\omega_0 = 4.7$ kHz). A precise real-time control system based on a static and dynamic characterization of the system is constructed. By coupling the FTS control to the machine tool control, the patterns can be produced according to the main spindle's rotational angle [7,8]. Both the actuated tool and the technology described later are the subject of patent applications and have already been put into practical use in industrial applications [9].

Thus the classic flat milling process is supplemented by the two controlled variables, structuring frequency f_s and structuring amplitude a_s. The main variables influencing surface and micro pattern quality were determined on the basis of a static testing plan. An approach for measuring surface roughness along the tooth cutting arc is developed in order to determine the micro pattern quality and the effect of axial excitation on topography. A confocal microscope with suitable algorithms for profile distortion correction is applied. The surface roughness along the tooth path decreases with increasing cutting speed and structuring frequency. An increasing rake or clearance angle has a comparable effect. In feed direction, the influence of the structuring amplitude on the roughness is 2.5 times higher than the influence of the structuring frequency. Feed per tooth and clearance angle are significant influencing factors in this context. The results correlate with the existing findings without axial excitation [10,11].

The resulting data density is restricted by the kinematic relations. The number of patterns along the tool's engagement is determined by the relation between the structuring frequency f_s and the rotational speed n. The number of patterns in feed direction is determined by the structuring amplitude a_s and the feed per tooth f_z. In order to generate patterns with defined depth and length, an exact control of the axial tool movement is essential [12–14].

In order to demonstrate the process kinematics and the achievable quality of the patterns, comparative tests are carried out on Al7075, C45, 42CrMo4, and TiAl6V4. For this purpose, the actuator is stimulated with a sinus oscillation every 15 rotations during finish milling in order to generate defined patterns in the surface. To change the pattern's length and depth, frequency and amplitude are varied on two levels. Cutting speed and feed rate are set according to a finish milling depending on the particular material. The process parameters are summarized in Table 2.1.2.1.

Furthermore, two cutting edges are used that differ in their width between 250 and 500 μm. Fig. 2.1.2.6 shows the cutting edge's geometry.

Table 2.1.2.1 Process parameters depending on the workpiece material

Material	Rotational speed [min^{-1}]	Feed per tooth [mm]
Al7075	3500	0.08
C45	3000	0.06
42CrMo4	3000	0.06
TiAl6V5	2500	0.04

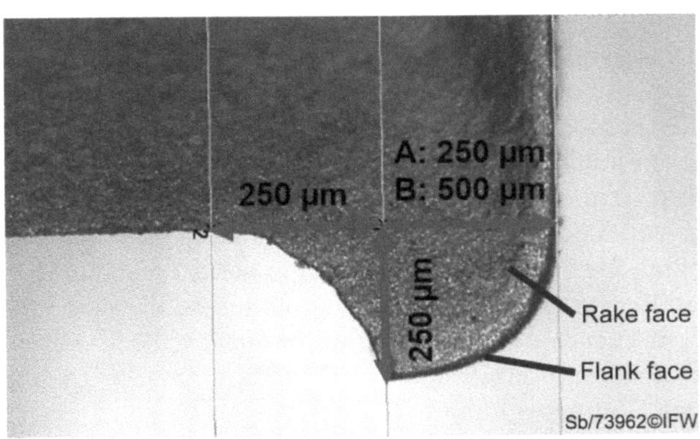

Fig. 2.1.2.6 Cutting tool for binary structuring.

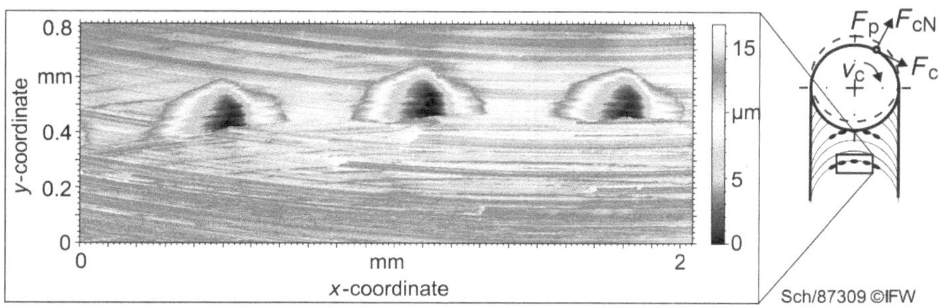

Fig. 2.1.2.7 Surface topography of milled micro patterns in TiAl6V5.

By using these cutting tools it is possible to generate defined patterns in the workpiece surface during the finishing process. As an example, generated patterns in TiAl6V4 are shown in Fig. 2.1.2.7. The figure is made with a confocal white-light microscope.

The characteristics of the generated micro patterns vary for the four different materials. These variations occur in form of geometry and the distance between the patterns. Furthermore, the surface roughness varies depending on the workpiece material and the corresponding process parameters.

2.1.2.4 Structure geometry depending on the workpiece material

The depth of the micro patterns is affected by the engagement depth of the tool, which results from the oscillation amplitude a_s, specified by the given electrical voltage U_{cp}. For the experimental evaluation, the tool's oscillation amplitude is varied on two levels at 10 and 20 µm. However, these are the oscillation amplitudes that appear on idle mode without engagement in the workpiece surface. Although the pattern depth can be increased by increasing the oscillation amplitude, the measured pattern depths are lower than the given ones. The given amplitude $a_s = 10$ µm can be reached more easily than the amplitude of 20 µm. Fig. 2.1.2.8 depicts the depth of the micro patterns depending on material, amplitude, and frequency for cutting edge width of 250 µm. A cutting edge width of 500 µm shows comparable tendencies. The pattern depth does not depend on the workpiece material. Machining of Al7075, which is the softest of all four materials, generates the deepest structures. Furthermore, pattern depth does not show a dependency on oscillation frequency either. As shown in Fig. 2.1.2.8 in the lower diagram, a change of the frequency from 1000 to 1500 Hz decreases

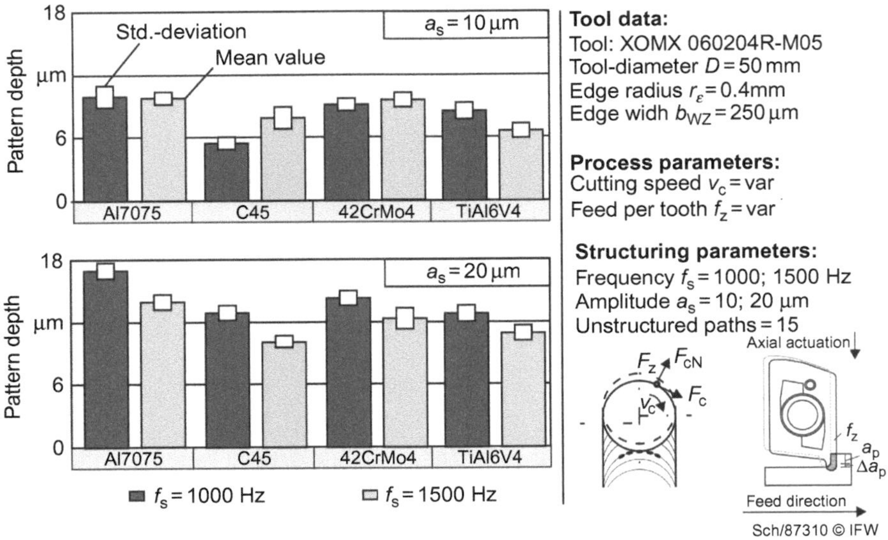

Fig. 2.1.2.8 Depth of micro patterns depending on material, amplitude, and frequency.

the pattern depth for a given oscillation amplitude of 20 μm. For an amplitude of 10 μm this cannot be confirmed. As the amplitudes during engagement differ from the idle amplitudes, further investigations have to be conducted concerning the engagement conditions of the cutting edge in the workpiece's subsurface.

The geometry of the generated micro patterns is examined with respect to their length and width. Fig. 2.1.2.9 shows the length depending on the oscillation amplitude for both frequencies, 1000 and 1500 Hz. According to the results, the variation of the oscillation amplitude between 10 and 20 μm shows no significant influence on the pattern length. An increase of the frequency from 1000 to 1500 Hz leads to a reduction of pattern length. For all workpiece materials, a higher frequency leads to a decrease of the pattern length. Machining of TiAl6V4 generates the shortest patterns. This can be explained with the fact that this material shows the highest resistance to machining due to its material properties. The main reason can be found in the low Young's modulus of 110 GPa and the high yield strength of the used alloy TiAl6V4. This leads to a high amount of springback and mechanical deformation resistance during tool engagement [15].

Besides the pattern geometry, the density of the generated patterns can be investigated. For this, the four workpiece materials are machined. During machining, the oscillation frequencies of 1000 and 1500 Hz are compared. In order to link the

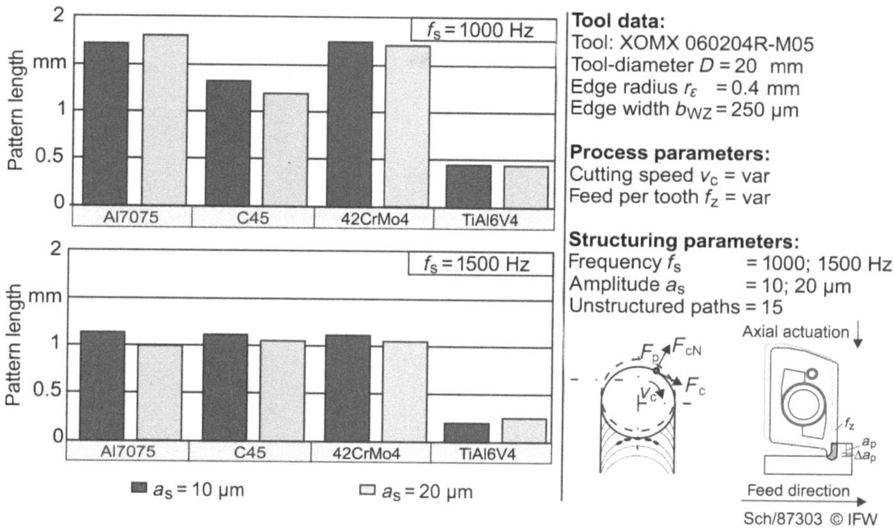

Fig. 2.1.2.9 Length of micro patterns depending on material, amplitude, and frequency.

Table 2.1.2.2 **Number of patterns per revolution for a width of cut $a_e = 20$ mm and material-depending finishing parameters**

Material	Rotational speed [min^{-1}]	Feed per tooth [mm]	Number of patterns/f_z	
			1000 Hz	**1500 Hz**
Al7075	3500	0.08	10	14
C45	3000	0.06	8	14
42CrMo4	3000	0.06	8	14
TiAl6V5	2500	0.04	42	60

investigation to the practical application, the machining tests are conducted with the material-depending finishing parameters. The finishing parameters and the obtained results are summarized in Table 2.1.2.2. Thus it can be deducted which density of binary information can be reached for each material with the recommended finishing parameters. TiAl6V4 shows the highest density of patterns because the lowest rotational speed n and lowest feed per tooth f_z are recommended for finishing.

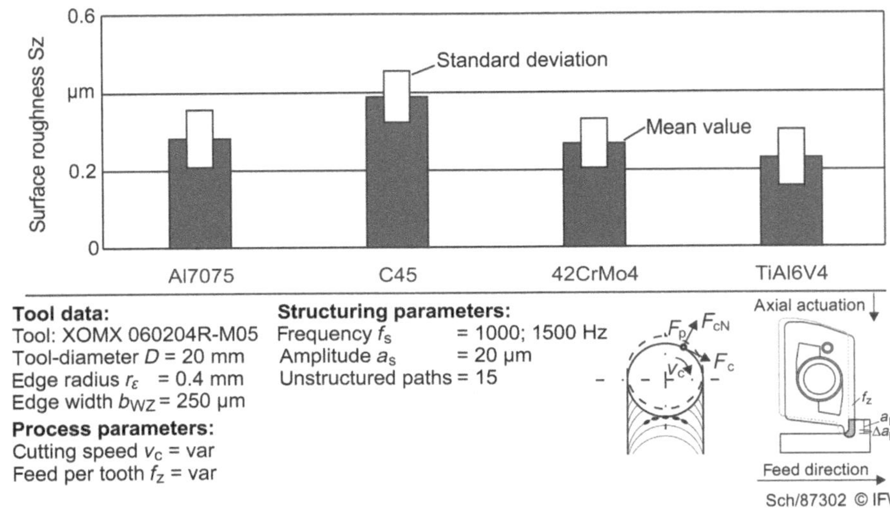

Fig. 2.1.2.10 Mean roughness in the finished surface without micro patterns.

Next, the roughness after machining different workpiece materials is shown. Using the finishing parameters depending on the workpiece material, the roughness in the area without micro patterns differs. The highest roughness value can be found after machining C45 as shown in Fig. 2.1.2.10. The other three machined materials show no significant difference in their surface roughness.

The surface roughness depending on the tool's cutting edge width is shown in Fig. 2.1.2.11. It can be seen that for all materials and for both operation frequencies a higher surface roughness results from the use of the tool with a cutting edge width of 250 µm. The reason for this can be found in the fact that the wider cutting edge takes away material from the prior cutting path and thus smooths the previously cut area. This leads to a lower surface roughness.

2.1.2.5 Process forces during cutting of micro patterns

Besides the generated geometries, the forces during the cutting process are examined with analogy experiments. They consist of single toothed face milling with an excited sinusoidal oscillation. Forces of the binary structuring process machining Al7075, C45, 42CrMo4, and TiAl6V6 are determined. In analogy to the binary patterns, the generated patterns are locally isolated and do not overlap.

The process parameters cutting speed, feed rate, cutting depth, oscillation frequency, and oscillation amplitude are varied in two steps. The force measurement

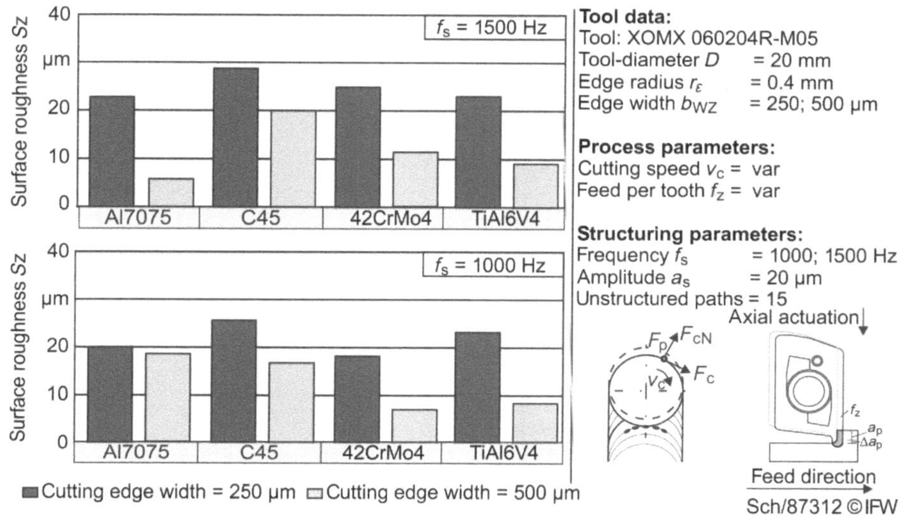

Fig. 2.1.2.11 Surface roughness in the finished surface without micro patterns depending on the tool width.

is adjusted by the self-oscillation rate of the force measurement plate based on the response characteristics. The structuring forces are clearly separable from the milling forces and identifiable as additional loads in cutting and passive directions (Fig. 2.1.2.12). The characteristics of the additional force components F_s are only influenced significantly by the structuring amplitude a_s and neither by the structuring frequency f_s nor by other process parameters. The structuring forces can be considered separately and expressed by a linear behavior of a defined incremental factor (Eq. 2.1.2.1) depending on the structuring amplitude. This factor depends on the material and the process conditions, e.g., specific cutting force. Additional tests with further amplitudes do not disprove the linear characteristic. This described behavior is proven for all investigated materials. Differences only exist in the magnitude of the structuring forces. The process forces in cutting and passive direction can be added up to the amount of the forces out of conventional face milling and structuring (Eq. 2.1.2.1) [16,17].

$$F_{i,\text{total}} = F_{i,\text{regular}} + k_{i,\text{struc}} \cdot a_s(S(\varphi)) \quad \text{with } i = c, z \tag{2.1.2.1}$$

$F_{i,\text{regular}}$ represents the cutting forces without structuring. The constant $k_{i,\text{struc}}$ is the gradient coefficient, which is calculated as a product of the specific cutting force

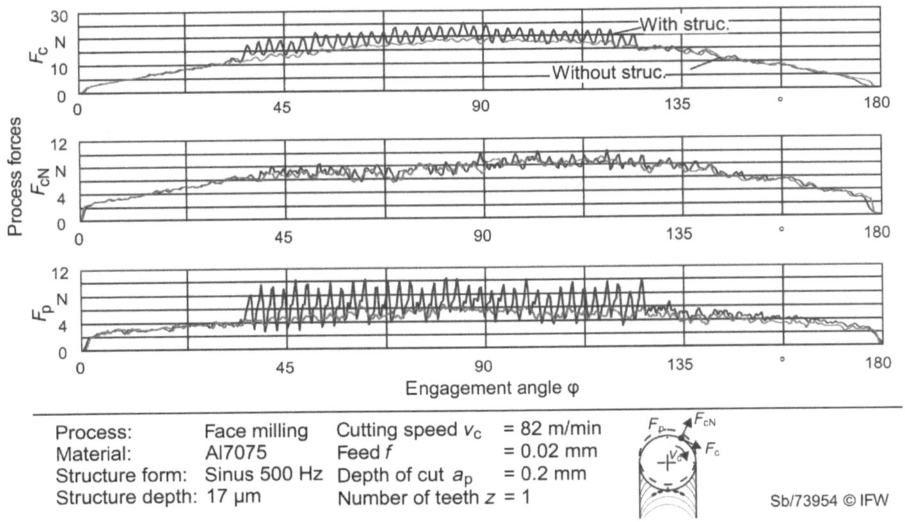

Fig. 2.1.2.12 Process forces during milling with sinusoidal stimulation.

constant and the equivalent chip thickness. The structuring depth depends on the deflection signal $S(\varphi)$, which in turn depends on the engagement angle φ.

It has been demonstrated which surface topographies result from cutting of micro patterns depending on process parameters and workpiece material. In the following, the process of reading the generated micro patterns is investigated.

2.1.2.6 Reading information-carrying micro patterns

The reading algorithm for turned micro patterns [1] proved suitable to be adapted to micro patterns produced by flat face milling. To that end, the existing read out process was expanded by automatic distortion correction. That involves nonparametric arc curvature estimation utilizing the periodic structure of the image [8]. The main wavelength of the image intensity values in feed direction, which is robustly determined using the ESPRIT (Estimation of Signal Parameters via Rotational Invariance Techniques) [18] spectral estimation method, represents the track separation. The track deviation from an assumed array of parallel tracks is equivalent to the local phase offset of the signal. The latter is determined based on the estimated frequencies. However, the phase still has to be unwrapped due to its 2π ambiguity. Subsequently, the image undergoes geometrical distortion correction and rectification such that tracks are sufficiently parallel (Fig. 2.1.2.13).

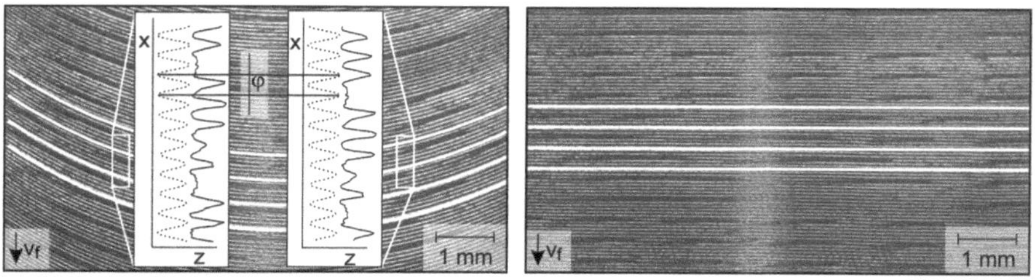

Fig. 2.1.2.13 Geometrical distortion correction and rectification of micro patterns created by flat face milling process.

2.1.2.7 Robustness of information-bearing micro patterns

The technology described in the preceding sections for the storage of information on workpiece surfaces using micro patterns was investigated with regard to data security in the case of near-realistic surface damage. The initial bases for the investigation were car drive shafts and CV joints after 10, 15, and 20 years of use on the vehicle. Damage characterization reveals a dominance of surface change due to corrosion. Mechanical damage occurred to a minor degree, while no thermal damage was found.

For a controlled damage simulation, information-carrying specimens were corroded in salt-mist spray chambers. As corrosion increases, the mean uncorrected error rate (proportion of incorrectly detected bits relative to total number of bits before any error correction) increases from 2×10^{-3} to 9×10^{-2} after 6 h and 2×10^{-1} after 9 h due to local concentrations of heavy corrosion. Beyond 15 h, data reading ceases to be possible because the algorithm for track synchronization fails. In addition, investigations were carried out concerning mechanical surface damage. Therefore scratches were incrementally introduced to the surface on a lathe. In these cases, the bit error rate increases to 9×10^{-3}. Those two errors, which simultaneously affect multiple bits of a symbol, correspond to a two-dimensional burst error in communication technology terms. To increase the robustness of the process and the security of the data against surface changes, the results were used as the basis for designing forward error correction (FEC) according to the expected surface damage (Fig. 2.1.2.14). For that purpose, a Reed-Solomon (RS) block code was used. The resistance to burst errors was increased by the development of a block-interleaving scheme that swaps the bits of a data block according to a pseudorandom permutation pattern after encoding and before application to the surface. Due to the use of RS FEC and the interleaver, it was possible to achieve quasi-errorless reading of the data with scratched and corroded specimens. At the same time, an information density of 2 kbit/cm^2 was achieved.

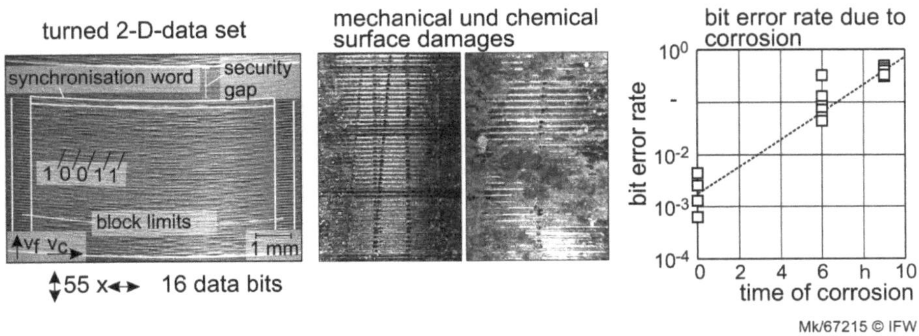

Fig. 2.1.2.14 Left: 2D data block; center: mechanical and chemical surface damage; right: effect of surface damage on bit error rate.

The effect of subsequent coating on data readability was determined. The PVD coating process shows no negative and sometimes even a positive effect on the detection of structures, regardless of coloring. That also applies to different layer types and thicknesses. The property of the process to amplify the roughness peaks of the topography has a positive effect on data readability. The reading system only had to be slightly adjusted to the different reading conditions [19] as the existing image processing methods already demonstrate adequate immunity to topography changes. Smoothing coating processes such as CVD, chemical wet deposition coating, or the application of paints had a negative effect due to the reduction of pattern depth and resulted in increasing data loss even at small layer thicknesses.

To improve data reconstruction and signal quality, two methods of sensor data fusion were evaluated. The use of RGB color cameras instead of monochrome cameras did not show any advantages over the existing process. Nor was any improvement in data reconstruction achieved by the analysis of stereoscopic images created by a pair of cameras because the substantial reflections from metallic surfaces did not allow a sufficiently accurate registration of both stereo images. Significant improvements were achieved, however, by the fully automatic reading of information-bearing micro patterns. For that purpose, a large number of optical, geometrical, machining technology, and signal parameters are estimated. Data reading algorithms use experimentally optimized parameter sets as a priori knowledge. Thus the number of degrees of freedom, i.e., the scanned area can be reduced, which means an improvement of computing efficiency. In that way, improvements of two orders of magnitude were achieved so that the execution time is now around 0.5–1.5 s depending on the applied parameters.

Besides the workpiece's surface, the tool properties have an influence on the quality of generated micro patterns. So, tool wear has been investigated additionally. Due to the fluctuating machining cross-section and fluctuating contact

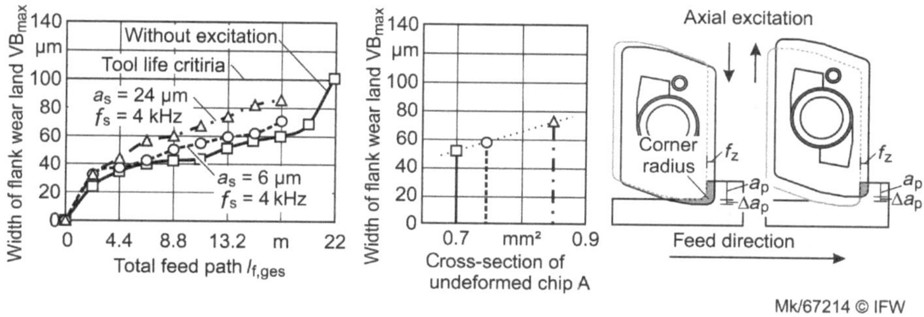

Fig. 2.1.2.15 Effect of axial tool excitation on flank wear of the end-cutting tip.

length, excitation has a negative effect on wear. In particular, the flank wear on the end-cutting tip increases with rising amplitude (Fig. 2.1.2.15).

As shown, the quality of generated micro patterns is important for the readability. Thus it can be of advantage to predict the micro patterns' quality before the machining process is performed. One method is a simulation model which may help to predict the generated micro patterns. This will be shown in the following section.

2.1.2.8 Process design for generating and simulating binary structures

In order to generate a method for the process design, a surface simulation applying a dexel-based surface simulation software CutS is constructed. It allows the generated surface to be mapped based on kinematic connections (Fig. 2.1.2.16).

Furthermore, the theoretical roughness depending on the cutting edge geometry and based on the process parameters can be determined. The area requirements for a defined binary pattern can also be determined simulatively. The simulation model is suitable for mapping the generated surface after cutting of micro patterns binary patterns into a workpiece surface. However, there are differences between the simulated and the actually measured surface roughness. As an example, this is shown for the roughness value S_a in Fig. 2.1.2.17.

In order to improve the model's results, this simulation model should be expanded to consider the tool's behavior shown in this paper. Also, the influence of different workpiece materials on the binary pattern geometry could be integrated into the simulation to generate results that are closer to the measured surfaces after the machining process.

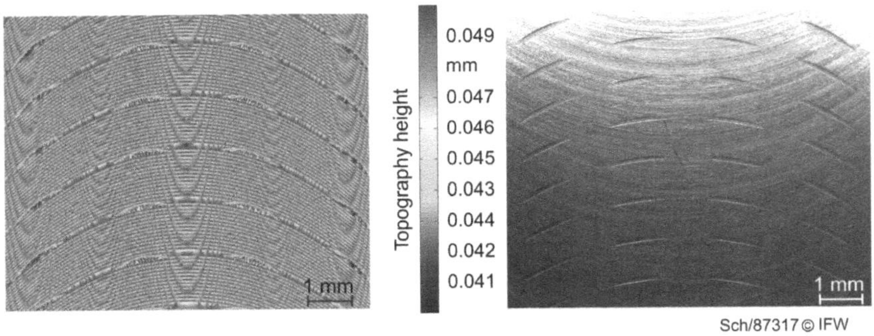

Fig. 2.1.2.16 Surface simulation with defined micro patterns.

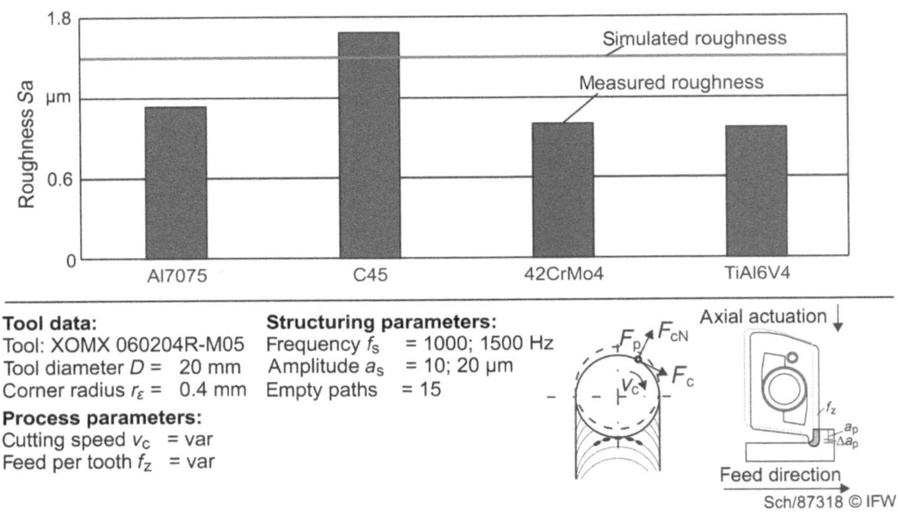

Fig. 2.1.2.17 Comparison of measured and simulated surface roughness in nonstructured surface area.

2.1.2.9 Possibilities and limitations of the structuring process

To demonstrate the potential of the kinetics of actuated face milling with one cutting edge for specific surface topography manipulation a tool prototype with a slip ring for energy transmission is used. Due to the slip ring energy transmission, this prototype is

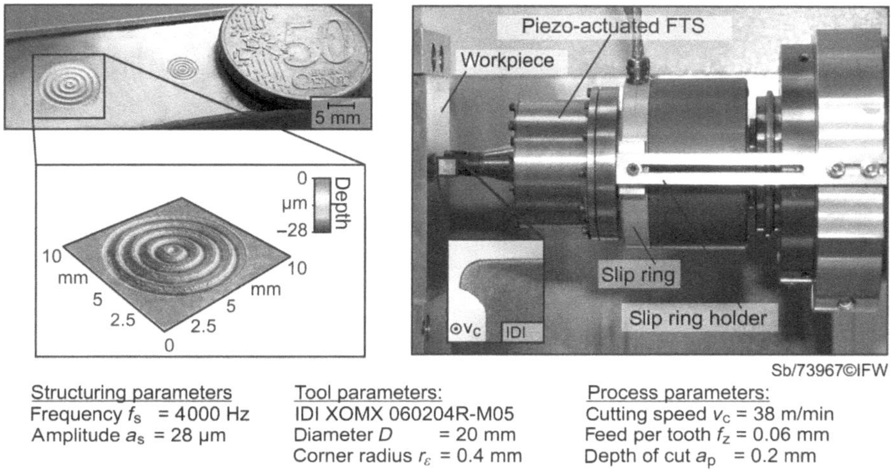

Fig. 2.1.2.18 Structuring exemplary patterns with actuated oscillated milling tool.

not suitable for converting a trigger signal into a movement of the piezo actuator. The main reason for this can be found in the physical fact that the inductive energy transmission cannot transfer DC voltages but only changes in the voltage. Therefore a constant deflection of the piezo actuator is not possible. However, after the extension and adjustment of the tool control the generation of any patterns and structures in one milling path is feasible, as shown in Fig. 2.1.2.18 [14].

First, based on any chosen gray scale image, a structuring matrix is generated as a preparation step. This matrix contains the tool's deflection signals depending on the position in feed direction and the momentary rotation angle of the cutting edge. The size of the matrix is predefined by the maximum amount of deflections per rotation (structuring frequency), the cutting speed, and the feed velocity. Next, together with the information of the spindle's rotation position, the matrix is processed by a Matlab/Simulink model. It is necessary to handle the signals in real time. For this reason, a dSPACE real-time computer system is used for the model. Its sampling rate is $f = 40$ kHz to discretize the maximum structuring frequency of $f_s = 4$ kHz into 10 gradations. This is necessary to guarantee an approximately homogeneous movement of the tool tip. A position control is not realized in this control solution. The system's trigger signal is amplified by a high-voltage amplifier and transmitted to the piezo actuator. The control principle of the tool system is given in Fig. 2.1.2.19.

Furthermore, this control principle allows the generation of several large-scale textures or patterns of various milling paths next to each other, as to be seen in Fig. 2.1.2.20. With the help of a trigger signal, which is activated by the rotary encoder of the feed drives, it is possible to start each pattern at the same position so that the

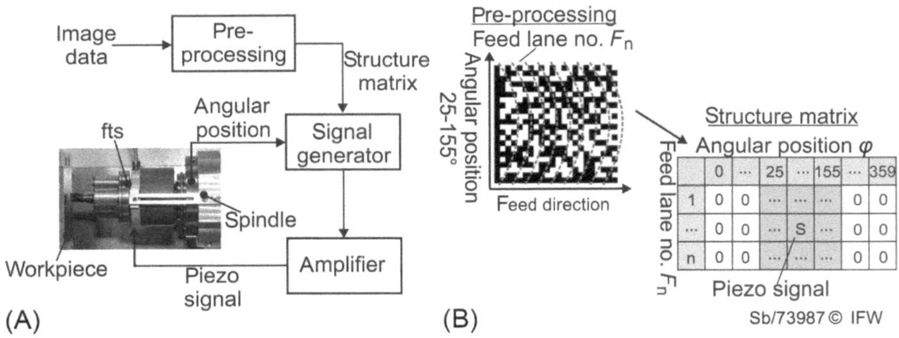

Fig. 2.1.2.19 Control principle (A) and image-processing (B) of the surface-structuring process.

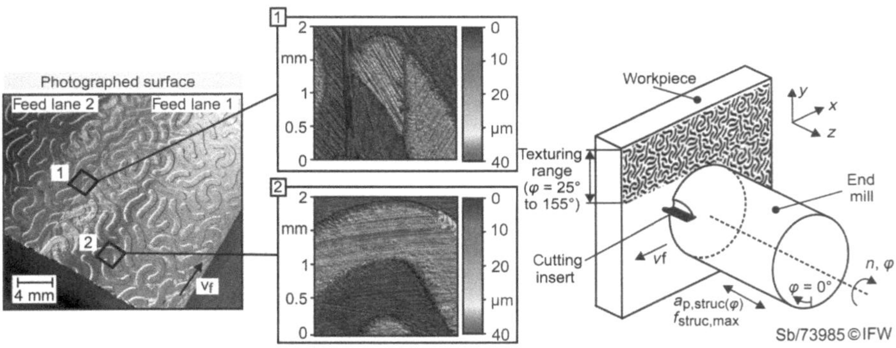

Fig. 2.1.2.20 Generation of connected textures on adjacent milling paths.

generated patterns are equally positioned. This way two patterns may be generated on two adjacent milling paths without an offset between them. So far, a small material web remains between the milling paths. However, this is an issue that can be solved in the course of further optimizations. By using this control principle, it becomes possible to mill textures and patterns with medium feed velocity within a short amount of time (about 5 s/cm^2).

The conducted tests show that an actuated fast tool servo is capable of generating defined micro patterns. Furthermore, the subsurface is affected by the axial tool excitation in terms of residual stresses. This opens the opportunity to define not only the workpiece's topography but also the residual stresses. The investigations of inducing residual stress by actuated cutting and hammer peeling are described in Chapter 3.3.

2.1.2.10 Summary

It was successfully demonstrated how data can be introduced directly into a workpiece surface by means of piezo-electrically actuated turning and milling processes in machining. For that purpose, the internally developed tool concepts were characterized according to their behavior in use and suitable operating conditions were derived. By the use of the binary number system, it was possible to encode information on the component surface and read it back using an optical process. The use of the actively affected deterministic surface component now makes it possible to record process and product-related information directly on the component. Thus it is conceivable that, in the future, products will be able to self-authenticate and independently control their own path through the production and/or maintenance processes. The application of the process to different materials was investigated and a simulation approach described to predict surface topography after the actuated machining process. The used methods were tested for robustness by means of corrosion tests and mechanical damage.

References

[1] R. Dragon, C. Becker, B. Rosenhahn, J. Ostermann, Reading from scratch—a vision-system for reading data on micro-structured surfaces, in: Proc. 31st DAGM Symposium, Springer, 9.-11.9.2009, 2009, pp. 402–411.
[2] B. Denkena, D. Boehnke, D. Spille, R. Dragon, In-process information storage on surfaces by turning operations. CIRP Ann. Manuf. Technol. 57 (1) (2008) 85–88, http://dx.doi.org/10.1016/j.cirp.2008.03.058.
[3] B. Denkena, L. de Leon, I. Steinbrenner, The influence of machining parameters and tool geometry on the quality of information carrying microstructures, in: Proc. 24th Annual Meeting, ASPE, 4.-9.10.2009, 2009, pp. 148–151.
[4] R. Dragon, J. Ostermann, B. Denkena, I. Steinbrenner, Machining microstructures by means of a high-dynamic fast-tool-servo, in: Proc. 9th Euspen (2), 2009, pp. 451–454.
[5] B. Denkena, J. Kästner, B. Wang, Advanced microstructures and its production through cutting and grinding. CIRP Ann. Manuf. Technol. 59 (1) (2010) 67–72, http://dx.doi.org/10.1016/j.cirp.2010.03.066.
[6] R. Dragon, J. Ostermann, B. Denkena, B. Breidenstein, T. Mörke, Data storage in gentelligent components: a new way for self-authentication, in: Self-X in Engineering, MV-Wissenschaft, 2010, pp. 1–13.
[7] B. Denkena, J. Köhler, O. Gümmer, T. Mörke, High performance cutting of micro patterns. in: Proc. 5th CIRP Conference on High Performance Cutting, 4–7.6.2012, Elsevier, 2012, pp. 144–149, http://dx.doi.org/10.1016/j.procir.2012.04.024.
[8] B. Denkena, J. Ostermann, B. Breidenstein, T. Mörke, B. Spitschan, Advanced machining and accessing of information by means of gentelligent micro patterns, in: Proc. 1st Joint International Symposium on System-Integrated Intelligence, 27–29.6.2012, PZH Verlag, 2012, pp. 78–81.

[9] Planfräsverfahren und Planfräskopf, German patent, DE 10 2012 002 140.2, patent application 27.01.2012.
[10] F. Betz, Die Beeinflussung der Schnittflächenrauheit durch die Mikrogestalt der Werkzeugschneide, Ann. CIRP 19 (1971) 625–631.
[11] H. Saglam, F. Unsacar, S. Yaldiz, Investigation of the effect of rake angle and approaching angle on main cutting force and tool tip temperature, Int. J. Mach. Tool Manuf. 46 (2006) 132–141.
[12] B. Denkena, J. Köhler, A. Seibel, Experimental analysis of cutting forces in actuated face milling of micro patterns, Procedia Technol. 15 (2014) 424–430.
[13] B. Denkena, H. Henning, L.E. Lorenzen, Genetics and intelligence: new approaches in production engineering, Prod. Eng. Res. Dev. 4 (1) (2011) 65–73.
[14] B. Denkena, T. Grove, A. Seibel, Direct part making by vibration assisted face milling, Procedia Technol. (2016) 185–191.
[15] T. Grove, Hochleistungszerspanung von Titan, Dr.-Ing. Dissertation Leibniz Universität Hannover, Berichte aus dem IFW, PZH Verlag, Garbsen, 2015.
[16] M. Magnevall, M. Lundblad, K. Ahlin, G. Broman, Frequency measurements of cutting forces in milling by inverse filtering, Mach. Sci. Technol. 16 (4) (2012) 487–500.
[17] J. Köhler, A. Seibel, FTS-based face milling of micro structures, Procedia CIRP 28 (2015) 58–63.
[18] B. Denkena, B. Breidenstein, T. Mörke, J. Ostermann, R. Dragon, Ascertainability of coated microstructured surfaces, in: Proc. 9th THE-"A" Coatings Conference, Editions ZITI, Thessaloniki, 3.-5.10.2011, 2011, pp. 19–26.
[19] R. Roy, T. Kailath, ESPRIT-estimation of signal parameters via rotational invariance techniques, IEEE Trans. Acoust. Speech Signal Process. 37 (7) (1989) 984–995.

2.2 Employment of subsurface properties

In this chapter it will be demonstrated how a component's subsurface can be used as an information source for experienced loads, furthermore different possibilities of using a component's subsurface as a data storage unit will be presented.

Imagine mass produced components had sensory properties. Which opportunities would arise if millions of wheel carriers or passenger car drive shafts could give detailed information on mechanical loads during their life cycle? Which impact would safety-relevant components such as engine mounts have if they reported on the maximum load stress they have experienced since the last inspection? Rather than attaching sensors to a component, its subsurface properties can be used to provide information on loads during its life cycle. Future mass production components that give feedback during their application represent an invaluable benefit for maintenance planning as well as for design evolution regarding tailored adaptation to load.

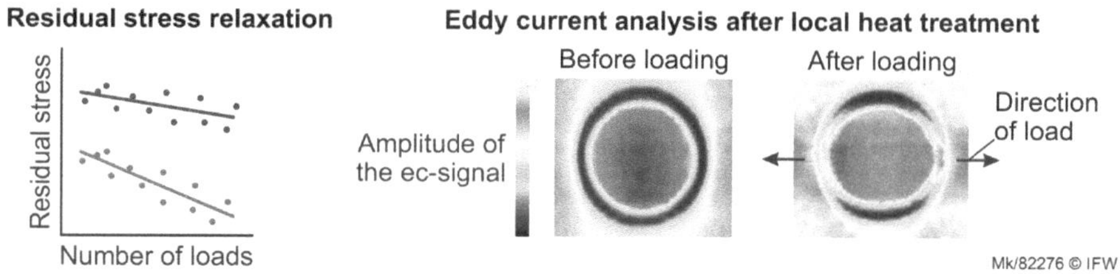

Fig. 2.2.1 Load detection by monitoring subsurface properties. Left: Residual stress relaxation due to mechanical load. Right: Load-induced changes of microstructural properties.

In the following, two approaches are presented to confer sensory abilities to metal components by monitoring changes of a component's subsurface properties (Fig. 2.2.1). If one correlates the initial subsurface condition with the subsurface properties during its life cycle, this offers the possibility to determine loads due to effects such as residual stress relaxation, change of dislocation density, or work hardening (see Chapter 2.2.1). Furthermore, the CRC 653 uses microstructure transformation by local heat treatment of cold-formed metastable austenitic steels to locally manipulate the material's yield strength. Its exceedance upon an external loading causes the formation of deformation-induced martensite within locally tempered regions (see Chapter 2.2.2).

A second not yet assessed feature of selective subsurface and microstructure modifications is the storage of product and production data (Fig. 2.2.2). Since the demand for product traceability throughout its production constantly increases, the unique identification of technical components is of growing importance. Using material properties realizes an inseparable merge of data and product. The technologies of the CRC 653 enable components to inherently carry information and to meet the requirements of tomorrow's production. The local heat treatment, which has been previously introduced with respect to load detection, bears an additional potential regarding component identification. For steel and light metal, the use of laser technology enables the

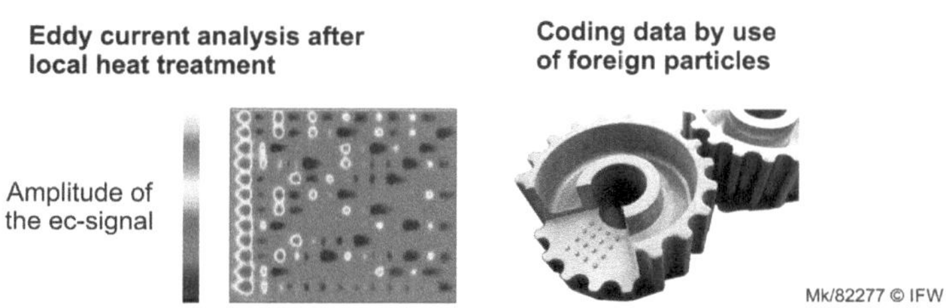

Fig. 2.2.2 Data storage by selective manipulation of subsurface properties. Left: Data matrix code stored by local heat treatment. Right: Data pattern realized by foreign particles in metal powder matrix.

coding of data by means of local microstructure transformation. Thus it enables an inherent component labeling and data storage (see Chapter 2.2.3). Besides machining and laser application, powder metallurgy represents a promising manufacturing technology to suspend the physical detachment of component and corresponding data. When filling the die with base powder, foreign powder particles are applied in order to store information within the bulk material of the component (see Chapter 2.2.4).

See the following chapters for further information

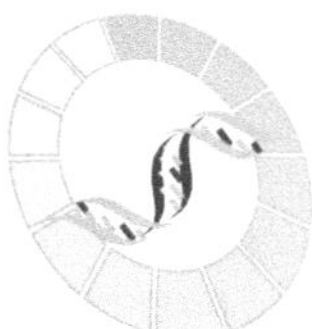

2.2.1 Estimation of mechanical loads by changes of subsurface properties
2.2.2 Storing the load history of a component in the subsurface region
2.2.3 Data storage within the subsurface of a component by local heat treatment
2.2.4 Data storage and component labeling within bulk material by sintering

2.2.1 Estimation of loads by changes of subsurface properties

Bernd Breidenstein, Tobias Mörke, Rolf Hockauf
Institute of Production Engineering and Machine Tools, Leibniz Universität Hannover, Garbsen, Germany

2.2.1.1 Initial level of knowledge and scientific issues

At the beginning of our investigation, it was already known that residual stresses relax components due to mechanical and thermal loads [1]. Our initial attempts using reversed bending loads on AISI 1060 specimens resulted in proving the following literature findings: residual stresses relax if the equivalent stress, as proposed by von Mises, and calculated as the sum of residual and load stresses, reaches the yield strength of the material [2]. Furthermore, the different relaxation behaviors of residual stresses, at three heat treatment states, became evident. For the purpose of our investigation, the generation of residual stresses by turning, as a result of cutting speed and feed rate, is examined systematically. In a follow-up step, the relaxation behavior in torsion and bending tests of steel with a carbon content of 0.3 and 1%, respectively, is demonstrated. The final scientific issue is the development of a method to estimate the loads experienced by determining the residual stress relaxations. This method is applied in the demonstration scenario of a racing car where the relaxation of the residual stresses of a pushrod is measured by eddy current analysis. Before that can take place, an approach to quantify the relaxation behavior from an eddy current analysis has to be found.

2.2.1.2 Generation of residual stresses

2.2.1.2.1 Residual stress depth distributions due to turning of AISI 1060

Before beginning investigation of the relaxation behavior of residual stresses, an investigation of how residual stress depth distributions are generated by turning of AISI 1060 due to cutting speed and feed rate is performed. To that end, alternating bending specimens of AISI 1060 are machined in a longitudinal turning process on a Gildemeister CTX 520 L CNC lathe, in accordance with German standard DIN 50113 (smallest diameter 7 mm). The results are shown in Fig. 2.2.1.1. It can be observed that feed rate is the main factor that affects the generation of residual stress by turning. The depth of induced residual stresses especially increases with a higher feed rate. The range of surface residual stresses moves from approximately −400 MPa compressive residual stress at $v_c = 300$ m/min and $f = 0.01$ mm to nearly 600 MPa tensile residual stress at $v_c = 300$ m/min and $f = 0.5$ mm in axial direction. Furthermore, a low cutting speed shows a tendency to cause a lower compressive residual stress level in terms of distance to the surface. The depth of the maximum compressive residual stress and the absolute values of the maximum tensile residual stress both increase with an increasing feed rate. The maximum influenced depth of the subsurface by turning is 200 μm.

2.2.1.2.2 Generation of residual stresses depending on the heat treatment of AISI 1060

In addition to the process parameters, the material properties have a great influence on the induced residual stresses. In order to determine this influence, specimens of AISI 1060 at two different heat treatment states are turned at a cutting speed of $v_c = 30$ m/min and a feed rate of $f = 0.01$ mm. The result of the heat treatment is shown in Fig. 2.2.1.2A. The ultimate strength correlates with the hardness of the workpiece material and with the process forces in the turning process. As a result, the induced residual stresses increase with increasing ultimate strength (Fig. 2.2.1.2B). The repeatability of the induced residual stresses is rather poor. For example, the axial residual stresses of the quenched and tempered specimens at 200°C have a range from -950 MPa $< \sigma_{yy} < -1700$ MPa. A reason for the poor repeatability lies in the deviations in the turning process, such as vibrations, irregular chip formation, or tool wear. Due to these reasons, a process has to be found which would enable the generation of residual stresses with a good repeatability. Therefore deep rolling of fully annealed and hardened AISI 1060 with a yield

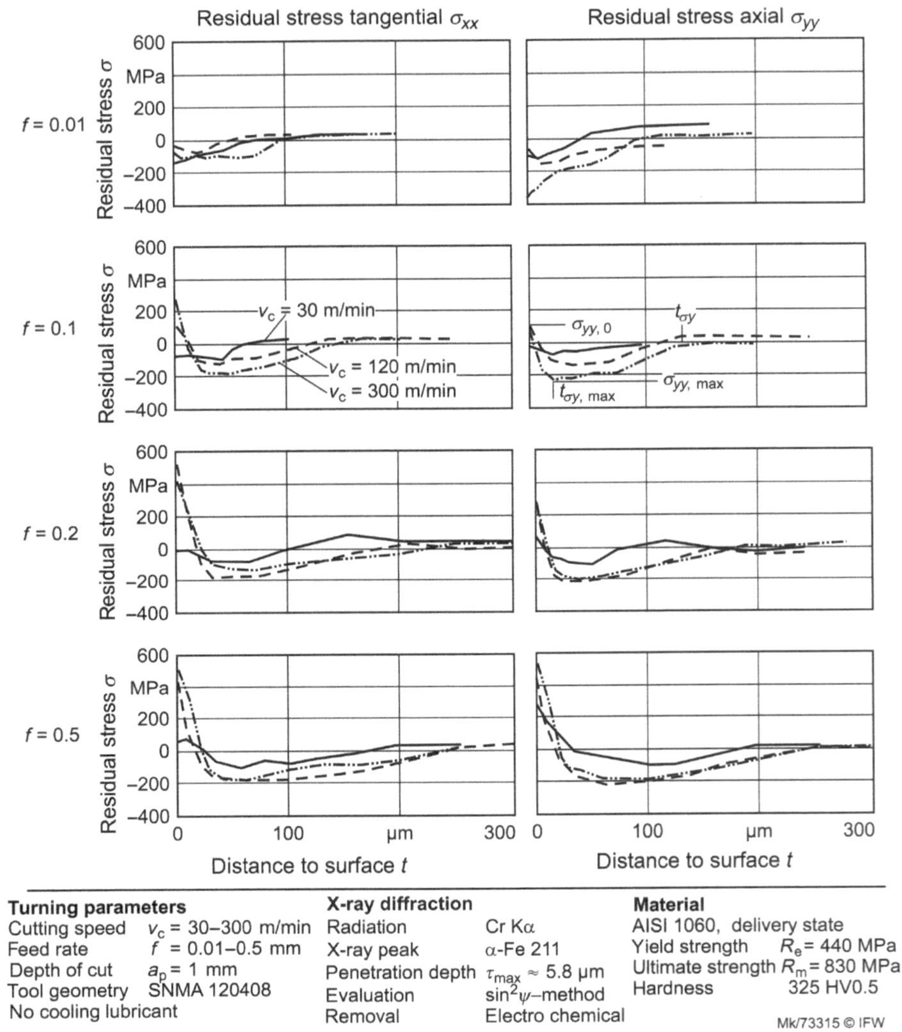

Fig. 2.2.1.1 Generation of residual stresses in AISI 1060 by turning [3].

strength of 330 and 1606 MPa is investigated [4–7]. Here, the repeatability of the axial residual stresses moves into a range of 20 MPa $< \sigma_{yy} <$ 200 MPa for the fully annealed and 50 MPa $< \sigma_{yy} <$ 250 MPa for the hardened steel. A stable deep rolling process with constant contact conditions results in good repeatability. There is no influence from tool wear or a deviation in depth of cut or feed rate.

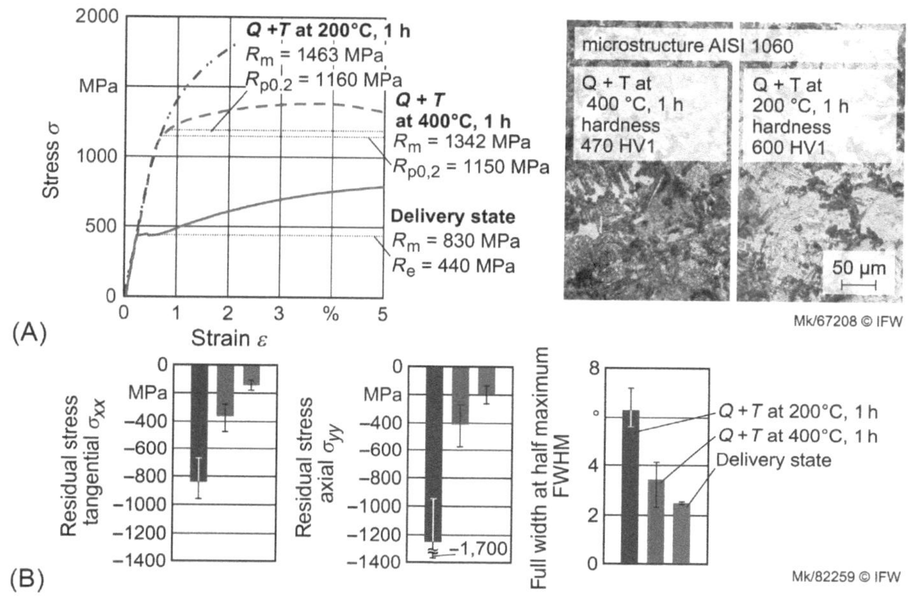

Fig. 2.2.1.2 (A) Mechanical properties and microstructure of AISI 1060 and (B) residual stress state by turning.

2.2.1.2.3 Generation of residual stresses depending on the carbon content of the steel

For general knowledge about the relaxation behavior of residual stresses it is necessary to investigate different kinds of steel, for example, in terms of their various carbon contents. Therefore alternate bending specimens of AISI 1035 and 52100 steel are turned with two process parameter sets and the initial residual stress states are compared to AISI 1060 (Fig. 2.2.1.3). The superposition of thermal and mechanical strains during the cutting process induces the surface residual stresses shown in Fig. 2.2.1.3. All specimens turned with a cutting speed of $v_c = 120$ m/min and a feed rate of $f = 0.5$ mm have high tensile residual stresses in the upper subsurface. The content of carbon is not the main influence, but an increase in carbon content results in a decrease of residual stresses. The yield strength of the material might also be responsible for this effect, however. In such a case, the AISI 1060 specimens should have the highest tensile stresses. Due to the small feed rate of $f = 0.01$ mm and a cutting speed of $v_c = 30$ m/min, the surface residual stresses, detached from the carbon content, are at a low to medium compressive level. There is no uniform chip formation at this small feed rate so

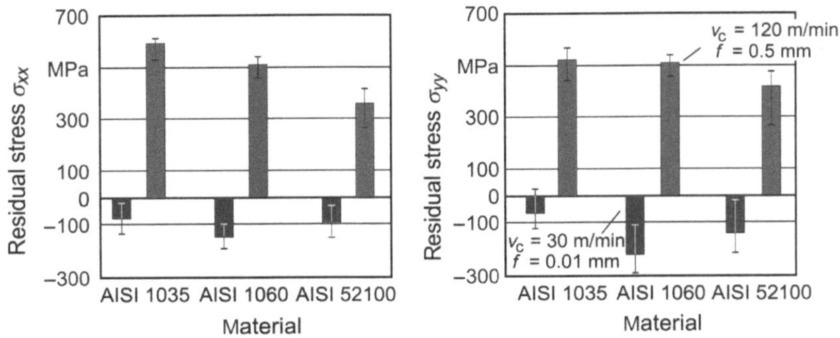

Turning parameters		**Materials**	**X-ray diffraction**	
Cutting speed	v_c = 30/120 m/min	AISI 1035, 1060, 52100 steel	Radiation	Cr Kα
Feed rate	f = 0.01/0.5 mm	9 specimens per set	X-ray peak	α-Fe 211
Depth of cut	a_p = 0.5 mm	Yield strength	Penetration depth	$\tau_{max} \approx 5.8$ µm
Tool geometry	SNMA 120408	R_e = 417, 440, 385 MPa	Evaluation	$\sin^2\psi$–method
No cooling lubricant		Ultimate strength		
		R_m = 630, 830, 635 MPa		

Hoc/80991 © IFW

Fig. 2.2.1.3 Comparison of surface residual stresses in AISI 1035, 1060, 52100 after turning [3].

that the material is squeezed under the cutting edge and the subsurface is plastically deformed. The carbon content is not the primary factor for the formation of residual stresses. Yield strength and material also need to be considered for future investigations of induced residual stresses in steel with different carbon contents, yield strength and material hardness. If the values of yield strength and hardness prove to be the same for all carbon contents, the description of the influence due to the carbon content will be more significant. The interaction between material properties should not be disregarded either. For the generation of residual stresses, the entire system of workpiece properties and tool and contact conditions needs to be taken into account.

Similar to the investigations of residual stress inducement, the deep rolling process is used to induce residual stresses with a good repeatability. The result can be seen in Fig. 2.2.1.4 compared to the result of deep rolling AISI 1060.

There is also no significant influence of the carbon content on the residual stress state after deep rolling. In general, the residual stresses in axial direction of the workpiece are higher than those in tangential direction. When increasing rolling pressure, no appreciable increase in compressive tangential residual stresses for all three materials can be observed. In radial direction of the shaft, pressure has a significant influence on the maximum value and the depth of the compressive residual stress. Compared to the turning-induced residual stresses, the deviation in

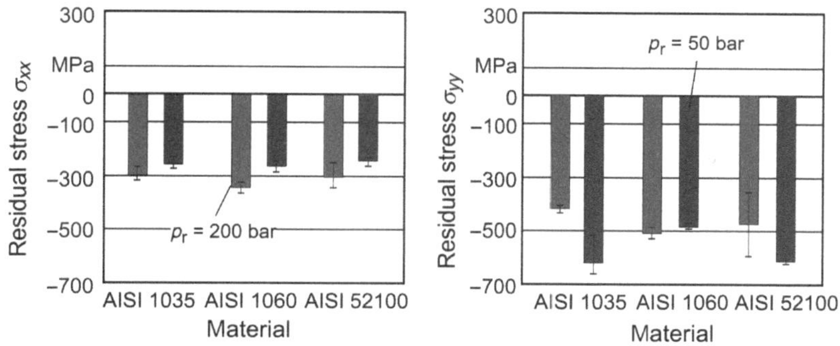

Deep rolling parameters
Rolling speed v_r = 100 m/min
Feed rate f = 0.05 mm
Rolling pres. p_r = 50/200 bar
Tool diameter d_t = 6.35 mm

Materials
AISI 1035, 1060, 52100 steel
9 specimens per set
Yield strength
R_e= 417, 440, 385 MPa
Ultimate strength
R_m = 630, 830, 635 MPa

X-ray diffraction

Radiation	Cr Kα
X-ray peak	α-Fe 211
Penetration depth	$\tau_{max} \approx$ 5.8 µm
Evaluation	$\sin^2\psi$–method

Hoc/80991a © IFW

Fig. 2.2.1.4 Comparison of surface residual stresses in AISI 1035, 1060, 52100 after deep rolling [5,6].

measuring values is smaller and the initial state can be adjusted more confidently by deep rolling.

2.2.1.2.4 Generation of residual stress gradients

In addition to the investigation of how equally distributed residual stresses and the subsequent relaxation behavior are generated, the generation of locally induced residual stresses is also of interest. It is assumed that high gradients of residual stresses cause a more significant relaxation behavior. Therefore the reconstruction of enduring loads needs to be more accurate. First, the generation of bordered residual stress areas is analyzed by turning. Shafts of AISI 52100 with a diameter of 10 mm are turned at a constant feed rate and with different cutting speeds. It is assumed that surface roughness would have no influence on the measuring results because of the identical feed rate. A ceramic tool and a carbide tool are used to investigate the influence of the tool material. In Fig. 2.2.1.5, the results of the residual stress measurements in feed direction of two shafts with a measuring point distance of 2 mm can be seen. The results of the shaft turned by a ceramic tool show high tensile residual stresses. The carbide tool causes rather compressive residual stresses. No significant gradient can be observed between the turned sections with increasing and those with decreasing cutting speeds. Furthermore, it is impossible to generate a homogeneous residual stress

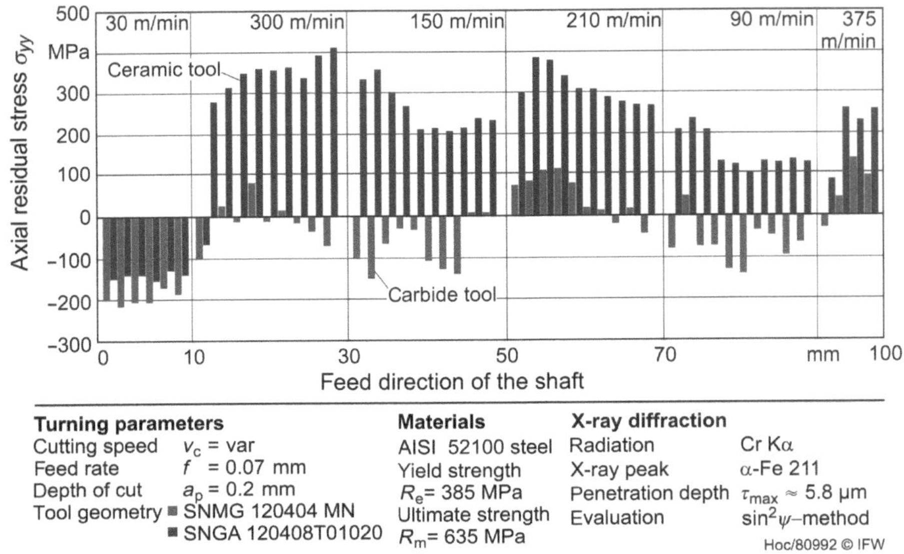

Fig. 2.2.1.5 Gradients of residual stresses by turning with two different tools.

state over a 20 mm area by means of turning. Due to the acceleration and delay of the spindle, the cutting conditions never reaches stability. This, among other characteristics, is manifested in the resulting undulating surface.

The AISI 1035 and 52100 deep rolling specimens from chapter 1.3 are used for the relaxation investigations. The gradients between turned and deep-rolled surfaces are significantly steep enough. In the turned surface, both materials show about 600 MPa tensile residual stress, whereas the deep-rolled surface shows about −500 MPa compressive residual stress. Before the influence of steep gradients can be tested, however, an investigation of the relaxation behavior of equally distributed residual stresses is necessary.

2.2.1.3 Relaxation of residual stresses

After investigating the generation of residual stresses in steel at different heat treatments and with different carbon contents, the relaxation behavior under alternating bending loads is tested. Alternating bending creates a cyclic stress in the specimen by turning it in a bent state, whereby the subsurface relaxes uniformly around the circumference. For the investigation of the relaxation behavior, the 54 specimens listed in Fig. 2.2.1.3 are tested at different loads to find the critical load for relaxation initiation. The alternating bending experiments are carried out using a rotating bending device, type Zwick/Roell UBM 200 tC. A constant bending torque is applied to each specimen. The setup is self-centering; in that way, insufficient linearity of the specimens can be compensated. The highest load stress is located on the specimen's surface. If the superposition of load and residual stresses following Eq. (2.1.2.1) exceeds the yield strength of the material, the residual stresses relax due to yielding. Following a common approach from literature [8], the description of residual stress relaxation is separated into a quasi-static (number of loads $0<N\leq 1$) and a cyclic phase. The latter describes the relaxation for a number of load cycles $N>1$. In both phases, an equivalent stress σ_v larger than the quasi-static or cyclic yield strength of the material (R_e^{stat} or R_e^{cycl}) leads to residual stress relaxation. As a consequence of the machining process, a plain residual stress state is present which has to be taken into account for the determination of the equivalent stress. Since the load stress is applied in axial direction of the specimen, the superposition of residual and load stresses requires a transformation of the principal residual stresses into the specimen's coordinate system. Therefore normal and shear stresses have to be taken into account. Providing a uniaxial load stress σ_{load} in axial direction (y) is applied, the equivalent stress is described by following the von Mises yield criteria equation (Eq. 2.2.1.2).

$$\sigma_v = \sqrt{\sigma_{xx}^2 - \sigma_{xx}\left(\sigma_{yy} + \sigma_{load}\right) + \left(\sigma_{yy} + \sigma_{load}\right)^2 + 3\tau_{xy}^2} \qquad (2.2.1.1)$$

The critical load stress results for $\sigma_v = R_e^{stat}$ or R_e^{cycl} to:

$$\sigma_{crit,load} = \frac{\sigma_{xx}}{2} - \sigma_{yy} \pm \frac{\sqrt{4 \cdot R_e^{i2} - 3 \cdot \sigma_{xx}^2 - 12 \cdot \tau_{xy}^2}}{2} \qquad (2.2.1.2)$$

Considering the residual stresses of the machined AISI 1060 specimens (Table 2.2.1.1), the critical load stresses are equal to $\sigma_{crit,\ load}^{stat} = -115\,\text{MPa}$ and +549 MPa as well as $\sigma_{crit,\ load}^{cycl} = -204\,\text{MPa}$ and +617 MPa, respectively.

In both cases, the compressive load will cause relaxation. Fatigue specimens are produced and the residual stress is determined according to the description earlier. Subsequently, fatigue experiments are carried out applying three different load stress amplitudes. Three identically machined specimens are used for each load stress amplitude in order to determine the influence of load stress and the number of load cycles on residual stress relaxation. Load stress amplitudes above the critical static stress (load stress 1) and above the critical cyclic stress (load stress 2) are applied. Additionally, a load stress of 80% of the material's yield strength (load stress 3) is analyzed (see Table 2.2.1.2). The residual stresses are determined after $N=1$, 10, 100, 1000, and 10,000 load cycles.

Fig. 2.2.1.6 depicts the residual stress relaxation with an increasing number of load cycles for the various load stresses. A load stress slightly higher than the critical quasi-

Table 2.2.1.1 Results of residual stress measurements directly after machining

Residual stresses type I								Types II & III	
σ_{xx} [MPa]		σ_{yy} [MPa]		τ_{xy}[a] [MPa]		σ_v[b] [MPa]		FWHM [°]	
μ	s	μ	s	μ	s	μ	s	μ	s
−286	16	−348	32	−86	15	238	27	2.1624	0.054

μ = mean value; s = std. dev. (sample).
[a]According to Dölle [9].
[b]According to von Mises [10].

Table 2.2.1.2 Applied load stresses

Load stress 1	$\sigma_{load} = \pm 132$ MPa
Load stress 2	$\sigma_{load} = \pm 220$ MPa
Load stress 3	$\sigma_{load} = \pm 352$ MPa

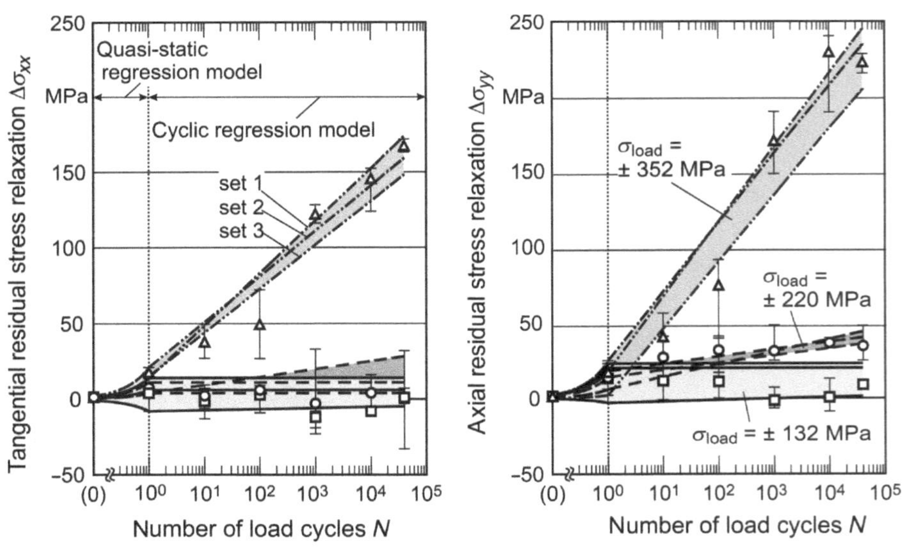

Fig. 2.2.1.6 Residual stress relaxation in tangential and in axial directions due to mechanical loading.

static load leads to a relaxation after the first load cycle. Successive load cycles do not cause any significant further relaxation. The residual stress relaxation in tangential (x) as well as in axial direction (y) increases with the increase in number of load cycles for loads above the critical cyclic load stress. When the load increases, quasi-static and cyclic relaxation increase as well. The quasi-static and cyclic relaxations show a linear dependence on the number of load cycles and the load stress. The full width at half maximum (FWHM) describes the width of the peak at half of its maximum intensity on the 2-theta scale in degrees. It correlates with grain size and dislocation density. As grain refinement does not matter in the performed experiments, all observed peak broadenings are attached to changes of dislocation density and, thus residual stresses of type II and type III. FWHM and residual stresses of type I show different behaviors due to fatigue loads. Fig. 2.2.1.7 shows that AISI 1060 exhibits no significant change of FWHM due to the load stress amplitudes below $\sigma_{load} = \pm$ 352 MPa $\approx$ 80% R_e. A linear decrease leading to ΔFWHM $=$ 0.3° after $N =$ 100,000 load cycles results. The difference between the residual stress and FWHM behaviors makes the consideration of FWHM feasible, thus creating a model for reconstructing the loads of a component [11].

The described relaxation behavior can also be observed for the alternating bending specimens of AISI 1035 and 52100. As shown in Chapter 1.3, the carbon content has no influence on the induced residual stresses due to the unbalanced yield strength. A similar conclusion in terms of relaxation behavior is drawn: carbon content and material properties exert no significant influence on relaxation.

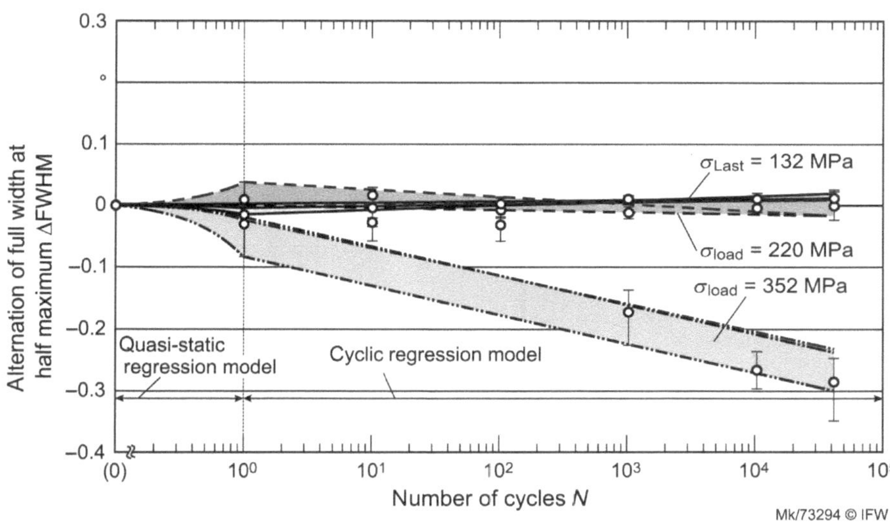

Fig. 2.2.1.7 Alteration of FWHM due to mechanical loading.

In order to test the hypothesis from Chapter 1.4, in particular the fact that steep gradients of residual stresses cause a more significant relaxation behavior, the deep-rolled alternating bending specimens of AISI 1035 and 52100 are loaded in the same way as the turned specimens (Fig. 2.2.1.6). According to Eq. (2.2.1.2) and the superposition theory, a low load stress of ± 77 to ± 175 MPa for the AISI 52100 and ± 25 to ± 190 MPa for the AISI 1035 specimens, referring to the initial state of residual stresses, should be enough to cause a relaxation. After the first and tenth cycle, however, no relaxation is measurable. To force a relaxation of residual stresses, each specimen is overloaded with a load stress of 250 MPa. This means that the von Mises equivalent stress on the surface exceeds the ultimate strength for all tested specimens. This load is applied for another 1, 10, 100, and 1000 load cycles on the same specimens. The result of this relaxation test is shown in Fig. 2.2.1.8. The residual stresses of the turned specimens with a deep-rolled area in the middle section do not relax at all.

This effect is explained as follows: Due to deep rolling, the grains of the microstructure are squeezed and the hardness is, thus increased from 225 to 308 HV1. This reduces the ability of the microstructure to transfer meaning that the relaxation of residual stresses is blocked.

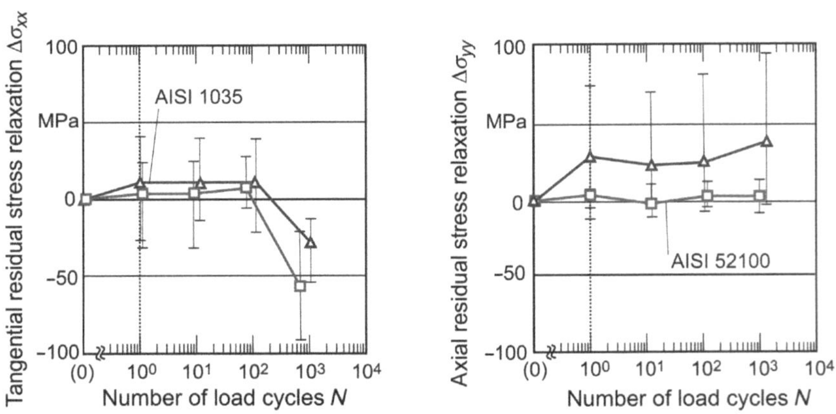

Deep rolling parameters
Rolling speed v_r = 100 m/min
Feed rate f = 0.05 mm
Rolling pres. p_r = 200 bar
Tool diameter d_t = 6.35 mm

Materials
AISI 1035, 52100 steel
9 specimens per set
Yield strength
R_e= 417, 385 MPa
Ultimate strength
R_m = 630, 635 MPa

X-ray diffraction
Radiation Cr Kα
X-ray peak α-Fe 211
Penetration depth $\tau_{max} \approx 5.8$ µm
Evaluation $\sin^2\psi$–method

Hoc/80991 © IFW

Fig. 2.2.1.8 Relaxation of deep-rolled axial and tangential residual stresses after alternating bending load.

2.2.1.4 Harmonic analysis of eddy current signals

Residual stresses are usually measured with X-ray diffraction. For applications in a shop floor environment, a method for determining residual stresses using eddy current is to be developed. Mechanical as well as electromagnetic material characteristics are defined by alloy composition, crystalline structure, grain size, along with the material's strain. In ferromagnetic materials, eddy current signals depend on the characteristics of the magnetic hysteresis. Due to the nonlinear dependency on magnetization and a magnetic field, the measuring frequency results in higher harmonics, which is, in turn, characteristic for the state of the material [12]. Mechanical and electrical properties of each specimen are determined by the 1st, 3rd, and 5th harmonic responses. The 3rd and 5th harmonic responses are regarded as providing qualitative information about residual stresses [13]. Details on the experimental setup are summarized in Table 2.2.1.3. Due to the skin effect, the penetration depth δ of the presented method is limited. It is estimated by using the excitation frequency f_p, electrical conductivity σ_{elec}, vacuum permeability μ_0, and the magnetic permeability μ_r, as found in Eq. (2.2.1.3) [14]:

$$\delta = \sqrt{\frac{1}{\pi \cdot f_p \cdot \sigma_{elec} \cdot \mu_0 \cdot \mu_r}} \tag{2.2.1.3}$$

The penetration depth is defined as the distance from the specimen's surface, where the eddy current density drops to $1/e$ (approximately 36.8%), compared to the value at the surface. In addition to the experimentally determined electrical conductivity $\sigma_{elec} = 7.9\,1/(\Omega\text{m})$, the magnetic permeability μ_r needs to be determined. Based on measurements for AISI 1035 low carbon steel, a bandwidth of μ_r 400 ± 200 should be taken into account. Accuracy can be increased by a specific measurement of μ_r. Since μ_r changes depending on the strength of the magnetic field, a bandwidth for AISI 1060, similar to AISI 1035, is expected. According to Eq. (2.2.1.3), when applying the vacuum permeability $\mu_0 = 4 \cdot \pi \cdot 10^{-7}$ V s/(A m), an excitation frequency of $f_p = 3.2$ kHz causes a penetration depth of $\delta = 177 \pm 48$ µm. Fig. 2.2.1.9 illustrates the penetration depth of eddy current measurements and X-ray diffraction regarding the measured residual stress depth profiles.

Table 2.2.1.3 **Details of eddy current measurements**

Probe diameter d **(mm)**	**Core material**	**Current** I **(A)**	**Voltage** U **(V)**	**Excitation frequency** f_p **(kHz)**
12	ferrite	0.6	4	3.2

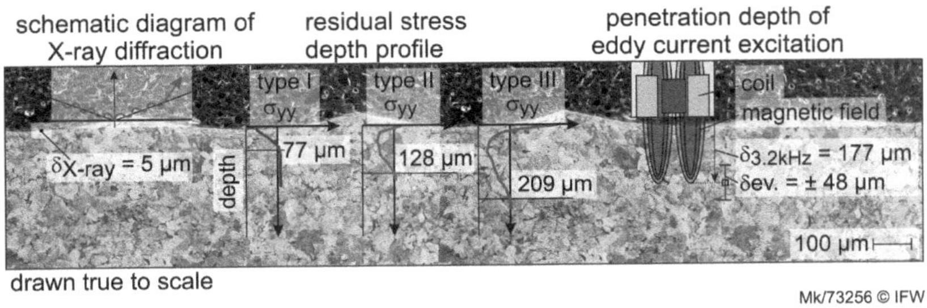

Fig. 2.2.1.9 Penetration depth of eddy current measurements and X-ray diffraction.

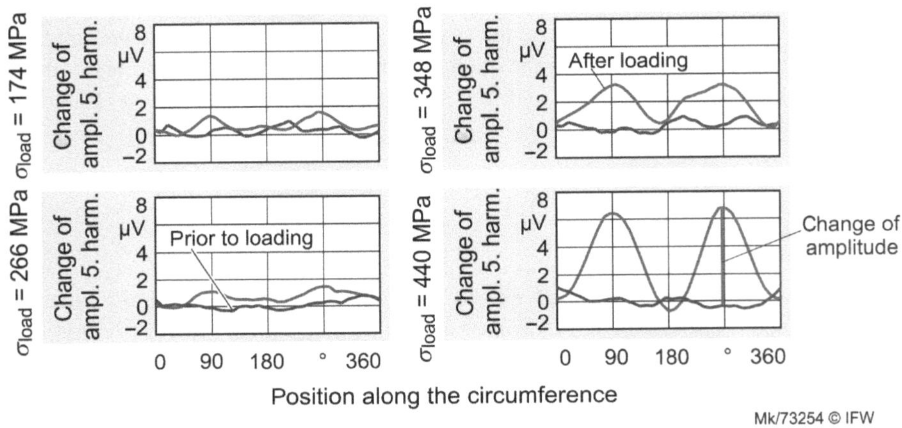

Fig. 2.2.1.10 Amplitude change of 5th harmonic response.

Since the depth analyzed by eddy current is 62%–290% of the depth affected by residual stresses, the reduction of the penetration depth leads to better comparability. The X-ray diffraction results of AISI 1060 presented here correlate with the changes in the 5th harmonic response of the eddy current measurement. Fig. 2.2.1.10 illustrates the amplitude changes for eddy current analyses along the circumference of the specimens. This is carried out for all specimen types and load stress amplitudes. Each diagram illustrates the measurement before and after load stress application. Furthermore, the phase change is also analyzed. The neutral axis of the load stress is located at 0° and 180° along the circumference. For specimen type I (residual stresses in load direction $\sigma_{yy} = -145$ MPa), the critical load is exceeded for $\sigma_{\text{load}} > 298$ MPa. Residual stresses of type I show changes of $\Delta\sigma_{\text{v}} = 31$ MPa for $\sigma_{\text{load}} = 348$ MPa, at which point the 5th amplitude of harmonic response shows significant changes. These increase with the load amplitude. The two peaks along the circumference can be explicitly associated with the maximum and neutral stress axes. For high load stress amplitudes ($\sigma_{load} > 266$ MPa), changes in the 5th harmonic amplitude can be determined for the specimens (residual stresses in load direction $\sigma_{yy} = +65$ MPa). A significant change in phase, though weak in amplitude, can be determined for $\sigma_{load} = 266$ MPa. In this case, $\sigma^*_{v,crit} \approx 30\%\ R_m$ (242 MPa) is slightly exceeded in terms of tensile stress. Current research includes a reduction of the penetration depth of eddy current in order to focus on the near surface area affected by residual stresses. According to Eq. (2.2.1.3), a doubling of the frequency to $f_p = 6.4$ kHz decreases the penetration depth to $\delta = 125 \pm 34$ μm, which represents 44%–205% of the depth affected by residual stresses.

In conclusion, the eddy current measurements show promising results when correlated to X-ray diffraction and seem to be qualified for use as an alternative method in terms of shop-floor application. The measuring of exact values, as carried out using X-ray diffraction, is not possible. Correlation quality, however, can be increased by reducing the eddy current penetration depth [15].

2.2.1.5 Reconstruction of loads

The common equations introduced by Kodama [8] are combined and resolved to determine the possible combinations of load stress and number of load cycles that lead to a given residual stress and to FWHM alterations. This approach is divided into quasi-static and cyclic behavior. At first, the results depicted in Figs. 2.2.1.6 and 2.2.1.7 are described according to Eq. (2.2.1.4). The quasi-static residual stress relaxations in both axial and tangential directions caused by the first load cycle ($0 < N \leq 1$) are approximated linearly to the load stress (Eq. 2.2.1.4).

$$\sigma_i(N) = \sigma_i(N=0) + b_i \cdot \sigma_{load} \cdot N \qquad (2.2.1.4)$$

$\sigma_i(N=0)$ represents the initial residual stress value. The coefficient b_i is determined in order to describe the relaxed stress per load cycle depending on the amplitude of the load stress. Exceeding the critical quasi-static load stress, b_i increases linearly to the load stress amplitude (Eq. 2.2.1.5).

$$b_i(\sigma_{load}) = p_1 \cdot \sigma_{load} + p_2 \qquad (2.2.1.5)$$

The corresponding influence of the cyclic load $\left(1 < N \leq 10^6\right)$ is quantified by a logarithmic function presented by [8]. Based on the condition after the first cycle $\sigma_i(N=1)$, the relaxation is described in terms of linear dependence vis-à-vis the logarithm of the number of load cycles (Eq. 2.2.1.6).

$$\sigma_i(N) = \sigma_i(N=1) - A_i \cdot \sigma_{load} \cdot \log(N) \qquad (2.2.1.6)$$

The coefficient A_i is determined in order to describe the cyclic relaxation per load cycle, depending on the load stress amplitude. Exceeding the critical cyclic load stress, A_i increases linearly over the load stress amplitude (Eq. 2.2.1.7).

$$A_i(\sigma_{\text{load}}) = q_1 \cdot \sigma_{\text{load}} + q_2 \tag{2.2.1.7}$$

In addition to the relaxation of the residual stresses in both axial and tangential directions, the change in the full width at half maximum is quantified according to the same procedure.

Once the initial condition of a component is given, the residual stress relaxation and the alteration of FWHM can be determined. Presuming a number of load cycles $N > 1$, the combination of Eqs. (2.2.1.4) and (2.2.1.6) leads to Eq. (2.2.1.8) and can be applied.

$$\sigma_i(N) = \sigma_i(N=0) + b_i \cdot \sigma_{\text{load}} - A_i \cdot \sigma_{\text{load}} \cdot \log(N) \tag{2.2.1.8}$$

Next, the dependencies of b_i and A_i on load stress are considered by substituting the coefficients with Eqs. (2.2.1.5) and (2.2.1.7). Since Eqs. (2.2.1.4) and (2.2.1.6) depend on two unknown elements, neither the individual equations nor the combination can be resolved. The infinite combinations of load stress and number of cycles can lead to specific changes. To quantify any possible solutions, the resulting equation is formulated (Eq. 2.2.1.9).

$$N(\sigma_{\text{load}}) = \exp\left(\frac{\sigma_i(N=0) - \Delta\sigma_i + \sigma_{\text{load}}(p_2 + \sigma_{\text{load}} \cdot p_1)}{q_2 + \sigma_{\text{load}} \cdot q_1}\right) \tag{2.2.1.9}$$

The previous equation is used for the residual stress in the axial and tangential directions (σ_{yy}, σ_{xx}) as well as for the FWHM. To illustrate the procedure, the initial conditions shown in Table 2.2.1.4 and changes in subsurface properties are assumed.

Fig. 2.2.1.11 illustrates all possible combinations of load stress and number of load cycles that can cause the given changes in axial and tangential residual stresses (solid and dashed lines) and in FWHM (dashed/dotted line). The intersections represent

Table 2.2.1.4 Changes in subsurface properties underlying Fig. 2.2.1.11

Residual stresses in tangential direction	$\Delta\sigma_{xx} = 113$ MPa
Residual stresses in axial direction	$\Delta\sigma_{yy} = 189$ MPa
Full width at half maximum	ΔFWHM $= 0.16°$

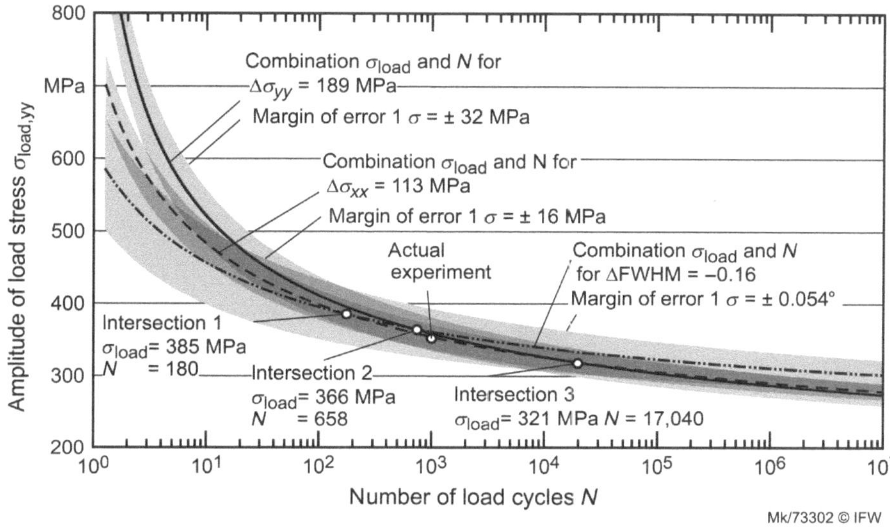

Fig. 2.2.1.11 Possible combinations of load stress and number of load cycles causing any given change in residual stresses in axial and in tangential directions as well as in FWHM.

possible combinations of interest since the solution must be true for all properties. The experimental parameters to be determined by the model are $\sigma_{load} = 352$ MPa and $N = 1000$. Three intersections are found at $\sigma_{load} = 385$ MPa and $N = 180$, $\sigma_{load} = 366$ MPa and $N = 658$, and $\sigma_{load} = 321$ MPa and $N = 17{,}040$. The model therefore either overestimates or underestimates the load stress by +33 or −30 MPa. Due to the logarithmic character of the underlying equations, the number of load cycles is overestimated or underestimated by +16,040 or −820.

Certain inaccuracies have to be taken into account in order to estimate the significance of the possible load cases determined. The deviation in the initial subsurface condition of the machined specimens is summarized in Table 2.2.1.2. The measured deviation 1σ between 16 and 32 MPa is individually considered and illustrated in Fig. 2.2.1.11 by gray shading. The range of load cases expands from $\sigma_{load} = 540$ MPa and N = 5 to $\sigma_{load} = 268$ MPa and N = 107. Due to the high uncertainty, the possible combinations cover almost the full range of load cycles. The load amplitude is narrowed down to a range of about 272 MPa [16].

The presented method is capable of determining loads above the critical load stress, as observed here. The estimation of possible load stress amplitudes is significantly more precise than for the number of cycles. In addition to the level of relaxation, the margin of error given by the reproducibility of the initial subsurface condition after machining is substantial for the model's accuracy.

References

[1] B. Scholtes, Eigenspannungen in mechanisch randschichtverformten Werkstoffzuständen, Ursachen-Ermittlung-Bewertung, DGM-Informationsgesellschaft, Oberursel, 1990.

[2] E. Macherauch, H. Wohlfahrt, Eigenspannungen und Ermüdung, in: Ermüdungsverhalten metallischer Werkstoffe, DGM-Informationsgesellschaft Verlag, 1985, pp. 237–283.

[3] T. Mörke, Randzonenanalyse zur Bestimmung mechanischer Belastungen im Lebenszyklus spanend gefertigter Bauteile, Leibniz Universität Dissertation Hannover, Berichte aus dem IFW, PZH-Verlag, Garbsen, 2016.

[4] B. Denkena, A. Abrao, B. Breidenstein, T. Mörke, Surface characterization of components subjected to deep rolling for cyclic loading applications, Lecture Notes in Production Engineering, Springer, 2013, pp. 47–53.

[5] B. Denkena, A. Abrao, B. Breidenstein, T. Mörke, Surface and subsurface alterations induced by deep rolling of hardened AISI 1060 steel, Prod. Eng. Res. Dev. (WGP) 8 (5) (2014) 551–558.

[6] B. Denkena, A. Abrao, B. Breidenstein, T. Mörke, The influence of deep rolling on the surface integrity of AISI high carbon steel, Procedia CIRP 13 (2014) 31–36.

[7] B. Denkena, A. Abrao, B. Breidenstein, T. Mörke, Alterações superficiais e subsuperficiais induzidas no aço alto carbono C60 pela operação de roleteamento, in: Proceedings of the VIII Congresso Nacional de Engenharia Mecânica, Uberlandia, MG, Brazil, vol. 10, 2014.

[8] S. Kodama, The behavior of residual stress during fatigue stress cycles, in: Society of Materials Science Japan (Ed.), Mechanical Behavior of Materials—Fatigue of Metals, vol. 2, 1972, pp. 111–118.

[9] H. Dölle, V. Hauk, Röntgenographische Spannungsermittlung für Eigenspannungssysteme allgemeiner Orientierung, Härterei-Tech 31 (1976) 165–168.

[10] R. von Mises, Mechanik der festen Körper im plastisch deformablen Zustand, Göttingen, Nachr.: Math.-Phys. Klasse 1 (1913) 582–592.

[11] B. Denkena, J. Köhler, B. Breidenstein, T. Mörke, Surface integrity—an inherent load sensor, Adv. Mater. Res. 797 (2013) 679–684.

[12] B. Heutling, Zerstörungsfreie Online-Materialcharakterisierung von Stahlfeinblechen mittels Harmonischen Analyse von Wirbelstromsignalen, Dissertation Leibniz Universität Hannover, PZH-Verlag, Garbsen, 2004.

[13] H. Stroppe, K. Schiebold, Wirbelstrom–Materialprüfung, Castell Verlag, Wuppertal, 2005.

[14] DIN EN ISO 15549:2011-03: Non-destructive testing—Eddy current testing—General principles (ISO 15549:2008), Deutsche EN ISO 15549:2010, Beuth, Berlin, 2010.

[15] B. Denkena, B. Breidenstein, W. Reimche, G. Mroz, T. Mörke, H.J. Maier, Changes of subsurface properties due to fatigue determined by $\sin^2$psi-method and harmonic analysis of eddy current signals, Procedia Technol. 15 (2014) 503–513.

[16] B. Breidenstein, B. Denkena, T. Mörke, T. Hockauf, Estimation of load history by residual stress relaxation, Procedia CIRP 37 (2015) 236–241.

2.2.2 Storing the load history of a component in the subsurface region

Sebastian Barton, Gregor Mroz, Wilfried Reimche, Hans Jürgen Maier
Institute of Material Science, Leibniz Universität Hannover, Garbsen, Germany

2.2.2.1 Introduction

Despite continual enhancements in the field of sensor technology, there is still lack of reliable sensors that allow for assessing the integrity of safety-relevant, highly loaded components. Thus abrupt breakdowns frequently occur due to unforeseeable component failure despite employing simulation-based designs, structural durability computations, and periodic inspections of technical equipment. This situation leads to high follow-up costs and, moreover, a risk exists of endangering people and/or the environment. Clearly, corrective measures are needed, and ideally the component should autonomously collect and provide information about its loading history experienced during its use phase. This information should be easy to retrieve at any time.

Given this scenario, the objective was to develop a technology that allows for continuously and permanently monitor the status of safety-relevant, highly loaded components during their life cycle. With the current approach, the load sensors are directly integrated into the component in order to detect and store the loading history by the component itself. Specifically, the component-inherent loading sensors are based on deformation-induced martensite of metastable, austenitic materials and a local laser-induced heat treatment of the formed martensite to set the yield stress values in small regions of the material. The type of phase locally present after the usage is then retrieved using modified nondestructive retrieval technologies. This in turn provides the information about the local loading history.

2.2.2.2 Qualifying metastable austenitic steels as sensor materials

The properties of metastable austenitic steels and the occurrence of deformation-induced martensite during plastic deformation have been widely studied in the scientific literature, e.g., [1–4]. The steel's face-centered cubic crystal lattice is transformed into a distorted, tetragonal body-centered martensitic lattice during plastic

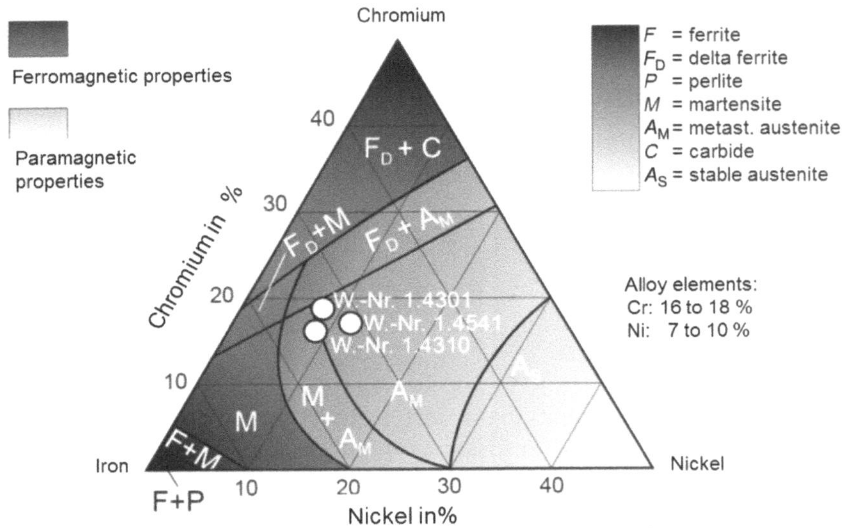

Fig. 2.2.2.1 Influence of alloying elements on the microstructural constituents present in CrNi steels.
Recompiled from H. Dietrich, W. Heimann, H. Strom, Einfluss der Kaltverformung auf die mechanischen und magnetischen Eigenschaften von nichtrostenden Stählen, Thyssen Edelstahl Technische Berichte TEW (1976) 61–69 with permission from Outokumpu.

deformation. Both the fraction of paramagnetic ε-martensite increases as well as that of the ferromagnetic α′-martensite. Due to the retention of the hard and brittle martensite in the soft austenitic microstructure, the toughness and the Young's modulus are reduced while the strength values are increased.

The austenitic steels that are widely used in engineering typically feature a nickel content of 9–16 wt.% and a chromium content of 16–21 wt.%. In particular, austenitic steels having a nickel content of 7–10 wt.% and a chromium content of 16–18 wt.% demonstrate pronounced metastable properties, cf. Fig. 2.2.2.1. Such steels qualify for use as high strength spring steel due to the onset of martensite formation during cold forming [5].

In addition to the mechanical properties, which are substantially altered due to the onset of martensite formation during plastic deformation, the electrical and magnetic properties are also changed upon the phase transformation from austenite to martensite [6]. Thus this effect can be exploited to monitor the phase transformation. In the present study, both the high-frequency eddy current technology and the magnetic-inductive testing method along with the harmonic analysis of the eddy current signals were employed to nondestructively assess the microstructural state.

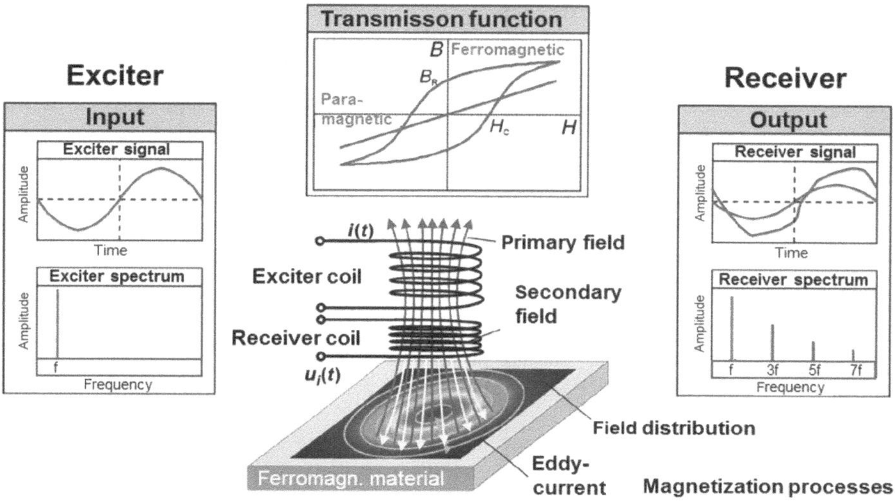

Fig. 2.2.2.2 Functional principle of the harmonic analysis of eddy current signals [7].

For the harmonic analyze of the eddy current signals (Fig. 2.2.2.2), a transmitter coil generates a sinusoidal primary alternating magnetic field that interacts with the material. Depending on the test frequency f, the permeability μ, the conductivity σ, and the test piece geometry, magnetizing processes and eddy currents develop in the material, which in turn lead to the emergence of a secondary magnetic field. The superposition of both magnetic fields is detected using receiver coils. Specifically, the induced measured voltage is distorted due to the nonlinear relationship between the magnetic field strength H and the magnetic flux density B. Owing to this distortion, the measured signal contains higher harmonic signal constituents, which can be analyzed in terms of their amplitudes and phase values, or their real and imaginary parts [5,8–10]. By selectively acquiring and evaluating higher harmonic data from eddy current signals, both the change in magnetic parameters can be sensitively detected in ferromagnetic materials as well as the phase transformation in austenitic materials owing to the formation of martensite. To quantify the martensitic phase fractions, different eddy current sensing probes were developed according to the inspection task and qualified for the application. Besides characterizing materials, the harmonic analysis of eddy current signals is also suitable for establishing the stress distribution in ferromagnetic materials. Hence, both residual stresses as well as loading stresses can be detected. These effects were employed for determining the loading stress in a tie rod of the Formula Student racing car RP09 (see Section 2.2.2.5.1) [5].

To qualify a suitable material for registering and storing loading data, various metastable materials were characterized with respect to their tendency to form as large a fraction of ferromagnetic ά-martensite as possible upon static loading. For this purpose, specimens of 1.4301, 1.4310, and 1.4541 were loaded in 5% strain steps up to a total strain of 35% and subsequently the fraction of deformation-induced martensite were determined with a local probe using the harmonic analyses of eddy current signals. In the 1.4310 material the largest changes in ferromagnetic phase content were obtained. Light microscopy analysis of the three materials at different strain steps revealed that the 1.4301 and 1.4310 materials exhibited particularly pronounced fractions of lathe-shaped α′-martensite formation following plastic deformation. Moreover, the martensite content increased continuously with increasing plastic deformation up to 35% permanent strain, as shown in Fig. 2.2.2.3. These results were also confirmed by scanning electron and transmission electron microscopy [5,11].

To better estimate the phase fractions present during deformation, specimens of the 1.4301 and 1.4310 materials in their undeformed and deformed states were also examined using X-ray diffractometry. Noticeably large differences occurred in the ratios of peak heights of the γ-phase and of the α′-phase as a function of the applied strain. The resulting peaks of the α′-phase clearly show the formation of deformation-induced martensite with plastic deformation, particularly for the 1.4310 material, cf. Fig. 2.2.2.4.

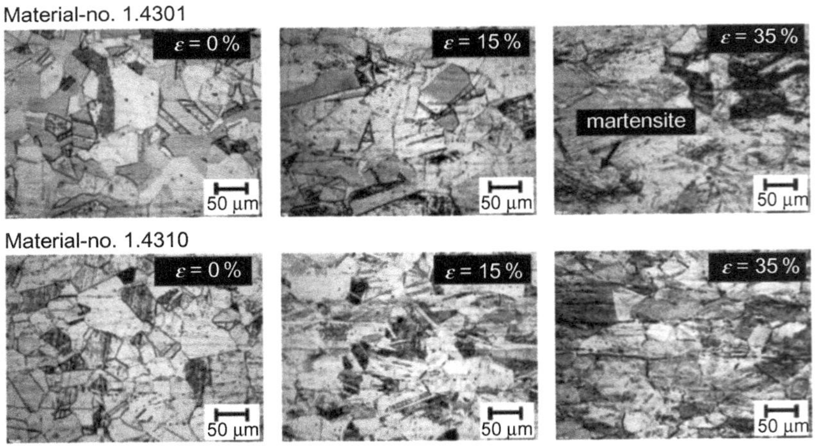

Fig. 2.2.2.3 Martensite formation, in 1.4301 and 1.4310 steel upon plastic straining. Recompiled from G. Mroz, Entwicklung von Techniken zur bauteilinhärenten Informationsspeicherung und Erfassung von Bauteilbelastungen, Dissertation, Leibniz Universität Hannover, Berichte aus dem IW, PZH Verlag, Garbsen, 2013.

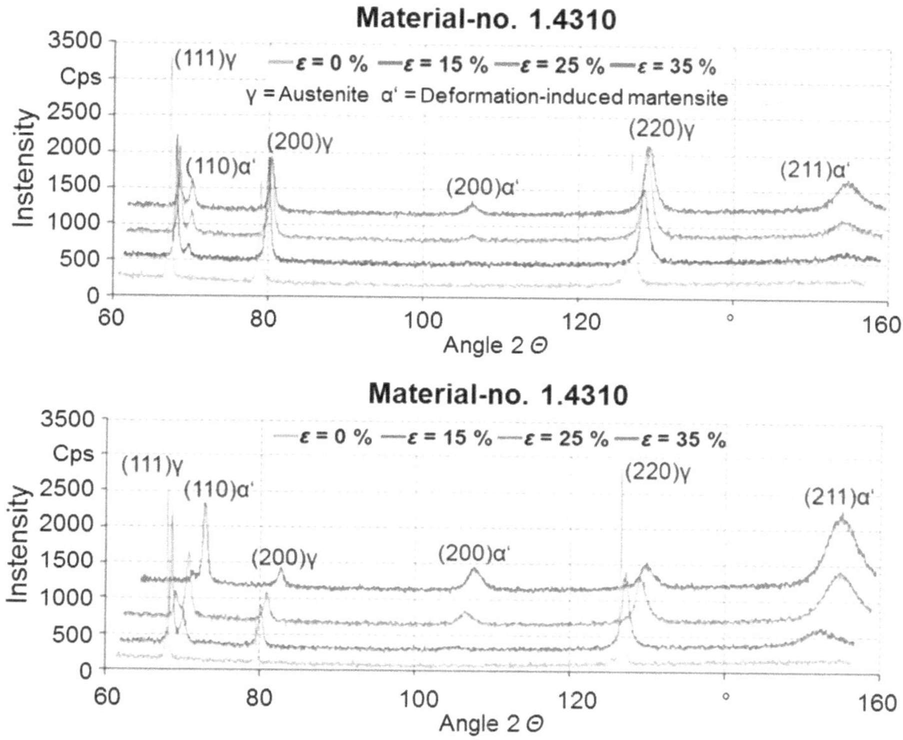

Fig. 2.2.2.4 X-ray diffraction diagrams used for estimating the phase fractions [5].

In order be able to understand the formation of martensite in the metastable austenitic material upon fatigue loading, cylindrical specimens were cycled in torsion and bending using a fatigue test rig. The specimens were continually probed in gauge section using an enveloping eddy current sensor. The formation of martensite was evaluated based on the harmonic analyses of the eddy current signals. Thus the stress level and the amount of plastic deformation at which the martensite formation initiated could be established. While lower loading stresses, which only resulted in elastic deformations, did not result in martensite formation, stresses above the yield stress triggered the onset of the phase transformation. The strain-induced formation of martensite also resulted in a material exhibiting enhanced ferromagnetic properties [5,6]. The increase in the ferromagnetic properties correlated with an increase in the amplitude of the 3rd harmonic, as shown in Fig. 2.2.2.5.

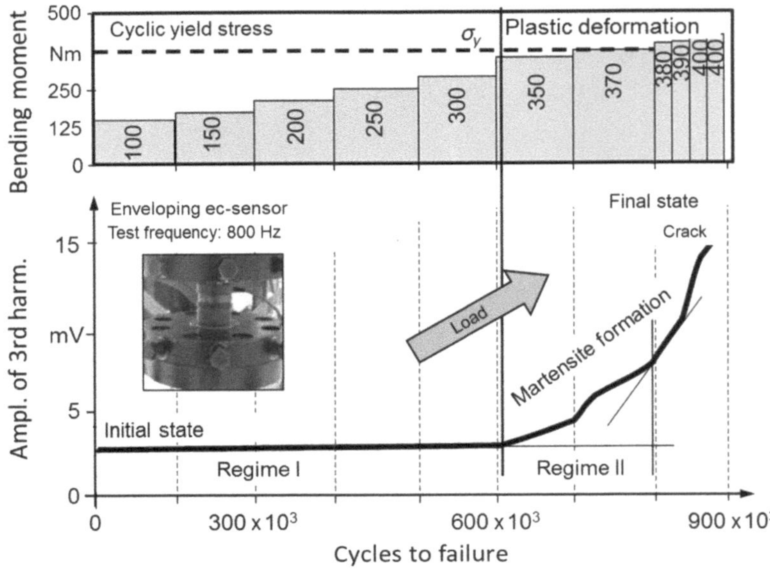

Fig. 2.2.2.5 Formation of martensite in cylindrical specimens subjected to cyclic loading [5].

In order to detect the formation of martensite in the near surface region of the cylindrical specimens of the 1.4541 material as a function of the loading scenario, the specimens were measured at a test frequency of 200 Hz following a specified number of load reversals using a high-resolution eddy current probe. The change in the eddy current signal's 1st harmonic is depicted in Fig. 2.2.2.6 as a function of the bending and torsional loading moment for different numbers of load reversals.

A significant increase in the formation of α'-martensite occurs in the cyclically loaded bending specimen at a bending moment of ± 350 N m, or above. Particularly at the 90° and 270° orientations, high tensile and compressive stresses alternately occurred which led to plastic deformation and the formation of deformation-induced martensite. The eddy current signal amplitude in the specimen's local eddy current scans demonstrates the increase in martensite formation with increasing number of load reversals (LR). At 772 kLR, crack initiation occurred that subsequently led to fracture of the specimen.

Under cyclic torsional loads of ± 430 N m, a local change in the material properties due to martensite formation can also be identified with increasing number of

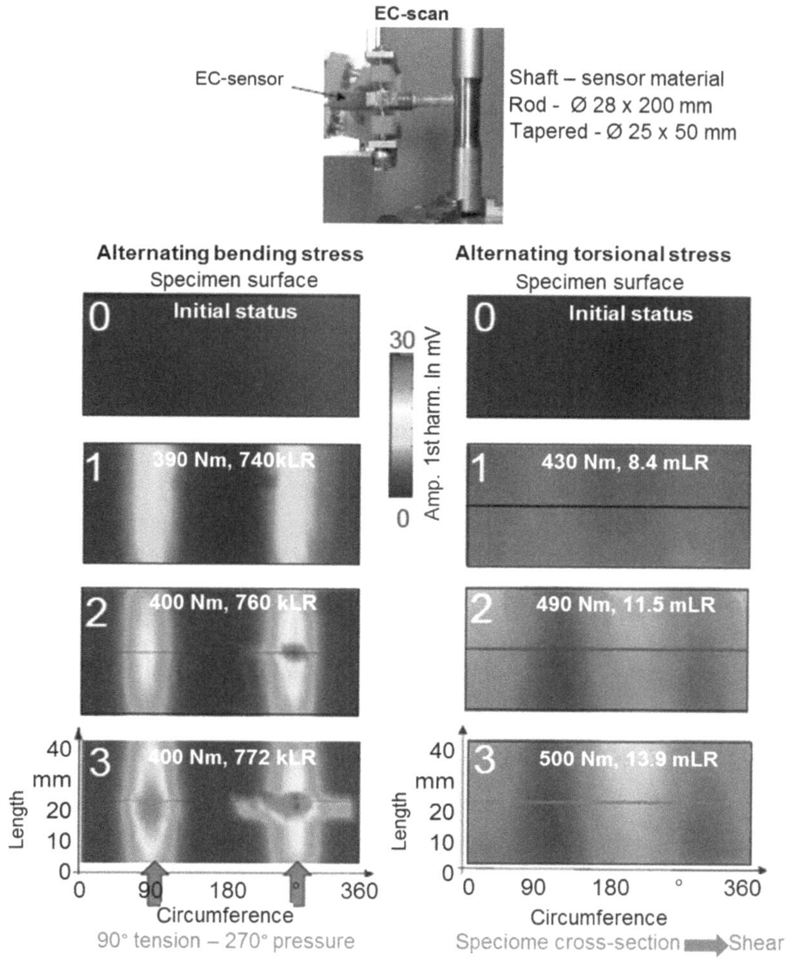

Fig. 2.2.2.6 Changes in the EC's 1st harmonic's amplitude as a function of the cyclic bending and torsion loading, 1.4541 material [5].

load reversals. In contrast to the bending loading, the cross-section of the specimen's material is subjected to shear stresses under torsional loading. In this loading case, the loading stresses are homogeneously distributed in the specimen's subsurface region and produce relatively uniform martensite formation with increasing load and number of load reversals [5].

2.2.2.3 Development of inherent, directionally responsive yield stress sensors

The investigations described in the previous section demonstrated the suitability of metastable austenitic steels, particularly the 1.4310 material, for obtaining deformation-induced martensite upon plastic straining. This steel, in contrast to the 1.4301 and 1.4541 materials, features lower Ni and Mn contents and, owing to its pronounced metastable state, a substantial tendency to martensite formation. The martensitic transformation sets in upon plastic straining and increases with increasing plastic strain. By tempering the martensitic constituents, the material's yield stress can be lowered, and thus the fraction of metastable austenite is raised. In these regions, lower loading levels already lead to plastic deformation and to the onset of martensite formation in the sensor material. However, with a global approach this would also change the mechanical properties of the entire specimen or component.

By contrast, a low yield stress value present only within a small region hardly influences the components overall load carrying capacity. Depending on the geometry and/or the alignment of such regions, the local plastic deformation can correspond to the magnitude and direction of the mechanical loading, whereas the surrounding regions are only elastically stressed. Such a behavior can be achieved by a local heat treatment using a laser. In the present study, an ytterbium-doped high-powered fiber laser was employed using a downstream optical focusing system, which resulted in a good beam quality (diffraction index $M^2 < 1.1$) and a small focal spot ($\geq$15 μm). This laser allowed for a sharply defined region to be shortly, locally heat treated [12]. The laser beam's power density as well as its energy input into the component is characterized by the three parameters: beam power, focusing, and emission duration. By adjusting these parameters, the type of heat treatment in the volume of interest can be controlled, and thus the resulting microstructural transformation behavior can be tailored.

The martensite present initially in the metastable austenitic steel is relieved, tempered, and partially transformed into metastable austenite by the intensity of the laser energy. Thus lower yield strength values can be set locally, and the heat-treated regions fulfill a sensor function. In other words, the martensitic transformation is triggered in these sensory regions whenever they are subject to high loading accompanied by plastic straining, and the transformation can be detected by the ensuing changes in ferromagnetic properties of the material. The intensity of the tempering process, and thus the yield stress, can be differentially specified by stepwise introducing a defined energy, for details see [5,13].

The sheet material used in the present study possessed an initial yield stress of 1100 MPa, which was set by cold working. By locally tempering the material in steps using a laser power from 110 to 130 W, the local yield stress could be set in steps from 950 MPa down to 800 MPa, cf. Fig. 2.2.2.7.

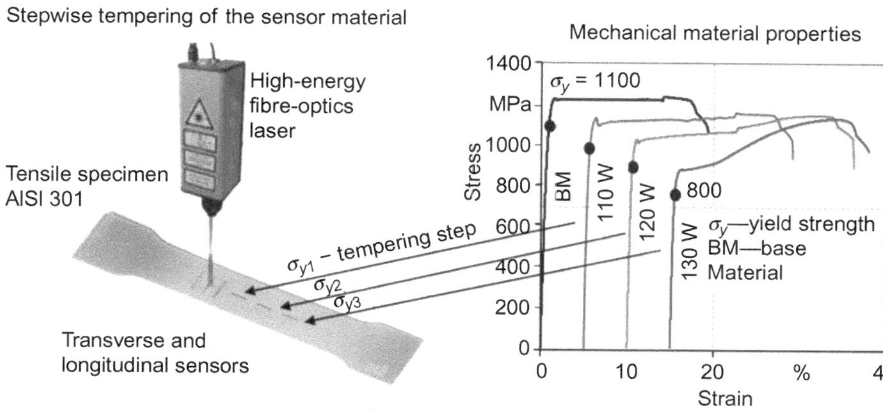

Fig. 2.2.2.7 Matching the sensor's strength to the component's loading stresses. Recompiled from G. Mroz, Entwicklung von Techniken zur bauteilinhärenten Informationsspeicherung und Erfassung von Bauteilbelastungen, Dissertation, Leibniz Universität Hannover, Berichte aus dem IW, PZH Verlag, Garbsen, 2013.

The tensile specimens' local heat treatment was carried out in lines along two orientations. In this way, directionally dependent yield stress sensors resulted that can register the loading longitudinally or transversely. This orientation dependence results from constraint of the surrounding matrix. Specifically, one set of the sensors is only responsive to stresses which act perpendicular to the loading direction since, in the parallel direction, the surrounding higher strength material reinforces the tempered region and carries the loading stress. This is different for the tempered lines lying transverse to the loading direction. In this case, the load stress is carried by the tempered region.

In order to produce sensors in the subsurface region which, apart from the loading magnitude, can also indicate the loading direction, circular sensors were introduced into the component. Here, these tempered regions again function as sensors and can be individually set to a response level according to the parameters used during the tempering step.

2.2.2.4 Retrieving the loading information

In order to retrieve the loading information inherently stored in the component's subsurface region due to the microstructural transformations, nondestructive retrieval technologies were developed that should allow for an easy retrieval of the component's loading history.

2.2.2.4.1 Eddy current technology for retrieving the stored loading data

The eddy current technology is widely employed as a nondestructive inspection method for characterizing materials by using absolute sensors and for detecting defects by using differential and multidifferential sensors. Using this inspection method, it is possible to characterize material inhomogeneities in the component's subsurface. According to the field of application, a suitable testing rate and resolution can be obtained by means of employing adapted inspection sensors and frequencies. The information can be retrieved from the loading sensors by automatically guiding the testing sensor using multiaxis systems or an industrial robot.

In order to optimize the measuring technology with respect to monitor local microstructural changes, high-resolution eddy current tactile sensors were employed. In addition to the probe's material properties, the sensor's design, and the test frequency, the formation of the field is governed by the austenitic and ferritic probe shielding. For low electrical conductivity of the austenitic steels, small eddy currents are formed in the austenitic shielding by which the primary field is only slightly damped. If a ferromagnetic material having a high permeability is employed for the shielding then the magnetic flux is intensively concentrated by the shielding. By employing a ferromagnetic shield in combination with a coil core diameter of 0.5 mm, an intensively concentrated field can be generated which provides for an increased local resolution [5].

Fig. 2.2.2.8 shows the effect of the different tempering steps and orientations of the line-shaped sensors along with a comparison between the eddy current scans of a nonloaded specimen and a specimen longitudinally loaded to 800 MPa.

The EC scans in Fig. 2.2.2.8 show the distribution of the eddy current amplitudes along the specimen's surface. In the nonloaded state, the three longitudinal and three transverse sensors can be clearly detected. These sensors were introduced into the component using laser powers of 110–130 W. The yield stress values in the lineal sensors were thus set to 950, 900, and 800 MPa. Due to the stepwise tempering, changes occurred in the eddy current amplitudes of the ferromagnetic, martensitic constituents in the metastable austenitic material. The plotted measured profiles show an increasing eddy current amplitude as a function of laser power and the associated changes in the material properties. Following the specimen's mechanical loading to a loading stress of 800 MPa, strain-induced martensite is formed in those sensors oriented transversally to the loading direction. Owing to the correlation of the eddy current amplitudes to the local fraction of martensite, the eddy current amplitudes also show a significant change after mechanical loading, which approaches the level of the cold-worked martensitic microstructure of the substrate material. It is clear that martensite only forms in those sensors lying transverse to the loading direction, whereas the sensors placed longitudinally exhibit no change following loading. In addition, it is clear that the transverse sensor introduced using 130 W demonstrates a significantly bigger change than the transverse sensors produced using 120 and 110 W. Loading the specimen further to 1000 MPa led to a more

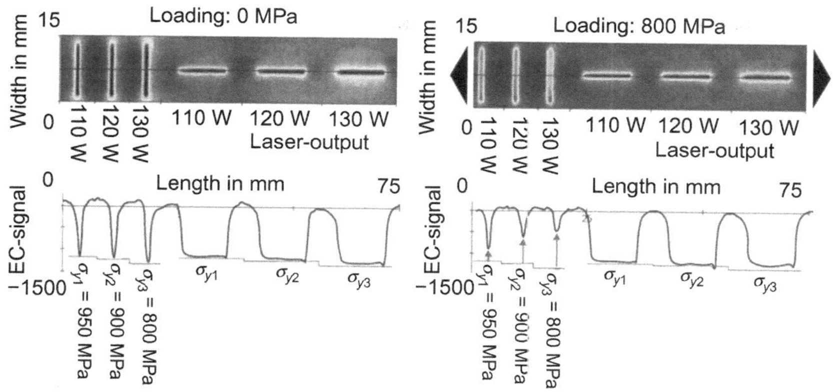

Fig. 2.2.2.8 Yield stress sensors in the transverse and longitudinal directions prior to and following loading.
From G. Mroz, Entwicklung von Techniken zur bauteilinhärenten Informationsspeicherung und Erfassung von Bauteilbelastungen, Dissertation, Leibniz Universität Hannover, Berichte aus dem IW, PZH Verlag, Garbsen, 2013.

markedly pronounced formation of martensite in all three sensors. Thus according to the field of application and to the sensor's orientation, the sensor can be configured to given loading steps and different loading directions [5,11,13].

In order to create sensors in the subsurface region that sensitively respond to all possible loading directions, circular sensors were introduced into the component, cf. Fig. 2.2.2.9. For these sensors the same laser parameters were used for the linear shaped ones.

From Fig. 2.2.2.9, it is also clear that there is only formation of martensite in those regions of the sensors, which lie transverse to the loading direction. The sensor's elements lying parallel to the loading direction are reinforced by the surrounding, stronger microstructure and only elastic strains occur in the high-strength substrate microstructure and in the nearby part of the yield stress sensors.

In order to adapt the technology to actual component geometries, the sensor region was further minimized for registering several load increments. For this purpose, five different tempering steps were combined in one single lineal sensor. According to the response characteristics of the individual sensor's segments, the experienced loading can thus be inferred [14].

The signal shown in Fig. 2.2.2.10 was recorded from a 6 mm long sensor that featured six temper increments (laser power: 100–140 W). Upon loading the specimen to a stress of 700 MPa, newly formed martensite, and therefore a change in the eddy current amplitude, is seen only in the sensor segment that was tempered using a laser power of 140 W. At a stress of 900 MPa, changes in the segments 2 and 3 become evident. The fourth segment (110 W) only responds at a loading of 1100 MPa, whereas segment 5 (100 W) still shows no response.

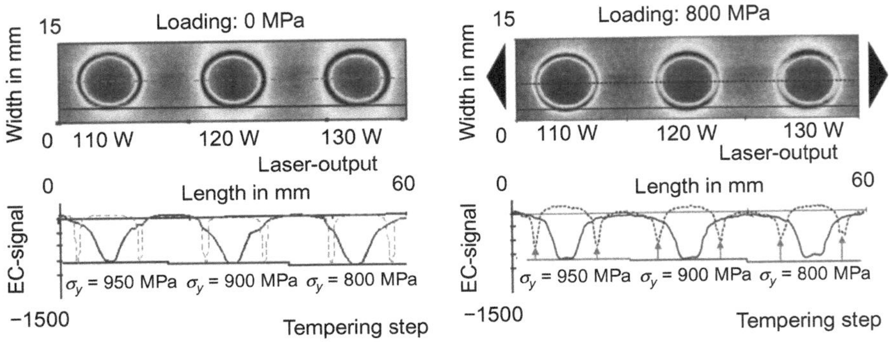

Fig. 2.2.2.9 Data obtained from circular yield stress sensors prior to and following loading [5].

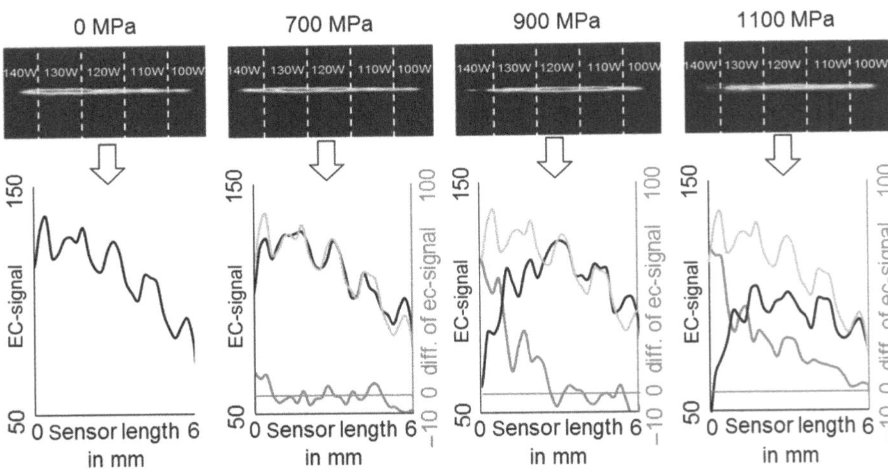

Fig. 2.2.2.10 EC amplitudes of multiincrement yield stress sensors following loading [14].

2.2.2.4.2 Development of induction thermography for retrieving the stored loading data

Since a locally resolved surface scan using the eddy current technology is time consuming, inductively excited thermography with pulse excitation was developed parallel to the eddy current technique in order to retrieve the information from the yield stress sensors at a high acquisition rate.

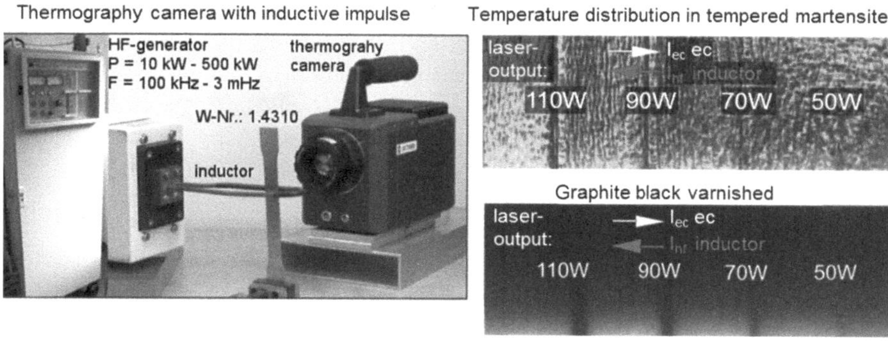

Fig. 2.2.2.11 Material characterization of yield stress sensors using induction thermography, see main text for details.
Recompiled from G. Mroz, W. Reimche, W. Frackowiak, O. Bruchwald, H.J. Maier, Setting discrete yield-stress sensors for recording early component loading using eddy-current array technology and induction thermography, in: Proc. 2nd International Conference on System-Integrated Intelligence: Challenges for Product and Production Engineering, SysInt 2014, July 2nd–4th, Bremen, Germany, 2014, pp. 484–493.

This global image processing recording procedure possesses a high lateral resolution and is based on the dependence of the evolution of temperature fields on the material properties when subject to transient, high frequency (HF) excitation. A controlled impulse from a HF generator connected to an inductor is used to locally introduce a high energy into a component, cf. Fig. 2.2.2.11. Due to the transient pulse excitation, the dynamics and sharpness of the temperature field increase in comparison to conventional longer-term excitations or heating methods, while at the same time the component's global heating remains low. In this way, damage to the component by overheating can be avoided. Besides the spatial analysis of the temperature field, the time and frequency ranges of the temperature-time profile can also be analyzed. In the experiments, an excitation and delay period of 50 ms provided best results. This corresponds to an impulse excitation frequency of 10 Hz and a duty factor of excitation and measuring cycle of 50%. The temperature-time curve's profile of an image point depends on the eddy current distribution at the observed location in the component. This eddy current distribution does depend strongly on the physical material properties, the excited region's distance from the inductor, and the geometry of the component. In order to optimize the signal, the inductor needs to be positioned such that the region with the sensor is uniformly heated. Fig. 2.2.2.11 shows the temperature distribution at the sensor that featured different tempering increments. Owing to the locally changed electrical and thermal material properties, which depend on the laser power employed, various intensities of the excited temperature field occur in the sensor region due to the transient, high-frequency impulses. Fig. 2.2.2.11 demonstrates that high-frequency, impulse induction thermography allows for discriminating the differently tempered segments, and thus the amount of martensite formation in the yield stress sensors can be rapidly retrieved [15]. It should be noted it was possible to reduce spurious effects resulting from reflections by coating the component's surface with black graphite.

2.2.2.4.3 Optical evaluation of the surface in the sensor region following plastic deformation

Another approach to nondestructively detect plastic deformation is to analyze the changes in the surface topography resulting from dislocation slip in the material. These topographic changes in the component's surface can be very rapidly probed using a portable digital microscope. Specimens, in which different yield stress sensors were introduced, were loaded in increments within the range of 5%–40% plastic strain and their surface topographies were optically recorded. When subjected to low loads, i.e., within the range of elastic strains, the surface topography remained largely unchanged. Upon exceeding the local yield stress value, the roughness increases as shown in Fig. 2.2.2.12. To assess the permanent deformation as well as to unambiguously evaluate the exceedance of the yield stress values due to the loading, both the roughness values R_a and R_z as well as an image comparison of the surface structure are pertinent, Fig. 2.2.2.12, left. Thus by optically surveying the surface topography in the region of the sensor, this additional retrieval method can also provide for information about the component's loading history [5,11].

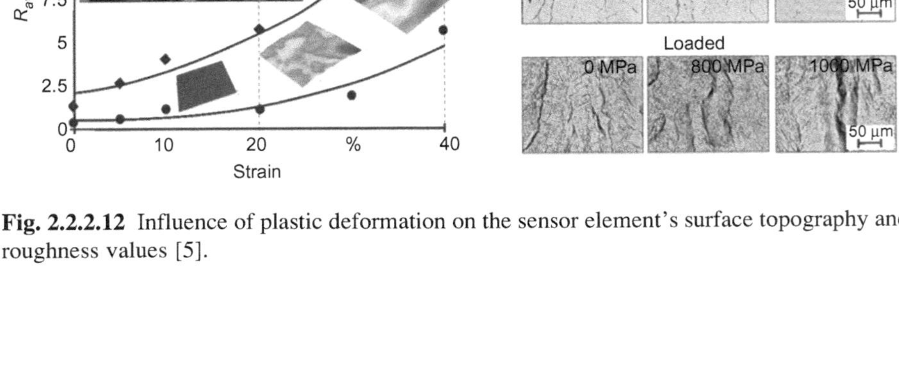

Fig. 2.2.2.12 Influence of plastic deformation on the sensor element's surface topography and roughness values [5].

2.2.2.5 Application of yield stress sensors

2.2.2.5.1 Demonstrator: Tie rod

In order to demonstrate the capabilities and potential applications of inherent, directionally dependent yield stress sensors, a tie rod of the Formula Student racing car RP09 was furnished with yield stress sensors having different sensitivities. Thus it is possible to retrieve the loading history from this safety critical component, which it has experienced during its use phase. Such data could be used to extend the racing car's service life or prevent a component from being replaced prior to it reaching its actual service life.

In order to correlate the stored loading history with online recorded data, loadings of the tie rod, in which the yield stress sensors were also introduced, were recorded and evaluated using harmonic analyses of the eddy current signals. Here, the Villari effect was exploited. Specifically, the changes in the nonlinearity of magnetic hysteresis upon mechanical loading, which led to changes in the amplitudes of the higher harmonics, were exploited. In addition, load measurements were also carried using harmonic analyses of eddy current signals on components made of the magnetic magnesium material developed in the Collaborative Research Centre 653 (Fig. 2.2.2.13) [16,17].

The load data obtained during the use phase of components using these two novel methods were employed within the CRC 653 with focuses on the development of the next component generation adapted to the loading profile (see Chapter 4.2) and on optimization of maintenance activities (see Chapter 4.3).

2.2.2.5.2 Combination of design and sensor materials for creating gentelligent components

Clearly, the metastable austenitic steels are suitable materials for the sensor design proposed here. In order to furnish components manufactured from other structural materials with yield stress sensors, a different approach is needed to create a responsive component. The challenge here is to suitably join the sensor and the other structural material. As a rule, joining processes using welding technology involve local thermal interaction with the component. Thus notch effects can occur, which weaken the component. Joining processes based on bonding or soldering technologies are, for the most part, incapable of transmitting the required stresses. Possible joining technologies that produce minimal heat-affected zones, result in a fine grained microstructure with a low microstructural notch effect, and provide for a large joint strength include laser beam, electron beam, and TIG welding as well as certain brazing technologies.

Within the framework of the CRC 653, a Cardan drive shaft of 1.0535 (C55) material was combined with the 1.4310 sensor material. Specifically, a sleeve manufactured from the sensor material was joined to the shaft by means of laser beam

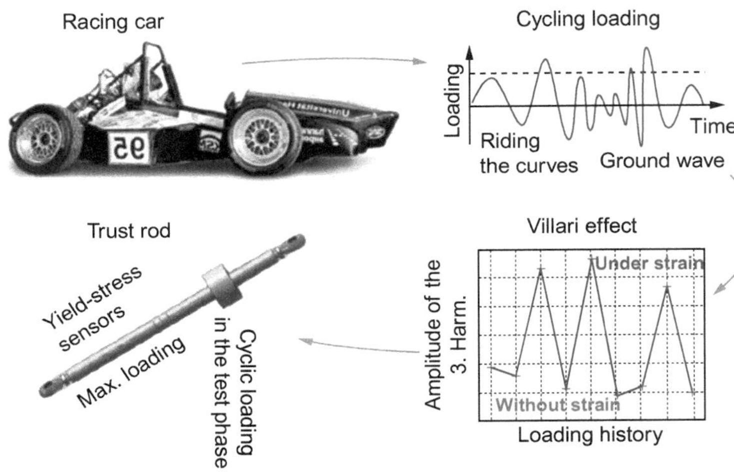

Fig. 2.2.2.13 Loading measurements by means of yield stress sensors and the harmonic analyses of eddy current signals [15].

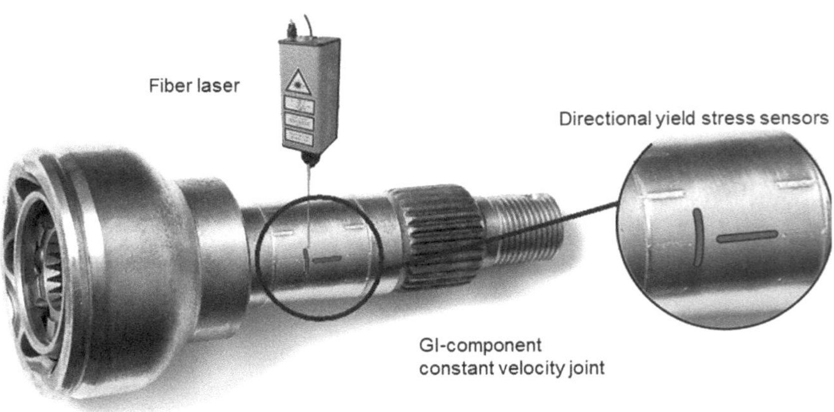

Fig. 2.2.2.14 Combination of sensor and structural materials in a Cardan drive shaft. Recompiled from G. Mroz, W. Reimche, Fr.-W. Bach, Frühzeitige Erfassung erhöhter Bauteilbeanspruchung mit lokalen Dehngrenzwertsensoren, in: DGZfP-Jahrestagung "Zerstörungsfreie Materialprüfung": 30. Mai - 1. Juni 2011, Bremen, Berichtsband, DGZfP, Berlin, 2011, p. 31.

welding and the yield stress sensor was introduced into the sensor material. To elevate the operating loading stress in the sensor region, the wall thickness of the sleeve was made thinner in the sensor region. The sensors were longitudinally and transversally oriented, as shown in Fig. 2.2.2.14.

2.2.2.6 Summary

Within the CRC 653, a technology was successfully developed and tested for sustained sensory monitoring of highly loaded components. This approach is based on the global cold working of metastable austenitic materials and subsequent local heat treatment of the formed martensite for setting the yield stress value in a small region of the material. High component loadings are inherently registered by and stored within the sensory material inseparably from the component. The sensors can be set to defined component load limits and loading directions for specific applications via the local heat treatment parameters.

In addition, nondestructive retrieval technologies were developed that allow for retrieving the information stored in the material. These data can then be used to shed light on the component's loading history.

With the aid of this novel technology, components can also be permanently monitored for to detect overloads during their life cycle. Thus damage can be detected early on and component failure be avoided.

References

[1] H.-J. Bassler, Wechselverformungsverhalten und verformungsinduzierte Martensitbildung bei dem metastabilen austenitischen Stahl X6CrNiTi1810, Lehrstuhl für Werkstoffkunde, Kaiserslautern, 1999.

[2] R. Sellger, Influence of strain rate on deformation behaviour, austenite structure and kinetics of the gama/alfa transformation of steels: Final report, EUR-OP, Luxembourg, 2003.

[3] T. Tsuta, R.J.A. Cortes, Flow stress and phase transformation analysis in austenitic stainless steel under cold working, JSME Int. J.: Ser. A (1993) 63–72.

[4] M. Smaga, F. Walther, D. Eifler, Deformation-induced martensitic transformation in metastable austenitic steels, Mater. Sci. Eng. A 483–484 (2008) 394–397.

[5] G. Mroz, Entwicklung von Techniken zur bauteilinhärenten Informationsspeicherung und Erfassung von Bauteilbelastungen, Dissertation, Leibniz Universität Hannover, Berichte aus dem IW, PZH Verlag, Garbsen, 2013.

[6] G. Mroz, W. Reimche, M. Diebel, Fr.-W. Bach, Erfassung diskreter Bauteil- und Belastungsinformationen in der Bauteilrandzone mit innovativer Sensortechnik, in: Tagung Werkstoffprüfung, Neu-Ulm, 2007, pp. 85–94.

[7] M. Bernard, W. Reimche, Fr.-W. Bach, Zerstörungsfreie Randhärte- und Einhärtungstiefenbestimmung an Verzahnungen mit der Harmonischen-Analyse von Wirbelstromsignalen, in: Tagung Werkstoffprüfung, Bad Neuenahr, 2006, pp. 415–422.

[8] W. Reimche, M. Bernard, R. Duhm, K.L. Feiste, B. Heutling, Fr.-W. Bach, Zerstörungsfreie Materialcharakterisierung zur Kennwertermittlung und Inline-Prozessüberwachung, in: DELTA-TEST Symposium, 2002.
[9] K. Schiebold, T. Knöll, Fehlerprüfung mit Wirbelstrom, Castell-Verlag, Wuppertal, 2006.
[10] M. Maaß, J. Nehring, Oberwellenanalyse bei der magnet-induktiven Wirbelstromprüfung in der Praxis, in: DGZfP-Jahrestagung "ZfP in Forschung, Entwicklung und Anwendung": 17-19. Mai 2004, Salzburg, Berichtsband, DGZfP, Berlin, 2004. v25.
[11] G. Mroz, W. Reimche, Fr.-W. Bach, Frühzeitige Erfassung erhöhter Bauteilbeanspruchung mit lokalen Dehngrenzwertsensoren, in: DGZfP-Jahrestagung "Zerstörungsfreie Materialprüfung": 30. Mai - 1. Juni 2011, Bremen, Berichtsband, DGZfP, Berlin, 2011, p. 31.
[12] G. Mroz, W. Reimche, Fr.W. Bach, Acquisition of discrete component and loading information in the component's edge region using innovative sensor technology, in: 4th I*PROMS 2008 Virtual International Conference on Innovative Production Machines and Systems, 2008, pp. 431–437.
[13] G. Mroz, W. Reimche, Fr.-W. Bach, The use of component's edge region as inherent information carriers and loading indicators, in: Proc. 1st Joint Symposium on System-integrated Intelligence: New Challenges for Product and Production Engineering, SysInt 2012, June 27th–29th, Hannover, Germany, 2012, pp. 158–160.
[14] S. Barton, G. Mroz, W. Reimche, H.J. Maier, Inherent load measurement and component identification by multi-dimensional coded data in the component's subsurface region, in: Proc. 3rd International Conference on System-Integrated Intelligence: New Challenges for Product and Production Engineering, SysInt 2016, June 13th–15th, Paderborn, Germany, 2016, pp. 537–543.
[15] G. Mroz, W. Reimche, W. Frackowiak, O. Bruchwald, H.J. Maier, Setting discrete yield-stress sensors for recording early component loading using eddy-current array technology and induction thermography, in: Proc. 2nd International Conference on System-Integrated Intelligence: Challenges for Product and Production Engineering, SysInt 2014, July 2nd–4th, Bremen, Germany, 2014, pp. 484–493.
[16] C. Klose, C. Demminger, G. Mroz, W. Reimche, F.-W. Bach, H. Maier, K. Kerber, Influence of cobalt on the properties of load-sensitive magnesium alloys, Sensors 13 (2013) 106–118.
[17] C. Klose, C. Demminger, S. Zwoch, W. Reimche, F.-W. Bach, H.J. Maier, K. Kerber, Measurement of static and dynamic loads utilizing sensory magnesium components, Euro Intelligent Materials, Kiel, 2013, p. 57.

Further Reading

[1] W. Reimche, G. Mroz, Fr.-W. Bach, Bauteil, Verfahren zum Einbringen von Informationen in ein Bauteil und Verfahren zum Ermitteln einer Belastungshistorie eines Bauteils, Deutsches Patent, DE102009056584A1, 2010.
[2] H. Dietrich, W. Heimann, H. Strom, Einfluss der Kaltverformung auf die mechanischen und magnetischen Eigenschaften von nichtrostenden Stählen, Thyssen Edelstahl Technische Berichte TEW (1976) 61–69.

2.2.3 Data storage within the subsurface of a component by local heat treatment

Sebastian Barton, Gregor Mroz, Wilfried Reimche, Hans Jürgen Maier
Institute of Material Science, Leibniz Universität Hannover, Garbsen, Germany

2.2.3.1 Introduction

Currently, metallic components are identified by marks which are molded, stamped, machined, or bonded to the component using letters, numbers, or bar codes, which can be optically read out [1]. These mechanical-optical techniques lead to irregularities in the component's surface, the information is limited due to the lack of space or only partly forgery-proof. In addition, there is always a risk that the information carrier becomes detached from the component, the information be changed by thermal or mechanical environmental influences, or the optical retrieval of the data is prevented due to soiling. Laser technology has, for many years, been employed in the field of surface technology as a modern tool for treating surface layers or for characterizing products. Currently, laser markers operate using a focused laser beam and produce an optical identifier or an engraving in the components surface [2,3]. Owing to the resulting notch effect, such a marking is always accompanied by a loss in component strength. In addition, a change in the surface's optical properties occur. Another technique for identifying components is radio frequency identification (RFID). An RFID system consists of a transponder, which is located either in or on the component and incorporates an identifying code, as well as a reading device to retrieve this code. Whereas early technologies were based on externally attaching the RFID transponder to metallic components, new technologies allow to introduce it into the component by either molding or additive manufacturing [4,5]. For storing data and for identifying sinter manufactured aluminum and magnesium components a technology was developed within the framework of the CRC 653 for the purpose of protecting against plagiarism. This was carried out by specifically introducing foreign particles into the metal powder. The data introduced into the sintered component can be nondestructively retrieved by means of computer tomography or radiographic technique [6].

Given the lack of suitable, universally usable techniques for securely identifying high performance metallic components, one objective of the CRC653 is to develop a technology to multidimensionally store information by locally modifying microstructural properties in the component's subsurface region.

By employing this innovative technology for marking components and storing data, the component's own subsurface region becomes the data store. By means of inherently combining the data with the microstructure of the component's subsurface, it is ensured that this information is inseparably bound to the component during its entire life cycle, the information remains readable owing to an outstanding resistance to environmental influences, and it provides for a high degree of forgery protection. Moreover, by varying the local introduction of energy, it is possible to exploit the degree and depth of the heat treatment as an additional dimension of storing data and thus to significantly increase the data density. The data and marking introduced by locally modifying the microstructural properties do not lead severe notch effects and are therefore suitable for highly loaded, safety-relevant components. Such marks can be read using nondestructive eddy current technologies in order to rapidly identify a component during its entire life cycle.

2.2.3.2 Retrieving data stored in the component's subsurface region using high-frequency eddy current technology

The data described in this section, which were introduced by locally heat treating the component's subsurface region, are visually inconspicuous and can therefore only be securely detected by means of a retrieval technology, which is not based on an optical method. The high-frequency eddy current technology is a suitable inspection technology for component defect testing as well as for evaluating the local microstructural properties in the component's subsurface region.

In the eddy current technology, an alternating magnetic field is generated by a sinusoidal alternating current in an excitation coil. The generated field interacts with an electrically conducting and magnetically permeable material, as shown in Fig. 2.2.3.1.

The developing eddy currents produce a further alternating magnetic field, which is superposed on the excitation field and induces a voltage in the receiver coil. The developing eddy currents are influenced by the probed material based on its permeability μ, electrical conductivity σ, and geometry. Local geometric and physical inhomogeneities in the material modify the distribution of the developing eddy currents and lead to a change in the measured signal [8].

To obtain a high measuring sensitivity and local resolution of the material's properties, high-resolution eddy current sensors having a core diameter of 0.5 mm and with austenitic as well as ferritic shielding were developed and tested within the frequency range from 10 kHz to 1 MHz.

Using an automated x-y scan of the component's surface with an eddy current sensor and subsequently evaluating the signal components in the eddy current signal, the introduced markings can be displayed. To perform such a surface scan, the sensor or

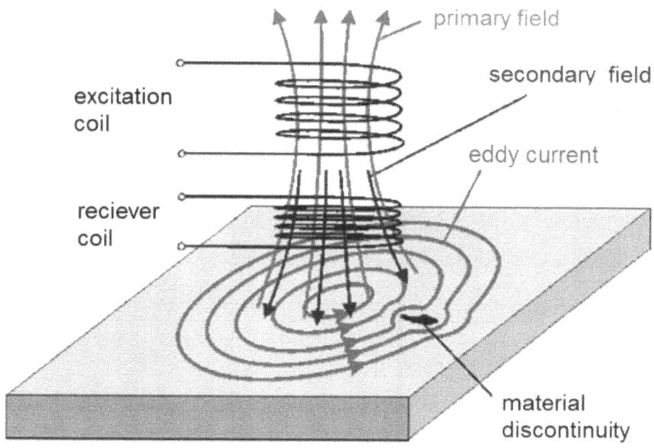

Fig. 2.2.3.1 Measuring principle of eddy current testing [7].

component was, depending on the component geometry, automatically traversed with the aid of a multiaxial system or a 6-axis robot.

Performing the surface scan by employing a single eddy current sensor requires traversing several tracks depending on the surface to be scanned and the employed sensor's diameter. In order to significantly increase the number of traversed tracks and therefore the retrieval rate, eddy current probe arrays were developed as opposed to using a single probe. Instead of traversing the surface using a single sensor, these EC sensor arrays possessed 4 or 8 sensor heads. The sensor field's focusing and the operating range of the coil was locally adjusted by means of austenitic and ferritic shielding. These can be reliably operated within the range from 60 to 300 kHz, and thus cover the frequency spectrum necessary to retrieve the marks. Parallel operation of the sensors in a common sensor field was accomplished by a multiplexer, as shown in Fig. 2.2.3.2 [9].

By adjusting the testing parameters, the eddy current test enables to differentially separate the wanted measured from the unwanted spurious effects depending on the type of probe, the probes shape, and the substrate material's electrical conductivity and magnetic properties. For example, it was possible to obtain a phase difference of almost 90° between the measured and spurious effect by employing a testing frequency of 300 kHz. Here, the measured effect, which is produced by the local change in microstructure, is separated from the spurious effect of the component's geometric discontinuities, such as drilled holes. To retrieve the data, only the signal component of the eddy current's signal is considered in which the relevant information is solely located. Thus the influence of spurious effects on the retrieval of the data can be minimized, as shown in Fig. 2.2.3.3 [10].

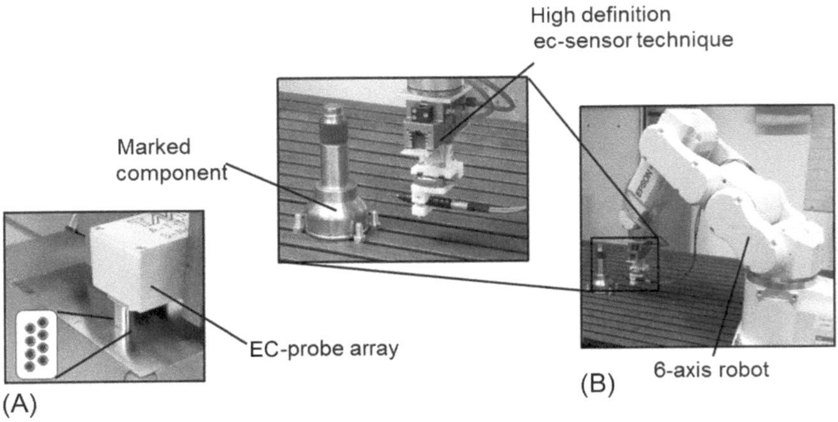

Fig. 2.2.3.2 (A) Multichannel eddy current sensor array and (B) 6-axis testing robot for rapidly detecting component information.

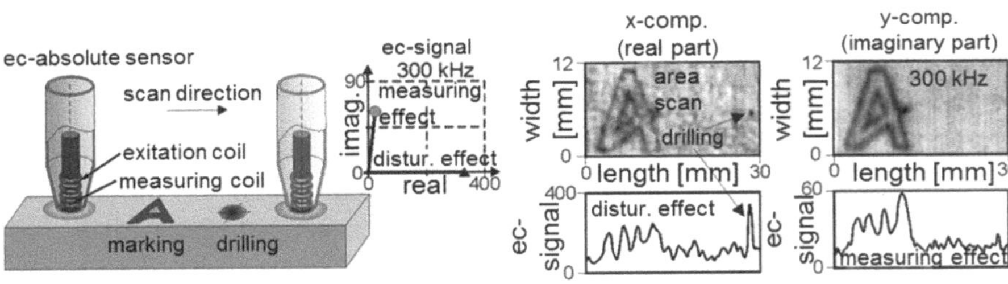

Fig. 2.2.3.3 Separation of the actual data from spurious effects with the phase separation procedure using EC absolute probes [10].

2.2.3.3 Information storage in the component's subsurface region

In order to demonstrate the applicability of the developed technology for a broad spectrum of technologically relevant materials, data were introduced into both steel materials as well as lightweight metals.

2.2.3.3.1 Information storage by modifying the microstructure

To mark and store component data in steel components, a local heat treatment of the component's subsurface region was carried out. The microstructural changes led to a change in the permeability and the electrical conductivity.

For hardened steels and work-hardened metastable austenitic materials, it is possible to set various microstructural states by stepwise tempering the martensite. In general, heat treating a work-hardened microstructure leads to stress relieving, tempering effects, and to possible recrystallization of the microstructure. Following rapid heating above the austenitizing temperature and rapid cooling via heat conduction by the material or a stream of coolant, heat treatable carbon steels having elevated carbon contents form a martensitic hardened microstructure.

These local, physical inhomogeneities produced in the component's subsurface region can be depicted with the aid of eddy current testing. The thermal energy introduced into the component's subsurface region is decisive for the extent to which material properties are changed. To introduce the marks, a ytterbium doped high-powered fiber laser was employed using a downstream optical focusing system characterized by a good beam quality (diffraction index $M^2 < 1.1$) and a small focal spot ($\geq$15 μm) thus enabling a sharply defined region to be locally heat treated [11], Fig. 2.2.3.4. The laser beam's power density as well as its energy input into the component is specified by three parameters: beam power, focusing, and emission duration. By adjusting these parameters, a direct influence can be exerted on the type of heat treatment in the irradiated volume. In order to avoid oxidation and decarburization during the heat treatment process, a shielding gas (argon) was employed.

Different variations of structured data stored in the component's subsurface region and unambiguous component marking are depicted in Fig. 2.2.3.5. To multidimensionally store and code data in the component's subsurface region, the so-called IW code was developed. This code creates new possibilities for

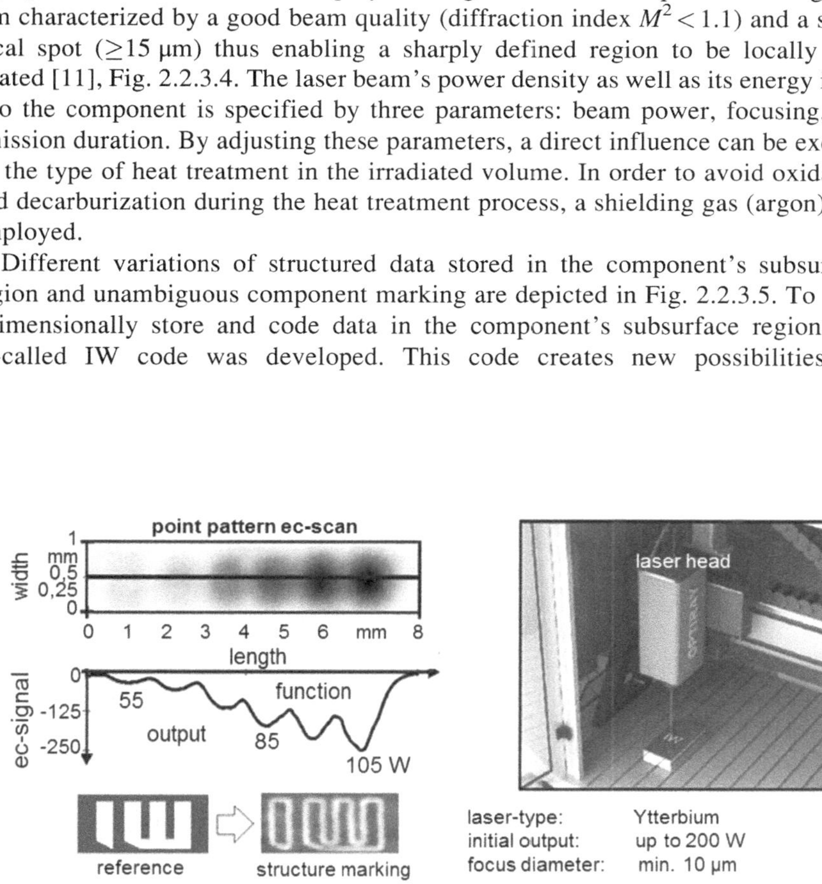

Fig. 2.2.3.4 Visually inconspicuous, inherent information storage in the component's subsurface region using a high-powered fiber laser.

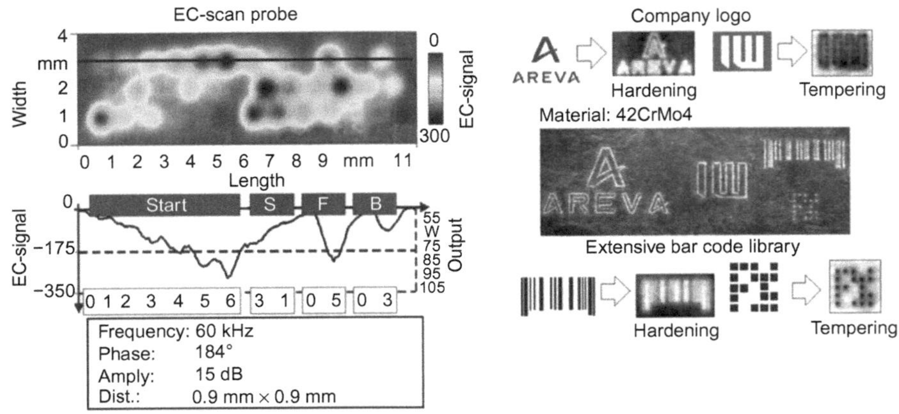

Fig. 2.2.3.5 Storing component-inherent, multidimensional information via the degree of microstructural changes (hardening/tempering) [10].

multidimensionally and inherently storing discrete component data in the component's subsurface. This double-digit marking code enables to store 49 conditions at two positions via seven material states. Each of these states was assigned a letter, a number, or a punctuation character in a code matrix. Thus it is possible to code data, introduce it into a component and, after retrieval, to decode it. By optimizing the process parameters and with a marking pitch of 0.9 mm, a specimen of 1.4310 material was inscribed with the character set "SFB 653 TP-S3." The marks, which were introduced on a 2 cm^2 surface, are visually inconspicuous compared with the pristine component surface, but it was possible to easily read them using an eddy current absolute probe having a coil diameter of 0.5 mm and ferritic shielding, as shown in Fig. 2.2.3.5 left. With the aid of the IW code, inscriptions of up to 126 marks were also successfully accomplished. Fields of application were also demonstrated for employing two-dimensional mark variants. It was possible to introduce different company logos, bar codes, and dot codes into components of various steels, as shown in Fig. 2.2.3.5 right [10–12]. As an example, serial numbers in plain text were also introduced and clearly read.

In order to be able to rapidly and securely identify components along with a high data density, a coding procedure was implemented that is based on the principle of color QR codes. Here, several, same sized, different colored QR codes were superposed on each other and the resulting mixed colors were used as the third dimension for increasing the data density [13]. The principle was transferred to data matrix codes and bar codes by a stepwise changed setting of the microstructural states. Thus 4-step bar codes were introduced into the components of 1.4310 material. The steps of the tempering process were obtained by employing

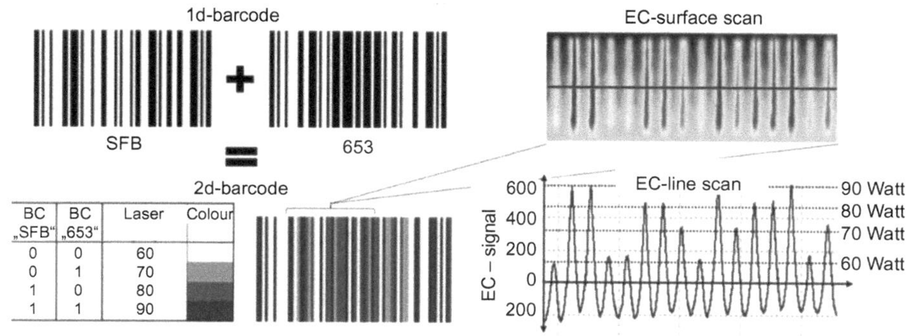

Fig. 2.2.3.6 Generating and reading a 2D bar code.

different laser powers in the range of 60–90 W in the lines of the bar codes. In this way, one line can show four steps, as states which can be unambiguously differentiated from each other using the eddy current technology. With this additional dimension of stored information, it is possible to combine two-dimensional bar codes and to double the data density as compared to conventional bar codes. In addition, the information from the component, in contrast to dot codes, can be read in a line scan and does not have to be scanned using a flat scanner in several tracks. This enables the stored data to be read at a significantly higher rate, as shown in Fig. 2.2.3.6.

Depending on the field of application, the optimal variant for storing information can be selected in order to identify the component or obtain an appropriate data density.

The formation of the affected zone as well as the penetration depth of the marks in the component's subsurface region was investigated by carrying out microstructural analyses and micro-hardness testing. The formation of the heat-affected zone depends on the thermal energy introduced. With increasing laser power and decreasing operating speed or longer emission duration, the heat-affected zone becomes larger at a point. The laser's focal spot diameter on the component's surface also influences the spreading of the heat-affected zone. For the laser employed here including the optical focusing system, this diameter was ≈ 15 μm. Owing to the beam's cone used here, the focal spot diameter varied with increasing focal distance to the component along with decreasing the laser beam's power density. It is not suitable to directly use the focused system for introducing the marks since, depending on the beam's power, this can lead to a too high power density and therefore to local melting of the sample. In order to avoid overheating, a focal length in the range from 1 to 4 mm was chosen for introducing the marks.

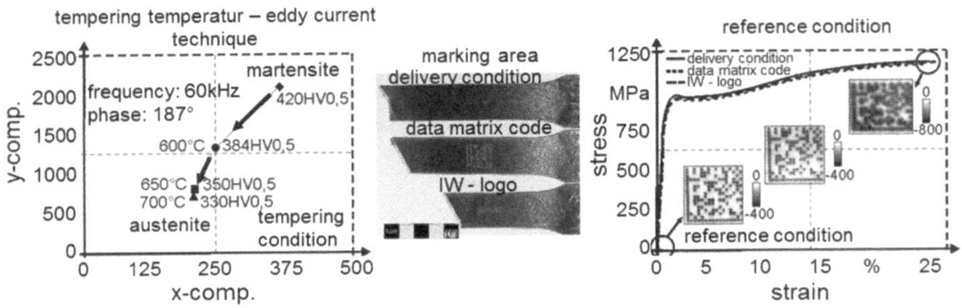

Fig. 2.2.3.7 Influence of the tempering temperature on the micro-hardness and EC signal amplitudes, influence of the marks on the component's mechanical properties, and the change in readability of the marking upon plastic deformation [10].

To evaluate the influence of the heat-affected zone on the material's mechanical properties, specimens of 1.4310 material were examined. Penetration widths from 0.2 to 0.5 mm and penetration depths from 0.3 to 0.6 mm were employed. Within the temperature range from 500°C to 700°C, the microstructure transformed from martensite into austenite and the eddy current signal reached a minimum value. In the heat-affected zone, the hardness values continuously decreased toward the mark, starting from 450 HV0.1 in the substrate material and attained values between 400 HV0.1 and 300 HV0.1 in the mark's mid-region. By means of micro-hardness measurements, it was confirmed that the martensite was tempered and the austenite was partially transformed by the local introduction of energy and the local temperature gradient in the area of component's subsurface region, as shown in Fig. 2.2.3.7 left. Tensile tests were carried out to globally evaluate the component's properties regarding changes in the strength and deformation characteristics due to a local heat treatment. In comparison with the as-received work-hardened condition, a local temper led to only small changes in the yield stress and the fracture strain. This contrasts with the ablative marking technologies which significantly weaken the component locally due to the resulting notch effects. Even after considerable straining and deforming of the material, the inscription can still be easily detected, read, and identified following component failure, as shown in Fig. 2.2.3.7 right [10].

Fig. 2.2.3.8 depicts the development of the microstructural changes in the surface's heat treated zone and the penetration depth as functions of the varied process parameters. Owing to the local heating of the component's surface by means of a fiber laser, the temperature profile constantly decreases with increasing distance from the position of the energy input. Owing to the dependency of the heat treatment on the local temperature-time profile, a decrease in the level of heat treatment occurs with increasing distance from the position of the energy input.

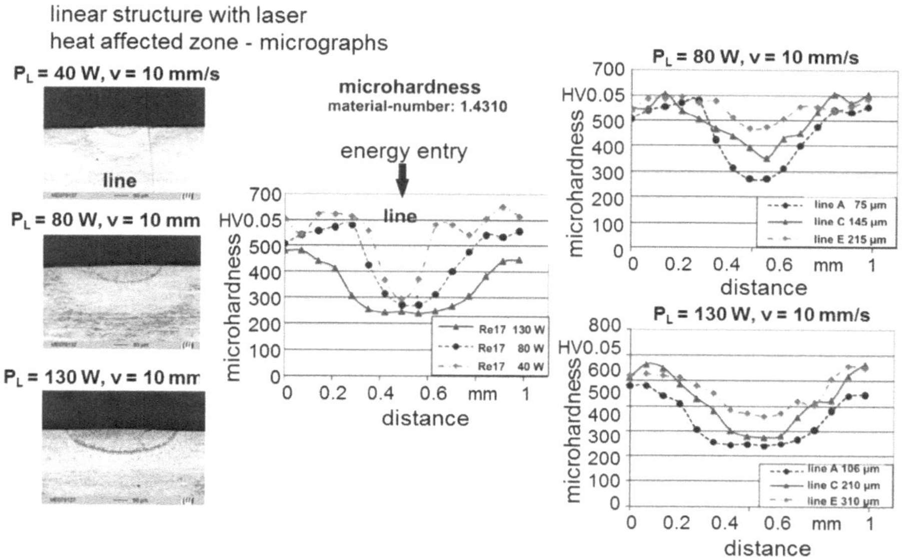

Fig. 2.2.3.8 Graded material properties in the region of a heat-treated line structure [14].

For laser power values of 40, 80, and 130 W, linear tempered regions were produced having a line thicknesses from 0.3 to 0.8 mm on the component's surface. The micro-hardness of the work-hardened substrate material was approx. 500 HV0.5. By locally tempering the substrate material, the micro-hardness values continuously decreased in the region of the heat-affected zone and reached their minimum values of 300 HV0.5 (40 W), 270 HV0.5 (80 W), and 240 HV0.5 (130 W) in the mark's mid-region. Besides the width and the degree of microstructural changes in the material, significant changes in the depth of the heat-affected zone occurred depending on the process parameters. By employing a laser power of 80 W, a tempered region having a width of approx. 0.3 mm and - a minimum hardness of about 480 HV0.5 was obtained at a depth of 215 µm. With a laser power of 130 W, this region was significantly more pronounced having a width of 0.6 mm and a minimum hardness of 280 HV0.5 at a depth of 210 µm.

Regarding the broad field of application in mechanical engineering, a C60 carbon steel and 42CrMo4 steel were chosen as materials for storing information in the component's subsurface region. In its normalized state, the C60 steel features of a soft ferritic/pearlitic microstructure. If the material is heated above the austenitizing temperature of 810°C, the iron carbides are dissolved and the body-centered cubic lattice of the ferrite transforms into the face-centered cubic lattice of austenite. The kinetics of the austenitization are dependent on the local

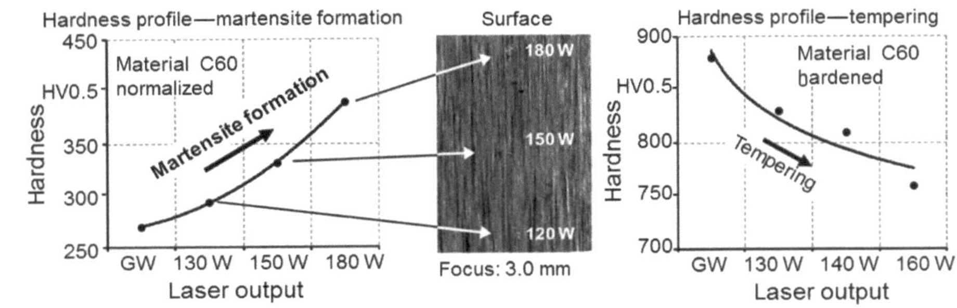

Fig. 2.2.3.9 Influence of the laser power on the micro-hardness, C60 in the normalized and hardened states [10].

temperature-time profile in the austenite region. In the normalized state, the C60 material has a hardness of 270 HV0.5. For a C content of 0.6 wt.%, the transformation starts in the γ-phase at a temperature of approx. 810°C. If the austenitic microstructure is sufficiently rapidly cooled, then this leads to a transformation into the martensitic state. By locally treating the C60 material using a laser power of 180 W at a nonfocused 3 mm operating spot and for a processing time of 0.2 s, a subsurface hardness of approx. 390 HV0.5 was obtained.

Following a short heating time of the C60 material in its hardened state, the martensite was locally tempered and stress relieved in the subsurface region. By locally treating the C60 material using laser powers from 130 to 160 W at a nonfocused 3.3 mm operating spot and for a processing time of 0.2 s, it was possible to significantly reduce the hardness in the marks' mid-region. Starting from the substrate material, the hardness values were reduced from 880 HV0.5 to a hardness of approx. 760 HV0.5 for a laser power of 160 W, as shown in Fig. 2.2.3.9.

This local hardness increase in the component's subsurface region can, as previously mentioned, be verified with the aid of the eddy current technology and used for classification. It was possible to precisely read the coded, multidimensional inscription. Similar to the carbon steels, different microstructural states can be set via stepwise tempering of the martensite in the 42CrMoV4 alloy and other variants in their hardened states [10].

2.2.3.3.2 Storing information by means of dispersing hard magnetic particles

In addition to steel components that have ferromagnetic properties, nickel-based alloys or lightweight components of aluminum or magnesium alloys are widely employed, particularly for high thermal loading. These materials have diamagnetic or paramagnetic properties which, also similar to the electrical conductivity, only slightly vary as a function of the material state. Thus it is not possible to directly mark components made from these materials by locally heat treating the component's subsurface region and then reading these marks using eddy current technology.

To achieve secure information storage, it is therefore necessary to introduce additional information carriers into the component's subsurface region having different properties as compared to the substrate material. One approach is local dispersing of magnetic materials, such as Fe_2O_3, CrO_2, SmCo, $Nd_2Fe_{14}B$ possessing particle diameters of 40–120 μm, in the component's subsurface region. The dispersal of particles in the component's subsurface region depends on the material matching of the particle with the substrate material and on the laser processing parameters.

To avoid oxidation in the region of the energy input, the fiber laser was operated with a constant flow of argon shielding gas, which flowed around and enveloped the laser beam. In order to prevent the loose particles on the component's surface from being scattered by the flow of the shielding gas, various methods were tested for anchoring the particles. It was possible to obtain highly adherent particles using a thin layer of paste and by using the thermal spraying, as shown in Fig. 2.2.3.10.

The larger the particles, the lower the probability is of complete melting. For a focused operating spot and a high power density, both particle vaporization and substrate material ablation can occur. It was possible to obtain good bonding between the component's substrate and the introduced particles by selecting a favorable operating spot and an appropriate energy input per unit line length. Thus it was possible to inherently store extensive bar code libraries and company logos in the component's subsurface region. The marks and inscriptions introduced into the components surface by dispersing hard magnetic particles could be successfully read using the eddy current technology, as shown in Fig. 2.2.3.11 [10,15].

2.2.3.4 Environmental impact resistant, robust data storage in the component's subsurface region

Unambiguous identifiability, i.e., reliable data storage, assumes that the marking cannot be deleted by external influences or the data changed during component use. The component marks that were introduced by using a stepwise defined, laser-induced local change in the microstructure, can be read by a high-resolution eddy current technology at any time. Since this retrieval method is not based on optical technologies, it is also possible to read multidimensional identifiers when the surface has been mechanically damaged or painted. Since only temperatures of several hundred

Fig. 2.2.3.10 Attaching the magnetic layer to the substrate material [10].

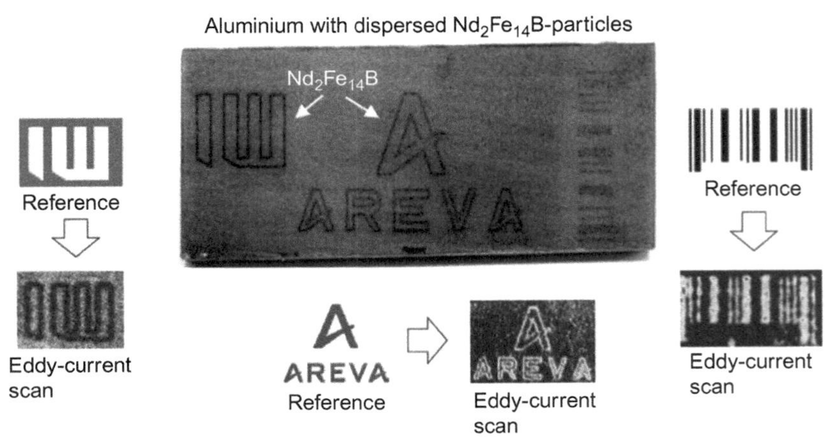

Fig. 2.2.3.11 Verification of the particle distribution using a high-resolution eddy current probe—aluminum specimen containing Nd2Fe14B particles [10].

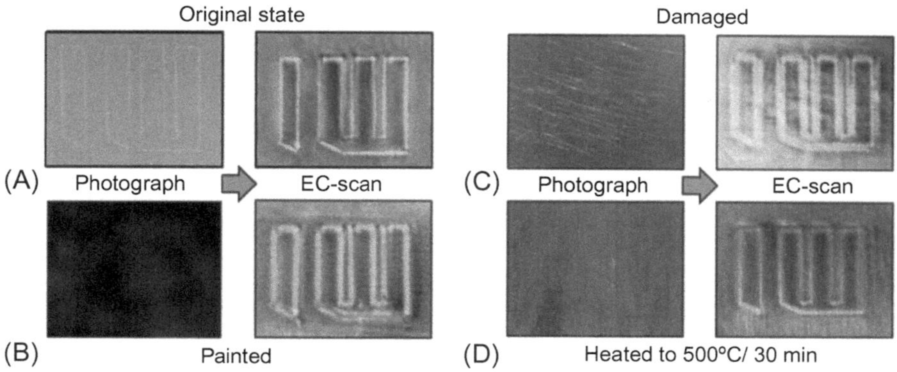

Fig. 2.2.3.12 Environmental impact resistant, robust component marking [16].

degrees can change the microstructure of steel components, the introduced marks are also resistant to substantial component heating during the use phase.

In Fig. 2.2.3.12A, a 12.5 mm × 9 mm mark in the form of a logo is depicted, which was introduced into a C45 steel (material no. 1.0503) component. The mark was introduced into the component using a laser power of 100 W and a nonfocus of 1 mm. Owing to the material's local heating above the austenitizing temperature and the subsequent direct cooling in air, a hardness increase occurred in the heated region. The mark, introduced by the local heat treatment, can be clearly read after its introduction using the high-resolution eddy current technology.

Painting of the surface leads to complete coverage of the mark and prevents the mark from being visually read. Using the eddy current technology, the mark nevertheless can be precisely identified. However, due to the thickness of the paint layer, a liftoff effect of the eddy current sensor from the material occurs. Associated with this is a weakening of the secondary field compared to testing the nonpainted component, this leads to a lower eddy current amplitude in the measured signal. By adjusting the measured signal strength, the mark could also be clearly read beneath the paint layer, Fig. 2.2.3.12B.

Also, mechanical damage to the surface does not render the logo illegible, since the marks are not exclusively stored on the surface by the local heat treatment but are also stored in the subsurface. However, damage to the surface also exerts an influence on the eddy current signal. These effects can be filtered by appropriately separating the signal from the spurious effects. This can be realized by suitably selecting the testing frequency and by adjusting the phase rotation, Fig. 2.2.3.12C.

Even if the component was heated to 500°C and held at this temperature for 30 min, the mark was retained and could be read although the component showed incipient effects and some scale formation despite the use of a shielding gas atmosphere.

At a temperature of 600°C and a holding time of 30 min, the logo resulting from a local heat treatment could no longer be unambiguously read. This tempering processes substantially diminished the differences in the magnetic and electrical properties between the mark and the substrate such that these could no longer be clearly differentiated using eddy current testing [16].

2.2.3.5 Summary

The CRC 653 successfully developed and tested a technology for inherently marking components and storing data in the subsurface region of metallic components. This method is based on the local heat treatment of a steel component's subsurface region and the associated local changes of the physical material properties. By tailoring the laser-induced heat treatment, it is to significantly increase the data density. The versatile applicability of this method was verified for different steels. In addition, light metals, such as aluminum and magnesium, were successfully marked by dispersing magnetic particles in the component's subsurface region.

By developing suitable retrieval technologies, such as the high frequency eddy current technology, the data stored in the material can be nondestructively read and/or the component can be unambiguously identified. It was possible to significantly increase the retrieval rate, in comparison to employing conventional eddy current technologies, by developing an eddy current array technology.

It was demonstrated that the data introduced using this innovative technology are robust with respect to mechanical and thermal environmental influences. By employing a retrieval technology which is not based on optical effects, data can also be read when they are, for example, covered by a layer of paint.

Using this innovative technology, data can be multidimensionally and robustly stored in the component's subsurface region. The component's material itself functions as the data store, and thus enables a component to be unambiguously identified. For instance with this method, marked components can be protected against piracy. The method for inherently storing data and component marking described in this article is a registered patent DE102009056584A1 [17].

References

[1] K. Meissner, B. König, Die neue Gießuhr-vollflexibles Markiersystem für Gussteile, Giesserei 98 (6) (2011) 108–110.

[2] M. Brinkheinrich, Lasermarkiersysteme: Yb-Faserlaser im Vergleich zu anderen DPSS-Lasern, Photonik 38 (1) (2006) 42.

[3] R. Poprawe, Lasertechnik für die Fertigung: Grundlagen, Perspektiven und Beispiele für den innovativen Ingenieur, Springer, Berlin, 2005.

[4] J.F.I. Paz, J. Wilbig, C. Aumund-Kopp, F. Petzoldt, RFID transponder integration in metal surgical instruments produced by additive manufacturing, Powder Metall. 57 (5) (2014) 365–372.

[5] C. Pille, In-process-embedding of piezo sensors and RFID transponders into cast parts for autonomous manufacturing logistics, Smart systems integration, Como, Italy, Paper 20, 2010.
[6] B.-A. Behrens, Fr.-W. Bach, W. Reimche, E. Gastan, F. Lange, G. Mroz, Verfahren zur Einbringung und Wiedergabe von Daten in Sinterbauteilen, Metall - Internationale Fachzeitschrift für Metallurgie 62 (2008) 298–301.
[7] M. Bernard, W. Reimche, C. Scheer, F.-W. Bach, Nachweis von Härterissen im Verzahnungsbereich von Zahnrädern mit Thermografie und Wirbelstromtechnik, in: DGZfP-Jahrestagung "ZfP in Forschung, Entwicklung und Anwendung", 28–30. April 2008, St. Gallen, Berichtsband, DGZfP, Berlin, 2008.
[8] Norm DIN EN ISO 15549, Zerstörungsfreie Prüfung - Wirbelstromprüfung - Allgemeine Grundlagen (ISO 15549:2008); Deutsche Fassung EN ISO 15549:2010, Beuth, Berlin, 2010.
[9] G. Mroz, W. Reimche, W. Frackowiak, O. Bruchwald, H.J. Maier, Setting discrete yield-stress sensors for recording early component loading using eddy-current array technology and induction thermography, in: Proc. 2nd International Conference on System-Integrated Intelligence: Challenges for Product and Production Engineering, SysInt 2014, July 2nd–4th, Bremen, Germany, 2014, pp. 484–493.
[10] G. Mroz, Entwicklung von Techniken zur bauteilinhärenten Informationsspeicherung und Erfassung von Bauteilbelastungen, Dissertation, Leibniz Universität Hannover, Berichte aus dem IW, PZH Verlag, Garbsen, 2013.
[11] G. Mroz, W. Reimche, Fr.-W. Bach, The use of component's edge region as inherent information carriers and loading indicators, in: Proc. 1st Joint Symposium on System-integrated Intelligence: New Challenges for Product and Production Engineering, SysInt 2012, June 27th–29th, Hannover, Germany, 2012, pp. 158–160.
[12] W. Reimche, G. Mroz, O. Bruchwald, Fr.-W. Bach, Bauteilinhärente Belastungssensoren und Informationsspeicherung in der Randzone, Tagung Werkstoffprüfung 2009 - Fortschritte der Kennwertermittlung für die Forschung und Praxis, 2009.
[13] K.A.H. Nurwono, R. Kosala, Color quick response code for mobile content distribution, in: Proceedings of the 7th International Conference on Advances in Mobile Computing and Multimedia, December 14th–16th 2009, Kuala Lumpur, Malaysia, 2009, pp. 267–271.
[14] G. Mroz, W. Reimche, Fr.-W. Bach, et al., Setting of gradient material properties and quality control of high tension 3D NVEB-weld joints, Adv. Mater. Res. 137 (2010) 375–411.
[15] G. Mroz, M. Diebel, W. Reimche, Fr.-W. Bach, Entwicklung von Sensortechnik zur Erfassung diskreter Bauteil- und Belastungsinformationen in der Bauteilrandzone, in: DGZfP-Jahrestagung, Jahrestagung "ZfP in Forschung, Entwicklung und Anwendung", 14–16 Mai 2007, Fürth, Berichtsband, DGZfP, Berlin, 2007, p. 35.
[16] S. Barton, G. Mroz, W. Reimche, H.J. Maier, Inherent load measurement and component identification by multi-dimensional coded data in the component's subsurface region, in: Proc. 3rd International Conference on System-Integrated Intelligence: New Challenges for Product and Production Engineering, SysInt 2016, June 13th–15th, Paderborn, Germany, 2016, pp. 537–543.
[17] W. Reimche, G. Mroz, Fr.-W. Bach, Bauteil, Verfahren zum Einbringen von Informationen in ein Bauteil und Verfahren zum Ermitteln einer Belastungshistorie eines Bauteils, Deutsches Patent, DE102009056584A1, 2010.

2.2.4 Data storage and component labeling within bulk material by sintering

Bernd-Arno Behrens, Martin Bonhage, Irfan Yousaf Malik
Institute of Forming Technology and Machines, Leibniz Universität Hannover, Garbsen, Germany

2.2.4.1 Introduction

Lightweight materials like magnesium or aluminum provide a high potential for the weight reduction of components. In aerospace and automotive industries this is of special interest as a decrease in weight means less fuel consumption and a smaller environmental impact [1].

Powder metallurgy (PM) combines a large number of advantages in processing these materials. A notable economic advantage is the high utilization of material, which is particularly important for mass as well as near-net-shape production. Furthermore, the energy requirement is lower compared to casting as the sintering material does not have to be brought to its melting point. An important technical advantage is the possibility of generating alloys which cannot be produced by means of melting metallurgy [2]. In the following, a procedure which allows storing information inside sintered light alloy components is described. By this, component information can be stored decentralized. Another aspect is the prevention of plagiarism. Since the information is embedded in the material, manipulation is impossible and highly secure relevant components can be safely identified.

2.2.4.2 Background and results of prior research

The production of PM components with assimilated storage possibility is primarily similar to common PM production processes. The only difference is an additional process step to integrate foreign particles (e.g., different material and grain size) before pressing the powder into green compacts (Fig. 2.2.4.1).

First, a layer of the base powder is filled into the die. This layer is flattened to provide a platform for the positioning of foreign particles. These particles are placed onto the powder surface before the rest of the die is filled up with the base powder. Then the upper punch compresses the powder into the green compact which is ejected in the last

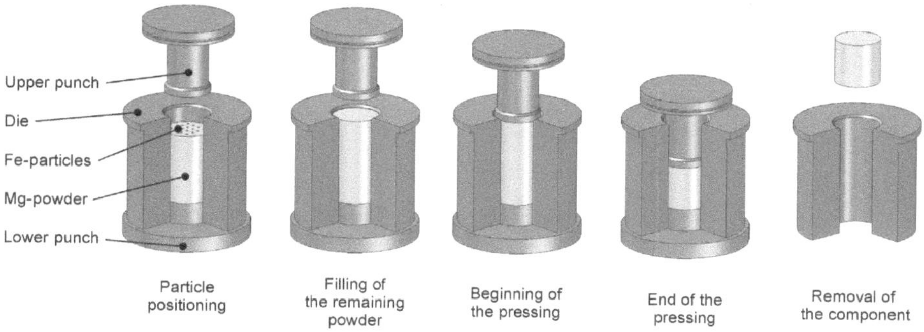

Fig. 2.2.4.1 Pressing process of foreign particles in a metal powder matrix [3].

step [3]. After densification, the specimens are sintered in a vacuum for 30 min at the specific sintering temperature of the base material.

2.2.4.3 Data storage by foreign particles

The storage of data and thus creating a part number within the volume of the produced part is achieved by means of foreign particles. The physical properties and diameter of these particles differ from the base powder material and can therefore be located by X-ray scanning. The information can be stored binary by arranging the particles in different manners, e.g., as Royal Mail 4 State Customer Code (RM4SCC) or in a square matrix (Fig. 2.2.4.2). By binary transformation, a 4×4 matrix is already able to store numerical values from 1 to 65,535 [4].

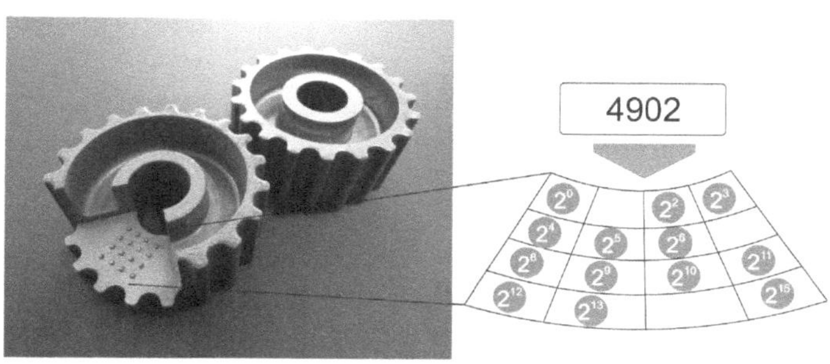

Fig. 2.2.4.2 Integrated foreign elements in a pulley wheel produced by metal powder [4].

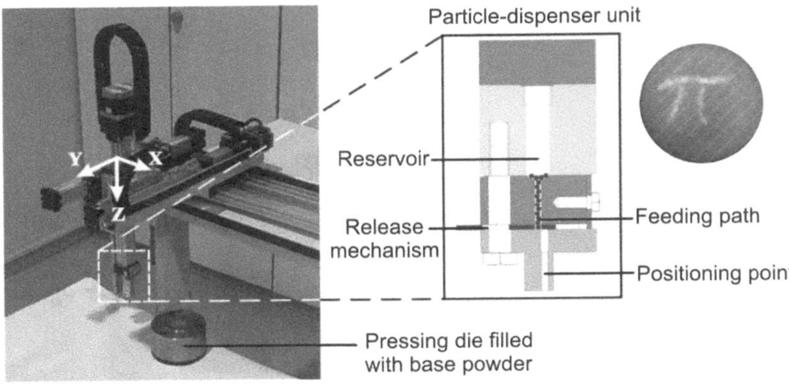

Fig. 2.2.4.3 Automated powder feeder for applying built-in symbols [3].

An alternative process chain was also examined which uses foreign powder written symbols instead of binary code in a metal powder matrix. In contrast to the previously described coding methods, the mean diameter is about the same as the base powder. A nozzle places a powder line where previously foreign particles were integrated. An automated system, as shown in Fig. 2.2.4.3, was tested for powder application. The flexibility and storage capacity of this method is mainly limited by the thickness of the powder line. Read out is done the same way as with the foreign particle approach, by means of X-ray technology. In Fig. 2.2.4.3 (right) the symbol π was applied [3].

2.2.4.4 Dynamic data storage

Another approach is the magnetic data storage in a PM component. Thus data can be complemented, changed, and deleted on demand. The applied process uses inductive magnetic recording heads to magnetize a small part of the components surface, similar to a hard disk drive.

Magnetizable powders consisting of Fe_2O_3 are used for data storage. These powders are distributed evenly in the basic magnesium powder. The volume distribution is 60% of magnesium and 40% of Fe_2O_3 by volume (Fig. 2.2.4.4). The component production follows the classical powder metallurgical production process. Special care has to be taken to ensure a homogeneous distribution of the magnetizable powder in the base powder. The chosen shape of the specimen allows investigating the writing on a plane as well as on a curved surface. For the read out of the stored data the magneto-optic Kerr effect (MOKE) is used. The MOKE read out process detects changes in the polarization of the light reflected by the surface of a magnetic body.

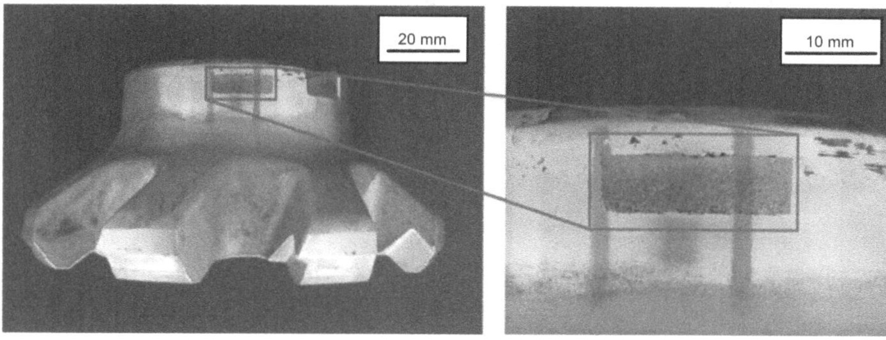

Fig. 2.2.4.4 Milling head made of AZ91 with integrated magnetic storage capacity ($Mg+Fe_2O_3$) [5].

Therefore the surface needs to meet two requirements. On the one hand, it has to possess homogeneous magnetism, and on the other hand, it has to reflect light. These requirements can be met with a soft magnetic keeper layer [5].

2.2.4.5 Integration of inherent data carriers in PM components

Another possibility of positioning data carriers in a metal powder (aluminum and steel) matrix is the use of magnetoelastic sensors consisting of NiFe layers applied to different substrates (aluminum, silicium, and aluminum oxide). As functional elements, magnetoelastic nickel-iron layers with a thickness of 500 nm on different substrate materials were embedded in sintered parts. The magnetoelastic NiFe layers on substrates such as aluminum, silicium, or aluminum oxide are especially appropriate for an online measurement of strains and resulting stresses within a part using an external eddy current sensor system. The variation of the resulting magnetic field in response to an elastic deformation can be correlated with the causing stress. The main point in integrating such elements is to achieve a high binding quality between the substrate material and the base powder. For this means, the substrate material should be properly paired with the basic material.

The experiments showed that the best result concerning bonding quality is achieved by integrating a NiFe layer on an aluminum substrate in an aluminum part. The substrates made of silicium or aluminum oxide are not appropriate due to insufficient ductility, so that the strains in the part cannot be fully transferred to the magnetoelastic layer. Concerning the signal strength, the base powder should consist of a paramagnetic material, for example, aluminum.

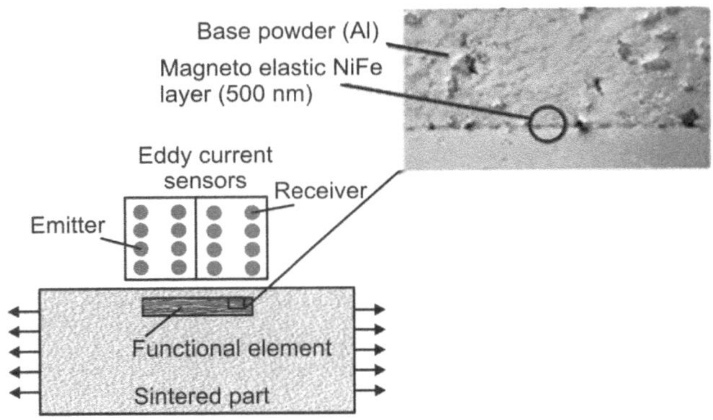

Fig. 2.2.4.5 A magnetoelastic NiFe element in a sintered part [6].

Ferromagnetic materials would interfere with the magnetic field of the eddy current sensor which results in inaccurate signals. Fig. 2.2.4.5 shows the principle of the loading measurement using an integrated magnetoelastic element and also a metallographic image of the bonding zone between the Al-substrate and the Al-base powder [6].

2.2.4.6 Numerical model

Cracks and distortions occurring during sintering are current problems of the PM industry. These problems are caused by an inhomogeneous distribution of the component density. Inhomogeneities occur during compaction by the interaction of foreign elements in a powder matrix. Here, finite element analysis (FEA) can assist in predicting the density changes and gradients caused by the positioning of foreign particles. The powder material is described by the Drucker-Prager Cap (DPC) model which originates from soil mechanics [7]. The DPC model describes the material as an elastoplastic continuum. Fig. 2.2.4.6 shows the model in the plane of hydrostatic pressure p and the von Mises equivalent stress q. The elastic region is bordered by the failure line F_s on the left and the elliptically shaped compaction cap F_c on the right. Stress states reaching the cap cause compression as well as an increase in strength. The FEA software used, Abaqus/Standard, includes a formulation of the DPC model independent of the relative density of the powder. To increase the accuracy of the simulation, the model is expanded by user-defined subroutines. The elastic modulus of the powder is also formulated as a function of the relative density. The particles are modeled as elastic bodies. The friction is modeled based on coulomb but is extended by the dependency on the relative density [8].

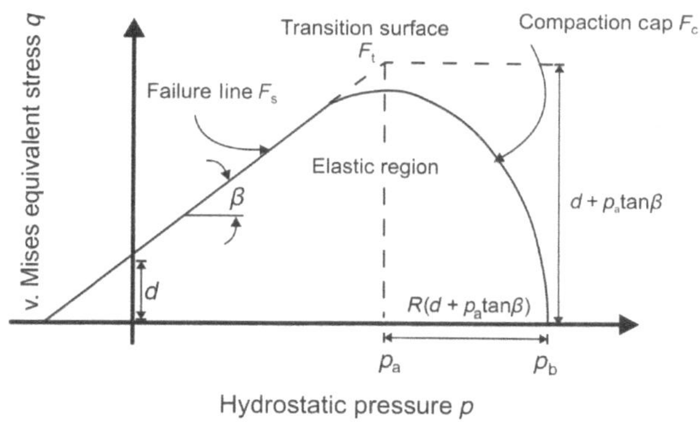

Fig. 2.2.4.6 Drucker-Prager Cap model (DPC) [7].

2.2.4.7 Numerical investigation of cold pressing process

During a double-sided compaction process, the friction between die and powder causes a density gradient symmetric to the mid-plane. The friction restricts the movement of the powder at the die surface. This phenomenon is observed at specimens with and without foreign particles. The influence of a foreign particle on the density is locally limited to an area about two times the size of the particle diameter. Friction hinders the movement of powder along the surface of the particles. This leads to a density gradient with the lowest density near the equatorial plane.

With the help of two models, one with and one without friction, the influence of friction on the density distribution is shown in Fig. 2.2.4.7. The left half shows the results of the model with friction whereas the right half shows the results of the

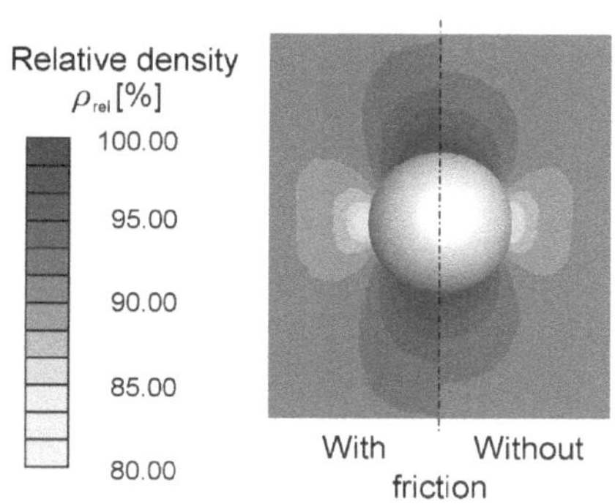

Fig. 2.2.4.7 Density distribution of metal powder matrix surrounding foreign elements [9].

one without friction. The areas of highest and lowest density are observed at the same locations in both models, respectively. The qualitative distribution is similar but with higher values for the model with friction. The results of the simulation show that the heterogeneous distribution of the relative density cannot be avoided when pressing the powder with inserted particles [9].

2.2.4.8 Numerical investigation of the sintering process

In order to estimate the effects of foreign particles on the sintering process, a 2D axial-symmetric model of a cylindrical sample has been taken as a basis along with a single introduced foreign particle. The sintering model uses the relative density and the net geometry as input parameters, which have been determined within the scope of the simulation for the pressing process. As the effect of foreign particles on the density distribution is limited to a small region, only a portion of the sample is examined. Focusing on the analysis, the question arises how far the density gradient around the foreign particle changes during a sintering process. The following states of the sintering process (sintering temperature 600°C) relate to the times indicated in Fig. 2.2.4.8. It can be determined that the density gradient decreases quickly around the foreign particle during the course of the sintering process. As shown in Fig. 2.2.4.8, for the time $t = 50$ min, practically no density gradient can be ascertained upon the equatorial plane toward the end of the sintering process as opposed to the surrounding area. On the basis of this value, it is to be presumed that only a slight weakening in the solidity of the component takes place [10].

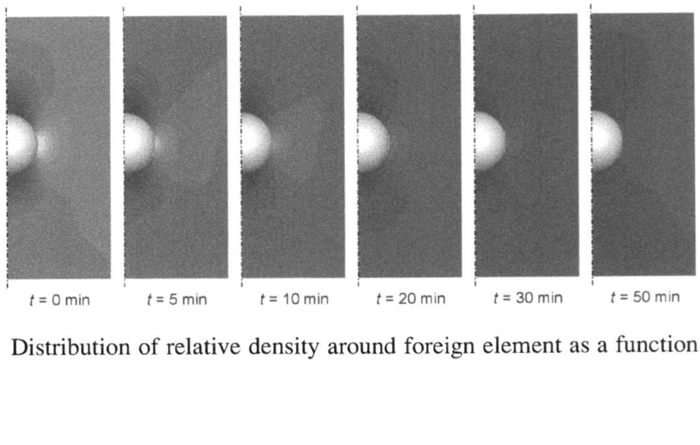

Fig. 2.2.4.8 Distribution of relative density around foreign element as a function of sintering time [10].

2.2.4.9 Summary

This brief overview shows the possibility of integrating foreign elements in a powder matrix with the objective of creating components with inherent properties. This research methodology possesses a diverse value. The increasing connection between production processes, material, and information flows requires components which fulfill these new challenges. By means of the integration of foreign particles, components become traceable over their entire lifetime. Furthermore, plagiarism is efficiently counteracted.

References

[1] K. Danilecki, M. Mrozik, P. Smurawski, Changes in the environmental profile of a popular passenger car over the last 30 years—results of a simplified LCA study, J. Clean. Prod. 141 (2017) 208–218.
[2] W. Schatt, K.P. Wieters, B. Kieback, Pulvermetallurgie Technologien und Werkstoffe, Springer-Verlag, Berlin, Heidelberg, 2007.
[3] B.A. Behrens, F. Lange, E. Gastan, N. Vahed, J. Mielke, Different approaches for static and dynamic data storage inside PM components, in: Proceedings Advanced Metal Materials and Technologies, AMMT 2009 8th International Scientific and Technical Conference, St. Petersburg, Russia, 2009, June 24th–26th, 2009, pp. 114–125.
[4] B.A. Behrens, F. Lange, E. Gastan, Production of PM components with integrated information, in: Proceedings 4th I*PROMS 2008 Virtual International Conference on Innovative Production Machines and Systems, Cardiff, United Kingdom, July 1st–14th, 2008, pp. 481–486.
[5] A. Belski, E. Gastan, N. Vahed, C. Klose, M. Rodman, F. Lange, M.C. Wurz, D. Bormann, L. Rissing, B.A. Behrens, F.W. Bach, Process principle for the production of sintered dynamic component-inherent data storage, Prod. Eng. Res. Dev. 5 (3) (2011) 233–240.
[6] B.A. Behrens, N. Vahed, M. Kammler, Functionalisation of PM components by Integration of inherent Data Carriers and sensory Elements, New Challenges for Product and Production Engineering, Hannover, Germany, June 27th–29th, 2012.
[7] B.A. Behrens, A. Bouguecha, Numerical and experimental investigations of aluminum powder compaction and sintering, Mater. Sci. Eng. 43 (2012) 481–556.
[8] F. Lange, B.A. Behrens, E. Gastan, G. Helms, Manufacturing of powder-metallurgical components with integrated information, in: Proceedings Metal Forming 2008, 12th International Conference, September 21st–24th, 2008, Steel Research International, Verlag Stahleisen, Krakow, Poland, 2008, pp. 486–492.
[9] B.A. Behrens, F. Lange, E. Gastan, Experimental and numerical analysis of a marking approach for sintered light alloy components, Powder Metall. 51 (3) (2008) 277–282.
[10] B.A. Behrens, E. Gastan, N. Vahed, F. Lange, Numerical analysis of the process chain for the production of PM components with integrated information storage, Prod. Eng. Res. Dev. 4 (5) (2010) 477–482.

2.3 Employment of magnetic effects in magnesium alloys

In this chapter it will be demonstrated how the lightweight material magnesium can be provided with magnetic properties and how these properties can be used as an information source for experienced loads as well as a data storage unit.

The enabling techniques described in this section are primarily based on magnetism. The generally known kind of magnetism is ferromagnetism, which only occurs in a few elements, namely, cobalt, nickel, and iron. These elements have a comparatively high density compared to lightweight materials like aluminum or magnesium. Hence, one vision of the CRC 653 is the development of new magnetic magnesium alloys to equip lightweight materials with ferromagnetic properties, which are usable for magnetic sensors or data storage applications (see Fig. 2.3.1).

The development and characterization of magnetic magnesium alloys is described in Chapter 2.3.1. By manipulating alloy elements, properties of the magnesium alloys can be tailored, such as magnetostriction, coercivity, and magnetic remanence.

The alloys developed within the CRC 653 enable the detection of loads using the inverse magnetostrictive effect and the harmonic analysis of eddy current signals (see Chapter 2.3.2). One advantage of this technique is that the measurement is contactless and thus can be applied in locations that are limited or even impossible for conventional sensors, for example, due to geometric restrictions.

Furthermore, the ferromagnetic properties enable a magnetic data storage to locally store information specific to an individual component obtained throughout its life cycle. High stability of the stored information is strived for and this, in turn, requires a high coercivity medium as well as a new recording technique, namely, a heat-assisted magnetic recording. In order to increase the density of the written data tracks and the reliability of the read/write head, the sensor element of the read head and the air gap of the write head had to be scaled down and

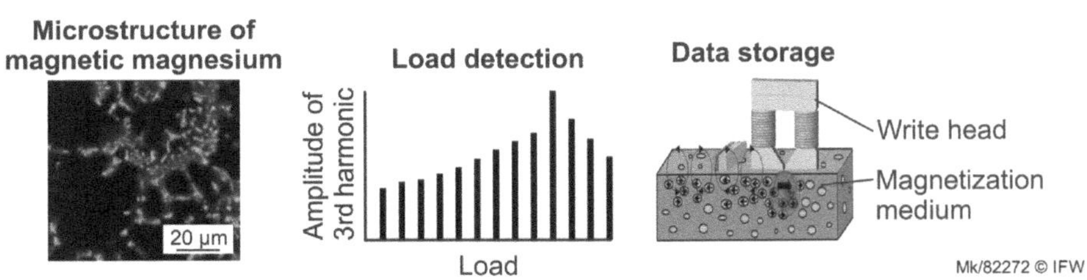

Fig. 2.3.1 Development of magnetic magnesium enabling load detection and data storage.

were therefore designed and manufactured on a flexible substrate using thin film technologies (see Chapter 2.3.3).

See the following chapters for further information

2.3.1 Development of magnetic magnesium alloys

Christian Demminger, Christian Klose, Hans Jürgen Maier
Institute of Material Science, Leibniz Universität Hannover, Garbsen, Germany

2.3.1.1 Alloying of magnesium with magnetic powders

The focus of the Collaborative Research Centre 653's (CRC 653) research on the alloying of magnesium with magnetic powders was on understanding the process-property relationship. Both the online measurement of the mechanical loads acting on the specimens and the magnetic labeling on the material's surface were important technologies developed within the CRC 653. The first funding period of the CRC 653 aimed at the development of casting processes for the production of magnesium alloys with ferromagnetic properties as well as a basic understanding of the properties of suitable alloying systems. Common magnesium alloys such as AZ- or AM-based alloys and pure magnesium are paramagnetic, hence ferromagnetic materials had to be added in order to provide the alloys with magnetic properties.

In the initial investigations, pure magnesium as well as standard magnesium alloys such as AZ91 was alloyed with commercially available magnetic powders based on NdFeB, AlNiCo, and SmCo in order to create ferromagnetic properties in the castings. Because it was assumed that the casting process would have a major influence on the materials' properties, the material composition and the microstructure were subsequently investigated depending on the cooling conditions present during casting. To achieve different cooling rates, both rotational molding and melt spinning were employed [1]. Gravity die casting was used as a reference. During the rotational casting, the acting centrifugal force caused the melt to form layers with a thickness of 5–6 mm on the inner side of the die. The cooling rates were calculated to reach up to 100 K/s. By contrast, the melt spinning process led to higher cooling rates values of up to 10^4 K/s. Here, the magnesium melt was poured into a water-cooled crystallizer with which thin sheets were produced. These castings had a thickness from 0.3 to 0.6 mm.

The local chemical compositions of the cast alloys were analyzed using X-ray spectroscopy (EDX) and electron probe microanalysis (EPMA). Light microscopic investigations were conducted in order to reveal the grain structure as well as the morphology of the precipitates in the microstructure. In addition, the ferromagnetic precipitates were characterized using transmission electron microscopy. The combination of the local chemical composition data from EDX with X-ray diffraction measurements (XRD) allowed for identification of the magnetic phases. The average grain size of the specimen produced by melt spinning was below 10 μm. On the other hand, the specimens manufactured with the rotational casting process had a greater grain size between 30 and 50 μm.

In order to determine the microstructure-property relationship of the magnesium alloys produced, the mechanical properties were characterized for different cooling rates. Thus tensile tests and Vickers hardness measurements (HV10) were carried out. A comparison of the stress-strain curves of four alloys of the type $MgSm_xCo_yCu_z$ is shown in Fig. 2.3.1.1A. Yield strength and elongation to fracture were improved by increasing the mass fraction of the SmCo powder. Fig. 2.3.1.1B demonstrates the results of the Vickers hardness tests. With

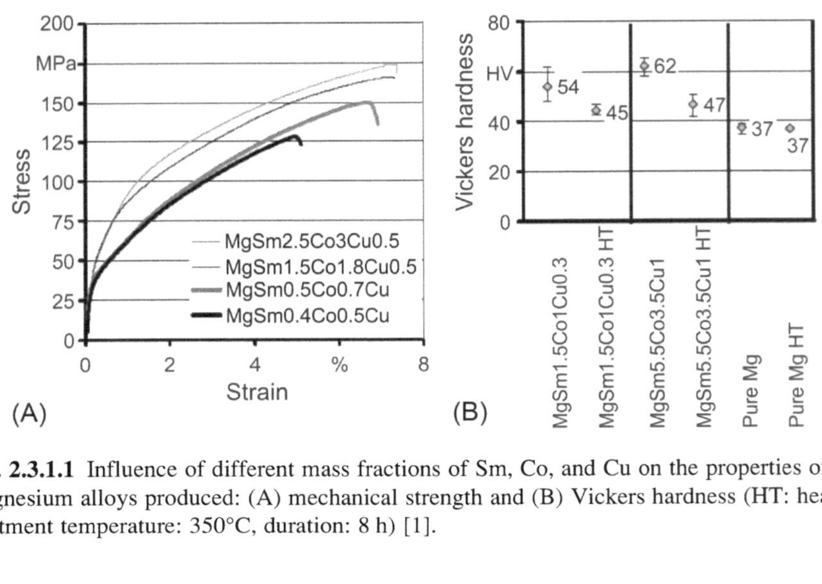

Fig. 2.3.1.1 Influence of different mass fractions of Sm, Co, and Cu on the properties of the magnesium alloys produced: (A) mechanical strength and (B) Vickers hardness (HT: heat treatment temperature: 350°C, duration: 8 h) [1].

increasing amount of magnetic powder, the measured values rose up to 62 HV10 compared to 37 HV10 obtained for the reference (pure magnesium). Additionally, a heat treatment at 350°C for 8 h was applied. As expected, this had no significant effect on pure magnesium, but the heat treatment decreased the Vickers hardness of the magnetic magnesium alloys by approx. 25%.

For magnetic data storage applications, the residual magnetic field strength after application of an external magnetization is essential. The magnetic remanence of the magnetic magnesium alloys depended on the composition of the ferromagnetic phases and precipitates formed in the magnesium matrix during the melting, casting, and solidification processes. The diffraction spectra of specimens made from pure magnesium and ferromagnetic powder containing the elements Co and Sm revealed the presence of the following intermetallic phases: $Mg_{41}Sm_5$, Co_2Mg, $CoMg_2$, Sm_2Co_{17}, $SmCo_5$, $Sm_{3-X}Mg_XCo_5$, and $Sm_{2(1-X)}Mg_{2X}Co_{17}$. Especially the phases $SmCo_5$ and Sm_2Co_{17} have strong ferromagnetic properties and are important for production of common sintered magnets based on Co and Sm.

Various alloys with different mass fractions of the commercially available magnetic powders SmCo, NdFeB, and AlNiCo were produced. The magnetic remanence of alloys made from pure magnesium and SmCo powder generally increased with increasing mass fraction of the magnetic powder. In contrast, alloys based on standard magnesium alloys, e.g., AZ91 or WE43, alloyed with SmCo, NdFeB, or AlNiCo did not exhibit any magnetic remanence. Based on the corresponding microstructural investigations, it can be assumed that the magnetic alloying elements mostly formed nonferromagnetic phases and precipitates due to their poor solubility in magnesium. Moreover, several phases were observed which consisted of ferromagnetic and nonferromagnetic elements, leading to the conclusion that the alloying process itself had to be investigated in more detail.

2.3.1.2 Development of advanced magnetic magnesium alloys based on pure elements

Different commercially available ferromagnetic powders were successfully alloyed to pure Mg. The materials showed magnetic properties which could be used for magnetic labeling tests; however, reproducibility of the alloys' compositions and the resulting properties were still an issue. In addition, the multitude of alloying elements in the magnetic powders did not allow for a systematic evaluation of the individual elements' effect on the magnetic properties. Therefore further investigations were necessary to establish the properties of magnetic magnesium alloys based on the pure elements.

For these investigations, pure Mg was alloyed with pure Co in order to determine the influence of varying amounts of Co on the magnetic and mechanical

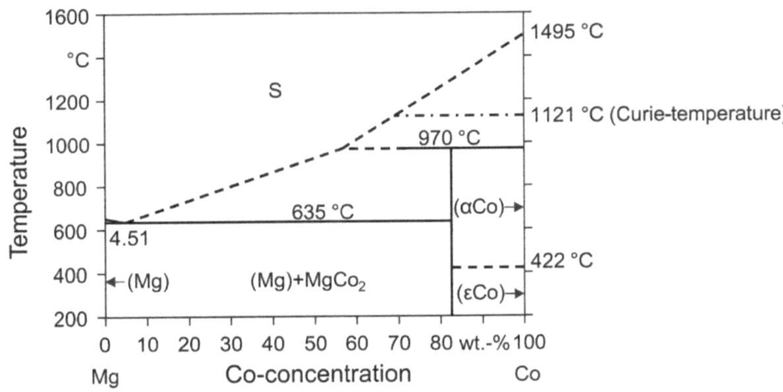

Fig. 2.3.1.2 Binary phase diagram of the Mg-Co system [2,3].

properties of binary Mg-Co alloys. Co is an unusual alloying element of Mg compared with Al, Mn, or Zn, but it has strong ferromagnetic properties and it is not as detrimental for the magnesium alloy's corrosion behavior as Fe or Ni. The binary phase diagram (Fig. 2.3.1.2) shows that in general, Co is hardly soluble in Mg, which was considered favorable for the precipitation of Co-rich ferromagnetic phases in the microstructure. There is also an eutectic composition that contains 4.51 wt.% Co. The substantial difference between the melting points of 1495°C (Co) and 650°C (Mg) places special demands on the alloying processes. Similarly, simple heat treatments were not promising due to the lack of a solid solution, and thus further alloying elements had to be used to increase the mechanical strength of these alloys.

The magnetic magnesium alloys were produced using a modified tilting furnace with electric resistance heating. To avoid oxidation of the Mg melt, the furnace was operated with a controlled shielding gas atmosphere of nitrogen doped with sulfur hexafluoride (N_2+0.3 vol.% SF_6). The pure magnesium ingots (acquired from Magnesium Elektron Ltd.) were melted in a boron nitride coated mild steel crucible and held at a temperature of 730°C during the casting process. The Co share was dispensed into the Mg melt at the beginning of the mechanical stirring and stirred for 90 min at 300 min^{-1}. The poor solubility of Co in Mg prevented the simple use of Co in the form of powder form or ingots for the alloying of the Mg melt. Hence, a proprietary master alloy had to be developed and produced by blending 40 wt.% of pure Co powder (Sigma Aldrich Co.) with 60 wt.% of pure Mg powder in a laboratory powder mixer for 150 min. The mixture was compacted to billets using a cold isostatic pressing method with a hydrostatic pressure of 200 MPa. The green compacts were then machined to billets with a diameter of 29 mm and a length of 80 mm.

Using a laboratory extrusion press, these billets were extruded to rods with a diameter of 6 mm at a temperature of 225°C to reduce the required forming forces by

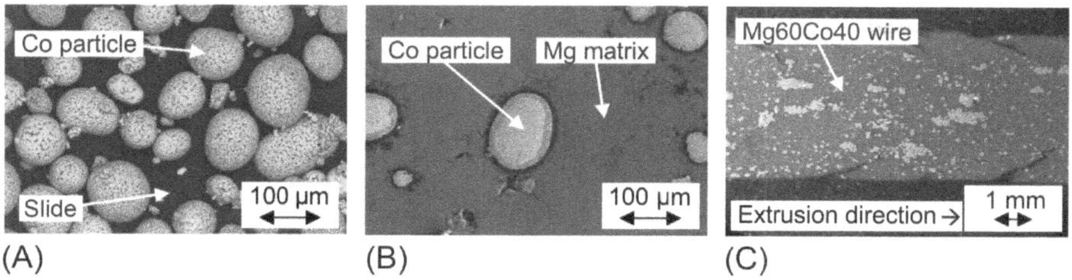

Fig. 2.3.1.3 Mg-40%Co master alloy production: (A) BSE image of pure Co powder, (B) SE image of Mg-40%Co green compact, and (C) SE image of extruded Mg-40%Co section [2].

activating pyramidal slip planes of the hexagonal densely packed crystalline structure. Fig. 2.3.1.3A shows a SEM image of the Co powder particles used in compositional contrast mode. An image of the green compact's structure is shown in Fig. 2.3.1.3B, as well as the microstructure of the extruded section in Fig. 2.3.1.3C. The spherical Co particles were not deformed during the master alloy's production. Although the particle distribution was not homogenous in the extruded rods and some agglomeration was observed, local inhomogeneity had no influence on the alloys' compositions because larger quantities of the Mg-40%Co rods were dispensed to the Mg melt during all casting experiments.

Conventional sintered magnets made from the elements Co and Sm have strong ferromagnetic properties based on the congruent melting phase $Co_{17}Sm_2$ as well as the high temperature phase $CoSm_5$ [3]. Usually, these phases are formed under high temperatures and pressures during a sintering process. To shed light on the phase formation occurring during the casting processes, a Sm-containing master alloy was produced similarly to the Co-containing master alloy. The Sm base material (Treibacher Industrie AG) was ground to powder using a ball mill with a resulting particle size from 30 to 150 μm and blended with 60 wt.% pure Mg powder afterward. In contrast to the Co particles (Fig. 2.3.1.3), the Sm particles shown in Fig. 2.3.1.4 were deformed and finely

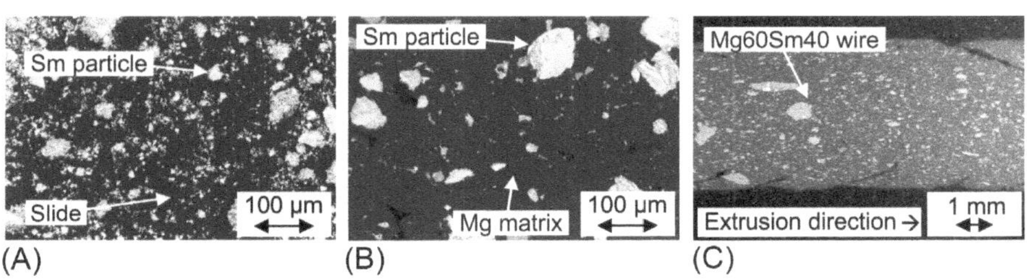

Fig. 2.3.1.4 Mg-40%Sm master alloy production: (A) BSE image of pure Sm powder, (B) BSE image of Mg-40%Sm green compact, and (C) SE image of extruded Mg-40%Sm section [2].

Table 2.3.1.1 Magnetic magnesium alloys produced and characterized, for further details see Ref. [2]

Group	Alloy	Basic material	Co share	Additional elements
Mg-Co	MgCo1	Mg, pure	1 wt.%	–
	MgCo2	Mg, pure	2 wt.%	–
	MgCo3	Mg, pure	3 wt.%	–
	MgCo4	Mg, pure	4 wt.%	–
	MgCo5	Mg, pure	5 wt.%	–
Mg-Co-Sm	MgCo4Sm	Mg, pure	4 wt.%	0.5 wt.% Sm
	MgCo4Sm1	Mg, pure	4 wt.%	1 wt.% Sm
	MgCo4Sm2	Mg, pure	4 wt.%	2 wt.% Sm
Mg-Co-Zn	MgCo4Zn	Mg, pure	4 wt.%	0.5 wt.% Zn
	MgCo4Zn1	Mg, pure	4 wt.%	1 wt.% Zn
	MgCo4Zn2	Mg, pure	4 wt.%	2 wt.% Zn
	MgCo4ZnREZr	ZEK100	4 wt.%	–
	MgCo4ZnRE	ZE10	4 wt.%	–
Mg-Co-Al	MgCo4Al2	Mg, pure	4 wt.%	2 wt.% Al
Mg-Co-Mn	MgCo4Mn	Mg, pure	4 wt.%	0.5 wt.% Mn

distributed by the extrusion processes. Several magnetic magnesium alloys, based on the elements Mg, Co, Sm, Al, and Zn, were alloyed and characterized in related studies [2,4–7]. The alloys studied are summarized in Table 2.3.1.1.

2.3.1.3 Casting geometries and subsequent processing

After melting and introduction of the alloying elements into the melt, the alloys were cast using several casting geometries in order to systematically study the effect of processing conditions on the mechanical and the magnetic properties. This included the use of different casting processes as the cooling rate usually has great influence on the material's properties. For instance, a sand casting process was used to manufacture a wheel carrier geometry. The wheel carrier in this case was the central part of the rear wheel suspension of a Formula Student race car that connected the wheel bearings with the spring-damper system.

The corresponding sand mold consisted of two core boxes, where the cavities were molded using a 3D-printed plastic model. Before casting, the mold was filled with the shielding gas mixture to prevent oxidation. The magnetic magnesium

Fig. 2.3.1.5 Comparison of reactions between melt and mold materials: (A) OBB. Sand S, (B) Bentomix-T, and (C) Rutile/Carbophen.

alloys based on pure Mg turned out to be more reactive as compared to conventional Al-containing magnesium alloys like AZ91. Hence, the initial casting experiments with oil containing mold sand failed. As demonstrated in Fig. 2.3.1.5A and B, the cylindrical sample geometry reacted intensely with molding sands of the types OBB Sand S and Bentomix-T, which is evident by the discolorations on the castings' surfaces.

In conclusion, conventional oil containing mold sands were not usable in combination with the magnetic magnesium melt. Therefore a mold made from rutile sand was manufactured using a cold box process. The sand mixture consisted of 97 wt.% rutile and a binder system based on resin and hardener. As the resin, Carbophen (Huettenes Albertus GmbH) was blended with rutile for 90 s in a sand mixer. The mold sand was then filled into the prepared core boxes. Afterward, the boxes were cured in a CO_2 gas atmosphere to initialize the cross-linking reaction. After casting and demolding, the surface of the casting showed only few reactions with the mold sand (Fig. 2.3.1.5C). This molding material was then used to manufacture the wheel carrier geometries from binary and ternary magnetic magnesium alloys for the Formula Student race car (Fig. 2.3.1.6).

However, thermal conductivity of the molding sand is generally poor compared to steel molds, thus the cooling rate is low, which in turn leads to a coarse-grained microstructure of the component. In other casting processes like die casting, the higher cooling rate increases undercooling of the melt, which results in a greater number of stable crystallization nuclei. This allows for a more fine-grained microstructure compared to sand castings [8]. Compared with the cast volume, the large size of the steel molds used in high-pressure die casting processes leads to an even higher cooling rate than in gravity die casting. Thus the relationship between the magnetic magnesium alloys' microstructures and

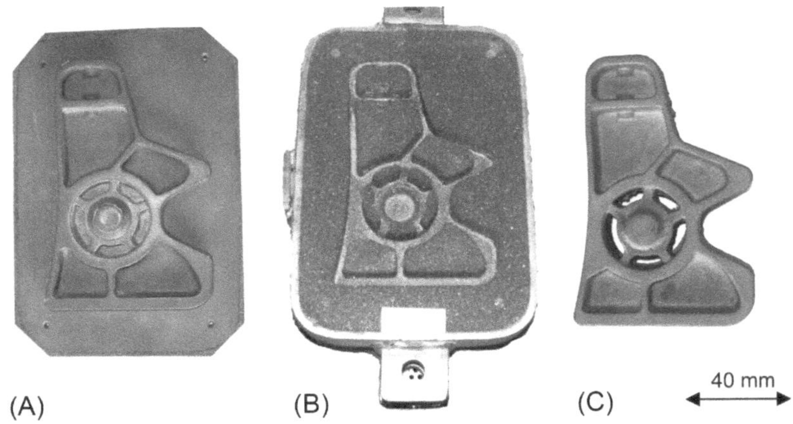

Fig. 2.3.1.6 Wheel carrier production process: (A) 3D-printed model, (B) core box, and (C) casting.

the cooling conditions was examined mainly using various die-casting methods. In contrast to the mold sands, the die casting molds made from steel and copper did not affect the surface quality of the specimens negatively.

The specimens for the mechanical testing were machined from bars, which were produced according to VDG-leaflet P 376 (Verein Deutscher Giessereifachleute e.V.). The die used for the casting of these bars consisted of EN-GJL-250. The die was equipped with a full-length feeder, whereas the bar geometry itself was located in the bottom area of the die and had a dimension of 22 mm in diameter and 230 mm in length (Fig. 2.3.1.7C).

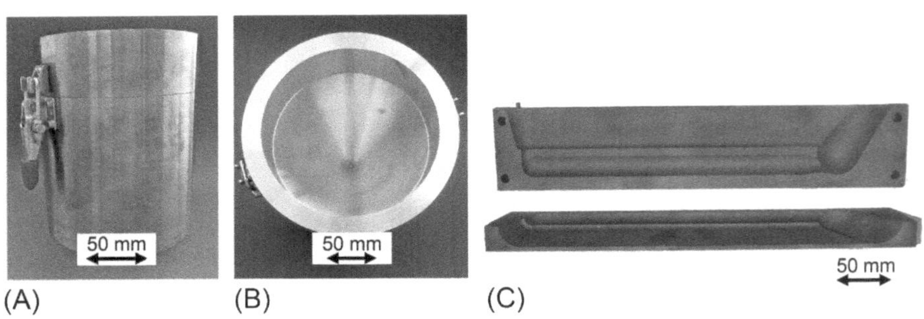

Fig. 2.3.1.7 (A) ETP-Cu mold side view, (B) ETP-Cu mold top view, and (C) EN-GJL-250 test bar mold.

In addition, a die with a conical inner shape was designed and manufactured from ETP copper (Fig. 2.3.1.7). The thermal conductivity of this material is about 394 W/m K [9] and the large wall thickness of the die enforced a short solidification time. This die allow for production of a cone-shaped casting with microstructures varying from very fine grained to coarse grained depending on the position the sample was extracted from. With the use of the casting simulation software Magma5 (Magma Gießereitechnologie GmbH) the casting, solidification, and cooling properties of the dies had been calculated prior to manufacture. Next, these simulations were validated experimentally, with four type-K thermocouples used for temperature monitoring during the actual process. The recorded temperature curves were then used to adjust the heat transfer properties for subsequent simulations.

2.3.1.4 Evolution of the microstructure depending on the cooling conditions

Fig. 2.3.1.8A shows the simulation results of the temperature distribution for the conical casting geometry based on an AZ91 alloy. In the cone's axis of symmetry four thermocouple-measuring points had been placed to monitor the temperature progression during mold filling and the solidification process. As expected, the cooling rate of the casting decreased with an increasing wall thickness from the tip of the cone to the

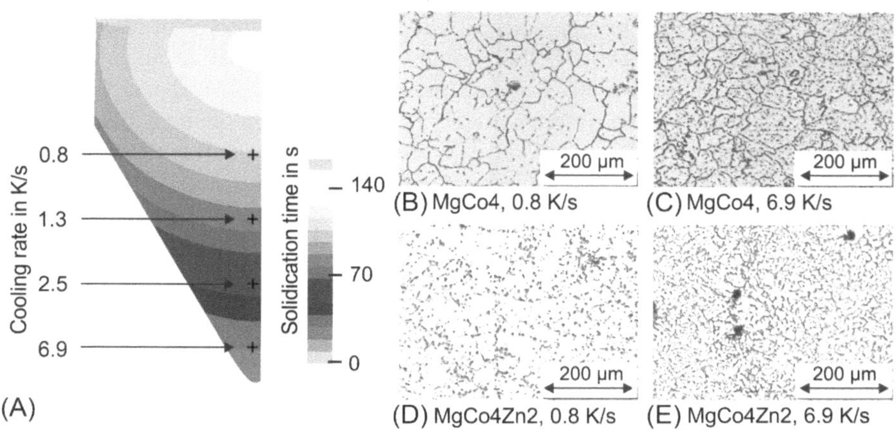

Fig. 2.3.1.8 (A) Simulation of the conical casting's solidification time; metallographic images of the corresponding microstructures: (B) MgCo4 (0.8 K/s), (C) MgCo4 (6.9 K/s), (D) MgCo4Zn2 (0.8 K/s), and (E) MgCo4Zn2 (6.9 K/s).

wide area of the cone. The associated specimens for SEM and metallographic investigations were cut by wire EDM. Next, the samples were ground and then polished down to a 1 μm grit size to remove the EDM-affected surface layer. In order to reveal the microstructure, the specimens were etched using a solution of 2 mL HCl, 3 mL HNO_3, and 95 mL C_2H_5OH.

For the sake of clarity, the investigated alloys will be designated in the following according to DIN EN 1753 with their nominal compositions in weight percent. The resulting microstructures of the binary MgCo4 alloy for cooling rates of 0.8 and 6.9 K/s are demonstrated in Fig. 2.3.1.8B and C, respectively. The Co-rich structures appear black because of the etching and the Mg-based material is more lightly shaded. These structures were mainly located in the areas of the interdendritic network and the grain boundaries, and a fine-grained microstructure led to a larger concentration of Co-rich phases. Based on the metallographic investigations, it could already be assumed that the ferromagnetic properties should be more pronounced with an increasing cooling rate.

Fig. 2.3.1.8D and E demonstrates the corresponding optical micrographs of the Zn-containing MgCo4Zn2 alloy. The addition of Zn significantly modified the microstructure of the magnetic magnesium alloy. The line-shaped Co-rich phase of MgCo4 changed to a more equiaxed microstructure in the area of the grain boundaries. In general, the grain size of the MgCo4Zn2 alloy was affected by the cooling rate similarly as in case of the MgCo4 alloy. The results of the EDX measurements performed in order to determine the alloy compositions are presented in Table 2.3.1.2. The values represent the mean values of 14 EDX measurements. Due to its poor solubility in Mg, alloying with Co proved to be a challenge, leading to generally lower amounts of Co in the alloy compared with the target concentration.

In addition, the microstructures of the alloys were investigated by electron probe microanalysis (EPMA) to characterize the local distributions of the alloying elements. In Fig. 2.3.1.9A, a BSE overview of a representative area is shown, where the elements with a high atomic number (i.e., Co, Zn) appear white and the lightweight Mg matrix is gray. In Fig. 2.3.1.9B–D the element distributions of Co, Zn, and Mg are illustrated. The Co share is clearly confined to the precipitations in Fig. 2.3.1.9B, whereas the Zn is also present in the matrix. However, the

Table 2.3.1.2 EDX spectra of specimens in Fig. 2.3.1.8B–D

Element	MgCo4 target	MgCo4 actual	MgCo4Zn2 target	MgCo4Zn2 actual
Mg in wt.%	96	97.09	94	95.05
Co in wt.%	4	2.91	4	3.10
Zn in wt.%	–	–	2	1.85

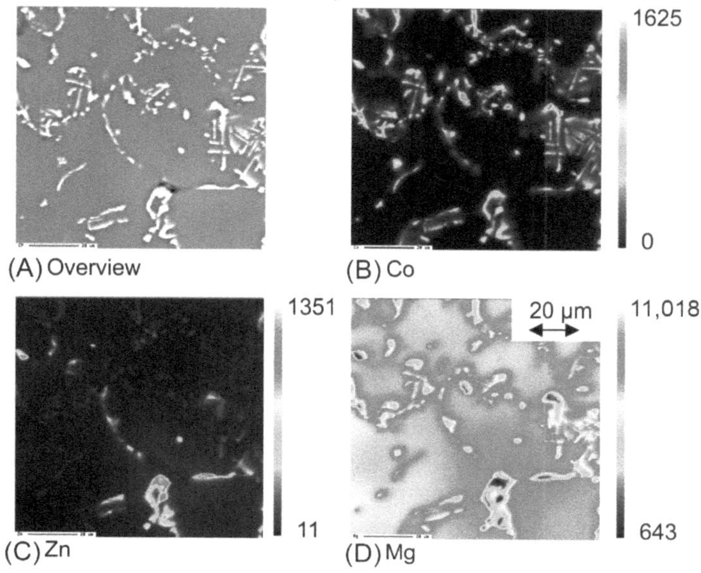

Fig. 2.3.1.9 EPMA scan of the magnetic magnesium alloy MgCo4Zn2.

distribution of Co and Zn is not completely identical, meaning that the precipitations consist probably of Co-Zn and Co.

In conclusion, this section described the development of magnetic magnesium alloys from adapted conventional available magnesium alloys to precisely designed alloys based on pure elements. The alloying processes and casting geometries have been demonstrated with respect to the influence of cooling parameters on the magnetic phases.

References

[1] F.-W. Bach, D. Bormann, Abschlussbericht SFB653, Teilprojekt E2, Magnetische Magnesiumlegierungen, 2008.

[2] C. Klose, Entwicklung magnetischer Magnesiumlegierungen mit sensorischen Eigenschaften, Dissertation Leibniz Universität Hannover, Berichte aus dem IW, PZH-Verlag, Garbsen, 2013.

[3] A.A. Nayeb-Hashemi, Co-Mg phase diagram, in: B. Predel, R. Börnstein, H. Landolt, O. Madelung, W. Martienssen (Eds.), Numerical Data and Functional Relationships in Science and Technology: New Series, Springer, Berlin, 1993, pp. 1–2.

[4] C. Klose, C. Demminger, H.J. Maier, Sensorische Werkstoffe - Erfassen und Nutzen mechanischer Beanspruchungsdaten im Lebenszyklus gentelligenter Bauteile, in: K. Bobzin (Ed.), 8. Aachener Oberflächentechnik Kolloquium 2013, first ed., Shaker, Herzogenrath, 2013.
[5] C. Klose, C. Demminger, S. Zwoch, W. Reimche, F.-W. Bach, H.J. Maier, Magnesium race car components with load-sensitive properties, Euro LightMAT 2013, Magnesium, Aluminium, Titanium - Science and Technology, DGM, Bremen, 2013.
[6] C. Klose, C. Demminger, S. Zwoch, W. Reimche, F.-W. Bach, H.J. Maier, K. Kerber, Measurement of Static and Dynamic Loads Utilizing Sensory Magnesium Components, DGM, Kiel, 2013, p. 57.
[7] C. Klose, C. Demminger, G. Mroz, W. Reimche, F.-W. Bach, H. Maier, K. Kerber, Influence of cobalt on the properties of load-sensitive magnesium alloys, Sensors 13 (1) (2013) 106–118.
[8] W. Seidel, F. Hahn, Werkstofftechnik: Werkstoffe - Eigenschaften - Prüfung - Anwendung, Hanser, München, 2012.
[9] Deutsches Kupfer Institut, DKI Werkstoff-Datenblatt Cu-ETP, 2016, https://www.kupferinstitut.de/de/werkstoffe/eigenschaften/kupfer.html (accessed April 2016).

2.3.2 Determination of loads with magnetic magnesium alloys

Christian Demminger, Christian Klose, Hans Jürgen Maier
Institute of Material Science, Leibniz Universität Hannover, Garbsen, Germany

2.3.2.1 Effect of cooling rate on the magnetic properties

The magnetic properties render the use of the alloys as materials for data storage possible. The magnetic properties of the alloys were analyzed by magnetic hysteresis measurements as a function of the cooling conditions using a vibrating sample magnetometer (VSM). Different cooling conditions were realized using different casting processes, i.e., gravity sand casting (GSC), gravity die casting (GDC), and high-pressure die casting (HP-DC). The comparatively high thermal insulation of the casting sand results in an overall low cooling rate of the casting. In comparison, the large tempered steel mold used in HP-DC processes lead to a high cooling rate and a short solidification time. The specimens were in each case extracted from step-plate geometries produced in the GSC and HP-DC processes. The HP-DC trials were conducted on a Frech DAK 350-40 cold chamber die-casting machine. The

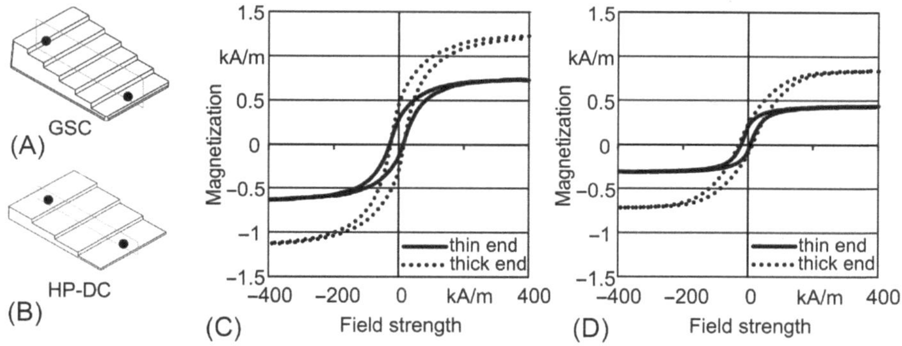

Fig. 2.3.2.1 Hysteresis (M(H)) curves of GSC (C) and HP-DC (D) specimens cut from the castings in (A) and (B).

cooling conditions of the step-plate geometry vary also depending on the material thickness.

Fig. 2.3.2.1A and B shows the locations of the specimens with dimensions of 5 mm × 5 mm × 2 mm, which were extracted from the thin and thick end of the step plates (Fig. 2.3.2.1C and D). The specimens' nominal ferromagnetic content was 4 wt.%. Thus the measured characteristics of magnetic properties are very low compared with literature values for bulk magnetic material. In general, the specimens produced by the GSC process showed a larger saturation magnetization than the HP-DC specimens. Furthermore, the thick ends of both step-plate geometries exhibited larger values for saturation and remanence magnetization than the thin ends (Table 2.3.2.1), i.e., the saturation magnetization increases with decreasing cooling

Table 2.3.2.1 Magnetic characteristics of the specimens in both casting processes

Casting process	Position	Coercive field strength in kA/m	Remanence magnetization in kA/m	Saturation magnetization in kA/m
GSC	Thin end	20.292	0.220	0.730
	Thick end	20.292	0.391	1.230
HP-DC	Thin end	20.193	0.144	0.438
	Thick end	20.292	0.176	0.836

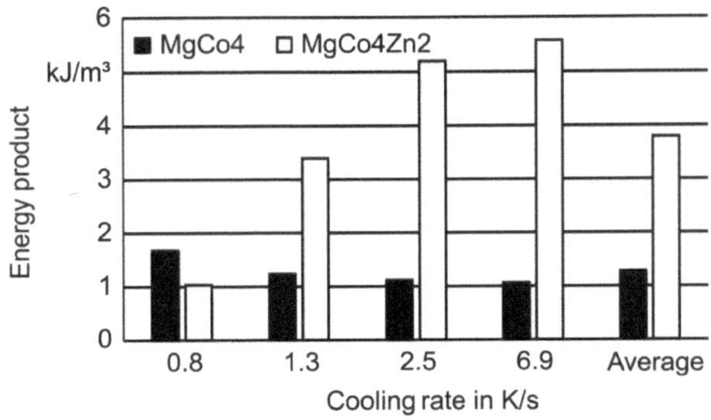

Fig. 2.3.2.2 Energy products of MgCo4 and MgCo4Zn2 based on measured hysteresis curves [1].

rate of the specimens. For comparison between the different alloys and cooling rates, the maximum energy product is given in Fig. 2.3.2.2.

2.3.2.2 Thermal stability of magnetic labels on magnesium alloys

To assess the microstructural and magnetic stability, magnetic write and read tests were performed. Using a magnetic write head several data tracks were written on the specimens made from the magnetic magnesium alloys. The influence of various cooling conditions on the magnetic labeling quality was analyzed using specimens extracted from different positions of the conical casting geometry.

In addition, the thermal stability of labeled magnetic data tracks was investigated. Therefore several die cast magnetic magnesium specimens were labeled magnetically and afterward heat treated at temperatures of 60°C, 80°C, 100°C, 120°C, and 140°C for 1 h in a convection furnace. The temperature was monitored with a thermocouple located in the specimens (Fig. 2.3.2.3).

The stability of magnetic labels is influenced by the quality of the labeling technique. If a magnetic label is used as a product protection feature to prevent plagiarism, the read out should be not negatively affected during the use of the component. Thus the thermal stability had to be investigated to obtain information about the technical application limits.

In Fig. 2.3.2.4 two SEM images of the microstructures of two different magnetic Mg alloys are displayed. In general, the image sections show a similar distribution

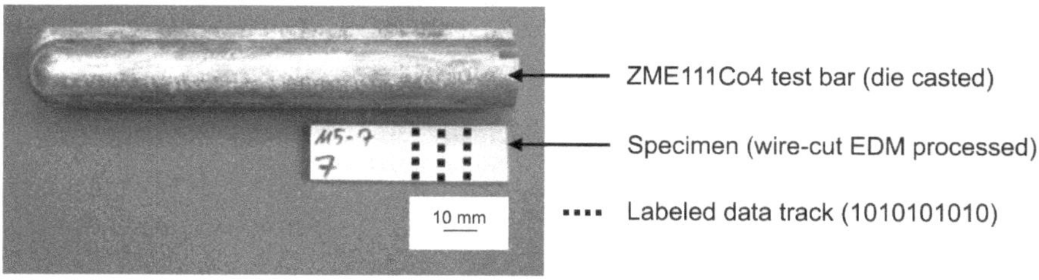

Fig. 2.3.2.3 Specimen geometry used for the magnetic labeling studies.

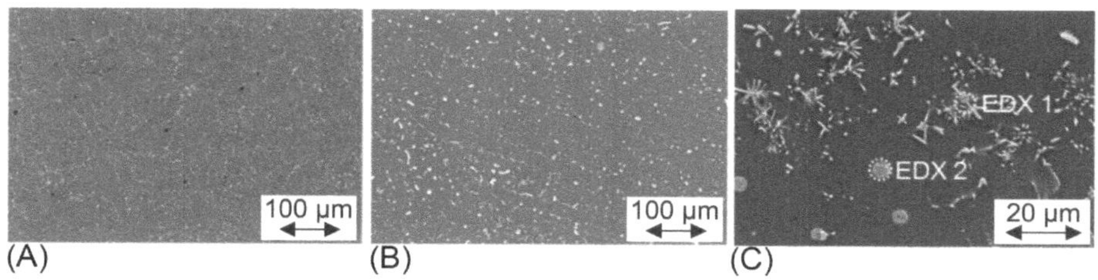

Fig. 2.3.2.4 BSE images of specimens made from (A) MgCo4Zn, (B) MgCo4Mn1RE1Zn1, and (C) higher magnification detail from (B) with EDX measurement spots.

of the white appearing Co-rich precipitates in the magnesium matrix. MgCo4Mn1RE1Zn1 featured larger precipitates than MgCo4Zn. The chemical compositions of the globular and star-shaped precipitates in Fig. 2.3.2.4C were determined with EDX measurements. The results are given in Table 2.3.2.2. The star-shaped precipitates consisted of mainly Mg, Co, and Zn but the globular-shaped precipitates

Table 2.3.2.2 EDX spectra of the locations marked in Fig. 2.3.2.4C

	Mass fraction in wt.%	
Element	**EDX 1**	**EDX 2**
Mg	38.5	54.1
Co	19.9	7.8
Zn	41.4	18.5
RE	0.2	19.6

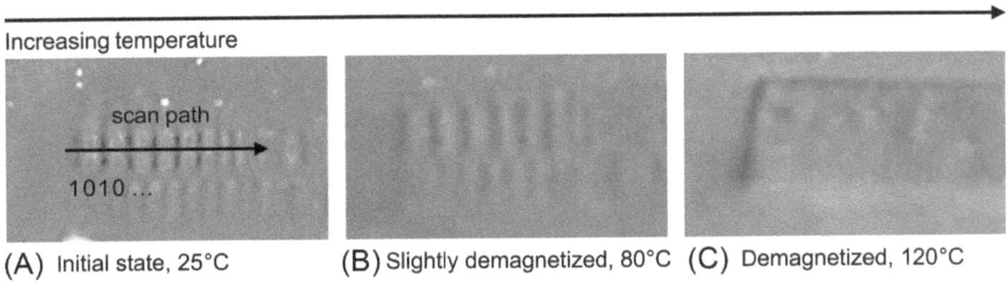

Fig. 2.3.2.5 Thermally induced degradation of the magnetic data tracks (see Chapter 2.3.3).

consisted of a similar mass fraction of Mg, a reduced Co and Zn content, and 20 wt.% of rare earth elements (RE).

All specimens manufactured from MgCo4Zn and MgCo4Mn1RE1Zn1 alloys were labeled with three data tracks, each with the bit sequence "1010101010," using the magnetic write head described in Chapter 2.3.3. After the writing process, each data track was read out in order to verify the label's quality. Next, the samples were heat treated and the read out was repeated. Fig. 2.3.2.5 shows an example of the thermally induced degradation of the magnetic data tracks.

Along the scan path marked in Fig. 2.3.2.5 the magnetic write head recorded the magnetization in the initial state and after the treatment. Depending on the materials' thermal stability, the magnetization decreased with increasing temperature and led to reduced measured amplitude values. Clearly, a correct allocation of the individual bits is mandatory for data storage applications. The only slightly decreased amplitude values resulting after 80°C heat treatment (Fig. 2.3.2.5B) are not critical with respect to data storage and product protection features. By contrast, the demagnetized specimen shown in Fig. 2.3.2.5C clearly demonstrates that such excessive temperature need to be avoided.

2.3.2.3 Load-sensitive properties of magnetic magnesium alloys

On the one hand, ferromagnetic magnesium alloys are usable for magnetic labeling techniques to store data directly on the surface of a component. Another application of these alloys is based on the sensory properties as mechanical loads can be monitored online by means of the harmonic analysis of eddy current signals. Here, a sinusoidal excitation voltage is generated and fed to the excitation coil integrated into the eddy current sensor, which generates an alternating magnetic field via the excitation current. The primary alternating magnetic field induces magnetic reversal processes and eddy currents which not only depend on the material's physical properties but also on the stress state. Thus the signal induced by the eddy current can be exploited to determine the loading condition. Specifically, analysis of the 3rd (and higher) harmonic of the eddy current signal is preferred for ferromagnetic materials as it is very sensitive to changes in the magnetic properties of the material tested.

If a ferromagnetic material experiences a mechanical force, the preferred magnetization direction changes due to the reversible deformation of the material's crystal lattice. This magnetoelastic effect leads to a macroscopic change of the magnetic properties, which can be monitored online by the harmonic analysis of eddy current signals. The application of magnetoelastic force and torque sensors has been reported in literature [2]. However, these sensor elements are often made from ferromagnetic materials like metallic glasses, which usually are available only as thin strips or wires in order to maintain an amorphous structure. By contrast, magnetic magnesium alloys allow for an online measurement of stresses acting on a structural component.

To compare the magnesium alloys' load-sensitive properties depending to the cooling conditions, tensile specimens were machined from Mg-Co-Zn-based alloys produced by gravity die casting (GDC) and high-pressure die casting (HP-DC). The GDC trials were conducted using the test bar geometry described before. The HP-DC specimens were manufactured with the cold chamber die-casting machine using a sextuple cavity according to DIN 50125-A 6×30. GDC and HP-DC tensile specimens were used as test pieces for the determination of the mechanical strength values in standard tensile tests. The results are summarized in Fig. 2.3.2.6. The additional Zn share led to a significant increase of 0.2% yield strength from 30 to 75 MPa in the GDC condition. The elongation at fracture and tensile strength were increased slightly. The mechanical properties of the HP-DC

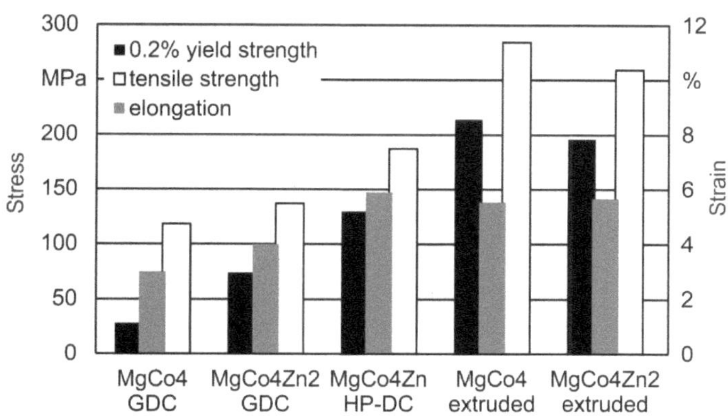

Fig. 2.3.2.6 Mechanical properties of magnetic magnesium alloys for different production processes.

processed specimens were further increased compared to the GDC processed specimens because of the comparatively high solidification and cooling rates introduced by the large HP-DC steel mold. The extrusion process led to the best mechanical properties for the cases studied. The 0.2% yield strength of the MgCo4 alloy was increased from 30 MPa in the GDC condition to 215 MPa in the extruded condition. In contrast to the strengthening effect of Zn in the GDC condition, the Zn content decreased the strength slightly, but increased the elongation at fracture from 5.5% to 5.7%.

Further samples of the Mg-Co-Zn alloy were subjected to loading tests using stepwise increasing loads from 200 to 1400 N in order to test whether correlations can be established between the forces applied to the specimens and their magnetic properties. During these experiments the magnetic properties were measured by the harmonic analysis of eddy current signals.

The results are shown in Figs. 2.3.2.7 and 2.3.2.8. The changes in the magnetic properties of all tested samples were sufficiently large to clearly distinguish between the loaded and the unloaded condition based on the amplitude of the 3rd harmonic. Moreover, a distinct correlation existed between the applied mechanical loads and the measured amplitude values. Following the test sequence in Fig. 2.3.2.7 from left to right, every value shown is the average result of three consecutive measurements under the same load. While a force of 200 N lead to a slightly greater amplitude than in the initial state, the highest load in this experiment (1400 N) caused the 3rd harmonic to increase significantly. In case of the GDC specimens (Fig. 2.3.2.7), there was a gradual increase of the amplitudes in the unloaded condition until they reached a

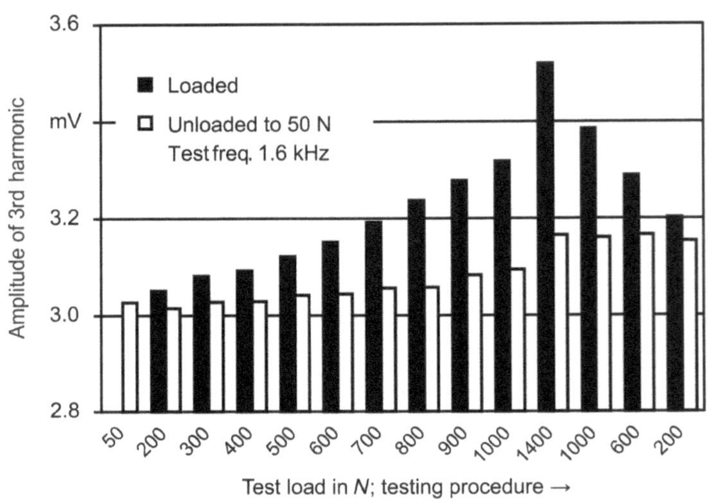

Fig. 2.3.2.7 Amplitudes of the 3rd harmonic measured in interrupted loading tests: gravity die casting condition [3].

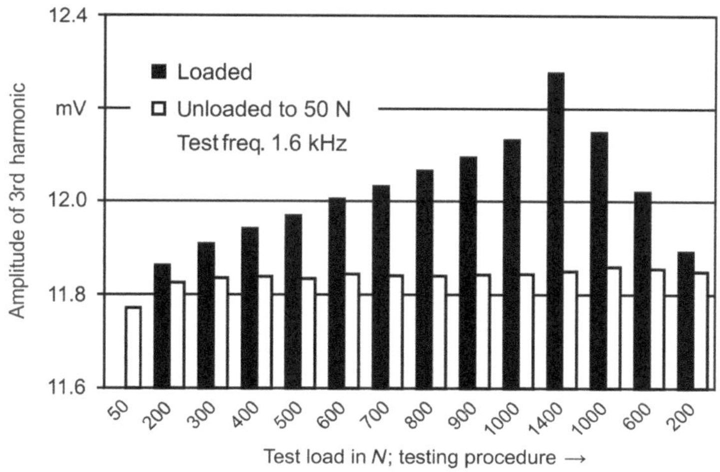

Fig. 2.3.2.8 Amplitudes of the 3rd harmonic measured in interrupted loading tests: high-pressure die-casting condition [3].

constant value after the application of the maximum load. This effect can be attributed to the onset of local plastic deformation, and thus is related to the relatively low yield strength. Still, the material clearly displayed load-sensitive behavior throughout the whole test sequence.

Compared with the GDC condition, the HP-DC specimens generally exhibited a much higher amplitude of the 3rd harmonic (Fig. 2.3.2.8). Basically, higher amplitudes are favorable because the actual signal is differentiated easier from the background noise during the measurements. In addition, the HP-DC alloy showed an almost ideal linear correlation between the measured amplitude and the actual load on the tensile specimen. Apart from the greater absolute values, the difference between the amplitudes under a load of 1400 and 50 N seemed to be virtually independent from the cooling conditions. In the current example, the difference was 0.49 mV for the samples in the GDC condition and 0.45 mV for the HP-DC specimens. In other words the sensitivities in the measurement of the mechanical loads were nearly the same for both materials.

2.3.2.3.1 Magnetic properties of plastically deformed specimens

As demonstrated in Figs. 2.3.2.7 and 2.3.2.8, plastic deformation can have a pronounced effect on the magnetic response of the material. To better understand the ramifications with respect to actual applications, the influence of a plastic

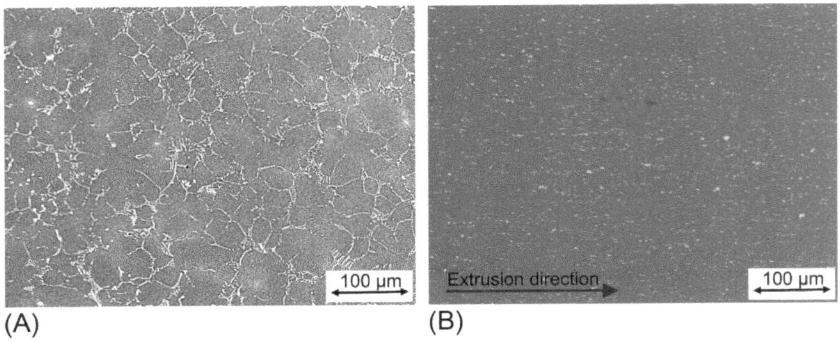

Fig. 2.3.2.9 BSE images of an MgCo4Zn2 alloy: (A) GDC condition and (B) extruded condition.

deformation on the material's load sensitivity needed to be studied in more detail. Therefore tensile specimens were manufactured from the alloy MgCo4Zn2. The alloy was produced by casting as described before; however, the cast billet with a diameter of 50 mm and a length of 150 mm was subsequently extruded to rods with a diameter of 10 mm each using a 2.5 MN extrusion press. The billets' and container's processing temperature was adjusted to 400°C and the ram speed was 0.2 mm/s during extrusion.

Two BSE images of the MgCo4Zn2 alloy manufactured using gravity die casting and extruding are displayed in Fig. 2.3.2.9. The extruded material's microstructure differs significantly from the as-cast condition. The white appearing Co-rich precipitations are located at the grain boundaries in the GDC condition. Upon extrusion the Co-rich precipitations and particles were oriented along the extrusion direction (from right to left in Fig. 2.3.2.9B). Because of the higher hardness of Co (c.1000 HV) compared with Mg (c.40 HV), the Co particles were hardly deformed but tended to agglomerate. In general the distribution of the Co in the Mg matrix appears to be more homogenous compared with the GDC condition.

In contrast to the interrupted loading experiments on cast material in order to determine its basic load-sensitive behavior, the extruded specimens were intentionally deformed permanently (Figs. 2.3.2.10 and 2.3.2.11). Prior to these experiments, standard tensile tests were performed to determine the material's yield point (cf. Fig. 2.3.2.6). Based on these values suitable tensile forces in the regimes of elastic and plastic deformation were selected. Given the specimen's diameter of 5.98 mm and the 0.2% yield strength of 196 MPa, substantial plastic deformation will set in at 5500 N. Thus the initial interrupted loading test sequence was performed in the elastic deformation regime, i.e., well below the

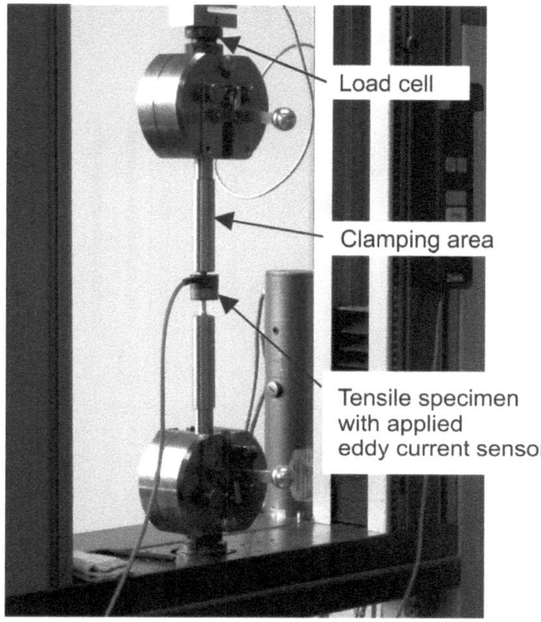

Fig. 2.3.2.10 Tensile testing with eddy current sensor.

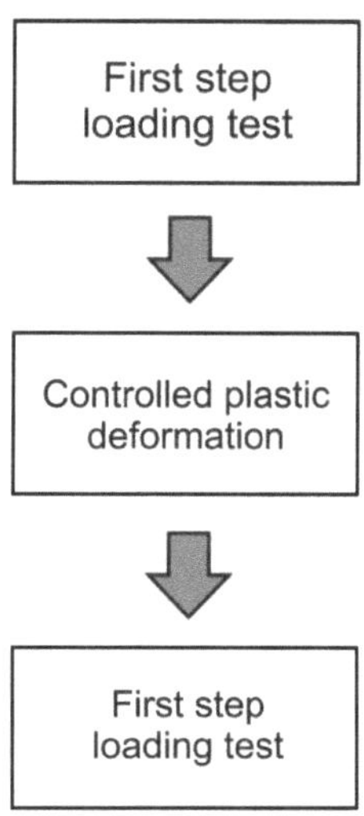

Fig. 2.3.2.11 Schematic of the test sequence.

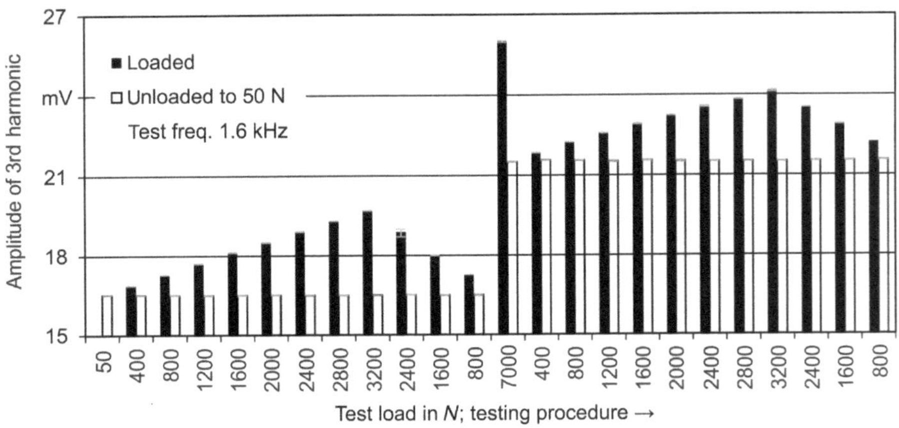

Fig. 2.3.2.12 Amplitudes of the 3rd harmonic measured in cyclic loading tests: MgCo4Zn2, extruded condition.

0.2% yield strength. Next, the sample was loaded with 7000 N to introduce a small degree of plastic deformation (Fig. 2.3.2.12). Finally, a second interrupted loading test was conducted. The relative sequence of applied forces was similar to those during the elastic tests (Fig. 2.3.2.7); however, the magnitude of the forces was adapted to the greater mechanical strength of the extruded material, cf. Fig. 2.3.2.12.

The open bars in Fig. 2.3.2.12 show the amplitudes of the 3rd harmonic of the eddy current signal upon unloading to 50 N. The full bars indicate the loaded condition. The constant level of the measured amplitudes (16.5 mV) in the unloaded state revealed that tensile loads up to 3200 N had no permanent effect on the samples. In accordance with the previous studies the amplitudes of the 3rd harmonic increased with increasing force. The maximum load of 3200 N resulted in an amplitude value of 19.7 mV and the signal difference between the loaded and unloaded condition, which describes the sensitivity of the system, is 3.2 mV.

The application of an intentional overload (7000 N) in the middle of the experiment led to a subsequent permanent increase of the amplitudes of the 3rd harmonic because of the introduced plastic deformation. In the unloaded condition, the signal level was increased permanently to 21.5 mV, while the linear load-sensitive behavior remained. Specifically, an amplitude value of 24.2 mV resulted at 3200 N and the sensitivity at 3200 N was 2.7 mV, which is decreased compared to the not deformed condition.

In conclusion, the material's sensory properties were not substantially degraded by the onset of plastic deformation. It was still possible to distinguish a mechanical load with the eddy current testing although the specimen had exceeded the yield point. In addition, the technique can also be exploited to detect an overload that resulted in plastic deformation if the original amplitude level is known.

References

[1] C. Demminger, C. Klose, P. Taptimthong, H.J. Maier, Material-inherent data storage using magnetic magnesium-cobalt alloys, in: Proc. 2nd International Conference on System-Integrated Intelligence: Challenges for Product and Production Engineering, SysInt 2014, July 2nd–4th, Bremen, Germany, 2014, pp. 188–197.
[2] S. Hesse, G. Schnell, Sensoren zur Erfassung mechanischer Größen, in: S. Hesse, G. Schnell (Eds.), Sensoren für die Prozess- und Fabrikautomation, Springer Fachmedien, Wiesbaden, 2014, pp. 185–231.
[3] C. Klose, C. Demminger, H.J. Maier, Microstructure and properties of cobalt- and zinc-containing magnetic magnesium alloys processed by high-pressure die casting, in: Magnesium Technology 2015: Proceedings of a symposium sponsored by Magnesium Committee of the Light Metals Division of The Minerals, Metals & Materials Society (TMS), 144th Annual Meeting & Exhibition, Orlando, Florida, John Wiley & Sons, Inc., Hoboken, New Jersey, 2015, pp. 451–454.

2.3.3 Magnetic data storage within a technical surface

Piriya Taptimthong, Marc Christopher Wurz*, Christian Demminger†*
*Institute of Micro Production Technology, Leibniz Universität Hannover, Garbsen, Germany, †Institute of Material Science, Leibniz Universität Hannover, Garbsen, Germany

2.3.3.1 Introduction

The current state of the development of inherent magnetic storage in the CRC 653 will be reviewed. This clarifies the concept of the technology, marks the achievements, and pinpoints the challenges which are the starting point of the research work done in the third funding period (2013–2017). This section is divided into two subsections. The first one is a short overview of the state of the art in magnetic storage technology. Only the minimum information on types of magnetic heads and media are given. Detailed implementations and mathematic treatments related to this report will be discussed in the following sections. The second subsection is a summary of the development of magnetic storage in CRC 653 done in the first and the second funding periods (2005–2009 and 2009–2013). The review indicates similarity to and adaptation of the state-of-the-art technology as well as challenges in inherent magnetic storage.

2.3.3.1.1 Overview of magnetic storage technology

In principle, magnetic storage consists of three main components, namely, a write head, a read head, and a medium. A simplified model of magnetic storage is depicted in Fig. 2.3.3.1. Information is stored into the medium by magnetization process, a process by which a magnetic field, called a fringe or stray field, from an inductive write head rearranges magnetic moment in the medium in such a way that the magnetic moment is parallel to the magnetic field. There are two types of inductive write heads, a ferrite head and a thin film head. An iron(III) oxide Fe_2O_3 (ferrite) compounded with NiZn or MnZn ($Ni_aZn_{(1-a)}Fe_2O_4$ or $Mn_aZn_{(1-a)}Fe_2O_4$) are common materials used as a magnetic core in ferrite head production. Coils are wounded around the core and an air gap is created at the bottom of the core. A rigid substrate, such as a compound of alumina and titanium carbide (Al_2O_3-TiC), is used in thin film head fabrication. A head structure is created via photolithography and electroplating. Considering only main structures, the thin film head consists of a planar Cu coil sandwiched between permalloy (NiFe81/19) cores and an air gap, usually alumina, structured between the cores at a region defined as a head pole. A detailed structure and fabrication of the thin film head can be found in [1,2].

There are two methods to read the stored information back. One is using an inductive write head to detect an induced voltage generated when the head moves past a magnetic transition in a medium. The magnetic transition is a region where magnetization reversal

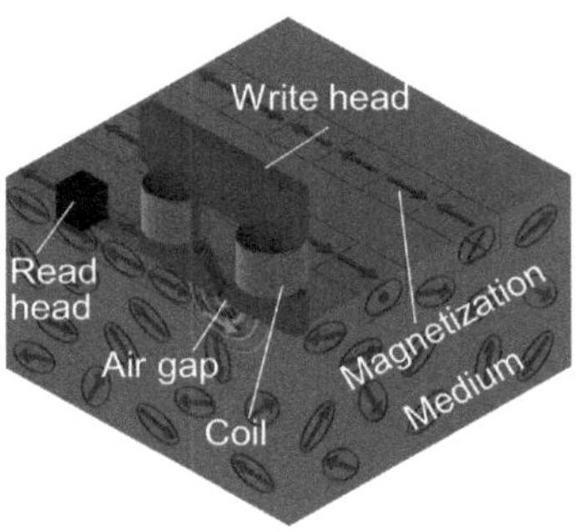

Fig. 2.3.3.1 Magnetic storage consists at least of a write head, a read head, and a medium. The write head emits a magnetic field from an air gap to magnetize the medium. The read head detects magnetization (the magnetic moment per unit volume) from the medium to recover stored data.

occurs. Another method is using a magnetoresistive (MR) sensor whose electrical resistance changes in the presence of a magnetic field. Magnetic media can be divided into two categories, particulate media and thin film media. The particulate media are used in magnetic tape recording and floppy disk technologies. For example, an audio magnetic tape composes of a polyester substrate, typically 20 μm thick. Acicular gamma iron(III) oxide (γ-Fe_2O_3) particles are dispersed in polymer binder, typically 11 μm thick and coated on top of the polyester substrate. The overall thickness of the audio magnetic tape is about 35 μm [3,4]. Thin film media, on the other hand, have a rigid disk structure and, due to their high magnetization and coercivity, are employed in hard disk drives. In general, a Ni-P plated aluminum disk served as a substrate is sputter deposited with a 50-nm Cr seed layer. The Cr under layer matches epitaxially to the 30-nm magnetic layer CrXY (X = Pt, Ta and Y = Co, Ni) deposited on top. The magnetic layer is covered by a 15-nm hard-sputter carbon overcoat and a 1-nm perfluoropolyether lubricant for protection from a head contact [1].

2.3.3.1.2 Inherent magnetic storage in CRC 653

In a context of the CRC 653, the first generation write head developed by Wu, Traisigkhachol and Gatzen [5,6] is made of a well-established MnZn-ferrite C-shaped cores with 0.52 T saturation magnetization. This head is comparable to a ferrite head described in the previous subsection. The manufacturing process of the head consists of four steps: profiling the C-cores by dicing, winding the C-cores, assembling the C-cores, and grinding the air gap. The dimensions of the head are 700 μm × 3.2 mm × 3.8 mm (W × H × L) and the air gap of the head is 100 μm long. Further optimization of a write head was done by Belski, Wurz and Rissing [7] in order to increase data density. The second head generation, shown in Fig. 2.3.3.2, is

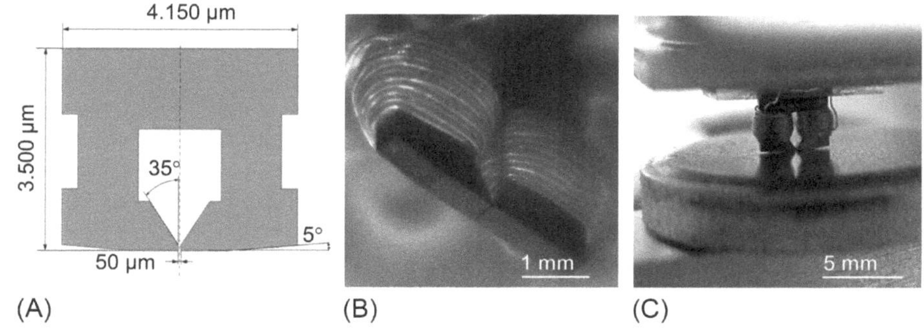

Fig. 2.3.3.2 (A) dimensions of the second generation write head, (B) the bottom of the write head revealing an air gap, and (C) the write head in contact with a sintered medium during writing.

slightly higher and longer (3.5 mm × 4.15 mm) but only 500 μm wide. The width is further reduced to the minimum value of 350 μm. An air gap is also shortened and measured 50 μm.

Instead of particulate or thin film media, Wu et al. [8] and Belski et al. [9] applied powder metallurgy to produce a sintered storage medium. Here, the standard magnesium alloy AZ61, as a base material, is blended with γ-Fe_2O_3 and compacted in a cylindrical die. Up to 40% by volume of γ-Fe_2O_3 is blended and successfully compacted. The medium is then sintered below melting point of Mg at 450°C. Computed tomography (CT) images indicate that diameters of agglomerates of γ-Fe_2O_3 are below 180 μm. The sintered media cannot be machined and thus are integrated into technical components. For example, a medium is inserted in a casting mold and casted into a component made of magnesium alloy AZ91. Klose et al. [10] also developed machineable media from a magnesium-cobalt (MgCo) alloy by means of gravity die casting. Components manufactured from these MgCo alloys are considered as technical components with inherent magnetic storage. Due to limited solubility of Co in solid Mg and large difference in melting points between Co and Mg, they developed a technique which makes alloying of Co in Mg possible. In short, Co powder is blended with Mg powder, compacted and extruded in a rod form. During casting process, the pure Mg melt is added with the extruded MgCo rods. The optimal alloy has a form of MgCo4 nominal wt%. Microstructures of this alloy indicate a finely dispersed structure of Co-rich eutectic precipitations in an interdendritic network and at grain boundaries. A high cooling rate causes more dense distribution of Co-rich precipitations.

Instead of using a write head or a MR sensor to read back stored information, a novel read back technique was proposed by Gatzen, Wu and Traisigkhachol [11]. The magneto-optical Kerr effect (MOKE) is exploited to read the stored information from a distance which is a main advantage of this technique. However, a thin soft magnetic layer called a keeper layer needs to be deposited on top of a medium. The keeper layer serves as a magnetic coupler that mirrors magnetization of the medium (the stored information) within its surface. The reflectivity of the surface of the keeper layer together with its magnetic properties and the magneto-optical (MO) rotation are the keys of the MOKE read out. Parallel to the magneto-optical technique, Belski et al. [12] adapted a conventional giant magnetoresistive (GMR) sensor as a read head and integrated the sensor into the write head. The sensor dimensions are 0.96 mm × 0.25 mm × 1.46 mm (W × H × L). They developed also a special signal detection technique to deal with inhomogeneous distributions of hard magnetic particles in both sintered and gravity die casted media.

In spite of the fact that the development of the keeper layer has been intensively done [13], an application of the keeper layer is still limited because of fabrication complexity. If a smooth and reflective surface of the keeper layer as well as an easy magnetic anisotropic axis in the plane of the keeper layer is not obtained, the observed Kerr rotation will be diminished resulting in low quality MO images.

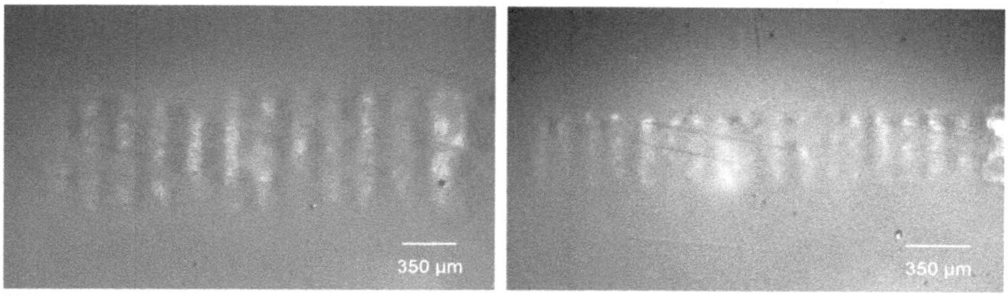

(A) (B)

Fig. 2.3.3.3 Magneto-optical images of data tracks written on a sintered medium (A) using the first generation head and (B) using the second generation head.

Data tracks on a γ-Fe_2O_3 sintered medium written by the first- and the second-generation heads were visualized with a MOKE microscope as shown in Fig. 2.3.3.3. Note that a magneto-optical indicator was placed above the media in order to make these high quality MO images, strictly speaking, through the Faraday effect. The indicator is a Bi-substituted ferrite garnet film grown on a gadolinium gallium garnet substrate and covered with a reflector and a protective film. An air gap of a write head determines the size of a data bit. Assume that a fringe field from the write head is concentrated in the air gap region and there is no bit overlapping. Thus, the bit size equals the air gap size. The minimum bit size is 7×10^{-4} cm^2 and 1.75×10^{-4} cm^2 for the first- and the second-generation heads. The corresponding calculated data density is 1.4 and 5.7 kbit/cm^2 respectively. However, an attainable performance of the magnetic storage depends on the interaction of the write head, the read head, and the medium. Due to the size of the GMR sensor and inhomogeneous distribution of hard magnetic particle in the media, the real maximum data density is limited to 100 bit/cm^2.

Head-medium interface plays an important role in real-world applications of the magnetic storage within technical surfaces. Principally in the case of hard disk drive, with a help of an air bearing, a head flies over a smooth medium at constant spacing (called fly height) on the order of nanometers in a clean and controlled environment [2]. The fly height determines an effective write field and thus a written data bit size. Such environment cannot be guaranteed in magnetic storage applications within technical surfaces. Moreover, changes in the fly height, e.g., due to vibration of a head fixture, cause variations of the effective write field which can lead to incomplete magnetization and bit overlapping. To ensure optimal writing, the write head needs to be operated in a contact mode (see Fig. 2.3.3.2C). This unfortunately leads to system susceptibility to mechanical shocks and vibrations. In response to these challenges, Belski et al. [14,15] proposed a flexible head design that can be operated in the contact mode without

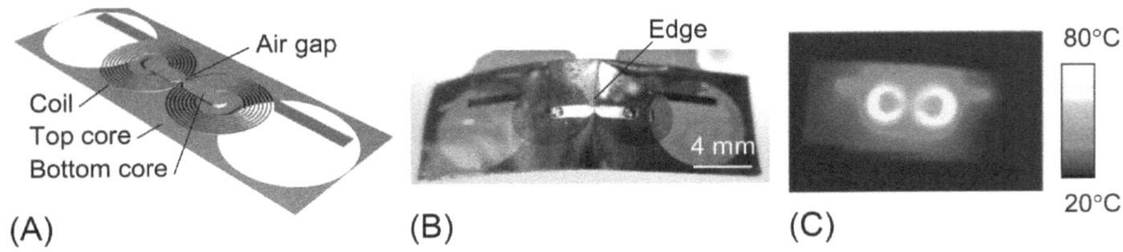

Fig. 2.3.3.4 A flexible write head: (A) a model, (B) a real head, and (C) a thermal image of the head during write test.

breaking. The main idea is the use of a copper-cladded polymer foil (42.5 μm thick) as a flexible substrate for the fabrication of a thin film head. On the backside, planar coils were structured directly on the copper layer (17.5 μm thick). This simplified the fabrication of the head and eliminated coil winding processes. After embedding the coils in an insulation layer, openings for contact pads were created and holes for core legs were drilled. The top core made of 50 μm permalloy foil was attached to the polymer foil by adhesive tape. On the front side, the bottom magnetic core (20 μm thick) and core legs were electroplated to form the complete head core (Fig. 2.3.3.4). Data tracks can be written on a Type I/IEC I cassette tape using this flexible head and thus the feasibility of the flexible head design is proofed.

However, the overall thickness of the flexible head limits the head flexibility in such a way that the head must be formed before usage and can be adapted only to a certain extent to surfaces of media. Furthermore, the maximum write field is low due to the head design and large air gap requirement. The low write field hinders the use of high coercivity media. Inhomogeneity of a magnetic medium seen as inconsistent contrast in Fig. 2.3.3.3 implies low data density because a large data bit size must be written to ensure adequate intensity of magnetization. If a data bit is too small, the bit may occupy a region where there is no magnetization thus resulting in data loss.

2.3.3.2 Flexible read/write head with HAMR capability

2.3.3.2.1 Flexible substrate as a key to flexibility and design simplification

Despite the fact that the flexible write head has been successfully developed [16], its design and use of multilayered construction constrain its flexibility. In addition, folding must be done to bring the head to its final form and this process is not trivial. If the folding forms an edge not in the air gap region or the edge angle is too small causing a sharp edge, a contact between the head and the medium will be suboptimal. In fact, only an area defining a head-medium interface, in this case the head poles, needs to be flexible. Hence, it is better, if only the head core and its poles are directly fabricated on a flexible substrate so that the overall thickness is kept low. At this point it is advisable to revise the head design. A design space of the write head depends on applicable thin film fabrication techniques, which in turn must be compatible with an available flexible substrate. Hence, an investigation of a material which serves as a required flexible substrate is the first task to be taken.

Elastic polymer materials are generally used as flexible substrates. Critical properties that flexible substrates must retain in MEMS production using thin film technology are primarily thermal stability, chemical stability, and vacuum compatibility. During physical vapor deposition (PVD) and etching, the flexible substrates experience high temperature stress and high vacuum. Cleaning agents like acetone and isopropyl alcohol are commonly used solvents in photolithography and are aggressive to polymers. Considering these factors, polyimides are appropriate as the flexible substrates because of their superior properties [17,18]:

- High thermal stability (up to 300°C)
- High chemical resistance to acetone and isopropyl alcohol.
- Low outgassing under high vacuum

Among other polyimides, Kapton is selected as the flexible substrate because of its properties and availability. Kapton is commercially available both in roll and in sheet format with standard thickness of 0.0254–0.127 mm. Fixation of the Kapton film on a carrier is necessary for thin film deposition and photolithography. This requires an adhesive layer between the Kapton film and the carrier. Such a layer should be flexible, compatible with both the film and the carrier, chemically stable, and dimensionally stable over a wide temperature range. Kapton films laminated with a pressure-sensitive adhesive (PSA) layer are commonly available. The PSA layer is made of an acrylic (polyacrylate) or a silicone (polysiloxane). The latter is preferable and is selected because of its better chemical resistance and higher thermal stability. Silicone withstands acetone and isopropyl alcohol and is thermally stable up to 180°C.

We proposed an idea [19] and also did baseline examinations to assert compatibility of the flexible substrate in thin film processes for microelectromechanical systems (MEMS) [20]. Further experiments on electroplating were performed as well. According to the results, a polyimide (Kapton HN) film coated with a silicone adhesive layer, e.g., 3M Polyimide Film Tape 5413, is selected as a flexible substrate. The film is laminated on a rigid carrier, which is in our case a silicon wafer, using a Bungard Elektronik RLM 419p laminator. Manual lamination by hand is also possible, even though there is a high chance of bubbles trapped between the film and the carrier. The PSA silicone layer provides adhesion through hydrogen bonding and has immediate tack. Nevertheless, the PSA silicone layer is semicured and the adhesion continues to build up over a longer time [21,22]. For this reason, the Kapton film laminated on the carrier is rested at ambient temperature up to 24 h. This process is called wetting and should be performed according to PSA silicone layer used by different film manufacturers. The dimensional stability of the Kapton film depends on a coefficient of thermal expansion of Kapton and residual stresses in the Kapton film. The latter causes the Kapton film to shrink on its first exposure to elevated temperature [17,18]. Prebaking process described in [20] is used to bring the film to a stable form before starting the fabrication. The prebaking process also helps removing residual substances in the PSA layer of the film reducing outgassing in vacuum chambers of equipment for thin film fabrication. Photolithography both with AZ 9260 positive and AZ 5214 negative resist from MicroChemicals GmbH is compatible to the flexible substrate. Due to the large difference in a thermal expansion coefficient between the Kapton and deposited metal films, the process temperature must be controlled. In PVD process, intermittent depositions are done at moderate power ($\leq$200 W). A deposition time and pause is set to 50% duty cycle. Films of NiFe and Au containing high aspect ratio structures are successfully deposited without crack. Examples of fabricated structures are shown in Fig. 2.3.3.5.

Unlike a conventional substrate where separation of an individual system after fabrication is done by sawing or dicing, a release process, here called debonding, is required to separate the flexible substrate from the carrier. Chemical etching or softening of the sacrificial layer and the complete etching of the carrier are conventional methods used in debonding. Since xylene softens a silicone adhesive layer, the chemical softening method is selected. First, the carrier is immersed in a bath of Roticlear, a harmless xylene-like solvent from Carl Roth GmbH+Co KG [23]. When the silicone adhesive layer softens, tweezers are used to peel off the flexible substrate from the carrier. Silicone residuals on the flexible substrate are gently wiped out. Finally, the flexible substrate is rinsed with acetone and isopropyl alcohol. A clean debonded film is shown in Fig. 2.3.3.6. With all the techniques and results presented in this

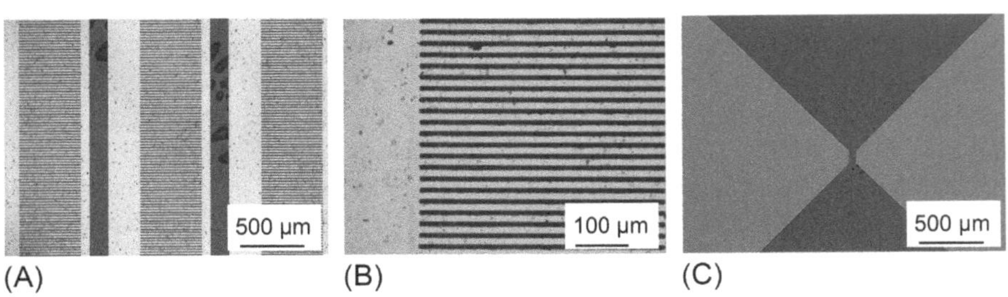

Fig. 2.3.3.5 (A) 100-nm-thick Au grating structures, (B) a close up revealing high aspect ratio grating structures with sharp edges, and (C) 500-nm-thick permalloy (NiFe81/19) magnetic flux guide.

(A) (B)

Fig. 2.3.3.6 (A) a flexible substrate with Au grating structures after a debonding process and (B) grating structures are intact and have no crack indicating successful debonding.

section, we ensure an application of a Kapton film as a flexible substrate that is compatible with thin film fabrication techniques.

2.3.3.2.2 AMR sensor on a flexible substrate

The current-generation head has been equipped with a read head adapted from a commercial giant magnetoresistive (GMR) sensor, the GF708 from Sensitec GmbH. The sensor measures 0.96 mm × 0.25 mm × 1.46 mm (W × H × L) and is relatively large comparing with the size of the air gap of the write head. An integration of a read head into a write head also requires cabling and fixing that inevitably make a large footprint of this read/write head. The dimensions of the sensor and the large footprint greatly reduce not only the flexibility of the system but also the data density. Since magnetic poles of a new flexible write head will be created on a flexible substrate, if a read head is fabricated on the same substrate, this will automatically resolve the integration problem and reduce the footprint of the read/write head. Anisotropic magnetoresistive (AMR) and GMR sensors are standard types of magnetic read head fabricated using thin film techniques on silicon wafers. On one hand, the GMR sensor delivers a larger magnetoresistive effect but at a cost of a complex multilayer thin film fabrication. The AMR sensor, on the other hand, has a simpler structure and can be made of the same material as a head core, namely, permalloy. Considering fabrication criterions, the AMR sensor is selected. However, there is still a challenge relating to the creation of a sensor element on the flexible substrate. Unlike, silicon wafer, which is a standard substrate for magnetoresistive (MR) sensors, the flexible substrate is rougher and has higher thermal expansion coefficient. Effects of surface topography of thermally oxidized silicon wafers and polymer substrates on magnetic properties of AMR sensors were studied in [24,25], respectively. Generally speaking, a substrate with

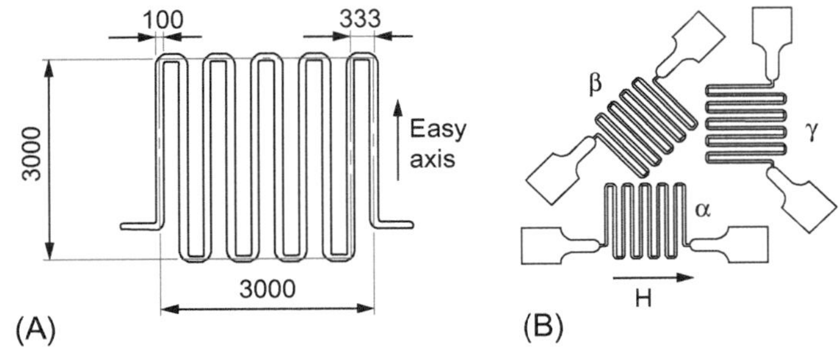

Fig. 2.3.3.7 (A) an AMR sensor in meander form (dimension in μm) and (B) alignment of sensors: perpendicular (α), parallel (γ), and 45° (β) to a measurement field *H*.

a smoother surface yields softer magnetic properties and better electrical surface conductivity. Both are the critical requirements for a large AMR effect. A large thermal expansion mismatch between the sensor layer and the substrate also induces film stress that affects sensor performance. To estimate the feasibility of the direct integration of the AMR read head into the write head, we carried out investigations on AMR sensors fabricated on various polymer substrates, including a Kapton HN film, and reported the results in [26,27]. Only important parts of the results relating to this work will be discussed here. The Kapton HN film 5413 form 3M [28] is used as a flexible substrate and laminated on a carrier (a silicon wafer) as described in the previous section. The AMR sensor is photolithographically fabricated into a meander form in different orientations (Fig. 2.3.3.7B). The form is selected to ensure strong shape anisotropy in a direction parallel to the length of the meander structure. A permalloy thin film with a nominal thickness of 100 nm is deposited on the meander structure using sputtering.

Direct examinations of the surface topography of a Kapton film are not simple due to its transparency and softness. Instead, the examinations are done on the created AMR sensors. Also, examinations of sensors created on thermally oxidized silicon wafer are performed and are used as references. With a reflective surface of a sensor structure, it is possible to measure surface topography as line roughness according to EN 25178 using a confocal laser scanning microscope (CLSM) VK-9700 from Keyence. Furthermore, curvature of the carrier was determined before and after fabrication by means of a tactile profilometer Veeco Dektak^{3}ST. The estimated radius of curvature is calculated from the measured curvature difference. Using Stoney's equation, the magnitude and orientation of film stress within the Kapton film are determined. A resistance change as a function of a magnetic field $\Delta R/R_0(H)$ is the most meaningful information regarding characterization of a magnetic sensor. During the characterization, AMR sensors are mounted on a fixture of a Helmholtz coil and connected to Wheatstone bridges and amplifier circuits developed in-house. The sensors are aligned perpendicular (α), parallel (γ), and at 45° (β) to a homogenous alternating magnetic field generated by the Helmholtz coil. The resistance change is calculated from output voltage signals recorded using an oscilloscope (Tektronix TDS 2014).

Calculations of film stress in the substrates are plotted in Fig. 2.3.3.8. Magnitude of stress is represented in relation to the estimated stress in the SiO_2 layer on a silicon wafer. Stress in a Kapton film depends on a direction of film lamination. Tensile stress occurs in a direction parallel to the lamination direction, while a compressive stress occurs in an orthogonal direction to the lamination direction. In Table 2.3.3.1, an arithmetic average of surface roughness (R_a) and an average distance between the highest peak and lowest valley (R_z) of AMR sensor on a Kapton film are 7.7 and 3.3 times higher than those on the SiO_2, respectively.

As expected, the higher surface roughness causes increasing of base resistance R_0 of sensors on Kapton films and negatively reduces the AMR effect. AMR characterization curves $\Delta R/R_0(H)$ of a sensor on Kapton film shown in Fig. 2.3.3.9B reveals three interesting things, i.e., peak shifts, reduced sensitivity, and unusual sensitive directions. The peak shifts (from the zero field position to around

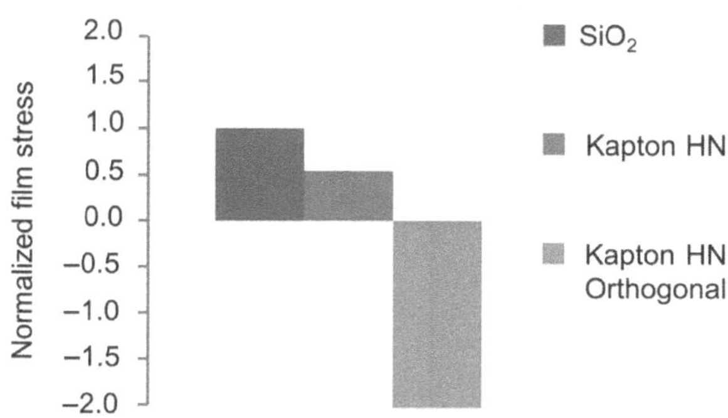

Fig. 2.3.3.8 Estimated normalized film stress in substrates.

Table 2.3.3.1 Surface roughness of AMR sensors fabricated on substrates

Substrate	Surface roughness R_a [μm]	Surface roughness R_z [μm]
SiO_2 layer on a silicon wafer	0.0088	0.0565
Kapton HN film	0.0675	0.1867

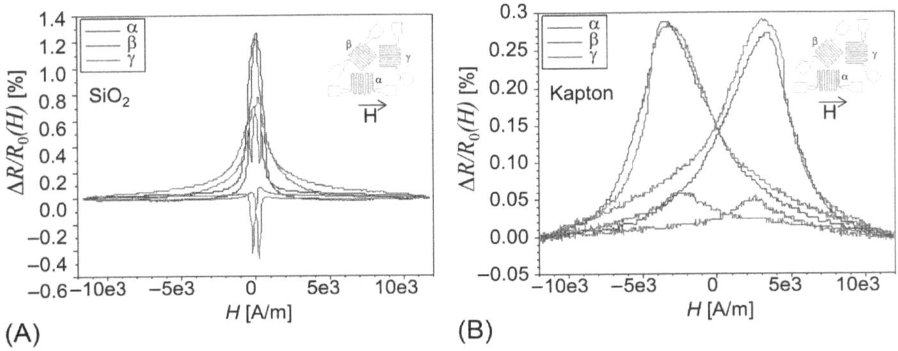

Fig. 2.3.3.9 AMR characterization curves $\Delta R/R_0(H)$.

Table 2.3.3.2 **Average values and standard errors of parameters of AMR characterization curves**

Substrate	Base resistance R_0 [kΩ]	Base resistivity ρ_0 [μΩ/cm]	AMR effect [%]	Peak shift [kA/m]
SiO_2 layer on a silicon wafer	2.12 ± 0.07	64 ± 0.02	0.89 ± 0.09	0.241 ± 0.016
Kapton HN film	2.66 ± 0.07	81 ± 0.02	0.35 ± 0.08	6.26 ± 1.11

± 5 kA/m) are results of hindrances to domain wall motion (also known as domain wall pinning) which is possibly due to the film stress and the rough surface of the Kapton film [29]. The reduced sensitivity is originated from high electrical surface resistivity of the sensor layer and also from the domain wall pinning. The unusual sensitive directions indicate domination of the stress in the Kapton film over the shape anisotropy of the sensor. A summary of parameters of AMR sensors is given in Table 2.3.3.2. To summarize this section, the following remarks on the AMR sensor fabricated on the Kapton film are given. The AMR sensor is suitable as a read head for detection of magnetic transition in a magnetic medium, even with the peak shifts and reduced sensitivity, given that magnetization representing a data bit is high enough. Lastly, film stress occurring during the production of the flexible read/write head can negatively affect the desired sensitivity, but a good signal detection circuit should solve the problem.

2.3.3.2.3 Hybrid design of a flexible write head with a long air gap

In general, high data density and high data stability are preferred and this requires a write head with a short air gap and a medium with high coercivity H_c. These requirements set design parameters of the write head which are summarized in Table 2.3.3.3. Both surface roughness and vibration limit the minimum fly height of the head. In this work, the fly height is set to 5 μm in order to account for a rough surface of a technical component. The current medium made of MgCo4 has coercivity equal to 23 kA/m and is magnetized in a direction parallel to the medium plane, also known as longitudinal magnetization.

A paradigm in design of an inductive write head roots in the Karlqvist's equation. The equation describes the simplest but yet most useful model of the inductive write head [30]. Assume that permeability of a magnetic core is infinite which means that there is no flux leakage. Also the air gap region is small compared with the size of the core and thus H_g is constant. A magnetic fringe field or a write field in x- and y-direction in the air gap region under the head shown in Fig. 2.3.3.10 is described by Eq. (2.3.3.1) and (2.3.3.2). Note also that the H_x component of the write field is relevant for longitudinal magnetization, while the H_y component is a matter

Table 2.3.3.3 Design parameters

Requirement	Parameter				
	g	H_c	H_g	ni	d
High data density	Small	High	–	–	Low
High data stability	–	High	High	–	–
High H_g	–	–	–	Large	–
Small air gap	–	–	High	Small	–

g: length of an air gap, H_c: coercivity of medium, H_g: fringe field (stray field) in an air gap, ni: magnetomotive force (MMF), d: fly height between an air gap and a medium surface.

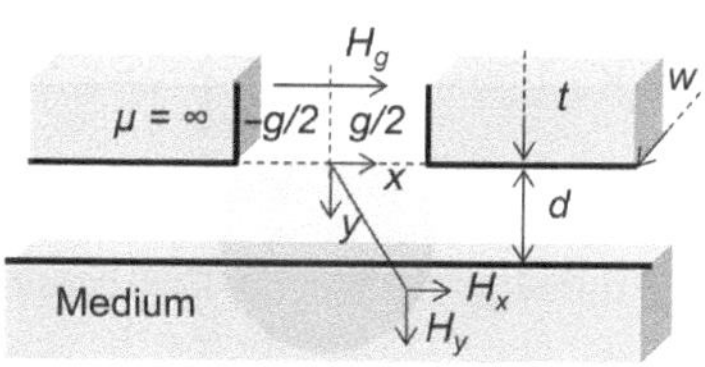

$$H_x(x,y) = \frac{H_g}{\pi}\left[\tan^{-1}\left(\frac{x+\frac{g}{2}}{y}\right) - \tan^{-1}\left(\frac{x-\frac{g}{2}}{y}\right)\right] \quad (2.3.3.1)$$

$$H_y(x,y) = \frac{H_g}{2\pi}\left[\ln\frac{\left(x+\frac{g}{2}\right)^2 + y^2}{\left(x-\frac{g}{2}\right)^2 + y^2}\right] \quad (2.3.3.2)$$

Fig. 2.3.3.10 The Karlqvist head and write field in a medium.

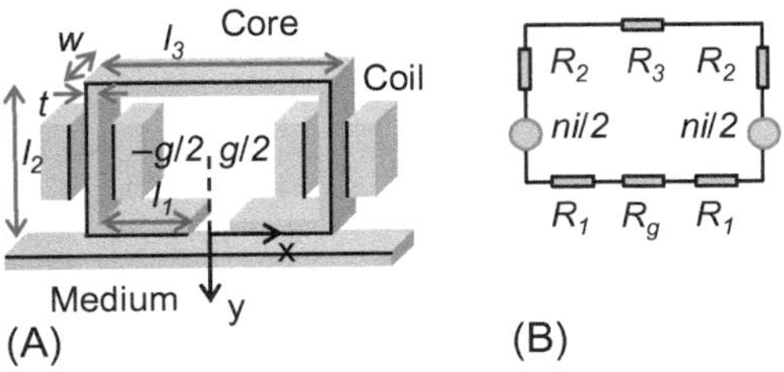

$$l_c = l_3 + 2\left(l_1 - \frac{g}{2} + l_2\right) \tag{2.3.3.4a}$$

$$R_c = \frac{l_c}{(\mu A_c)} \tag{2.3.3.4b}$$

$$A_c = A_g = wt \tag{2.3.3.4c}$$

Fig. 2.3.3.11 (A) A model of a write head and (B) its equivalent magnetic circuit.

for perpendicular magnetization, in which the write field is perpendicular to the medium plane.

If H_g in the Karlqvist's equation is known, a write field can be calculated. Using an equivalent magnetic circuit of a write head as depicted in Fig. 2.3.3.11B, H_g in the air gap of the write head can be calculated according to Eq. (2.3.3.3).

$$H_g = \frac{ni}{g + \left(l_c \frac{\mu_0}{\mu}\right)\left(\frac{A_g}{A_c}\right)} \tag{2.3.3.3}$$

ni : Magnetomotive force [A · Turn]
l_c : Length of a magnetic core [m]
μ_0 : Permeability in free space [$\mathrm{H \cdot m^{-1}}$]
μ : Permeability of a magnetic core [$\mathrm{H \cdot m^{-1}}$]
A_g : Area of an air gap [$\mathrm{m^2}$]
A_c : Area of a magnetic core [$\mathrm{m^2}$]

As previously discussed in Section 2.3.3.2.1 about a flexible substrate and design simplification, only a part of a magnetic core of the write head needs to be flexible. It is unnecessary to create magnetic coils directly on the flexible substrate. In fact, conventional wound coils can be used. Separating the coils from the flexible head core reduces dramatically the fabrication complexity. Later, the head assembly can be done easily as well because the flexible core can be folded, inserted through the coils, and wrapped into a complete head core. To ensure flexibility, the thickness of the flexible core is kept low under 10 μm. Because of the required deep penetration depth of the write field, the air gap length is set to 50 μm. Also, the head assembly process implies that the flexible core must be long enough to ensure good handling. For that reason, we set the length and width of the flexible core to 38 mm and 10 mm respectively. However, those requirements limit the usage of the Karlqvist's equation and the magnetic circuit because:

- The permeability of the real core is not infinite and the very long but thin flexible core yields an abnormal core-length-to-film-thickness ratio, e.g., 38 mm:4 μm (9500:1). This causes both considerable magnetic flux leakage and undesirable magnetic flux saturation at the core legs l_2 (see Fig. 2.3.3.11).
- The length of the air gap is large comparing with the core thickness and as a result, the fringe field H_g is not constant along the air gap. This will deform the fringe field in the air gap region.

To cope with the long core length, we proposed a hybrid design in Ref. [31]. The main idea is the use of a thick magnetic foil with high magnetic flux saturation B_s to create the core legs l_2 and the top core l_3 of the write head. Only the bottom core l_1 needs to be flexible and is fabricated on a Kapton film. The thick foil reduces the flux leakage and prevents magnetic saturation at the core legs, thus more magnetic flux is guided through the bottom core to the air gap. In our case, the thick magnetic foil is made of VACOFLUX 48 from VACUUMSCHMELZE GmbH & Co. KG [32]. Using numerical simulation based on a finite element method (FEM), the write field can be determined. Various head geometries and parameters as given in Table 2.3.3.4 are studied in order to determine both the form and the strength of the fringe field. The simulation is done with the finite element software ANSYS. The head is modeled with Solid 97 element in 3D.

The simulated results, e.g., from the head model M_1 as pictured in Fig. 2.3.3.12A, clearly show varying H_g due to the long air gap. The field strength is maximum near head poles and minimum at the middle of the air gap. Note that, this minimum should be considered as the maximum useful field strength attainable from the write head. Comparing the head model M_2 with M_3 in Fig. 2.3.3.12B, the magnetic saturation in the core legs of M_2 is eliminated when the thick magnetic foil is applied in M_3. As a result, more magnetic flux is guided to the air gap and thus the write field strength is increased. The field strength of M_3 plotted in Fig. 2.3.3.13 is two times higher that one of M_2. The field strength of M_5 is almost the same as the one of M_4 at head poles but lessens at the middle of the air gap due to its longer air gap. The characteristic curve M_{1a} is analytically calculated based on the head model M_1 using the Karlqvist's equation. The maximum field strength

Table 2.3.3.4 Models of a write head and their corresponding parameters

Parameter	Model				
	M_1	M_2	M_3	M_4	M_5
$l_1/l_2/l_3$ [mm]	6-g/2/7/12				
w [mm]	10				
g [μm]	10	10	10	10	15
t_1/t_2, t_3 [μm]	4/4	4/4	4/150	4/150	4/150
μ_1/μ_2, μ_3 [–]	280/280	N/N*	N/73	N/73	N/73
B_{s1}/B_{s2}, B_{s3} [T]	0.67/0.67	0.67/0.67	0.67/2.2	0.67/2.2	0.67/2.2
ni [A]	20	20	20	60	60

N: a measured BH curve of an electroplated permalloy (NiFe81/19) core.
M_5: a model of a real flexible write head with a hybrid design; the real write head is shown in Fig. 2.3.3.14.

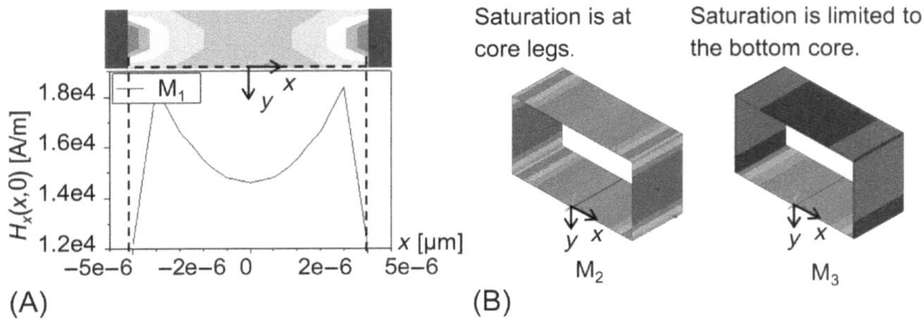

Fig. 2.3.3.12 (A) The H_x component of the fringe field H_g in the air gap of the write head model M_1 at $y=0$ μm; (B) Magnetic flux in the head core of the write head model M_2 (left) and M_3 (right).

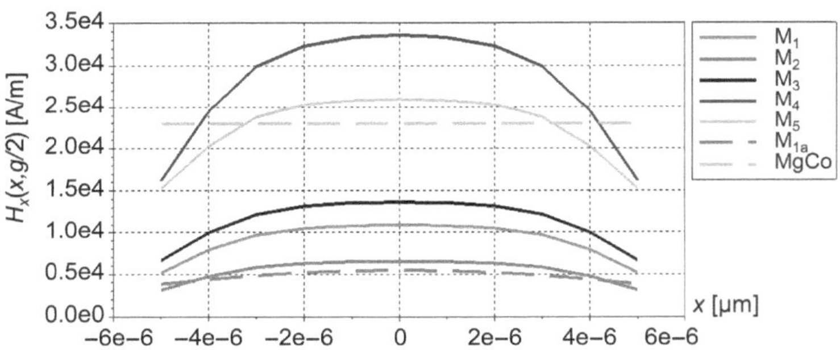

Fig. 2.3.3.13 The H_x component of the fringe field at 5 μm fly height ($y=5$ μm); M_4 and M_5 deliver the write field high enough to magnetize the MgCo4 alloy ($H_c=23$ kA/m).

of M_{1a} at $y=0$ is scaled to M_2 at $y=0$ in order to include flux leakage. It can be seen that both models yield similar results.

Verification of the hybrid design is done by a creation of a test head using the head model M_5. The assembly of the test head is straightforward and can be done as follows: The flexible bottom core is inserted through the magnetic coils and wrapped into the closed head core (Fig. 2.3.3.14A). A 150-μm thick magnetic foil (VACOFLUX 48) is folded into a U-shape core. To complete the magnetic circuit, the U-shape core is inserted through the coils to form the thick core legs and the thick top core. A Type I/IEC I cassette tape with coercivity equal to 32 kA/m is employed as a test medium. As shown in Fig. 2.3.3.13, the write field from the head model M_5 at 5-μm fly height is not sufficient to magnetize the tape medium. Hence, to gain a higher write field, the head and the medium are brought into contact and

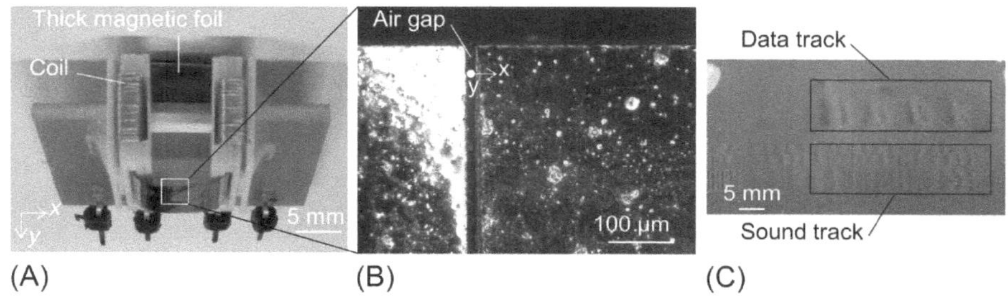

Fig. 2.3.3.14 (A) A flexible write head with a hybrid design; (B) the flexible bottom core and an air gap; (C) a Type I/IEC I cassette tape with an original sound track and a data track written using the flexible write head.

write tests are performed in a contact mode. We will close this section with an image of a flexible write head with a hybrid design (Fig. 2.3.3.14A). The successfully written data track is shown in Fig. 2.3.3.14C. It should be noted that sizes of transitions on the data track are not constant. This is because the write tests are done manually by hand without a head fixture and control electronics.

2.3.3.2.4 Flexible read/write head

We presented in Ref. [33] a complete design of a flexible read/write (r/w) head which is based on results obtained from Sections 2.3.3.2.1–2.3.3.2.3. The head is shown in Fig. 2.3.3.15 and consists of:

- A 7-µm flexible magnetic core of a write head and an AMR read head, both fabricated on a Kapton film by means of thin film techniques
- Two head coils fabricated via precision mechanical processes
- A U-shape 150-µm-thick magnetic foil made of VACOFLUX 48.

The head coils are attached on a fixture designed to be mounted on a read/write system. An inset of Fig. 2.3.3.15A reveals trapezoidal poles at the bottom of the write head and a read head above these poles. The read head consists of an AMR sensor and electrical conductors. The AMR sensor is a NiFe layer fabricated using a physical vapor deposition (PVD) technique and has a form of a meander structure with 200 nm nominal thickness as depicted in Fig. 2.3.3.15B. The Au conductors are 2 µm thick and can be folded to the top side of the r/w head where an electrical connection, e.g., via a ZIF connector, is possible. In Section 2.3.3.2.3, a prototype of a flexible write head has 15 µm air gap and no pole cap. The flexible r/w head presented here, however, has 50 µm air gap in order to accommodate the AMR read head and guarantee deep penetration depth. This long air gap causes a higher magnetic reluctance which in turn leads to a reduced magnetic field strength in the gap. To compensate the high

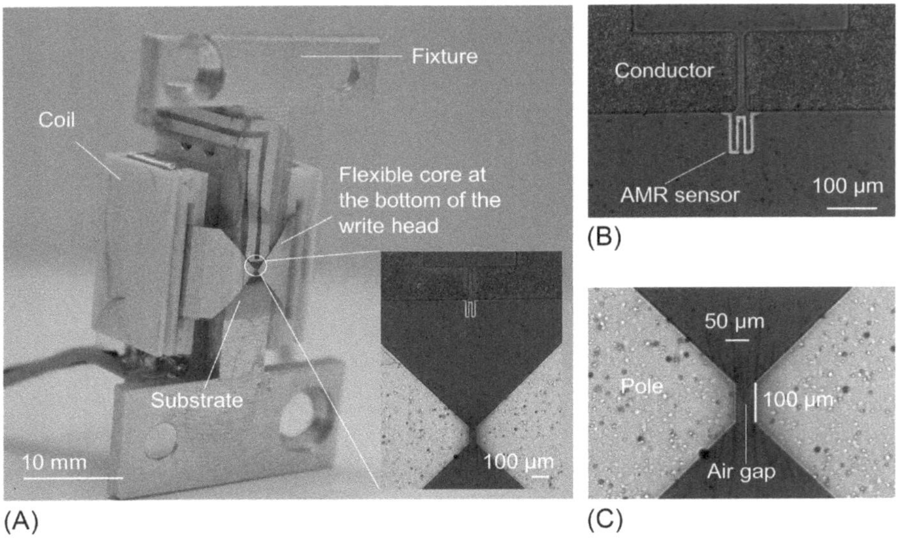

Fig. 2.3.3.15 (A) a flexible read/write head mounted on a fixture, (B) a read head, and (C) trapezoidal poles and an air gap.

magnetic reluctance, the trapezoidal pole cap with 100 μm pole width is employed to concentrate a magnetic field in the air gap (Fig. 2.3.3.15C).

A typical performance gauge of an AMR read head is a transfer characteristic of a resistance change. The resistance of the AMR element depends on the orientation between the path in the AMR element, through which an electrical current flows, and magnetization of the AMR element. In the simplest form as shown in Fig. 2.3.3.16, when an AMR element is subjected to an external magnetic field, there is a force that tries to rotate magnetization of the AMR element in a direction parallel to the external field. This results in a change of the orientation between the current path and the magnetization, thus causing the change in resistance. In order to design a signal detection circuit for reading back stored data, important parameters, namely, the maximum resistance change, deviations from a simple model, and a sensitivity range of an AMR read head must be known.

A characterization of the read head is done in the same way as explained in Section 2.3.3.2.2. In Fig. 2.3.3.17, a ratio of ΔR to an initial resistance ($R_{0|H=0}$ = appx. 200 Ω) is plotted as a function of an applied magnetic field H. Close examination of those responses reveals that the read head behaves in a low- and high-field region (>2.5 kA/m) distinctly. On one hand, in the low-field region, the responses resemble a resistance change of an AMR element with ε

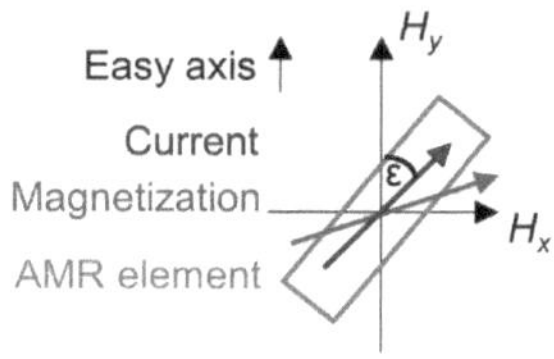

$$\frac{\Delta R}{R_0} = \frac{\Delta \rho}{\rho_0}\left(\begin{array}{c} \cos^2\varepsilon - \cos 2\varepsilon \left(\frac{H_x}{(H_y + H_k)}\right)^2 \\ + \sin 2\varepsilon \frac{H_x}{H_y + H_k}\sqrt{1 - \frac{H_x}{H_y + H_k}} \end{array} \right) \quad (2.3.3.5)$$

H_k : Anisotropy field
H_x : Magnetic filed normal to an easy axis
H_y : Magnetic field parallel to an easy axis
ε : Angle between an electrical current and an easy axis
ρ : Resistivity of an AMR element

Fig. 2.3.3.16 A simplified model of an AMR element in a homogeneous magnetic field and a corresponding equation of a resistance change as a function of an applied field [34].

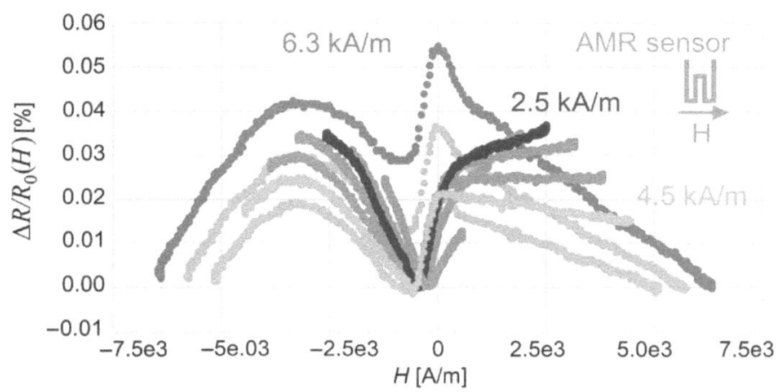

Fig. 2.3.3.17 $\Delta R/R_0$ of an AMR read head subjected to a homogeneous AC magnetic field.

equal to 90°. On the other hand, in the high-field region, the responses can be estimated via a superposition of resistance changes with ε equal to 0°, 45°, and 90°. By varying H_k and $\Delta\rho/\rho_0$ in Eq. (2.3.3.5), an approximation of $\Delta R/R_0$ of an AMR read head is done and the corresponding parameters are given in Table 2.3.3.5. The maximum $\Delta R/R_0$ in the low- and high-field regions are measured 0.035% and 0.055%, respectively.

The maximum write speed is limited by a resonance frequency of the write head. The resonance frequency can be calculated from a frequency response of the write head. An impedance of a flexible write head is measured using Fluke PM6306 RCL meter which can output up to 1 MHz test signal. Calibration of a test fixture is performed to minimize measurement errors before an impedance measurement. In a write test, the flexible read/write head is mounted on a rail system of a precision

Table 2.3.3.5 Estimated parameters of an AMR read head at various field levels

Parameter		H [kA/m]		
		2.5	**4.5**	**6.3**
H_k [kA/m]	0°	–	5.5	5.5
	45°	–	0.8	0.8
	90°	2.0	2.0	2.0
$\Delta\rho/\rho_0$ [%]	0°	–	0.03	0.035
	45°	–	0.03	0.04
	90°	0.03	0.012	0.012

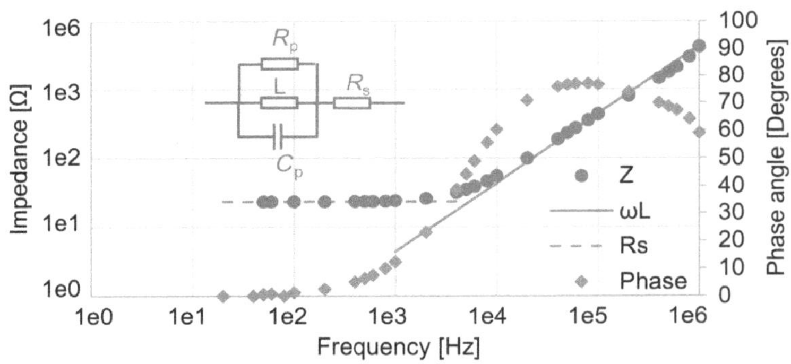

Fig. 2.3.3.18 Frequency response of impedance of a flexible write head.

XYZ translation stage developed in-house. The flexible read/write head is connected to a push-pull driver with internal clamp diodes (L293D) and driven with pulse test signals. Transient signals from the head are observed using Tektronix TDS 2014 oscilloscope.

The frequency response of a flexible write head is shown in Fig. 2.3.3.18. An impedance Z of the head can be modeled in a low frequency region below 0.1 MHz as a series L-R_s equivalent circuit, because the impedance of the head is in a low-medium range (<1 kΩ) [8]. The measured series resistance R_s is 23.6 Ω. The inductance L calculated from the frequency response is 700 μH. In a high frequency region above 0.1 MHz, an effect of a head parasitic parallel capacitance C_p between turns of coils and core loss modeled as parasitic parallel resistance R_p clearly appear and can be seen from a decreasing phase. By curve fitting technique, a calculated C_p and R_p are 2 pF and 10 kΩ, respectively. A corresponding self-resonance frequency (SRF) of the head is estimated to be at 4.25 MHz.

Transient responses of the flexible write head driven at various voltage levels are shown in Fig. 2.3.3.19. Rise times required for a current signal to change from 10% to 90% and from 0% to 95% of its maximum value are 44 and 64 μs, respectively. For comparison, a transient response is simulated using the head impedance and plotted in Fig. 2.3.3.19 as well. Note also that the calculated curve yields a slightly higher maximum current value. This is because parasitic resistance, e.g., from contact resistance and wire resistance, is not included in the simulation model.

To conclude this section, an image of data tracks written on a Type I/IEC I cassette tape is shown in Fig. 2.3.3.20. The image is made visible by means of a magneto-optical Faraday effect. Note that the image is created by merging many micrographs of the tracks along a write direction. Those micrographs are taken using a polarized light microscope that is equipped with a magneto-optical sensor (bismuth-substituted yttrium iron garnet, Bi:YIG). A length of data bits in the top data track is fixed at 400 μm and used as a reference scale. Lengths of data

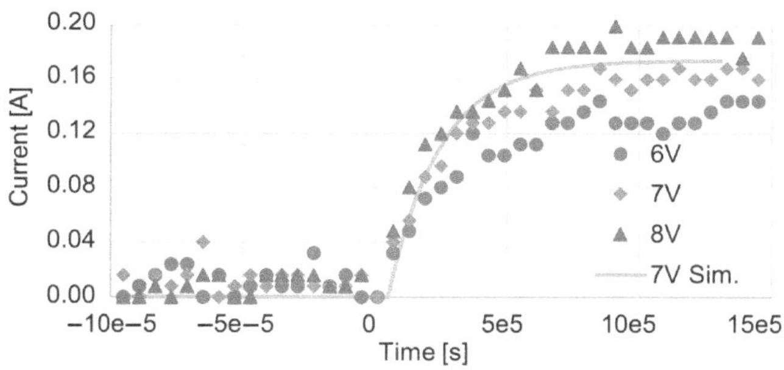

Fig. 2.3.3.19 Transient response of a flexible write head at various driving voltages.

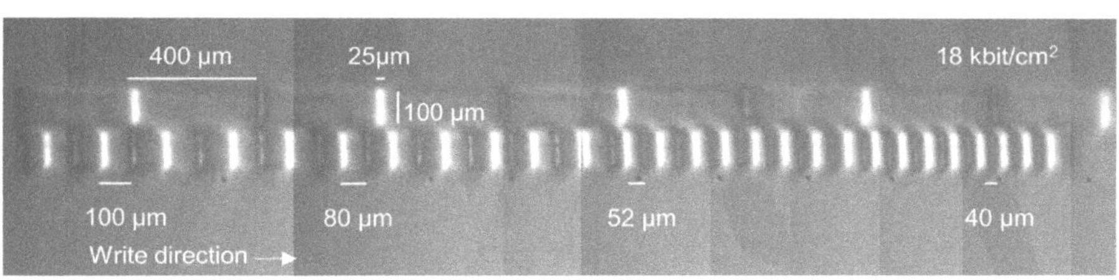

Fig. 2.3.3.20 Data tracks on a Type I/IEC I cassette tape written using a flexible write head.

bits in the bottom data track are varied in order to determine the minimum bit length. Black and white strips are transition regions between "0" and "1" bits. Down to a data bit length of 40 μm, the transition regions are distinct and clearly visible. A size of a transition region is measured 25 μm in length by 112 μm in width approximately. The transition length is 25 μm or half the air gap length. Note also that for a data bit length of 40 μm, the transition length is a little bit shorter, which means that data bits start to overlap each other. As a result of a side-fringing field, the transition width is slightly longer than the air gap width which is 100 μm. This additional width is called a side track and is measured 6 μm on both sides of a track. Considering a size of a read head, a 52 μm data bit size should be employed, thus yielding 18 kbit/cm^2 data density.

2.3.3.2.5 Heat-assisted magnetic recording

For high data stability, a storage medium with high coercivity H_c and high curie temperature T_c is employed. However, high coercivity means high write field, whose maximum strength is limited by a design and a soft magnetic material used in a fabrication of a write head. A solution to usage of high coercivity media is to temporarily reduce their H_c during the write process by means of energy-assisted techniques [35]. Among many techniques, heat-assisted magnetic recording (HAMR) shows high potential and is actively studied by many researchers from the hard disk drive domain [36]. We adapted HAMR to our work, equipped our flexible r/w head with a laser module and presented in Ref. [37]. In the following, some of those results will be discussed.

A laser module, Nano250 Series from Qioptiq, is employed as a heat source. The laser module can emit a collimated 532 nm green continuous wave (cw) laser beam. The laser module, a lens system mounted on a *XY* translation module, and the flex r/w head are assembled on a rail system and as a whole they become the HAMR module (Fig. 2.3.3.21).

Positioning of a laser beam through an air gap of a flexible write head is the first task in a HARM experiment. The HAMR module is mounted on a precision *XYZ* translation stage of a read/write system developed in-house as depicted in Fig. 2.3.3.22. The stage is equipped with stepper motors and can be driven in *x*- and *z*-direction with 4-μm step resolution. Due to the configuration of the HAMR module, a direct alignment using a photodiode is possible and, in our case, a standard photodiode, BPW 34 from Osram is employed. The photodiode is mounted on a focal plane of the precision *XYZ* translation stage, which is normal to the z-direction as shown on the top right of Fig. 2.3.3.22. Positioning the laser beam is done in two steps. The first step is focusing the beam on a focal plan, which is an interface between head poles and a medium (or a photodiode during an alignment process). A close-up image of the focal plane is placed at the bottom right of Fig. 2.3.3.22. The second step is aligning the focused laser

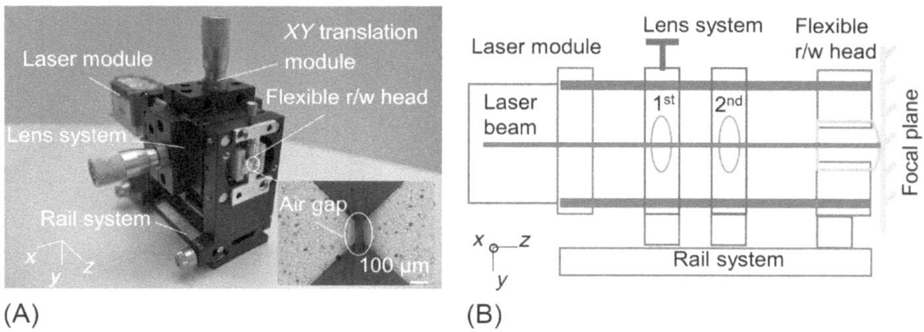

Fig. 2.3.3.21 (A) Heat-assisted magnetic recording (HAMR) module for inherent magnetic storage within surfaces of technical components and (B) its simplified layout.

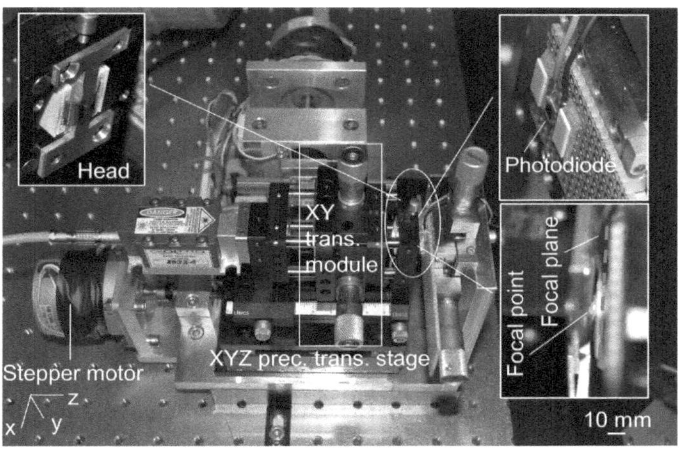

Fig. 2.3.3.22 Positioning a laser beam on a medium using a precision *XYZ* translation stage and a *XY* translation module of the lens system.

beam through the air gap. Fig. 2.3.3.23A illustrates a cross-section model of the flexible write head and reveals a through hole, which allows the focused laser beam to pass and reach the air gap at the bottom of the head. A region near the air gap can be scanned using the laser beam by manipulating the *XY* translation module of the lens system as schematically shown in Fig. 2.3.3.23B. An electrical signal from the photodiode decreases or increases whenever the focal point is blocked by the head poles or exposed to the photodiode, respectively. Using

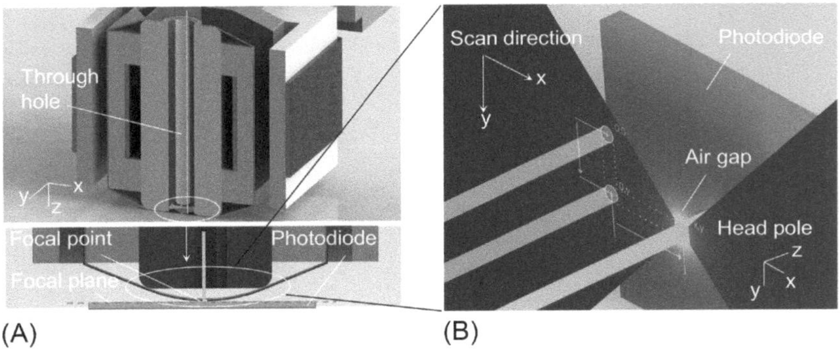

Fig. 2.3.3.23 (A) A cross-section model of a write head and (B) schematic procedure for aligning a laser beam through an air gap using a photodiode.

this information, a position of the air gap can be determined and used to align the laser beam. Since a Kapton substrate absorbs a laser beam, an opening through the substrate in the air gap region is needed. The HAMR module can be employed to create the opening by increasing power of the focused laser beam and performing a raster in the air gap region. There is no need to repositioning the laser beam before writing as long as the focal length to the focal plane is not changed.

Data tracks, consisting of a "101010..." bit pattern, are written in x-direction. The bit length is set to 100 μm, which is long enough to clearly identify a transition region between a "1" and a "0" bit. The strength of a write field is decreased to a lower level and tested to ensure that magnetization of the medium solely by the write field is not possible. During a write process, the head poles are in contact with the medium and the laser module is configured to output a continuous wave laser beam at 80 mW nominal power.

An image of a HAMR data track and an image of normal data tracks from Section 2.3.3.2.4 are shown in Fig. 2.3.3.24. An examination of these tracks reveals that data bits magnetized in the normal write process have a rectangular shape and their transition regions have a size comparable to the size of the air gap. For heat-assisted writing, it can be clearly seen that data bits have a circular shape and their transition regions have an approximated diameter equal to 36 μm. This is corresponding to a diameter of a focused laser beam at a focal plane. An increasing data density can be estimated from a ratio of the rectangular transition size to the circular transition size, which is 4.9. Since the heat tracks are furrowed into the medium, a contact between the medium and the magneto-optical sensor is not perfect. As a consequence, some data bits are blurred.

We will end this section with future works. To quantitatively describe the write process, a relation between overall effective write field (or write energy), which comes from both a write field and a laser spot, and a written data bit must be known.

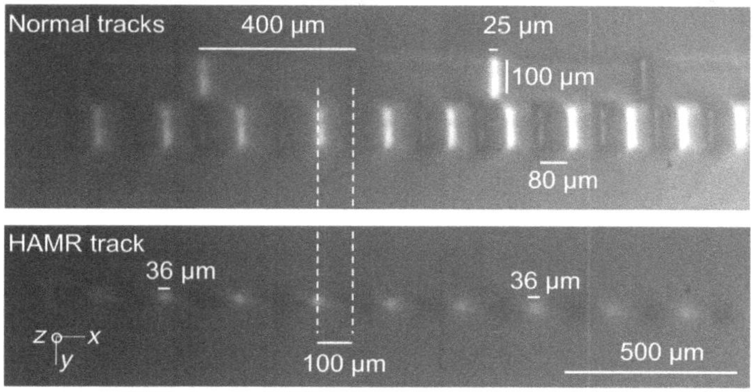

Fig. 2.3.3.24 Reference data tracks created by cool writing (top) and a data track created by HAMR (bottom).

However, the direct measurement of the energy is not trivial and poses a question on accuracy and precision. A numerical solution using FEM is more feasible and will be carried on in the future.

2.3.3.2.6 Inherent storage medium

We developed media from magnesium alloyed with Co and Zn by means of gravity die casting [38]. Magnesium is the lightest structural metal used in industrial applications. The density of pure magnesium at 20°C is 1.74 g/cm^3 and conventional magnesium alloys are usually paramagnetic [39]. However, our magnesium alloys possess ferromagnetic properties. These alloys enable two potentials of use. First, mechanical loads are measurable in a novel way utilizing the magnetoelastic effect. The magnetoelastic effect describes the change of magnetic properties in the magnetic field due to the reversible deformation of a ferromagnetic specimens crystal lattice [40,41]. By the employment of the harmonic analysis of eddy current, signals representing magnetization states is measured according to acting mechanical forces. Second, the ferromagnetic material enables magnetic data storage within a material surface. Therefore no additional labels are required to store information like a component ID, a manufacturing date or life cycle information. RFID or bar code labels are suitable to connect such information with a component but there are some disadvantages that should be considered. Pasted labels could be lost or polluted during manufacturing or operation. In contrast, a magnetic data track is written to a component surface inseparably and could not be lost or polluted. Additionally, the magnetic tracks could be applied as a protection against plagiarism using hidden data tracks on the surface.

Their inherent ferromagnetic properties of the magnetic magnesium alloys base on the element cobalt (Co), which is alloyed to the magnesium (Mg). Generally, the solubility of Co in Mg is negligible [42], therefore the alloying process needs to be modified using an extruded master alloys made from a powder mixture (60 wt% Mg and 40 wt% Co). The master alloy rods are added to molten pure magnesium held at temperature of 730°C in a boron nitride-coated steel crucible. To prevent oxidation during manufacturing, the resistance heated furnance a shielding gas atmosphere (N_2 and 0.3 vol.% SF_6). Furthermore, a mechanical stirrer is applied to the furnace to blend the alloying elements actively. The binary MgCo*X* alloys (*X*: 1, 2, 3, 4, 5 wt%) have low mechanical strength similar to pure magnesium in the range of 30 MPa yield strength [39]. By alloying Zinc (Zn) to MgCo4 the yield strength increases up to 74 MPa in case of 2 wt% Zn. In comparison with conventional Mg-based alloy, a yield strength of, e.g., AZ91 is 110 MPa in gravity die cast condition [39].

Comparing with the sintered magnetic component invented in Ref. [9] the casted magnetic magnesium alloys allow a machining process that is beneficial for sample preparation. As a part of the investigations the magnetic data storage properties are analyzed. Using the latest invented version of the magnetic write and read head, a problem occurs during the read back of the data tracks. Because the data tracks are written to the material in a denser way, some of the bits are not readable by the read head. Precipitation of Co depends on the cooling rate. Consider two SEM images of the alloy MgCo4Zn2 shown in Fig. 2.3.3.25. The matrix material is colored gray in the SEM images; the Co precipitates are colored white.

Due to the low solubility of Co in Mg the amount of Co is considerably higher in the precipitates than in the matrix. The indicated data tracks with a length of 50 μm demonstrate that some data bits are written to matrix areas with a low number of ferromagnetic precipitates and hence the magnetization produced by those areas is small and could not be read back completely. As a result, the stored data in those

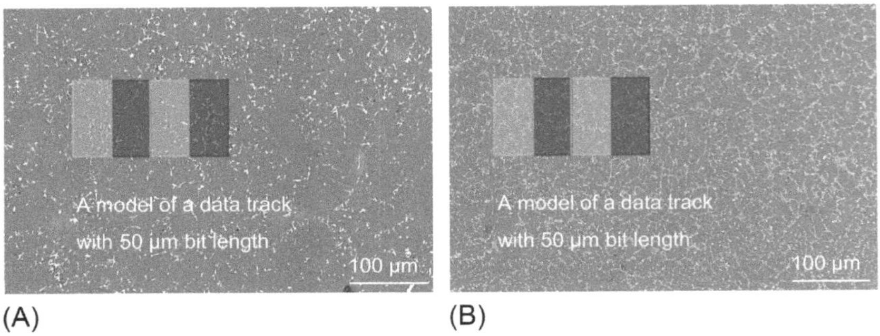

Fig. 2.3.3.25 SEM images of MgCo4Zn2 specimens manufactured with (A) low cooling rate, (B) high cooling rate; a model of a data track reveals that some data bits are in inhomogeneous areas containing an insufficient number of particles.

areas are practically lost. In contrast, the SEM image of the specimen manufacture with a higher cooling rate in Fig. 2.3.3.25B shows a smaller grain size and consequently a denser network of the relatively homogeneous distributed Co-rich ferromagnetic precipitates. On average, the spacing of precipitates is decreased from 45 to 21 μm by the rapid cooling conditions. Therefore the data tracks in these areas should yield a more probable successful read back.

References

[1] K.G. Ashar, Magnetic Disk Drive Technology: Heads, Media, Channel, Interfaces and Integration, IEEE Press, Piscataway, NJ, 1997.
[2] S.X. Wang, A.M. Taratorin, Magnetic Information Storage Technology, Academic Press, San Diego, London, 1999.
[3] J. Kadis, Foundation of Sound Recording Technology. Lecture 7: Magnetic Recording, 2014.
[4] Mediapedia, TDK AD60, Available at http:/mediapedia.nla.gov.au/browserecord.php?-action=browse&-recid=782 (accessed 18.03.16).
[5] K.-H. Wu, H.-H. Gatzen, Fabrication and assembly of a magnetic write head, in: Proceedings of the 10th Anniversary International Conference of the European Society for Precision Engineering and Nanotechnology, Zürich, Switzerland, EUSPEN, 2008, pp. 143–147.
[6] K. Wu, O. Traisigkhachol, H.H. Gatzen, A concept for magnetic data recording on genetically intelligent ("gentelligent") components, Phys. Status Solidi C 5 (2008).
[7] A. Belski, M.C. Wurz, L. Rissing, Redesign and fabrication of a magnetic head for Gentelligent™ products, in: Proceedings of the 11th International Conference of the European Society for Precision Engineering and Nanotechnology, Como, Italy, EUSPEN, 2011, pp. 191–194.
[8] K.-H. Wu, E. Gastan, M. Rodman, B.-A. Behrens, F.-W. Bach, H.H. Gatzen, Development and application of magnetic magnesium for data storage in gentelligent products, J. Magn. Magn. Mater. 322 (9–12) (2010) 1134–1136.
[9] A. Belski, E. Gastan, N. Vahed, C. Klose, M. Rodman, F. Lange, M.C. Wurz, D. Bormann, L. Rissing, B.-A. Behrens, F.-W. Bach, Process principle for the production of sintered dynamic component-inherent data storage, Prod. Eng. Res. Dev. 5 (3) (2011) 233–240.
[10] C. Klose, C. Demminger, G. Mroz, W. Reimche, F.-W. Bach, H. Maier, K. Kerber, Influence of cobalt on the properties of load-sensitive magnesium alloys, Sensors 13 (1) (2013) 106–118.
[11] H.H. Gatzen, K.-H. Wu, O. Traisigkhachol, Development of a recording head for magnetic storage of information on machine components, in: 210th ECS Meeting, Cancun, Mexico, October 29–November 3, 2006, pp. 191–202.
[12] A. Belski, P. Taptimthong, M.C. Wurz, L. Rissing, Integration of the read-after-write method for gentelligent applications, in: Proc. 1st Joint International Symposium on System-Integrated Intelligence: New Challenges for Product and Production Engineering,

SYSINT2012: Garbsen, Germany, June, 27th–29th, PZH Verlag, Garbsen, 2012, pp. 49–51.
[13] A. Belski, M.C. Wurz, L. Rissing, Process development of an electroplated permalloy layer on magnetic magnesium for gentelligent products, in: 218th ECS Meeting, Las Vegas, NV, ECS, 2011, pp. 93–104.
[14] A. Belski, P. Taptimthong, M.C. Wurz, L. Rissing, High permeability flexible bulk material for magnetic micro head applications, EPJ Web Conf. 40 (2013) 16003.
[15] A. Belski, M.C. Wurz, J. Rittinger, L. Rissing, Development, micro fabrication, and test of flexible magnetic write head for gentelligent applications, Microelectron. Eng. 110 (2013) 315–319.
[16] L. Rissing, A. Belski, T. Creutzburg, T. Griesbach, M. Kaiser, A. Wienecke, M.C. Wurz, Magnetic MEMS, SPbSPU J. Comput. Sci. Telecommun. Control Syst. 5 (205) (2014) 21–34.
[17] E.I. du Pont de Nemours and Company, DuPont™ Kapton® Summary of Properties.
[18] E.I. du Pont de Nemours and Company, DuPont™ Kapton® HN, 2011.
[19] P. Taptimthong, Biegsam statt starr: Daten speichern auf flexiblen Substraten, Available at http://www.phi-hannover.de/forschung/artikel/detail/biegsam-statt-starr-daten-speichern-auf-flexiblen-substraten/ (accessed 12.04.16).
[20] P. Taptimthong, J. Rittinger, M.C. Wurz, L. Rissing, Flexible magnetic writing/reading system: polyimide film as flexible substrate, in: Proc. 2nd International Conference on System-Integrated Intelligence: Challenges for Product and Production Engineering, SysInt 2014, July 2nd–4th, Bremen, Germany, 2014, pp. 230–237.
[21] B. Burkitt, B. Reilly, P. Danielle, New Technologies in Silicone Adhesives, Available at http://www.adhesivesmag.com/articles/91217-new-technologies-in-silicone-adhesives (accessed 31.03.16).
[22] NuSil Technology LLC, Adhesion and Sillicone Adheives: Application Note.
[23] Carl Roth GmbH+Co. KG, Roticlear®: Instruction for use, 2015.
[24] C.A. Grimes, Sputter deposition of magnetic thin films onto plastic: the effect of undercoat and spacer layer composition on the magnetic properties of multilayer permalloy thin films, IEEE Trans. Magn. 31 (6) (1995) 4109–4111.
[25] G. Choe, M. Steinback, Surface roughness effects on magnetoresistive and magnetic properties of NiFe thin films, J. Appl. Phys. 85 (8) (1999) 5777.
[26] J. Rittinger, P. Taptimthong, L. Jogschies, M.C. Wurz, L. Rissing, Impact of different polyimide-based substrates on the soft magnetic properties of NiFe thin films, in: SPIE Microtechnologies, Barcelona, Spain, SPIE, 2015, p. 95171R.
[27] J. Rittinger, P. Taptimthong, L. Jogschies, M.C. Wurz, L. Rissing, Magnetic microstructure analysis of sputter deposited permalloy thin films on a spin-on polyimide substrate, Microsyst. Technol. (2016).
[28] 3M Company, 3M™ Polyimide Film Tape 5413. Technical Data, 2014.
[29] B.D. Cullity, C.D. Graham, Introduction to Magnetic Materials, second ed., Wiley, Hoboken, NJ, Chichester, 2009.
[30] O. Karlqvist, Calculation of the magnetic field in ferromagnetic layer of a magnetic drum. No. 86, Transactions of the Royal Institute of Technology Stockholm, Sweden, 1954, pp. 1–27.
[31] P. Taptimthong, J. Rittinger, M.C. Wurz, L. Rissing, Flexibles magnetisches Lese-/ Schreibsystem: Design eines Schreibkopfes, in: VDI-Gesellschaft Mikroelektronik,

Mikrosystem- und Feinwerktechnik (Ed.), MikroSystemTechnik Kongress 2015, MEMS, Mikroelektronik, Systeme 26.-28. Oktober 2015 in Karlsruhe, first ed., VDE Verlag, Berlin, 2015, pp. 684–688.
[32] VACUUMSCHMELZE GmbH & Co. KG, Weichmagnetische Kobalt-Eisen-Legierungen: VACOFLUX 48·VACOFLUX 50VACODUR 50·VACOFLUX 17, 2001.
[33] P. Taptimthong, L. Rissing, M.C. Wurz (Eds.), P8.1—Characterization of a Read/Write Head, AMA Service GmbH, Germany, 2016.
[34] S. Tumański, Thin Film Magnetoresistive Sensors, Institute of Physics Pub, Bristol, 2001.
[35] K.H.J. Buschow, Handbook of Magnetic Materials, first ed., North-Holland, Amsterdam, Boston, 2011.
[36] M.H. Kryder, E.C. Gage, T.W. McDaniel, W.A. Challener, R.E. Rottmayer, G. Ju, Y.-T. Hsia, M.F. Erden, Heat assisted magnetic recording, Proc. IEEE 96 (11) (2008) 1810–1835.
[37] P. Taptimthong, J.F. Düsing, L. Rissing, M.C. Wurz, Flexible magnetic reading/writing system: heat-assisted magnetic recording, Procedia Technol. 26 (2016) 72–78.
[38] C. Demminger, C. Klose, P. Taptimthong, H.J. Maier, Material-inherent data storage using magnetic magnesium-cobalt alloys, Procedia Technol. (2014) 188–197.
[39] Kammer, Magnesium-Taschenbuch, first ed., Aluminium-Verl., Düsseldorf, 2000.
[40] A.S. Borovik-Romanov, H. Grimmer, Magnetic properties, in: T. Hahn (Ed.), International Tables for Crystallography, Wiley, Chichester, 1983, pp. 105–149.
[41] É. Du Trémolet de Lacheisserie, D. Gignoux, M. Schlenker, Magnetoelastic effects, in: E. Du Trémolet de Lacheisserie, D. Gignoux, M. Schlenker (Eds.), Magnetism: Fundamentals, Springer, New York, NY, 2002, pp. 351–398.
[42] B. Predel, Co-Mg (cobalt-magnesium), in: B. Predel, R. Börnstein, H. Landolt, O. Madelung, W. Martienssen (Eds.), Numerical Data and Functional Relationships in Science and Technology: New Series, Springer, Berlin, 1993, pp. 1–2.

2.4 Employment of electrical effects

In this chapter it will be demonstrated how sensors based on electrical effects and components merge to an inseparable entity, and which advances and possibilities are closely linked to these technologies.

The abilities to sense their environment, to monitor their status, and to gather data during their manufacturing process as well as during their life cycles are fundamental features of gentelligent components developed within the CRC 653 and requirements of the fourth industrial revolution.

To enable such capabilities, sensors are integrated into the components. Conventional sensors have their functional parts fabricated on rigid substrates, which are a few 50 μm thick (Fig. 2.4.1, left). They need to be carefully attached onto components. The accuracy of the measurements is, however, influenced by the substrates and the way the attachments are carried out. Moreover, in some situations sensors cannot be

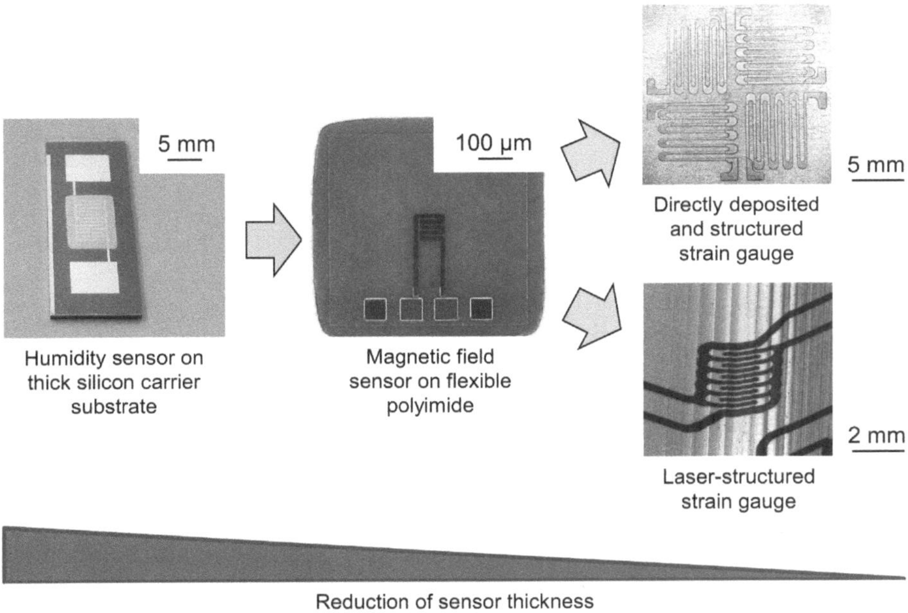

Fig. 2.4.1 The overall sensor thickness is reduced by using sensors manufactured on flexible polyimide carrier substrates, that are glued to the surface, or by directly depositing and structuring sensors onto components.

applied due to special requirements, such as the topology and tribology of component surfaces. Thus a new kind of technology for sensor integration is required in order to overcome those limitations. The CRC 653 has developed, for example, a very thin flexible substrate and a modular microsensor family to monitor temperature and strain of a component and the magnetic field in its environment (Fig. 2.4.1, center). By using this technology, the influence of the thick substrate on the measurands has been reduced, improving the sensitivity/accuracy of the sensor.

Jointly developed with an industrial partner, the application of the magnetic sensor has been extended and presented in Chapter 2.4.1. The next level of the sensor integration represents the complete elimination of sensor substrates and fabrication of sensors directly on components (Fig. 2.4.1, right). Since there is no substrate between a component and a sensor, the sensitivity and accuracy of a sensor are expected to increase. In addition, the direct fabrication means no need for adhesion and attachment, which have usage requirements, e.g., temperature range, and can deteriorate over time. As a result, higher stability and a wider application range are expected. Two prospective technologies have been developed. One is a direct deposition of sensors and thin layers directly onto the technical surfaces of arbitrary size (Chapter 2.4.2). The other is thin film patterning of deposited films with ultrashort pulsed lasers and fast laser scanners (Chapter 2.4.3).

See the following chapters for further information

2.4.1 Microsensors on film substrates

Lisa Jogschies, Marc Christopher Wurz
Institute of Micro Production Technology, Leibniz Universität Hannover, Garbsen, Germany

2.4.1.1 Introduction

The presented results are a spin-off from the group working on foil-based microsensors, discussed in Section 2.4.2, where a sensor family consisting of four different kinds of sensors has been developed: temperature, anisotropic magneto-resistance (AMR), strain gauge, and eddy current. The unique about this sensor family is its modular setup enabling the realization of all these sensors with only one set of photolithographic glass masks. The first sensor generation had been designed on a standard silicon wafer. In order to enhance their system integration, the silicon has been substituted by a polymer substrate. The following manufacturing process has been developed [1]:

1. The liquid, photosensitive Polyimide 7320 is spin coated on a 100 mm silicon wafer and is structured by photolithography in order to create spaces, in which contact pads can be grown by electroplating.
2. The polymer layer is cured in a hard bake process at 350°C for 1 h.
3. The sensor layers are deposited onto the polymer membrane and consist of the meander containing the functional layer, the four contact pads, and the feed lines. Only the eddy

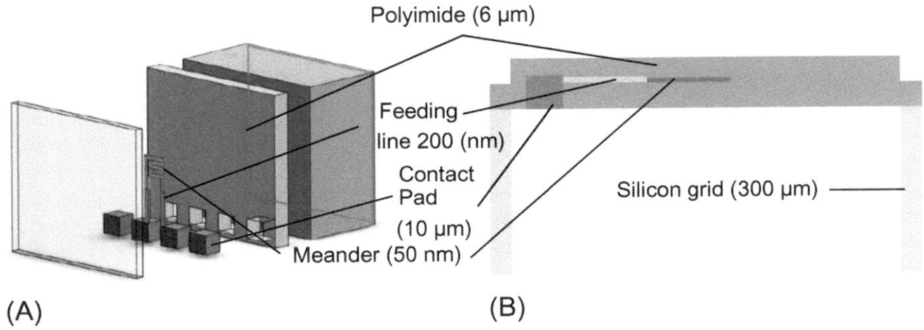

Fig. 2.4.1.1 (A) Exploded view of a technical drawing of a single sensor on the wafer and (B) cross-section of a single sensor on a wafer grid.

current sensor has an additional copper coil that is placed around the sensing layer and contacted at the two outer contact pads.

4. In a next step a second polyimide film is spin coated on top and cured as well at the same temperature.
5. The silicon substrate is partly removed in a deep reactive ion etching (DRIE) process leaving only a silicon grid carrying the polymer sensors. This last etching process is a key element in the fabrication of these sensors, as it creates very steep flanks inside the wafer grid and therefore realizes a high aspect ratio. Accordingly, a silicon grid with thicknesses of only 200 μm remains, leaving the membrane's backside as well as contact pads exposed.

Fig. 2.4.1.1A shows an exploded view of a single sensor element from the wafer assembly, while Fig. 2.4.1.1B depicts the cross-section of the element. As the silicon is removed, the polymer layer undertakes the task of serving as a substrate and as an insulating layer. To separate the sensors, a simple stamping tool can be used, without applying high forces. The ejected sensor elements have a rectangular form with an edge length of 1 mm and a thickness of less than 15 μm. The polyimide layer connected to the silicon and the silicon grid is referred to as the lower side of the sensor. The polyimide layer covering the sensor elements and is on top of the lower polyimide layer is referred to as the upper side of the sensor.

2.4.1.2 Preliminary investigation of the polyimide-based AMR sensors

The modular sensor design provides a single sensor with a footprint of 1 mm × 1 mm leading to an amount of approximately 5000 sensors per 4-in. wafer. The layout is depicted in Fig. 2.4.1.2. On the left-hand side, the layout for the AMR, temperature,

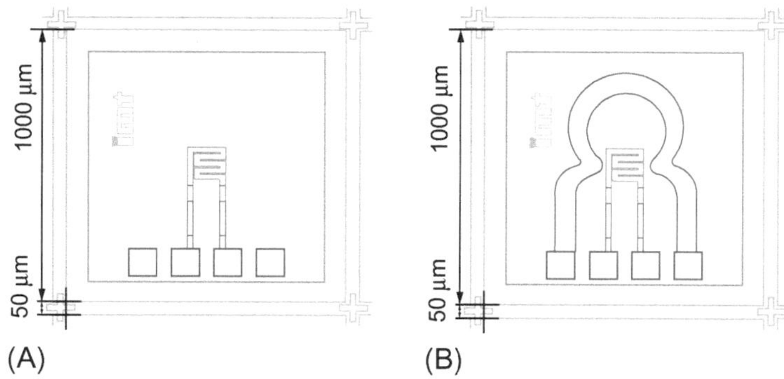

Fig. 2.4.1.2 (A) Layout of an AMR, a temperature, or a strain gauge sensor and (B) an eddy current sensor.

or strain gauge sensor can be seen; on the right-hand side the eddy current sensor is shown, which additionally consists of a copper coil.

The contact pads feature a size of 100 µm, making it impossible to solder the contacts. Consequently, this design is restricted to a bonding process with a gold wire concerning the packaging. However, within the CRC 653, only the AMR sensor has been developed further so only the layout on the left-hand side is concerned.

The AMR sensor is investigated to address two goals: First, the functionality of the sensor has to be ensured and second the ejection process from the wafer has to be introduced. To gain information about the behavior of the sensor during a separation process, a simulation has been done in ANSYS calculating the mechanic properties [2]. The simulation assumes a solid connection between polyimide and silicon as well as between the sensor components and polyimide. A previously measured friction coefficient of $\mu = 0.3$ between the steel stamping tool and the polyimide is taken into account as well. The silicon grid is assumed to be fixed and the model is supposed to act symmetric along the cross-section. Except for the meander-shaped functional layer, all materials of the sensor as well as the stamping tool are simulated as linear elastic materials. Only the meander-shaped functional layer is modeled for a better visualization. The simulation does not consider the actual detachment between polyimide and silicon. This would include the separation of nodes which would make the simulation much more difficult. Finally, the DRIE process for etching the silicon grid has to be considered by the simulation. It does not create ideal steep flanks. A slight undercut takes place at an angle of about 5 degrees. The silicon opening should measure 860 µm × 860 µm according to the design provided by the photolithographic mask. Furthermore, it does not create ideal sharp 90 degree edges but a radius of about 3 mm. This is depicted in Fig. 2.4.1.3.

Fig. 2.4.1.3 REM picture of a silicon grid etched by DRIE.

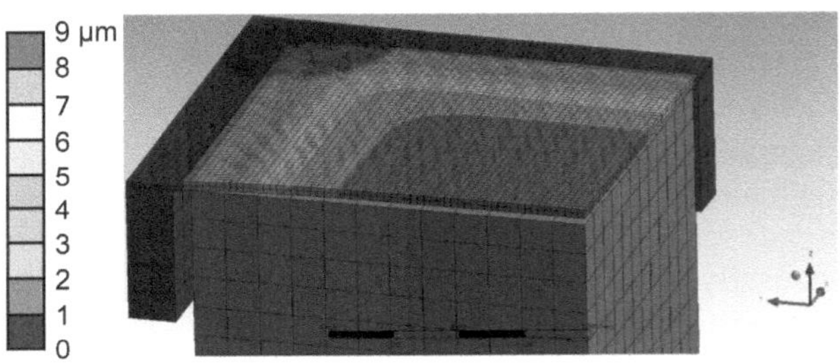

Fig. 2.4.1.4 Simulated picture of a quarter polyimide layer without sensor on a silicon grid during deformation from below by a planar steel stamping tool.

The carried out simulations take only the single sensor element connected with the silicon grid into account. As a reference, a simulation of a sensor without a sensor element is done and the results are depicted in Fig. 2.4.1.4. It shows the load resulting from a planar steel stamping tool pushing against the backside of the polyimide and moving upward for 6 μm. A constant deformation of the polyimide layer can be observed as expected with a maximum in the center of 9 μm. Because the deformation is symmetric in the sensor plane, only a separated quarter of the system wafer and tool is depicted.

Taking the sensor elements into account during the simulation the deformation behavior starts to differ as can be seen in the simulation results depicted in Fig. 2.4.1.5. The picture shows the sensor from above while the stamping tool is loading the sensor from the backside. It is a sectional view, so the contact pads are located on the left-hand side. It is interesting to note that the deformation becomes inhomogeneous as soon as the sensor elements are considered. The main deformation takes place right above the contact pads.

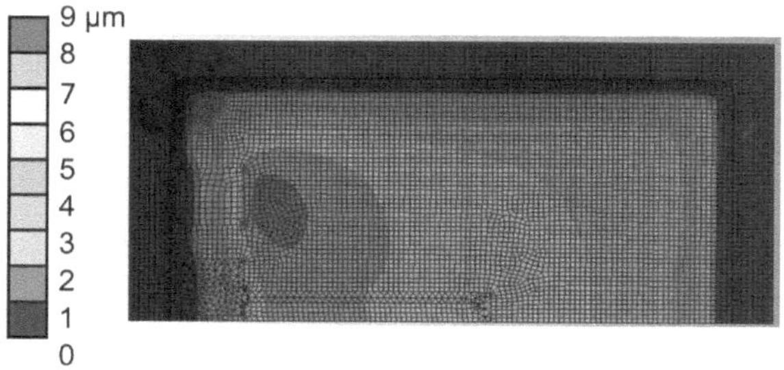

Fig. 2.4.1.5 Picture of simulated deformation of a half of a sensor during separation using a planar stamping tool.

In order to understand this phenomenon better, a measurement setup has been installed to gain measurement results that can be compared to the simulation. The wafer is clamped between two clamping rings and placed above a stamping tool. The clamp rings are screwed on four 75 mm long poles that maintain the necessary distance between wafer and the stamping tool. The tool is placed on top of a 3-axis precision translation which allows a three-dimensional movement of the tool. The stamping tool has a planar surface and a height of 1 mm, the same dimensions as used in the simulation. The whole setup, as shown in Fig. 2.4.1.6, is placed underneath a

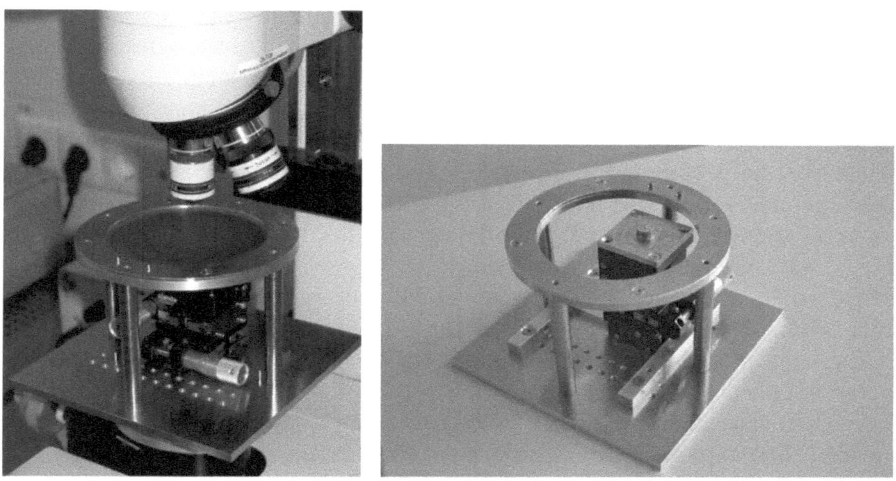

(A) (B)

Fig. 2.4.1.6 (A) Measurement setup with a wafer placed under a white light interferometer and (B) the measurement setup showing a stamping tool on a 3-axis precision translation stage.

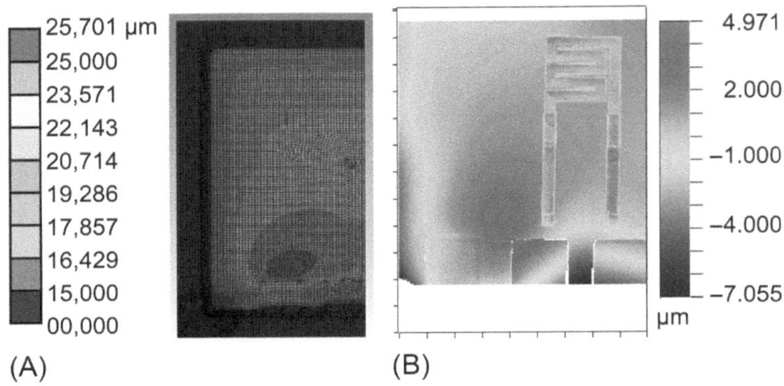

Fig. 2.4.1.7 Comparison of (A) simulated and (B) measured deformation during the ejection process.

white light interferometer to detect the deformation during the separation process of the sensor element. The white light interferometry is a contactless optical measurement. It uses interference of white light to create a three-dimensional picture of a reflecting surface.

Fig. 2.4.1.7 depicts the simulated and the measured deformation during separation and shows a very good conformity between simulation and measurement. In the simulation, an overall deformation of 10 μm was detected while the measurement delivered an overall deformation of 12 μm. The simulation assumes that the tool is being moved upward 12 μm, after reaching the contact to the sensor. Unfortunately, this cannot be compared to the measurement. The simulation calculated the silicon grid to be fixed, which is not practicable in the used measurement setup because it would prevent the view on the wafer. During the measurement the tool first moves the whole wafer upward before the sensor is separated. The spot of the maximum deformation is slightly shifted. The maximum deformation is calculated by the simulation located above the upper right corner of the outer contact pad (Fig. 2.4.1.7A). The measurement, however, showed the maximum deformation shifted to the location above the upper left corner of the inner contact pad (Fig. 2.4.1.7B).

The light microscope shot in Fig. 2.4.1.8 shows a wafer grid from the backside focused on the polyimide after the separation process that has been described earlier. All of them cracked above or through the contact pads approving the location of the highest deformation indicated by simulation and measurements.

An explanation can be found when having a look at the cross-section of the single sensor element. As the feed lines (Au) and the meander (NiFe 81/19 or permalloy) are very thin (200 and 50 nm), it can be assumed that these layers have no significant influence on the specific deformation behavior of the whole sensor element. The

Fig. 2.4.1.8 Destroyed sensor structures.

10-µm-thick contact pads are extended from the lower membrane into the upper membrane as depicted in Fig. 2.4.1.9. These pads most probably affect the polymer membrane causing the specific deformation during the ejection of a sensor.

To solve this issue, the stamping process has to be improved by using a supporter that will clamp the sensor during its separation. Several sensors could be successfully

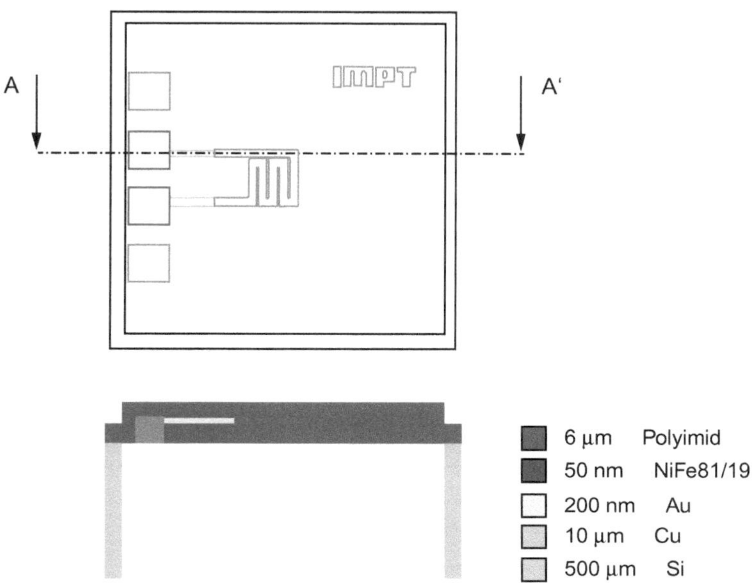

Fig. 2.4.1.9 Sectional view of a single sensor on a silicon grid.

separated without damage from the wafer with the stamping tool, when the supporter was applied.

2.4.1.3 Transferring the sensor into an application-oriented environment

2.4.1.3.1 Redesign of the sensor layout

In comparison to the presented sensor layout, an area of a sensor substrate can be expanded in the new layout, because the outer two contact pads are no longer necessary as the CRC 653 aims to further develop the AMR sensor and not the whole sensor family [3]. This makes the outer contact pads for the coil redundant. The new AMR sensor only consists of the lower and upper membrane, two contact pads, the feed lines, and the meander. To be flexible in the choice between soldering and bonding, the contact pads have to be enlarged to the dimensions of 300 μm × 300 μm with a spacing distance of 150 μm between each other. The feed lines have been enlarged to decrease their resistance so that their values become small and negligible compared with the resistance of the sensor element. The extension of the feed lines also increases the distance between the functional layer and the contact pads which reduces the risk of damaging the functional layer during the electric contacting. The meander-shaped sensing layer of the sensor is fabricated with more turns. This increases the electrical resistance of the sensor and the sensitivity in a direction parallel to the length of the sensor. The first redesign is shown in Fig. 2.4.1.10. The configurations have dimensions of 5 mm × 1 mm [4].

This sensor design was further investigated using a simulation, where the process of releasing the sensor from the silicon grid was estimated. To release the sensor, a planar steel tool with the dimensions of 4.8 mm × 0.8 mm and an edge radius of 3 mm is centered below the sensor element and moved upward. The von Mises stress in the sensor is calculated. In the simulation, the tensile strength of the

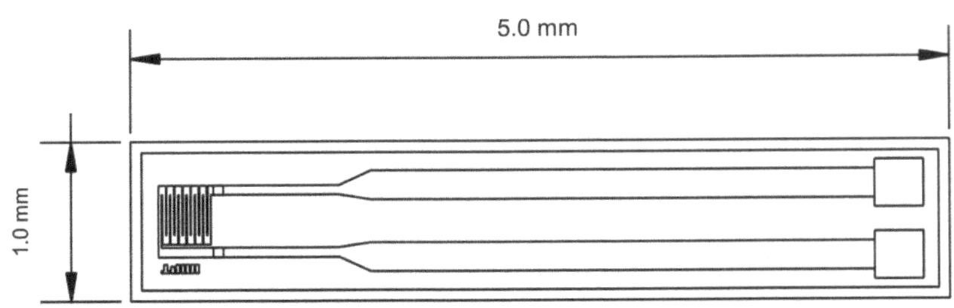

Fig. 2.4.1.10 Layout of the first redesign with a long meander-shaped sensor element and enlarged feed lines.

Fig. 2.4.1.11 Stress in the sensor simulated using the first redesign.

polyimide is set to 215 MPa, which is a value according to the manufacturer. Fig. 2.4.1.11 depicts the resulting stress in the sensor. The degree of stress is illustrated by different colors. Dark blue (or black) is the lowest and red (or light gray) is the highest. The color transition from orange (or gray) to red (or light gray) marks the tensile stress of the polyimide. All regions colored in red are likely to crack. Fig. 2.4.1.11 shows that the simulated deformation of the polyimide is sufficient to detach the polyimide from the silicon, which can be concluded from the red-colored edges of the polyimide. Also, there are no stress peaks near the contact pads, meaning that no cracks are to be expected in this area. Therefore, the distance between the contact pad and the edge of the polyimide is suitable and will be kept unchanged. The simulation also shows a stress peak at the transition area between feed line and functional layer. With this design, probably several sensors would be damaged during the stamping process. A higher radius in this area would solve this problem.

A second redesign has rounding transition areas between a functional layer to a feed line and from a contact pad to a feed line. Eventhough, the rounding transition from a contact pad to a feed line is in fact not mandatory, according to the simulation, but as a precaution it is included anyway. Moreover, the sensing element is enlarged and the meander is also turned by 90 degrees to use the space in a more efficient way. The 180 degree turns at the end of each section of the meander are widened in order to improve the sensitivity in the longitudinal direction. The turns have a curve structure to aid the liftoff process used in the fabrication of the sensor element. The second redesign results in the following sensor sizes as shown in Fig. 2.4.1.12.

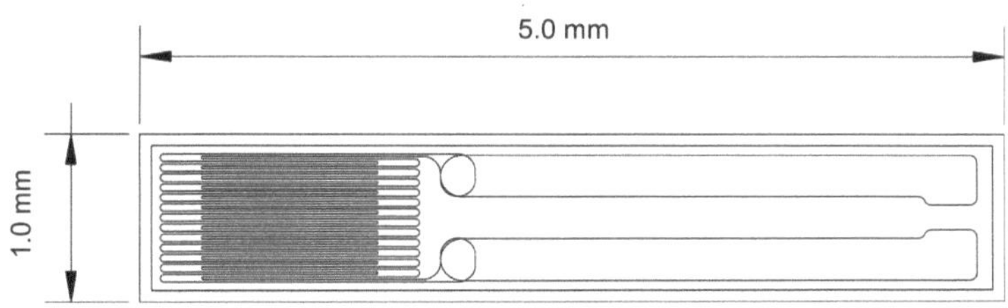

Fig. 2.4.1.12 Layout of the second redesign improving the sensor layout shown in Fig. 2.4.1.10.

Finally, the simulation has to consider that the DRIE process does not etch the whole wafer uniformly because the plasma distribution over the area varies. So, the cavities show a difference in their size from the middle of the wafer to its outer edge. The measurements done on the backside of the silicon grid show a mean value of the width of 879.1 μm (set value 860 μm) with a standard deviation of 16.45 μm and a mean value of length of 4891.2 μm (set value 4860 μm) with a standard deviation of 11.19 μm. Due to these negative impacts from the etch process, the resulting gap that occurs between the stamp tool and the silicon grid has to be taken into account. If the tool tip is 4.8 mm × 0.8 mm, the maximum gap calculated from the measurements is 86 and 117 μm for width and length, respectively. Simulating this worst-case scenario yields a heavily loaded area at the transition between the two polyimide layers. Fig. 2.4.1.13 shows a section of the sensor and the tool as simulated. The tool (lower left side) has been moved upward deforming the polyimide layers. On the right-hand side, the lower polyimide is still attached to the silicon grid.

An interesting finding from this simulation is that assuming the worst case, the stress exceeds the tensile stress of the polyimide at the corner of the upper polyimide and at the area next to it on the lower polyimide. Differently from the preceding simulation where the tool affects the sensor central, the sensor would crack most likely in the red marked area resulting in a high risk of damaging the sensor severely. Not only the polyimide layers are loaded when the worst-case scenario occurs but also the silicon grid is affected. This can be seen in Fig. 2.4.1.14 which shows the same situation as Fig. 2.4.1.11, but from the backside. Additionally, the stamping tool was redesigned to the geometry of the upper polyimide layer to reduce internal stress inside the microsystem during the ejection.

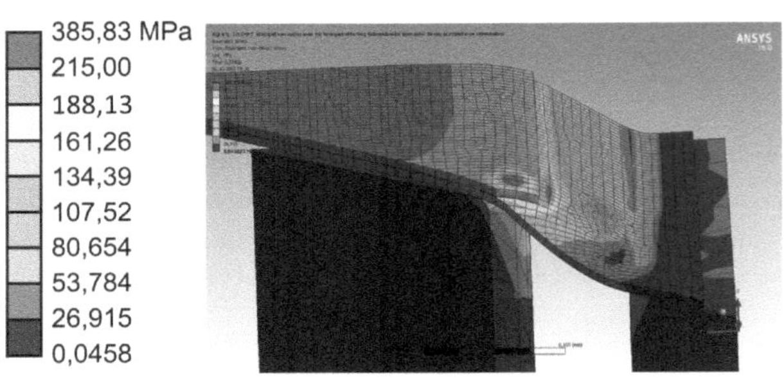

Fig. 2.4.1.13 High stress occurs at the transition between the two polyimide layers. Note also the position of the tool tip that is shifted from the optimal position due to a gap between the tool top and the silicon grid.

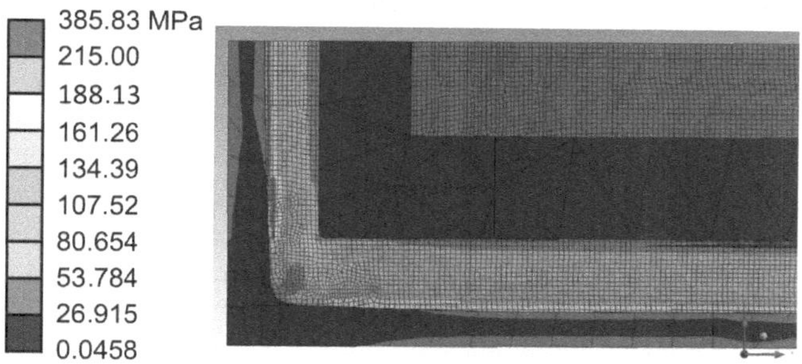

Fig. 2.4.1.14 Simulated stress inside the silicon grid.

First functional tests show that the polyimide-based AMR sensor is working properly, though it is still not good enough to be comparable to commercially available magnetic field sensors. A resistance change is measured using a Helmholtz coil and a Wheatstone bridge circuit. During the measurement an altering magnetic field is applied in the range between -10 and $+10$ kA/m, plotted on the x-axis and the change of resistance is measured and plotted on the y-axis as shown in Fig. 2.4.1.15.

Two disadvantages are revealed by this measurement. First of all, the maximum AMR signal is very low, being less than 0.4%. Second of all, the hysteresis causes peaks to shift from zero field to ± 4 kA/m and to have an unequal amplitude. In conclusion we can say that only the design modification was not sufficient to gain a fully functioning AMR sensor on a polyimide substrate.

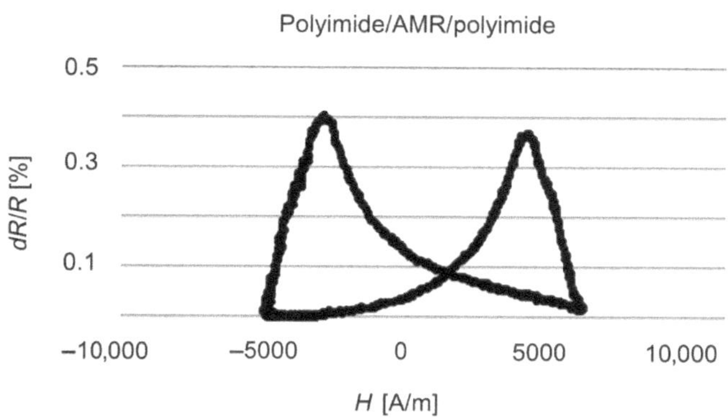

Fig. 2.4.1.15 Measurement of resistance change of an AMR sensor.

2.4.1.3.2 Stress release by buffer layers

A possible explanation for the misbehavior of the AMR sensor can be found by taking the internal stress of the polyimide into account. The stress induces in the layer itself because the liquid polyimide precursor is spin coated onto the wafer surface reaching a thickness of 12 μm. During the hard bake which is done for 1 h at 350°C, the polyimide cross-links and therefore shrinks to a thickness of 6 μm. It cannot change its size in the horizontal direction because it is attached to the silicon wafer. Furthermore, silicon has a coefficient of thermal expansion of 2.6 ppm/K, whereas the coefficient of thermal expansion of the polyimide amounts 55 ppm/K. All this leads to stress inside the polyimide which can be observed on a macroscopic level through the warpage inside the wafer that occurs when the sensor elements are exposed by the DRIE process. The remaining silicon grid can no longer support the necessary stability to keep the polyimide in plane. This can be depicted better when observing a single sensor element that has been released from the wafer assembly. In Fig. 2.4.1.16, a side view of a single sensor is shown. It has an enormous warpage which also hinders the handling of this system. It can be assumed that these internal stresses not only bow the polyimide mechanically but also are transferred inside the functional layer to the permalloy layer causing the untypical sensor behavior.

One possible way to release the stress from the permalloy layer is to integrate a buffer layer between the polyimide and the functional layer. Most suitable is a PECVD deposited SiO_2 layer. Depending on the N_2O/SiH_4 ratio which is used during deposition, a defined compressive or tensile stress can be realized within the SiO_2 layer [5]. So first of all we investigated the stress that develops inside the polyimide layer in order to be able to deposit an SiO_2 layer of defined stress of the same amount but with opposite sign so the stress state of both systems is compensated by each other.

Internal stress can be measured via X-ray diffraction (XRD) in crystalline. Because polyimide has not enough crystalline phases, this measurement cannot be used on this material directly. Instead, a polyimide layer was spin coated on a silicon wafer and hard baked. Subsequently, a 2 μm thick layer of permalloy

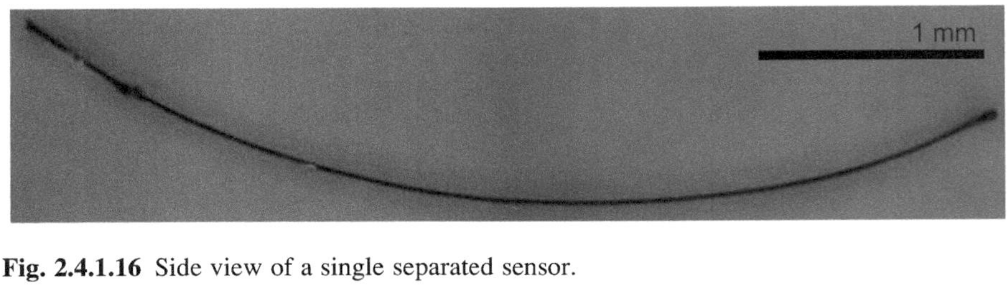

Fig. 2.4.1.16 Side view of a single separated sensor.

was sputter deposited on top. The measuring setup did not allow any thinner crystalline layers. The stress inside the polyimide will be transferred into the permalloy layer. Circles of 10 mm diameter were etched into the silicon, so the layer composition of polyimide and permalloy stood partly free. At these regions the XRD measurements took place. A tensile stress of 250 MPa was measured on this substrate. This value measured on the 2-μm-thick permalloy layer cannot be transferred to the 50-nm thick layer directly because the thickness is linked to the growing internal stress. Nevertheless, the measured value is a good indication on the range of the magnitude of internal stress.

To investigate the influence of a stress compensating buffer layer different kinds of sensors have been manufactured in one batch. First, a reference sensor was fabricated with the AMR structure deposited on a silicon dioxide wafer. Second, an AMR structure was deposited on a silicon dioxide wafer that was covered with a hard baked polyimide layer. Third, an AMR structure was sputter deposited on a silicon dioxide wafer and covered under a layer of PECVD deposited SiO_2 with a compressive stress of 250 MPa before the polyimide was spin coated on top and hard baked. In Fig. 2.4.1.17, the magnetization curves of these sensors are compared with each other and with the AMR sensor covered in hard baked polyimide from both sides.

In Fig. 2.4.1.17, it can be seen that the AMR sensor deposited on a silicon dioxide wafer (top left) works satisfactorily with a maximum AMR effect of 2% and no peak shift. It is shown that only adding a polyimide cover layer on top of the AMR structure (top right) already leads to a strong degradation of the AMR work performance. The

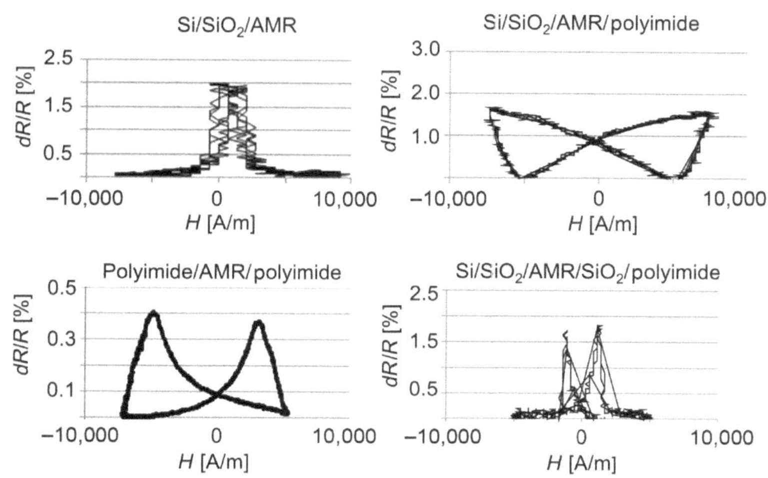

Fig. 2.4.1.17 Comparison of the AMR effect on different substrates.

peaks of the AMR effect shift to a value of ±5000 A/m and its value lowered to 1.5%. This strong shift of the two peak values is comparable to the result from the AMR sensor fully covered in polyimide (bottom left) but this AMR sensor only reaches an AMR effect of about 0.4%. An immense improvement could be achieved by adding a SiO_2 buffer layer between the AMR layer and the polyimide as can be concluded from the magnetization curve drawn in the bottom right of Fig. 2.4.1.17. Here, the distortion of the curve is reduced heavily. Therefore it can be argued that the tension inside the AMR layer causes these degradations of the AMR sensor performance. Nevertheless, the SiO_2 buffer layer is not enough to create an AMR sensor with an equal performance as an AMR sensor on a silicon dioxide substrate.

2.4.1.3.3 Choice of polyimide and stress influence

Another approach to improve the working performance of a flexible AMR sensor by removing internal stress is to have a closer look on the choice of polyimide. The used substrate influences the AMR layer and therefore its performance as a sensor due to its roughness and its internal stress as mentioned before. Therefore different spin on polyimides were coated on 500-μm-thick silicon dioxide wafers and hard baked before simple AMR structures were deposited on top of these. All the resulting polyimide layers were 6 μm thick. A silicon dioxide wafer without a polyimide layer was fabricated in the same stack as a reference probe. A drawing of the used geometries is shown in Fig. 2.4.1.18A. To evaluate the impact of the sensing element orientation relative to the substrate, the AMR structures consist of three single elements arranged in different angles as depicted in Fig. 2.4.1.18B.

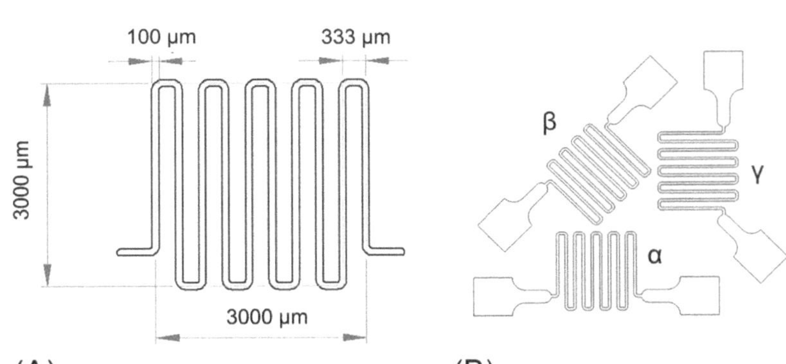

Fig. 2.4.1.18 Layout of (A) a single test structure and (B) the test structures oriented to each other.

Table 2.4.1.1 Surface roughness of different polyimide layers and the reference probe

Substrate	Surface roughness R_a [nm]
Silicon dioxide	8.8
Polyimide 7320	99.7
Polyimide 2611	98.2
Polyimide 10	43.6

Measurements of surface roughness R_a were performed on polyimide substrates with a sputtered permalloy thin film on top. The values of R_a are summarized in Table 2.4.1.1. The R_a value of the SiO_2 reference sample features the lowest surface roughness followed by the Polyimide 10. Worst results were obtained by the polyimide substrates Polyimide 2611 and 7320, which lie in a comparable range.

To gain further information about the different degrees of stress inside the polyimides and their influence on the AMR sensor, the radius of curvature of the silicon carrier wafer was determined before and after deposition of the polymeric layers and the permalloy thin film by means of a tactile profilometer. Using Stoney's equation, an estimation of the magnitude and orientation of film stress within the polymeric layers can be made. In Fig. 2.4.1.19, the magnitude of stress is presented in relation to the estimated stress in the SiO_2 reference specimen. As expected, all samples show tensile stress. The estimated film stresses of the Polyimides 10 (A) and 7320 (C) significantly exceed those estimated for Polyimide 2611. While the estimated film stress in Polyimide 2611 (B) only exceeds those estimated for SiO_2 by 1.3 times, the estimated

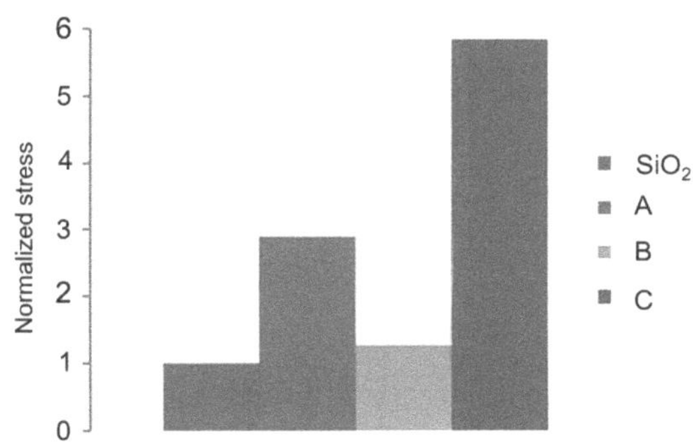

Fig. 2.4.1.19 Stress inside different substrates compared to those in silicon dioxide.

film stress in Polyimide 10 is 2.9 times higher and in Polyimide 7320 it is even 5.8 times higher.

During magnetic characterization, the homogenous AC field from a Helmholtz coils was applied in a direction parallel to the sample orientation labeled with γ as shown in Fig. 2.4.1.18B. Thus the samples experience the magnetic field at different angles, i.e., 90, 45, and 0 degrees for the sample orientation α, β, and γ, respectively. The percentage change of resistance as a function of the applied magnetic field was recorded simultaneously for each of the three sensor elements. The resulting magnetization curves are depicted in Fig. 2.4.1.20.

As expected, the best results were obtained on the SiO_2 substrate. The largest AMR effect and smallest shift of the AMR peaks along the magnetic field axis were determined. Thus the magnetization curves of the SiO_2-based sensor element is similar to what is predicted by theory when fields along the hard and easy magnetization axis of such single sensor elements are applied. Sensor element α, oriented perpendicular to the H-field, shows a typical transfer characteristic of an AMR permalloy thin film magnetoresistor [6]. Sensor element γ (parallel to the magnetic field) shows characteristics of the planar Hall Effect. The magnetization curves switch when the magnetic field orientation is rotated by 180 degrees. Sensor

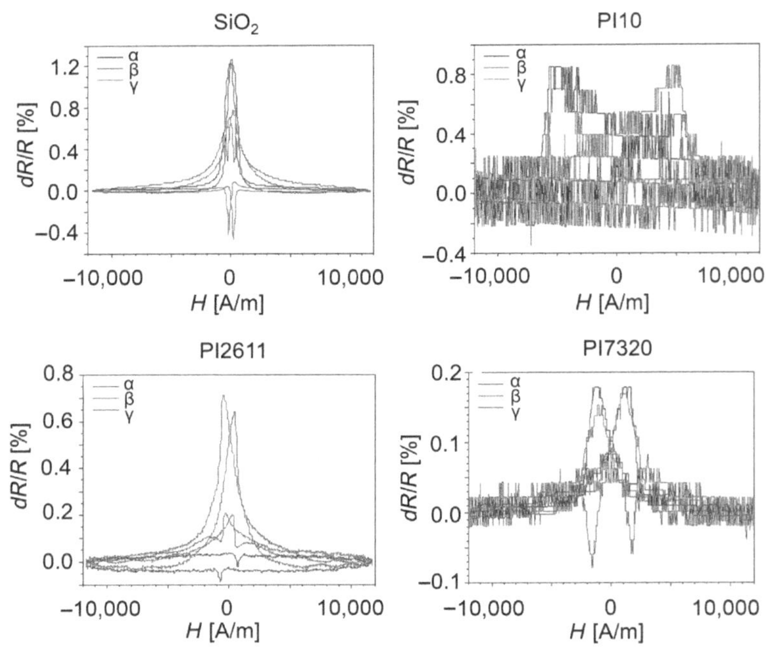

Fig. 2.4.1.20 AMR measurements on different polyimides and on silicon dioxide.

element β shows an overlap of the planar Hall effect with a lessened AMR effect, a typical behavior of a magnetoresistor in a magnetic field oriented at an angle of 45 degrees to its anisotropy axis.

All the polyimide-based AMR sensors show a worse performance when measured in a magnetic field, compared with the SiO_2 reference sample. However, great differences among these substrates could be shown. The maximum AMR effect is reduced significantly in all three cases and the distortion of the curve along the x-axis rises as well. Besides, the transfer characteristic of the sensor elements can no longer be purely attributed to shape anisotropy. While the maximum AMR effect of the sensor elements α, β, and γ on a SiO_2 substrate can be related to a specific orientation of the magnetic field, this is no longer possible for sensor elements on polyimide substrates [7,8].

Although Polyimide 10 obtained the best results concerning surface roughness and mediocre results concerning stress, the sensor element deposited on this substrate performed worst in respect of the maximum AMR effect and minimal AMR peak shift (Fig. 2.4.1.20). Even though by far highest film stress and mediocre surface quality were found out for Polyimide 7320, the average performance concerning magnetic characteristics of the sensor element was still better than those on Polyimide 10. Worst surface quality but lowest stress was identified for Polyimide 2611. In comparison to the other polyimide substrates, the corresponding sample clearly showed the best AMR performance. However, the magnetization curves of all samples deposited on polymeric substrates showed asymmetric transfer characteristics. According to literature, asymmetric transfer characteristics are due to wall state transitions that occur in multidomain structures. In other words, when the applied magnetic field is reversed, so-called Néel wall segments nucleate and expand which causes Barkhausen noise. Such Néel wall movement is known to be hindered by external factors such as surface roughness or film stress [9].

The Polyimide 2611 seems to be the promising candidate to investigate the various factors that affect magnetic thin films on polymeric substrates and will replace the before used Polyimide 7320. Consequently, several changes in manufacturing the sensor and its handling will appear. A main difference results from the fact that this polyimide is not photosensitive. In order to create the spaces inside the polyimide, where the contact pads can be grown the polyimide is structured by etching. To do so, a 4-μm thin negative photoresist is spin coated on top of the soft baked polyimide and is exposed through a mask that carries the forms of the contact pads. Afterward the specimen is developed in a TMAH-containing chemical. The TMAH not only develops the negative photoresist but also etches the polyimide. Because the polyimide is now structured by etching, the generated edges have a high undercut different to developed structures that feature a very steep edge. Thus it is crucial that the etch process is not executed too long. If the undercut becomes too big, the contact pads might lose contact to the polyimide as soon as the silicon carrier is removed. An adhesion promoter to enhance the

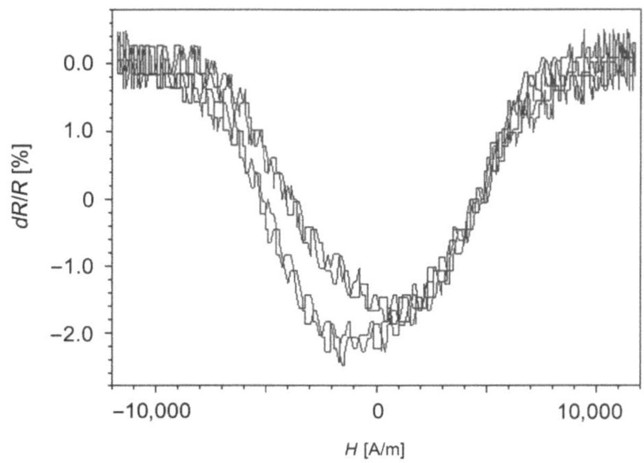

Fig. 2.4.1.21 Enhanced results in the AMR measurement.

contact between polyimide and silicon wafer becomes necessary which means an additional process step. This new polyimide is much more transparent than the previously used polyimide. This can be useful considering a stamping process, where the sensor structures can easily be seen from the camera to position a tool to the sensor element. If the aforementioned material and design modifications are combined, the magnetic behavior is further enhanced as can be seen from Fig. 2.4.1.21 where the AMR effect is plotted over the alternating magnetic field. Still a shift of the AMR effect maxima along the magnetic field has to be noted, although it is reduced. Furthermore, the maximum AMR effect almost reaches a value of 2%.

2.4.1.3.4 Stress reduction by optimizing the permalloy

By optimizing the sensor layout and minimizing the negative influence of the polyimide on the AMR performance, great improvements could be achieved. Still, the AMR performance is not fully comparable to commercially available ones, especially regarding the peak shift. A last possibility of reducing tensions inside the permalloy layer arises from the deposition process. The functional layer is deposited using a DC sputter process. Parameters that influence the internal stress of the growing layer are the power of the generator, the argon flux and the temperature during deposition. The sputtering process is always started at a pressure of 10^{-7} mbar. The layer is deposited out of a 200-mm target. In a separate heating chamber the specimens can be tempered under vacuum conditions and afterward can be moved into the deposition chamber without a break of vacuum. Argon plasma is used for sputtering.

During the deposition of NiFe on polyimide, stress can develop inside the metallic layer. The stress can be thermal or intrinsic. Thermal stress develops when the thermal expansion coefficient of both materials is different from each other. Consequently, when the deposition takes places at a certain temperature both materials experience different expansions but are in contact, so stress develops inside the structure. For this layer system an increasing tensile stress can be expected with increasing temperature. Intrinsic stresses can have a manifold of reasons, which can lead to either tensile or compressive stress. Grain boundaries and interstitial atoms are most likely to appear in this application. Chemical reactions between substrate and deposition layer would also lead to internal stress but are not taken into account. These two mechanisms are linked to each other and cannot be considered separately. Thermal stress will rise with increasing temperature as mentioned earlier while the intrinsic stress will decrease with rising temperature due to mechanisms like recovery and recrystallization that lead to bigger grains and logical to less grain boundaries and to an enhanced diffusion of atoms inside the layer which reduces the number of interstitial atoms. Here we aim to find the ideal temperature where a minimum of stress is reached. The mechanisms taking place are too complex to forecast the minimum exactly, it has to be determined experimentally. These correlations make clear that the temperature is a crucial factor for the stress reduction of a NiFe thin film.

The power of the generator influences the kinetic energy of the target atoms landing onto the substrate. In literature, it can be found that there is a critical energy that is sufficient to initialize diffusion processes of interstitial atoms in regions of energetically more favorable places [10]. It also has to be taken into account that an increasing power also enhances the amount of heat energy that is transferred into the system. The argon flux, which at the same time codetermines the pressure during sputtering, is crucial for the number of atoms that are released at once from the target. With an increase of the argon flux density and therefore an increasing partial pressure, the probability of collision of the target atoms among each other and between target atom and argon ion, respectively, rises. If collision takes place the kinetic energy of the atoms is reduced and also the angle in which they land onto the surface. With an increasing argon flux the deposition rate is enhanced as well, so a reasonable compromise has to be found here.

In a design of experiments (DoE) the three mentioned parameters have been investigated using the simplified AMR structures shown in Fig. 2.4.1.18 [11]. An ideal combination of parameters with respect to the maximum AMR effect was identified to be an argon flux of 20 sccm, a generator power of 750 W, and a substrate temperature of 150°C during deposition. This statistical analysis could show that the maximum AMR effect lies with a probability of 95% between 2.17% and 2.26%. Regarding the minimum peak shift, the ideal parameter variation is not the same. Still the shift is already so much reduced using the ideal parameters for the AMR effect that these parameters can be used as well as can be seen in Fig. 2.4.1.22.

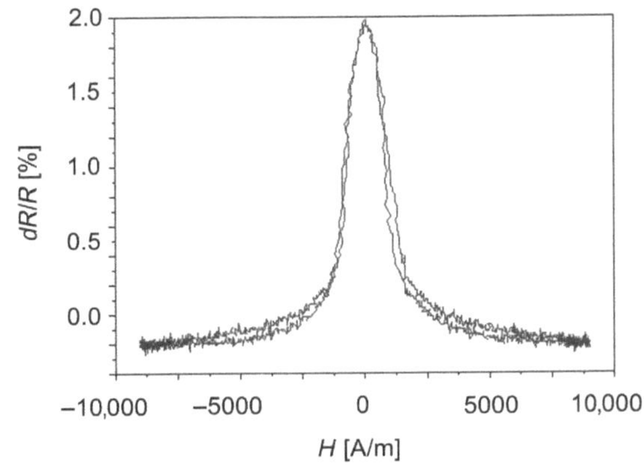

Fig. 2.4.1.22 AMR measurement with an optimized permalloy layer.

The necessary force to eject a sensor has been measured and simulated as well. The measurement setup was a Fineplacer, into which an adapter was integrated. The Fineplacer is a submicron bonder used for packaging issues with ultrahigh accuracy requirements. The adapter's tip was formed in the same way as the tool mentioned earlier. The Fineplacer positions the tool relative to the sensor and pushes against it with a defined force, which is increased by 1 N each time. A force of 6 N is found out to be sufficient to release the sensor, which agrees with the simulation that calculated a necessary force of 5.5 N.

Gathering all these information, a machine tool for the automated separation of the sensor assembly from the wafer can be designed. It is not planned to build a completely new machine. Instead, a stamping tool and a pick up tool are constructed and integrated into an existing die bonder with a wafer table that has a positioning accuracy of ±2–3 μm. The stamping tool is depicted in Fig. 2.4.1.23A. On the right-hand side, the complete stamping tool is shown. On the left-hand side and in the middle the tool is not fully assembled. The vacuum channels that can be seen on the right-hand side are installed to fix the wafer during the process. Fig. 2.4.1.23B depicts the pickup tool. Two vacuum channels can be seen on top that will be used to fix the separated sensor and transport it to its destination.

The wafer is laid down planar into the die bonder with the backside facing upward. The pickup tool, which is also working as the counterpart is placed above the wafer and moves into the cavity of it. The stamping tool is brought into contact with the sensor from the downside. The camera detection can only position the pickup tool relative to the sensor. Both tools are constructed in the way that they are centered to each other. When the sensor is clamped between the tools, the stamping process could work in both directions. The original idea was to eject the

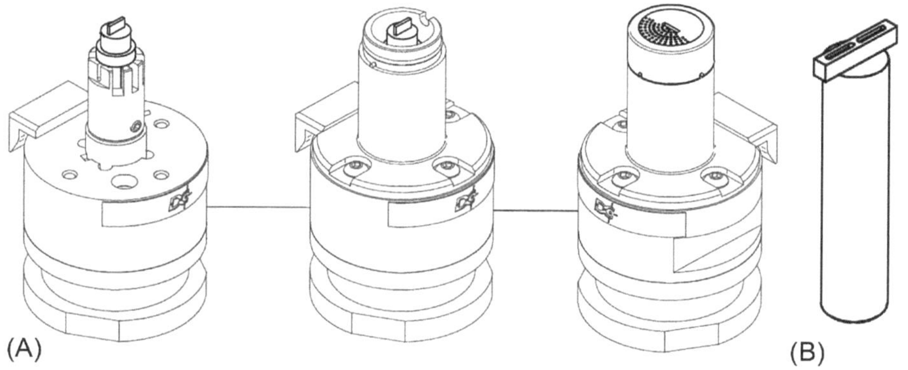

Fig. 2.4.1.23 Stamping tool (A) and pickup tool (B).

sensor from the bottom through the stable frame using the shear force out of the wafer grid. First tests showed that this process did not work as expected. The main troubles were due to the following points: The vacuum on the stamping tool side was not strong enough to hold the wafer down while the stamp moves upward. Sometimes it seemed as if the stamp or the pickup tool was blocked in the corner during the movement. Probably some silicon rests remain in some corners of the grid after the DRIE process. This could be observed before under the microscope as can be seen in Fig. 2.4.1.24. This is probably not easy to fix. With an increasing etching depth, the passivating layer becomes harder to remove. The etching process could be executed for a longer time, but it has to be considered, that the polyimide is being etched as well, though significantly slower.

If a separation process destroys a silicon grid at one point, the nearby sensor is also lost, because the stability is drastically decreased. A problem also occurred when the sensor was released from the grid at one of its long side but both of its short sides remained connected with its long side to the silicon. As a consequence, we tried to eject the sensor in the other direction, meaning that both tools after having clamped the sensor move downward to separate the sensor. Then the stamping tool remains and the pickup tool moves upward carrying the sensor. Though the sensor has to be moved through the grid, which bears the risk of stripping off the sensor at the grid, the ejection could be performed more reliable. Using this process, the shear force works against the cutout in the vacuum cap and the Si frame of the wafer can stay stable at the vacuum cap. The introduced stamping process successfully showed a way to automatically separate the sensors from the wafer grid. This is a huge step forward in the direction of transferring the

Fig. 2.4.1.24 Sensors on a wafer grid before ejection.

polyimide-based sensors into an industrial environment. Nevertheless, the wafer grid may have to be enlarged in order to increase process reliability.

2.4.1.4 Application of the magnetic field sensor

The flexible AMR sensor is aimed to be used to measure the rotational speed of a magnet wheel. Magnetic measurement methods are advantageous because they are robust against outer influences. In a first field test, a single sensor is glued onto a circuit board and placed above the magnetic wheel at a distance of 1 mm. A hard magnet NdFeB is placed behind the circuit board as depicted in Fig. 2.4.1.25A because the wheel is made out of a ferromagnetic material. Fig. 2.4.1.25B shows the magnet wheel in a technical drawing. It has a diameter of 120 mm and 100 poles each of a size of 2 mm. The hard magnet installed behind the sensor saturates the sensor while the moving ferromagnetic wheel changes the direction of the line of magnetic flux, thus changing the resistance of the sensor. The change of voltage is measured by an oscilloscope and can be transformed into a change of resistance. Simultaneous to the magnetic measurement the rotational speed was measured optically with a laser as a reference. Therefore a reflecting strip was attached to the wheel.

Fig. 2.4.1.25 (A) The magnetic wheel during the measurement and (B) in a technical drawing.

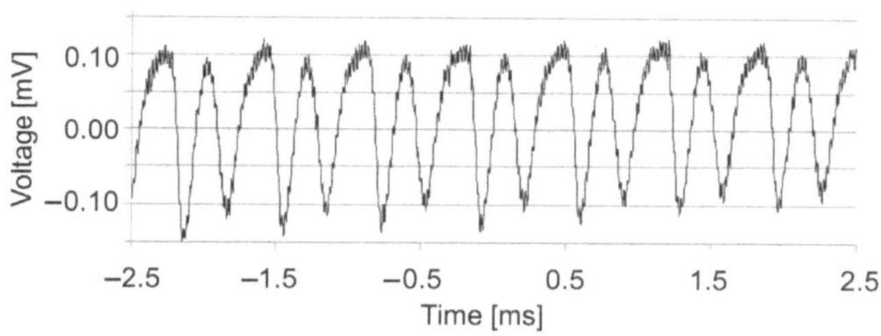

Fig. 2.4.1.26 Measurement of rotational speed with an AMR sensor.

Fig. 2.4.1.26 shows a graph of the change of voltage over the time of one sensor. Each pole could be measured by the sensor and the rotational speed was confirmed by the optical laser measurement.

The final application provides the use of two AMR sensors that are placed in a distance of 5.25 mm to each other in order to gain a 45 degree phase delayed sine signal that enables to detect the direction of rotation as well. In Fig. 2.4.1.27, a sheet of 40 circuit boards is shown that has been equipped with two AMR sensors each.

One of these circuit boards is placed in front and not above the magnetic wheel in the final application. Between the edge of the circuit board and the sensor lies a gap of 0.2 mm and between the circuit board and the magnetic wheel a space of 1 mm is left. Behind the sensors, a hard magnet is installed. A two-dimensional simulation done in FEM and depicted in Fig. 2.4.1.28 shows the change of

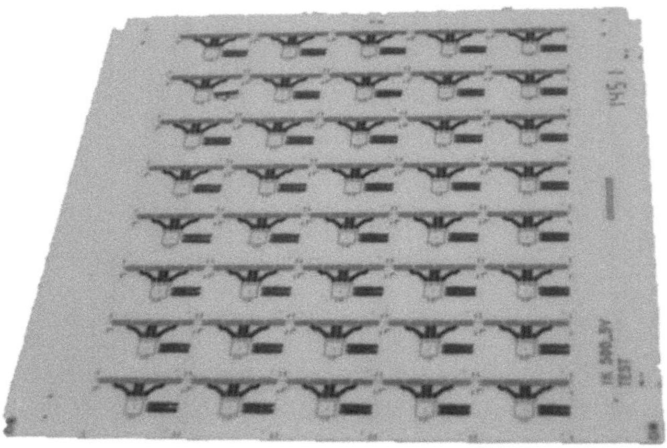

Fig. 2.4.1.27 Circuit board with AMR sensors.

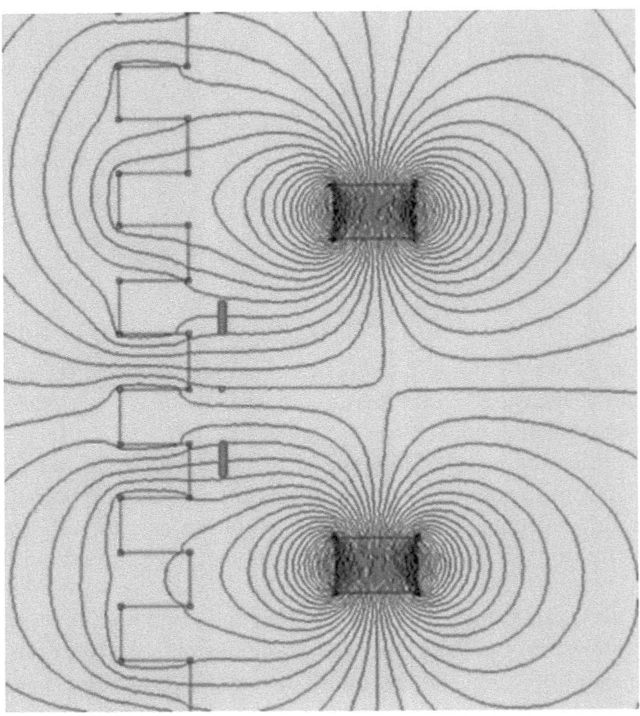

Fig. 2.4.1.28 Simulated magnetic flux from hard magnets flows through AMR sensors. The flux lines are guided by a magnetic wheel allowing a measurement of a speed and a direction of the wheel.

direction of the lines of magnetic flux of the two permanent magnets when the magnetic wheel (left-hand side) moves downward in this graphic. The red lines mark the position of the AMR sensors.

In an additionally made three-dimensional simulation, the absolute values for the provided magnetic field do not agree with the values found out in the two-dimensional simulation. The most probable explanation is that the field losses in radial direction (in the direction of the pole height) cannot be considered in a two-dimensional solution approach. Independently from this factor, the magnetic conducting functional layer of the sensor has not been considered so far in the simulation. This is probably not tolerable due to the small magnetic field that is provided in this application. But the used permalloy layer only has a thickness of 50 nm. Magnetic layers of such a low thickness can no longer be represented by simulation. Thus only approximate values can be found. A measurement to validate the simulation could be done successfully. Nevertheless, the signal-to-noise ratio was insufficient to meet the claim of a magnetic field sensor in an industrial application. Consequently, the material of the magnetic wheel was changed. Instead of using a ferromagnetic wheel, permanent magnets could be applied on the nonmagnetic wheel (Fig. 2.4.1.29), which would enhance the magnetic field force and its change of direction, respectively.

The simulated magnetic flux density depicted in Fig. 2.4.1.30A assumes that the permanent magnets are arranged as shown in Fig. 2.4.1.30B with the magnets on the wheel's front side in changing magnetic orientations with the sensors placed 2 mm away from the wheel. The orange dashed lines indicated the needed magnetic flux to gain a proper measurement, proving that the new measurement setup would deliver an easier measuring environment. Additionally, Fig. 2.4.1.30B shows the resulting change of angle during the measurement. A maximum difference of 10 degrees can be observed. A maximum change of

Fig. 2.4.1.29 Magnetic wheel with hard magnets.

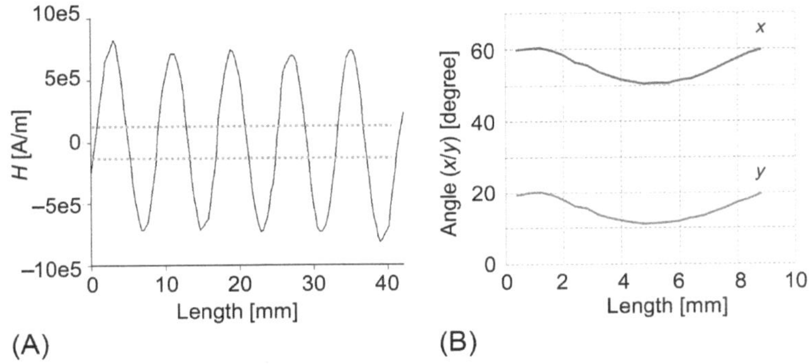

Fig. 2.4.1.30 (A) Simulated magnetic flux and (B) change of angle of the magnetic field during turn of the wheel for one tooth.

resistance in an AMR sensor will appear when the field direction is changed by 90 degrees. So a smaller signal can be predicted. Anyway, a proper measurement can be expected from the simulation.

References

[1] T. Griesbach, M.C. Wurz, L. Rissing, Development and fabrication of modular micro sensors on flexible polymer foils, in: 1st Joint Symposium on System-Integrated Intelligence: New Challenges for Product and Production Engineering, Hannover, Germany, 2012, pp. 66–68.

[2] L. Jogschies, D. Klaas, R. Kruppe, J. Rittinger, P. Taptimthong, A. Wienecke, L. Rissing, M.C. Wurz, Recent developments of magnetoresistive sensors for industrial applications, Sensors (Basel, Switzerland) 15 (11) (2015) 28665–28689.

[3] L. Jogschies, Ein Stempel für den Sensor, Available at http://www.phi-hannover.de/forschung/artikel/detail/ein-stempel-fuer-den-sensor/ (accessed 27.09.16).

[4] L. Jogschies, J. Heitmann, D. Klaas, L. Rissing, Investigations on strain behaviour of polymer substrates during a separation process, Procedia Technol. 15 (2014) 221–229.

[5] K.D. Mackenzie, D.J. Johnson, M.W. Devre, R.J. Westerman, B.H. Reelfs, Stress control of Si-based PECVD dielectrics, in: 207th Electrochem. Soc. Meet., May, 2005, pp. 148–159.

[6] S. Tumański, Thin Film Magnetoresistive Sensors, Institute of Physics Pub, Bristol, 2001.

[7] J. Rittinger, P. Taptimthong, L. Jogschies, M.C. Wurz, L. Rissing, Impact of different polyimide-based substrates on the soft magnetic properties of NiFe thin films, in: SPIE Microtechnologies, Barcelona, Spain, SPIE, 2015, p. 95171R.

[8] J. Rittinger, P. Taptimthong, L. Jogschies, M.C. Wurz, L. Rissing, Magnetic microstructure analysis of sputter deposited permalloy thin films on a spin-on polyimide substrate, Microsyst. Technol. 22 (7) (2016) 1627–1632.

[9] B.D. Cullity, C.D. Graham, Introduction to Magnetic Materials, second ed., Wiley, Hoboken, N.J., Chichester, 2009.
[10] C.A. Davis, A simple model for the formation of compressive stress in thin films by ion bombardment, Thin Solid Films 226 (1) (1993) 30–34.
[11] L. Jogschies, J. Rittinger, D. Klaas, M.C. Wurz, Stress reduction in sputtered thin NiFe 81/19 layers for magnetic field sensors, Procedia Technol. 26 (2016) 162–168.

2.4.2 Direct deposition of sensors on technical surfaces

Daniel Klaas, Marc Christopher Wurz
Institute of Micro Production Technology, Leibniz Universität Hannover, Garbsen, Germany

2.4.2.1 Introduction

Since many years, we are experiencing a rapid technological advance. Assistance systems within cars, ships, trains, and airplanes enable safer traveling; wearable electronics can detect the heartbeat; and modern technologies enable a micrometer precision manufacturing of components. Nearly all of these would be unthinkable without the use of sensors. They are used in high quantities within manifold applications [1]. Frequently, microsensors are used.

These microsensors are usually manufactured in a batch process on carrier substrates, such as silicon, glass, ceramic, or polymer substrates [2]. Afterward, the sensor elements are separated by, for instance, dicing; they are glued to the measurement point and contacted. As part of the Collaborative Research Center 653 (CRC 653), one research focus has been on the development of new sensor systems and sensor manufacturing technologies.

Within the first funding period of the CRC 653 (2005 until 2009) a tool box of sensor components has been developed. Each component was covering a specific basic function. By combining specific building blocks from the tool box, a modular sensor family was created. Such family includes sensors that are able to measure, for instance, strain, magnetic material properties, or temperature [3–5]. Such sensors have been manufactured on thick silicon carrier substrates.

Within the second funding period (2009 until 2013) a technology for replacing the relatively thick silicon substrates has been invented. Investigations showed that the use of polyimide in combination with a new manufacturing technology allows to reduce the overall sensor system thickness, to increase the level of system integration, and to attach sensors to nonplanar surfaces. The invented technology works as follows. Liquid polyimide is spin coated on a silicon wafer and cured. Afterward, sensors are deposited in a batch process on top of it. By using back side deep reactive ion beam etching a grid like silicon structure is manufactured. Such structure is used as a frame to span the sensors. By using a stamping tool, the sensors are separated [6]. By using this technology, strain and eddy current sensors have been manufactured [7,8].

The focus of the third funding period (2013 until 2017) is a deposition of sensors directly onto technical components of arbitrary size. Therefore, a new coating system and a technology for structuring sensors have to be developed. The development has the following advantages: By using direct deposition techniques, no thick intermediate layers such as the glue or the polymer foils are required. As a result, the negative influence of the layers on the measurands will be eliminated or reduced, enabling thinner sensors with a possibly higher measuring accuracy. Moreover, the limitation of the maximum operation temperature by glue or carrier substrates will be eliminated as well as the influence of the aging of the glue. As a result of the thin overall sensor thickness, the sensors can be used at new positions within machines or any other technical system. In the case of a direct deposition on electrically conductive surfaces, a thin insulation layer will be required. Fig. 2.4.2.1 gives an overview of silicon-, polymer-, and directly deposited sensors. The expected overall sensor thickness should be smaller than 10 μm.

A variety of technologies and processes are suitable for the direct deposition of sensors [10,11]. One very promising technology is sputtering (compare Section 2.4.2.5 for a detailed list of advantages). Usually, sensors are inserted into a vacuum chamber to enable sputter deposition. The size of such chamber is limiting the maximum size of the component on that a sensor can be deposited. Existing

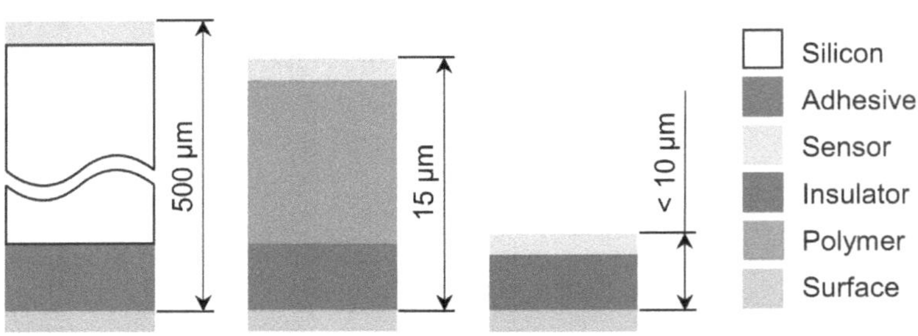

Fig. 2.4.2.1 The overall layer thickness is reduced to a minimum. From left to right: silicon-based, polymer-based, and directly deposited sensors [9].
Adapted from Klaas, D; Direct deposition of ultrathin sensor systems onto components of arbitrary size; In: Denkena, B. (Ed.); Cross Linked Production - Research Results for Industrial Applications; ISBN 9783959000574.

limitations have been exceeded by developing a system that enables a direct deposition on components of arbitrary size. Moreover, the existing sensor family has been extended by developing a direct depositionable humidity sensor.

The chapter is divided in different sections. First of all, research results concerning the development of the new humidity sensor element are presented. Afterward, a brief overview of the insulation layer development for directly deposited sensors is given. In the following, research results concerning the structuring of directly deposited sensors with shadow masks and the new sputtering coating system are presented. To proof the feasibility of this promising and powerful direct deposition technology, strain gauges are directly deposited onto a technical surface and test results are presented.

2.4.2.2 Development of a humidity sensor element

A variety of commercial humidity sensors are available on the market and are working on various measurement techniques. They differ, for instance, in size, design, measuring range, and the material used for detecting the humidity [12–14]. In order to extend the existing sensor family, a new humidity sensor element has been developed [15]. Compared to normal humidity sensors, it can be directly deposited on technical surfaces by using the new coating system (compare Section 2.4.2.5). As the system is using sputtering as a physical vapor deposition process, the used sensor material has to be sputter depositionable. Moreover, the sensor has to consist of as less layers as possible to enable a simplified manufacturing process.

Most of the various humidity measurement methods are described in the VDI-guideline 3514 [16]. Each method has various advantages and disadvantages depending on the required humidity range, the temperature during measurement, and the type of humidity that is measured, for instance, absolute humidity or the relative humidity. The resistive method appears to be very suitable for the new humidity sensor, as it can be easily miniaturized and cheaply manufactured. The working principle is as follows: The humidity causes a change of the electrical properties of a hygroscopic layer [16]. These changes can be measured.

2.4.2.2.1 Setup of the sensor

A sensor working on the resistive method [16] is consisting of two main parts, the hygroscopic material and an electrode structure. The developed setup is as follows: The interdigital structure is forming the electrodes of the developed sensor. Such electrodes are sputtered on a carrier substrate. As a material for these carrier substrates, electrically insulating silicon oxide or polyimide can be used. Alternatively, the sensor can be directly deposited onto a technical surface or, depending on the electrical conductivity of the technical surface, on a thin insulation layer. On top of

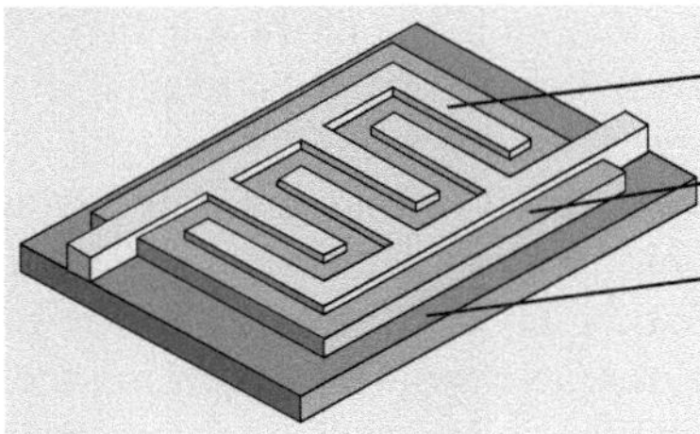

Fig. 2.4.2.2 Principle setup of the resistive sensor element [15].

these electrodes the hygroscopic layer (functional layer) is sputter deposited (Fig. 2.4.2.2). As a result, the hygroscopic material is in direct contact with the ambient air, enabling rapid response behavior.

As a material for the functional layer zinc oxide is used. The self-organization of the zinc oxide allows creating large, defined surface structures. Moreover, sputter deposited zinc oxide layers have a sufficient porosity to enable an adsorption of molecules and are therefore highly suitable as a hygroscopic layer.

2.4.2.2.2 Manufacturing of the sensor

A commercial sputter system Senvac Z550 is used to deposit the functional layers. Polished silicon oxide (525-μm-thick) has been used as a carrier substrate. This eliminates the influence of the surface roughness and enables a better investigation of deposited sensors concerning, for instance, the stoichiometry and layer thickness.

The setup of the humidity sensor element and the processing steps are illustrated in Fig. 2.4.2.3A. The electrodes have been structured by using photolithography and are made of 250-nm-thick chromium. Finally, 1-μm-thick solderable copper contact pads are deposited. In a third step, the functional layer is sputter deposited above the electrodes. Finally, the humidity sensor element can be separated by dicing forming the finished sensor element (Fig. 2.4.2.3B) [15].

A design of experiment has been carried out to detect the influence of the sputter power and duration on the sensitivity of the layer. Promising results have been achieved by using the following parameters. The sputter duration was 30 min at 100 W RF-power (radio frequency sputtering) with a substrate that was not extra heated.

To assure the stoichiometry, oxygen at a ratio of 1:1 was added to the argon process gas that had a flow rate of 10 sccm. The base pressure in the sputter chamber of the commercial sputter system Senvac Z550 before sputtering was at 1.9×10^{-4} Pa. The pressure during sputtering was at 4.9×10^{-1} Pa.

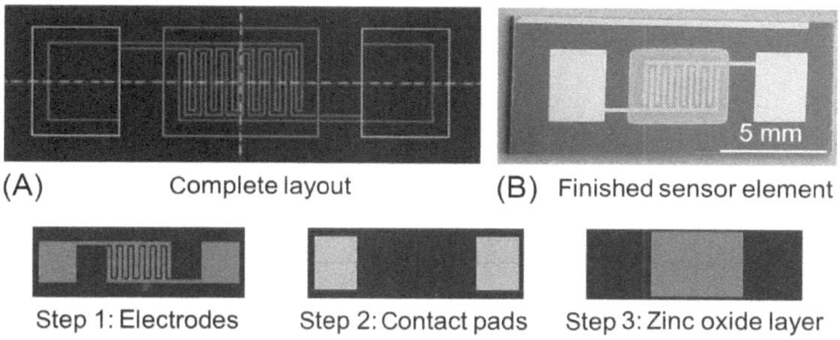

Fig. 2.4.2.3 (A) Setup of the sensor element consisting of the electrodes, the contact pads, and the zinc oxide layer (manufacturing step 1–3) and (B) an image of a manufactured humidity sensing element.

2.4.2.2.3 *Characterization of zinc oxide layers and sensor behavior*

Obtained zinc oxide layers were characterized by SEM (scanning electron microscopy) and XRD (X-ray diffraction) analyses [15]. Investigation showed that the crystallite size is increasing with higher sputter power and longer sputter time/thicker layers (compare Fig. 2.4.2.4A–D).

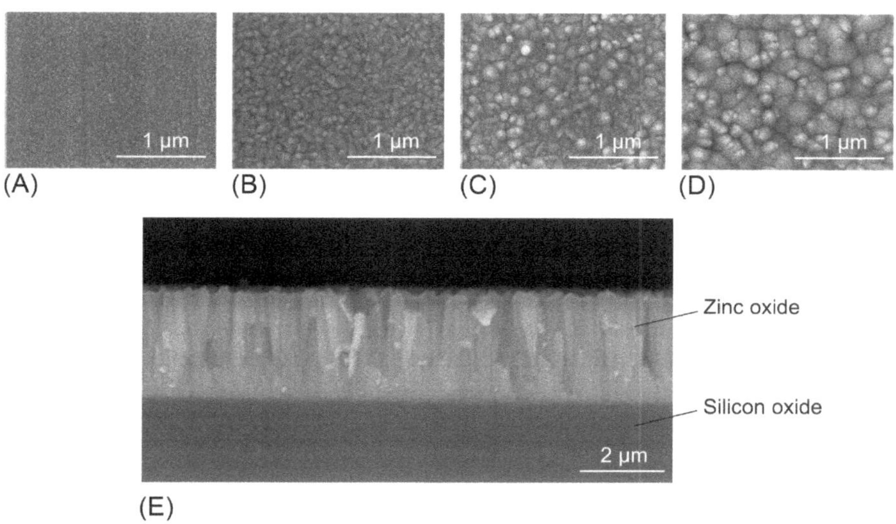

Fig. 2.4.2.4 Scanning electron microscope (SEM) images (A–D) of zinc oxide layers deposited at various sputter powers and durations: (A) 100 W, 120 min; (B) 300 W, 30 min; (C) 200 W, 75 min; (D) 300 W, 120 min. SEM image (E) of fracture plane of zinc oxide layer on a silicon oxide wafer.

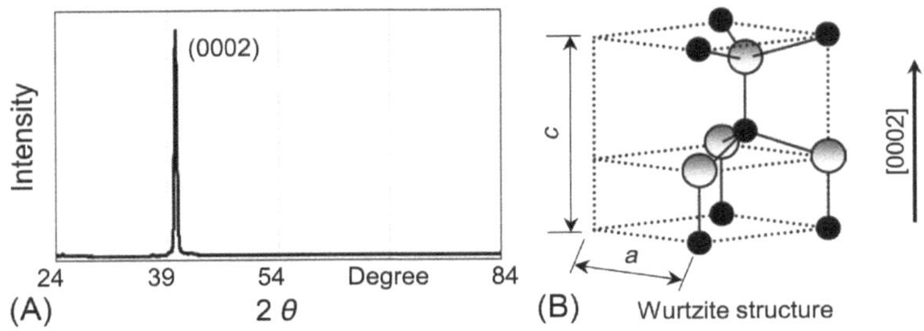

Fig. 2.4.2.5 XRD pattern (A) and crystallographic structure (B) [18] of sputtered zinc oxide (zinc atoms in gray and oxygen atoms in black) [15].

The fracture plane (Fig. 2.4.2.4E) shows columnar crystallites as a result of a preferred growth direction. As zinc oxide has hexagonal wurtzite structure, the velocity along the c-axis is larger than in other directions as a result of the different surface energies [17].

This effect causes a strong basal fiber texture reassured by XRD measurements (Fig. 2.4.2.5A). Corresponding to the (0002) plane of the wurtzite structure (Fig. 2.4.2.5B), all of the XRD measurements have one sharp (0002) peak. Static and dynamic measurement techniques have been used to characterize the sensor elements. For static measurements, the sensor element was placed inside a sealed chamber. Within this chamber a desired air humidity between 11% relative humidity (RH) and 85% RH was generated by using various saturated salt solutions. For dynamic measurements the use of salt solutions was inappropriate.

Therefore, a commercial climate chamber type ESPEC SH-241 has been used to investigate the electrical behavior of the sensors within a range from 25% RH to 85% RH. As direct current measurements might cause unwanted polarization effects, impedance measurement has been carried out by using a HAMEG HM8118 LCR-Bridge.

In both cases, static measurement as well as dynamic measurement, the sensors have been shielded against external influences such as light before measuring. The characteristic output curve (Fig. 2.4.2.6A) of the sensor element was created by using the LCR-Bridge in combination with LabVIEW and a commercial humidity and temperature sensor. Measurement results are showing the typical nonlinear behavior. The dynamic response of the resistance is presented in Fig. 2.4.2.6B showing a repeatable behavior.

2.4.2.2.4 Evaluation electronic concept

Up to now, the sensor element has been evaluated by using an LCR-Bridge. In order to develop a compact sensor system, an evaluation electronic concept has been designed and constantly improved to miniaturize the overall system. The latest system

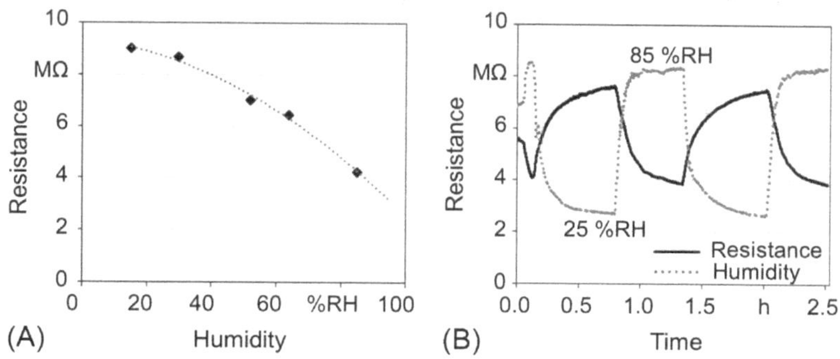

Fig. 2.4.2.6 Resistance values at specific humidity levels (A) and dynamic response of the sensor (B) [15].

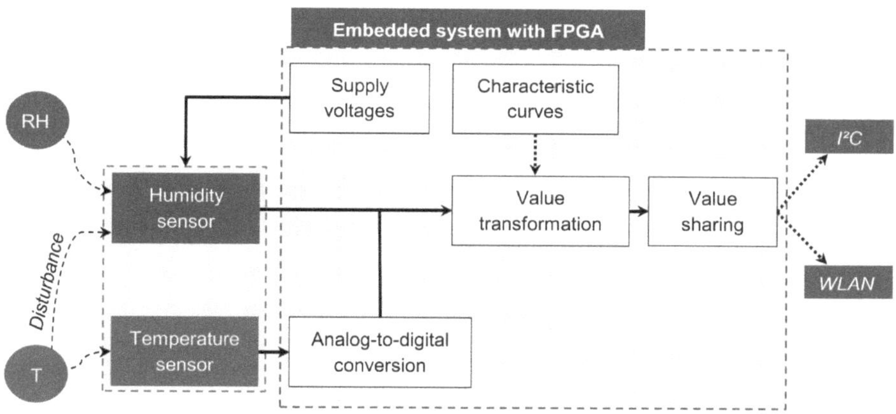

Fig. 2.4.2.7 Principle setup of the evaluation electronic. A field programmable gate array (FPGA) is part of the used embedded system. RH equals the external relative humidity and *T* stands for the temperature (compare [15]).

(Fig. 2.4.2.7) is consisting of three parts and is slightly different to former layouts (compare [15]).

As the relative humidity is depending on temperature, a platinum temperature element with an evaluation circuitry, consisting of a Wheatstone bridge, has been added to the system to detect the influence. Another circuitry has been developed for evaluating the humidity sensor element.

Such circuitry is creating a digital rectangular output signal that is changing in frequency depending on the measured humidity. A commercial embedded system consisting of a processor and a field programmable gate array (FPGA) is used for transforming the signals of both circuits to the actual humidity and temperature.

The developed circuits, the sensor elements, and the embedded system are connected by wire strands. The output signals of both sensor circuits are connected to the embedded system. The output of the analog voltage signal from the temperature sensor is converted to a digital signal; the frequency signal from the humidity sensor is directly forwarded. The characteristic curves of both elements are evaluated and saved within the embedded system. By using these curves, the values are transferred to the actual humidity and temperature and shared over a digital I^2C bus. Moreover, commands can be sent to the device to change parameters or to start a measurement. A direct access on the measurement results is also possible by getting access to the internal storage via wireless local area network (WLAN) from a mobile or tablet.

2.4.2.3 Development of an insulation layer

Besides the development of a humidity sensor, the overall aim is a deposition of sensors directly onto the surface of technical components. Most of these components are made of electrically conductive steel alloys or aluminum alloys, requiring a thin insulation layer between the sensor element and the technical surface. As the new sputter coating system was under development, first tests for depositing insulation layers and for investigating the influence of process parameters on the insulation layer quality took place in a commercial sputter coating system Senvac Z550.

One promising insulation layer material is high purity 4N-aluminum oxide (Al_2O_3). It has the advantages of a sufficient electrical insulation property, a high mechanical and chemical wear resistance, and it is stable over a wide temperature range. For characterizing the insulation layer quality, the average breakdown voltage and breakdown field strength have been evaluated by using a tera-ohmmeter [19]. A design of experiment has been used to investigate the influences.

Later on, the knowledge and experiences gained during the development of the insulation layer within the Senvac Z550 have been used to develop suitable insulation layers within the new sputter coating system. Therefore, another design of experiment has been carried out. Investigation showed an influence of various parameters on the insulation layer quality. Such are, for instance, the base pressure (pressure before sputtering), bias voltage, sputter pressure, sputter power, and amount of oxygen. The surface roughness has a high influence on the insulation capability why a polishing of the surface before sputtering is recommended. Depending on the material, a 10-min manual polishing with a felt polishing point can reduce the average mean roughness of up to 90%. Best insulation layer results have been achieved with the following parameters: argon flux of 15 sccm, sputter pressure of 1.14 Pa, and power of 275 W.

2.4.2.4 Structuring of sensors with shadow masks

Besides a technology for the direct deposition of layer materials, a method is required that enables a structuring of such functional layers. One very promising method is the use of shadow masks. Such masks are placed between the material source (in this case a sputter target) and the surface on that the material is deposited. By doing this, a thin layer of material is deposited and structured at the same time.

Shadow masks are used in a variety of manufacturing techniques, like chemical vapor deposition or physical vapor deposition processes [20]. Moreover, they are used to fabricate, for instance, optoelectronic devices, microelectrochemical systems, or sensors [21]. Recent research is, for instance, focusing on the fabrication of fine shadow masks to enable a structuring of emissive layers for the production of OLED displays [22].

Compared to conventional methods, like photolithographic structuring, shadow masks have the advantage that no photoresist is required. This means, process steps such as the spin-on of the resist, the exposure and liftoff are no longer required, making it a time-saving technique. Furthermore, it is possible to deposit and structure layers on nonplanar surfaces or to deposit layers on substrates that have a chemical incompatibility with the photoresist [23].

For manufacturing shadow masks, a variety of structuring methods are possible such as electrochemical machining or through wafer etching of silicon [2,24]. One promising technique for structuring shadow masks is laser machining. Lasering has the advantages that the beam can be focused to a very small diameter, it is a rapid process, and the laser spot can be reproducibly positioned, making accurate placed structures. Moreover, a micromachining of nearly all materials, especially metals and polymers, is possible [22].

2.4.2.4.1 Preliminary work to detect influences

For the fixation of such masks on the substrate, various concepts have been evaluated at the IMPT. In a first approach, shadow masks have been pressed with a blank holder onto a polished surface and centered by using locating pins (Fig. 2.4.2.8A). As a material for the masks, stainless austenitic sheet steel (1.4310) with a thickness of 50 μm has been used. Afterward, 100-nm thin layers of chromium have been deposited and the dimensions of the structured layers have been measured by using a microscope. Measurement results of deposited meander structures showed waved and serrated cutting edges with an average deviation of approximately 9% compared to the desired structural width. Maximum deviations have been evaluated to 22%, equaling a maximum deviation of approximately 70 μm for structures with a width of 300 μm (Fig. 2.4.2.8B).

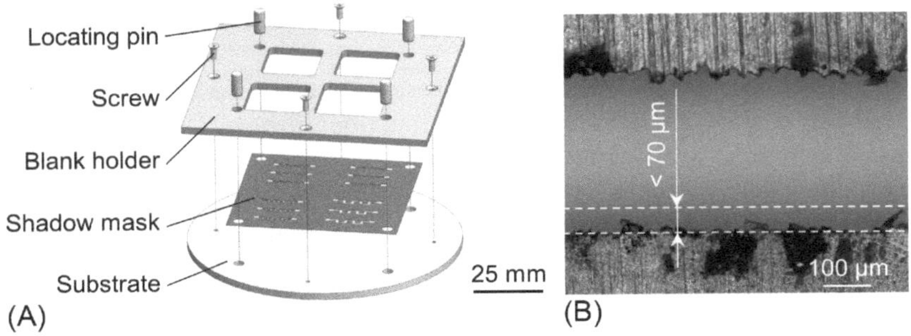

Fig. 2.4.2.8 (A) First approach: Shadow mask with meander structures pressed with a blank holder onto a polished metal substrate and centered by using locating pins. Screws are used for fixation [23]. (B) Example of waved and serrated cutting edges of a lasered shadow mask with a structural width of 300 μm showing a maximum deviation of 70 μm.

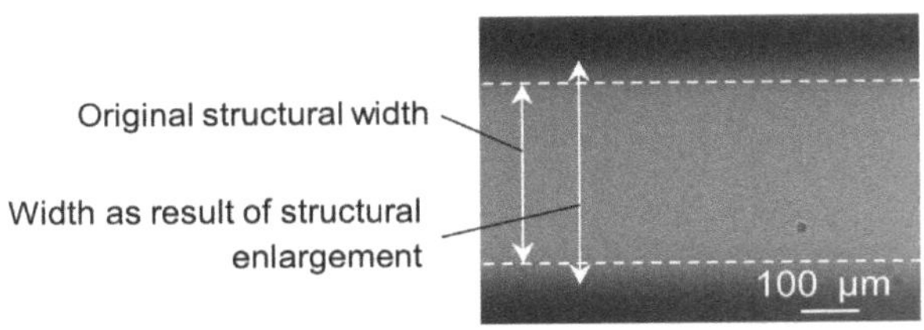

Fig. 2.4.2.9 A deposited structure showing the effect of blurring/structural enlargement.

These deviations are nearly exclusively a result of melted and resolidified material caused by the heat input during lasering. Investigations showed that melt is occurring on top of the meander structures or directly inside them [23].

Nearly all deposited structures showed a contour blurring that equals a structural enlargement (Fig. 2.4.2.9). This average structural enlargement was evaluated to approximately 19%.

The larger the distance between the shadow mask and the surface, the higher is the blurring effect. The average distance between mask and surface has been evaluated to be 50.9 ± 9.5 μm. A testing of the reproducible positioning showed an average deviation of 70 μm. Three main challenges have been identified: the resolidified melt, the blurring, and the reproducible positioning. In the following sections solutions for each of these challenges are listed [23].

2.4.2.4.2 Improvement of the laser process

The laser process has been improved to enable a reduction of the melted and resolidified material and to create more homogeneous linear cutting edges. Within the first approach (Fig. 2.4.2.10A), the shadow masks have been

Fig. 2.4.2.10 (A) Initial situation; (B) reduction of melt by using picosecond laser; and (C) best results by using a picosecond laser, an optimized overlap, and optimized process parameters.

manufactured by using a microsecond laser. The use of a picosecond laser showed a significant reduction of the resolidified material as the thermal impact on the cutting edges is reduced (Fig. 2.4.2.10B).

A reduction of the waved and serrated cutting edges was achieved by increasing the pulse repetition frequency and reducing the feed rate. As a result, the pulse overlap is increased (Fig. 2.4.2.11) and the shape of the cutting edge is improved to a nearly linear shape. A further reduction has been achieved by using a setoff in the feed direction for every new scan cycle. This was realized by preventing synchronization between the feed rate and the laser trigger within the controller. The result is shown in Fig. 2.4.2.10C.

For manufacturing the lasered shadow masks, a beam source from Coherent (type: RAPID) has been used with a wavelength of 532 nm and a focus diameter of approximately 27 μm. Optimized results have been achieved by using the following process parameters: The masks have been lasered with a pulse duration of 12 ps and a pulse repetition frequency of 100 kHz with an average power of 0.9 W. For a complete lasering of a mask a repetition rate of 550 is required. By using the optimized parameters the average deviation has been reduced from 55% to approximately 4% compared to the wanted structural width. The size of

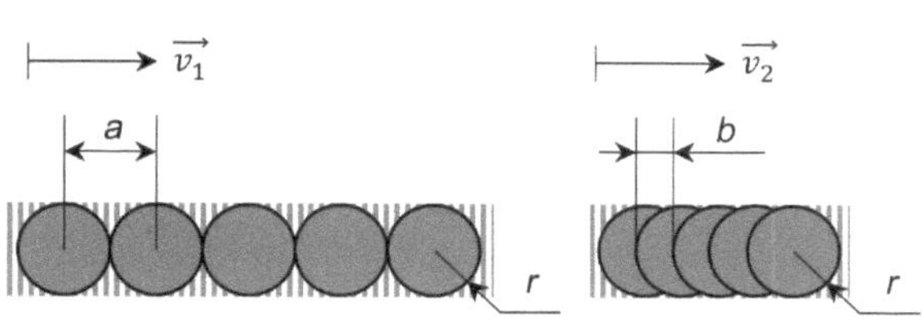

Fig. 2.4.2.11 Influence of the feed rates $\vec{v_1}$ and $\vec{v_2}$ $\left(\vec{v_2} < \vec{v_1}\right)$ on the overlap and the quality of the cutting edges with a and b distance of center points of the beam and r radius of the beam. Shaded are the expected shape of the lasered cutting edges [23].

the maximum deviations has not been reduced and equals approximately 70 μm. However, the amount of maximum deviations has been reduced by 90%, resulting in a more homogeneous cutting edge (compare Fig. 2.4.2.10C). The minimum structural width with respect to the used laser beam source and setup is approximately 50 μm.

This has two reasons: The first reason is the limited repeating accuracy of the galvanometer scanner. The second reason is the remaining material. As we are using contour lasering, material is left inside the lasered opening and has to be removed manually by using tweezers or ultrasonic cleaning. Investigations showed that underneath a structural width/length shorter than 50 μm the removing is impracticable [23].

2.4.2.4.3 *Optimization of the blurring*

The effect of the lateral blurring and lateral enlargement of the deposited layers compared to the actual size of the shadow masks is not erased by using the optimized laser process parameters. As already mentioned, the lateral enlargement is increasing with an increasing distance between the mask and the surface. This means a reduction can be achieved by reducing the distance between the mask and the surface. During research, various mask fixation techniques have been tested and evaluated. Best results have been achieved by using a mask that is fixed in a convex shape on a carrier and pressed onto the surface (Fig. 2.4.2.12).

Measurements of convex-shaped shadow masks pressed onto technical surface have been carried out with a confocal microscope. Assuming a constant material thickness of the mask, the distance between the surface and the mask has been reduced from 50.9 ± 9.5 μm to 29.2 ± 12.2 μm. This equals a reduction of 43%. By using the convex-shaped shadow masks, the structural enlargement of the deposited layers has been reduced from 19% to 11% [23].

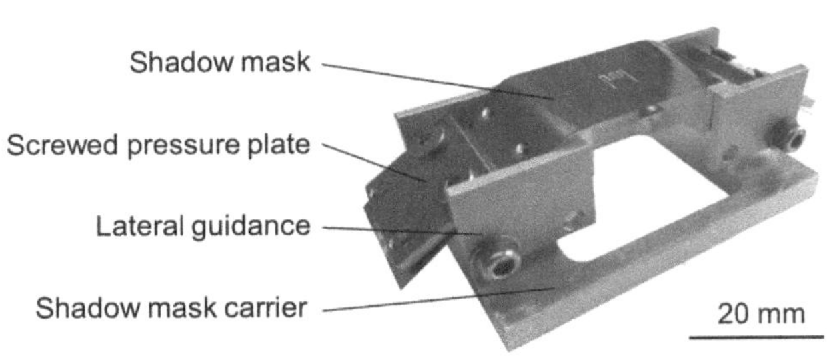

Fig. 2.4.2.12 Carrier with a fixed convex-shaped shadow mask [23].

2.4.2.4.4 Reproducible positioning of shadow masks

A high reproducible positioning of the shadow masks is required, especially in order to build multilayer systems. For this reason, a shadow mask carrier has been developed (Fig. 2.4.2.13). Such carrier enables a fixation of the mask as well as a reproducible positioning by using three locating pins. The structuring process is as follows: In the first step, a lasered shadow mask blank is fixed with four screws and two pressure plates on the carrier. The carrier is formed in such a way that the mask is fixed in a convex shape. In the second step, the carrier is placed upside down in an adapter, centered by the three locating pins and pressed onto a surface. Afterward, the laser structuring takes place. During the laser process the mask is centered by using the three locating pins and remains pressed onto the surface. This reduces structural deviations and positioning deviations.

Challenging are meander- or finger-shaped structures as they are springing back when pressed onto a surface. This results in a higher distance between the mask and the surface causing blurring. One solution is to divide the structure into two structures. These two structures are deposited after each other by using two masks, forming the required original structure (compare Section 2.4.2.6 and Fig. 2.4.2.20). The masks have to be cleaned on a regular base in order to use them for further deposition processes. It is not possible to set a specific time frame for the cleaning, as depositions/contaminations on the mask are depending on different factors, like the sputtered material or sputter power. For chromium, for instance, the layer adhesion is poor, resulting in high quantity of particles that have to be removed. It is recommended to use cotton swabs and a cleanroom wiper for cleaning in combination with isopropyl. The lasered structures should not be touched as they deform very easily. Moreover, particles should not be swept between the mask and the mask carrier as this might cause unwanted liftings. Investigations showed that the size reduction of the shadow mask structures, as a result of the deposition of material, is negligible.

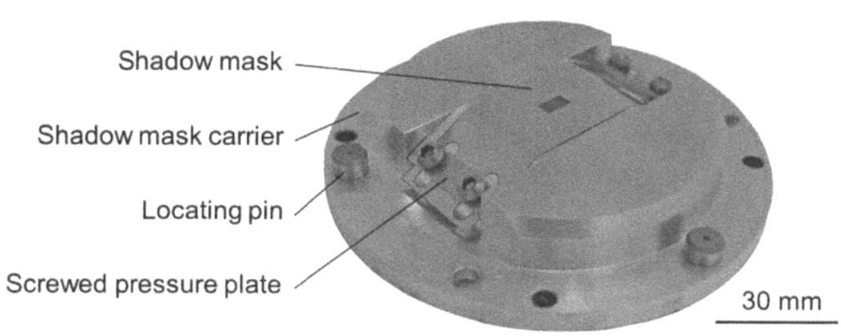

Fig. 2.4.2.13 Shadow mask carrier with fixed shadow mask and locating pins for a reproducible positioning.

2.4.2.5 The new sputter coating system

A new mobile coating system has been invented to enable a deposition of sensors and layers directly onto technical surfaces of arbitrary size. The system is working on the very promising principle of sputter deposition. As mentioned in Ref. [11] sputtering was first reported in 1852 [25]. Nowadays, magnetron sputtering is widely used in research, competing with the most important pulsed laser deposition techniques [11,26]. Many review articles [27,28] and literature [11,29,30] are available. Compared to the variety of other coating processes, sputtering has the advantage that the properties of the deposited layers are adjustable in a wide range depending on the deposition parameters like pressure or temperature. Moreover, the high kinetic energies of the target atoms ensure a good layer adhesion. Deposited layers are usually dense and defect free. Further advantages are the high homogeneity, the high purity, and the ability to adjust the layer thickness precisely and reproducibly. Furthermore, a variety of materials can be deposited, such as electrically conductive, nonconductive, and high-melting materials. The low thermal impact of the process on the substrate material allows coating a high number of materials such as metals, alloys, or polymers. Reactive sputtering allows to use metal targets for the deposition of compound semiconductors or insulators [31]. In addition, the stoichiometry of such thin layer can be adjusted. Moreover, sputtering is possible independently of the orientation of the system and it is relatively insensitive to vibrations and other external influences enabling a use in industrial production lines. Within commercially available sputter coating systems, components have to be inserted into a vacuum chamber to enable a sputter deposition. The size of such chamber is limiting the maximum size of the components on that functional layers are deposited. The new system is overcoming this problem, as the system is simply put on the surface of the component, making the surface one side of the chamber [32]. As a result, the system enables a deposition on components of arbitrary size. In cooperation with an industrial partner, a prototype was built [33].

2.4.2.5.1 The general setup of the sputter coating system

The sputter coating system is consisting of two main parts: the control station and the chamber rack (see Fig. 2.4.2.14). Both components are connected with a 3-m drag chain. The system can be used outside a clean room facility, for instance, directly in an industrial environment.

For depositing functional layers, the system is placed on a component by using a crane or robot. Alternatively, the system is put on a table and a component is placed in front of it. The deposition process is working independently of the orientation of the chamber rack. The system enables to deposit electrically insulating materials or conductive materials with a maximum diameter of 60 mm by using radio frequency (RF) magnetron sputtering or direct current (DC) magnetron sputtering with a maximum power of 300 W (RF-power) and 500 W (DC-power).

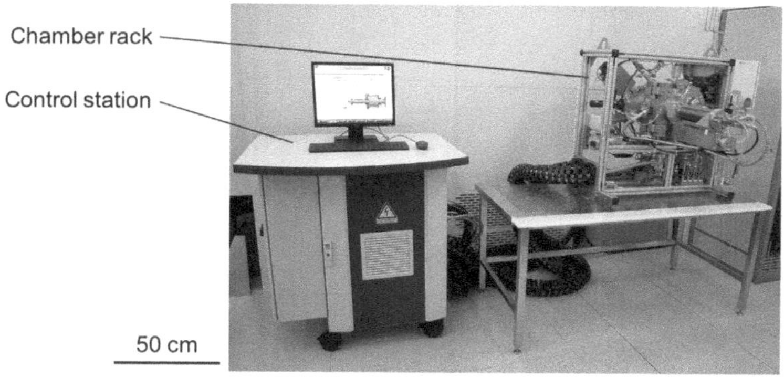

Fig. 2.4.2.14 The new sputter coating system is consisting of the control station and the chamber rack connected with a 3 m drag chain.
Modified from D. Klaas, J. Becker, M.C. Wurz, J. Schlosser, M. Kunze, New coating system for direct-deposition of sensors on components of arbitrary size, in: Proceedings: IEEE SENSORS 2016, October 30–November 2, 2016, Orlando, Florida, USA.

2.4.2.5.2 Setup of the control station

The control station includes nearly all electrical and mechanical components required to operate the system, such as two rotary vane pumps, power supplies, fuses, the RF- and DC-generators, the controller of the match box, a computer, and the I/O system. Moreover, the chamber rack is connected to the nitrogen, oxygen, power, and cooling water supply. The control station supplies the chamber rack with power, cooling water, compressed air, and vacuum from the rotary vane pumps. Just the process gases nitrogen, oxygen, and argon are directly connected to the chamber rack. The dimensions (height/width/depth) of the control cabinet are 1.1 m/1 m/0.9 m and the weight equals 250 kg. The chamber rack has dimensions of 0.9 m/1.1 m/1 m and a weight of 180 kg.

2.4.2.5.3 Setup of the chamber rack

The chamber rack is consisting of two chambers, process chamber 1 and process chamber 2, as shown in Fig. 2.4.2.15. Both chambers are separated by a vacuum gate valve. To enable a sufficient sealing of the vacuum chambers to the surface, an adapter plate is attached to process chamber 2. Such adapter plate can be replaced by other adapter plates to enable a deposition on nonplanar surface geometries.

Process chamber 1 is closed with a door at the front side of the system. The sputter source is screwed to the z-axis manipulator that is attached to the door. Such manipulator allows lowering the sputter source through process chamber 1 and 2 directly onto the technical surface. A structured deposition is possible by using a shadow mask. Such shadow mask is fixed in a shadow mask carrier (compare Fig. 2.4.2.13) that is attached to the sputter source. Alternatively, an unstructured deposition is possible by using a centering adapter in front of the

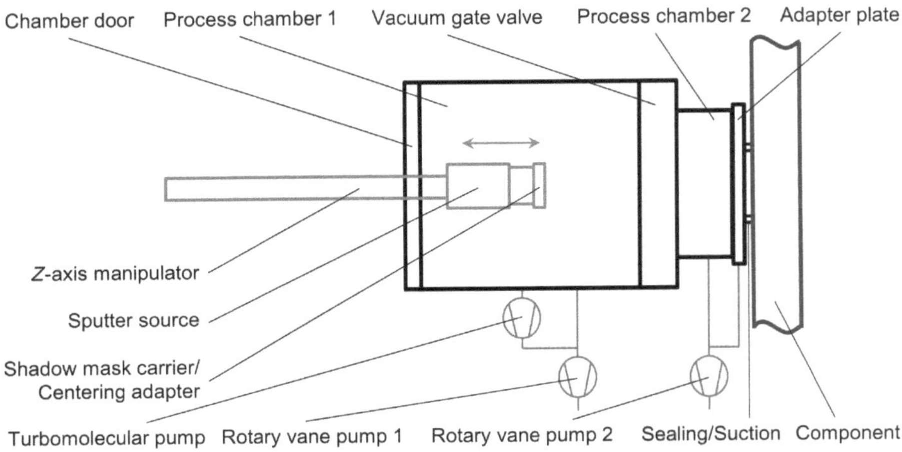

Fig. 2.4.2.15 Schematic of the new sputter coating system: The process chambers 1 and 2, the z-axis manipulator with sputter source, the vacuum circuit, and the component.

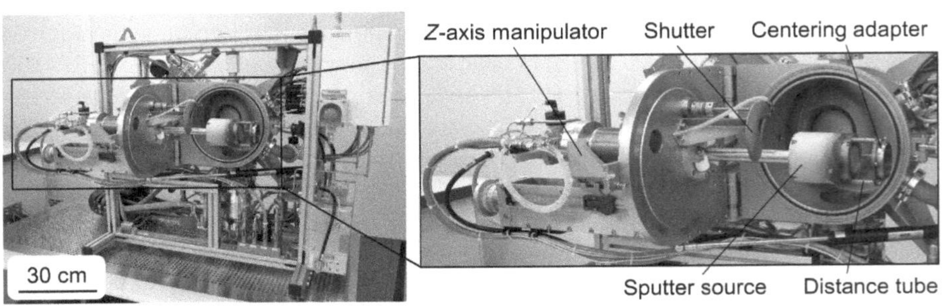

Fig. 2.4.2.16 Open process chamber 1 with detailed view on retractable sputter source with installed centering adapter.

sputter source instead of the shadow mask carrier (compare Fig. 2.4.2.16). The two-chamber system allows to open process chamber 1 for a change of the target or the shadow mask carrier/centering adapter without a complete vacuum loss in process chamber 2.

2.4.2.5.4 The process chambers

The open process chamber 1 has an inner diameter of 260 mm and is shown in Fig. 2.4.2.16. A shutter in front of the target allows a cleaning of the target surface before the actual sputtering process. The centering adapter/shadow mask carrier is

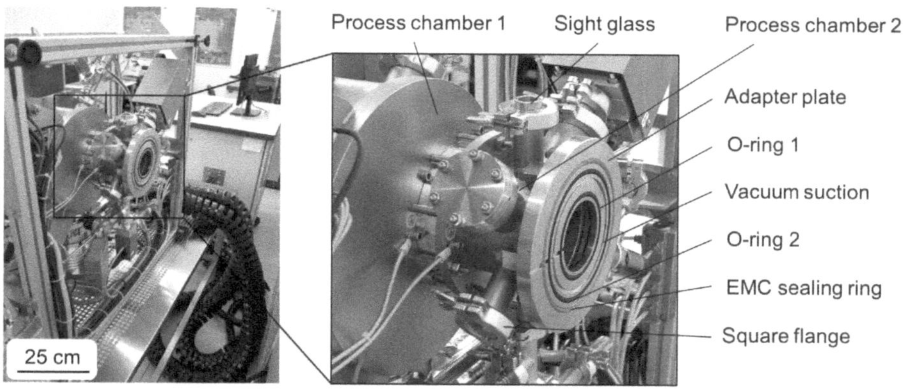

Fig. 2.4.2.17 View on the back side of the chamber rack with a detailed view on process chamber 2. The adapter plate is sealed with O-rings.

attached to the front of the sputter source. It enables a reproducible centering within the adapter plate, by using three locating pins.

The three locating pins are part of the centering adapter/shadow mask carrier (compare Fig. 2.4.2.13). Together with the center holes they are arranged in such a way that a contamination with sputter deposited material is avoided. The distance between the front side of the centering adapter/shadow mask carrier and the sputter target can be varied between 40 and 70 mm by using different distance tubes. The open process chamber 2 has an inner diameter of 100 mm and is located on the back side of the sputter system. A replaceable adapter plate is attached to process chamber 2 (see Fig. 2.4.2.17). The plate has two O-rings with an integrated vacuum suction between O-ring 1 and O-ring 2. This suction enables a better sealing of the chamber when the deposition takes place on surfaces with high roughnesses. An EMC sealing ring is used to fulfill electromagnetic compatibility (EMC) requirements and safety requirements. In order to preheat the surface of the component to enable, for instance, water desorption, a retractable lamp heater is attached to process chamber 2. In addition, process chamber 2 has a sight glass and a variety of square flanges.

2.4.2.5.5 Maintaining the vacuum

For maintaining the vacuum two rotary vane pumps and a turbomolecular pump are used (compare Fig. 2.4.2.15). Rotary vane pump 2 is attached to process chamber 2 and the vacuum suction of the adapter plate. Rotary vane pump 1 is a corrosive gas version and works as a backing pump for the turbomolecular pump that is attached to process chamber 1. During the deposition process, the turbomolecular pump is used for maintaining a vacuum within both process chambers.

2.4.2.5.6 *Operating the system and the deposition process sequence*

The system is operated with a computer located in the control cabinet. The user-friendly software gives an overview of the actual system status and allows to display and to set nearly all system parameters. New process sequences, so-called process recipes, are parameterized within the recipe editor, enabling the creation of cycles and subrecipes. The user is guided by instructions through the whole process. Each recipe is consisting of different recipe steps. One step can be a user instruction, such as "Open Shutter!," the switch-on of a rotary vane pump or a step that waits until a specific pressure is reached. Such steps are arranged after each other, forming a recipe.

After the sputter source is in position, the system is able to run in automatic mode requiring no interaction of the user. Moreover, a logging of all system parameters, databases for error, warning and status messages are implemented. In the following, a deposition process sequence is described. At the beginning of such a deposition process, chamber 1 is closed and evacuated. Process chamber 2 is also closed with a latch plate and evacuated. Moreover, a target is installed as well as a shadow mask carrier with a shadow mask. The shutter is closed, the vacuum gate valve is closed, and the sputter source is fully retracted and within process chamber 1. The deposition process sequence is presented in Table 2.4.2.1. After the deposition process sequence, further steps might be required to finish the deposition process. Such steps might be, for instance, a target change or the change of one shadow mask carrier with another one.

2.4.2.5.7 *Characterization of the system*

For characterizing the layer homogeneity of the system, electrically insulating (aluminum oxide) and conductive materials (nickel-chromium) have been deposited by using RF- and DC-sputtering. The layer homogeneity has been evaluated to better than approximately 4% (sigma/mean) in a square of 20×20 mm [33]. The evacuation/venting times of process chambers 1 and 2 to the required base pressure (freely selectable pressure before sputtering) and atmospheric pressure are listed in Table 2.4.2.2. The user is asked to open the vacuum gate valve between the process chambers when the pressure in chamber 1 equals 2×10^{-3} Pa and within chamber 2 at least 5 Pa. These values can be changed if required.

2.4.2.6 Manufacturing and characterization of sensors

For depositing strain gauges a full-bridge like structure, consisting of four quarter-bridges, has been developed (Fig. 2.4.2.18). The dimensions of the whole sensor structure are 15×15 mm. One single quarter-bridge is consisting of nine lines, each 150 μm in width, 4.95 mm long, and approximately 350 nm thick.

Table 2.4.2.1 **Deposition process sequence (compare [33])**

Step	Task
1	Vent and open process chamber 1.
2	Change target and install required shadow mask carrier/centering adapter.
3	Close and evacuate process chamber 1.
4	Vent process chamber 2 and remove latch plate.
5	Place new component in front of adapter plate and evacuate process chamber 2.
6	Open and start process recipe. The user is guided by specific instructions.
7	System waits until pressure in process chamber 1 is lower than 2×10^{-3} Pa and in process chamber 2 lower than 5 Pa. Then, open shutter and vacuum gate valve.
8	Lower sputter source until shadow mask carrier/centering adapter is in contact with component.
9	Confirm that sputter source is lowered to start sputtering. Wait until deposition is finished.
10	Retract the sputter source out of chamber 2 back to chamber 1 until the initial position is reached.
11	Close vacuum gate valve between both chambers.
12	Close the shutter.

Table 2.4.2.2 **Evacuation and venting times of the process chambers**

	Pressure	Time
Process chamber 1	From atmospheric pressure to 2×10^{-3} Pa	260 s
	Venting after sputtering	285 s
Process chamber 2	From atmospheric pressure to 5 Pa	360 s
	Venting after sputtering	40 s

Contact pads at the end of each sensor structure can be used for contacting each sensor element. The width of the connections to the contact pads and the connection arcs are enlarged to reduce cross-sensitivity. The basic resistance R of a sensitive line of one sensor is depending on the material and layer thickness and can be calculated with Eq. (2.4.2.1) in Fig. 2.4.2.19.

The use of four quarter-bridges has several advantages. The resistance of a sensor can be measured separately to get information about the resistance equality of all four sensors and the process homogeneity. Moreover, it is possible to characterize the sensor properties of four quarter-bridges manufactured in one single step. This is reducing the amount of required depositions. Moreover, the sensors can be connected to a full-bridge in two ways: externally or by using a sputtered connection layer, structured

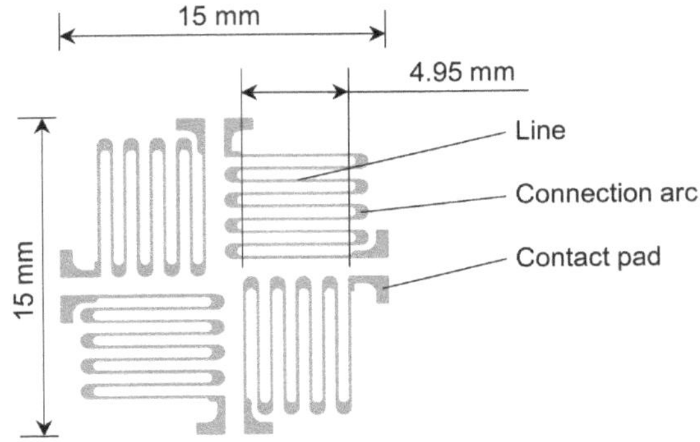

Fig. 2.4.2.18 Developed full-bridge like structure of the strain gauge consisting of four single quarter-bridges.

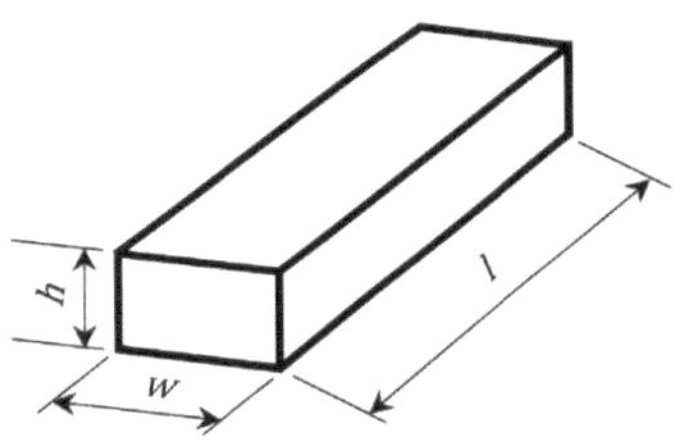

$$R = \rho \frac{l}{w\,h} \qquad (2.4.2.1)$$

ρ: Specific resistance
w: Width of line
h: Thickness of line
l: Length of line

Fig. 2.4.2.19 The basic resistance of one single line of the quarter-bridge can be calculated with Eq. (2.4.2.1).

with another shadow mask. When the complete sensor structure is sputtered with a shadow mask layout as shown in Fig. 2.4.2.18, the result is an enlarged sensor structure. This enlargement is a result of a gap between the mask and the surface causing the so-called blurring (compare Section 2.4.2.4.3). In this specific case, especially the long fingers from the meander structure are springing back. To overcome this problem, the sensor structure is divided as shown for one of the four quarter-bridges in Fig. 2.4.2.20.

By using mask 1, the horizontal lines of the sensor are structured, by using mask 2 the connection arcs and the contact pads. A small overlap between the different masks ensures a sufficient contact. Another mask can be used for depositing a solderable material just on top of the contact pads.

The shadow mask is made of stainless austenitic sheet steel (1.4310) with a thickness of 50 μm and is fixed in a convex shape in a carrier (compare Fig. 2.4.2.13). The lasering of the structure takes place as described in Section 2.4.2.4.2. This means the carrier with the mask is placed upside down in an adapter, centered by three

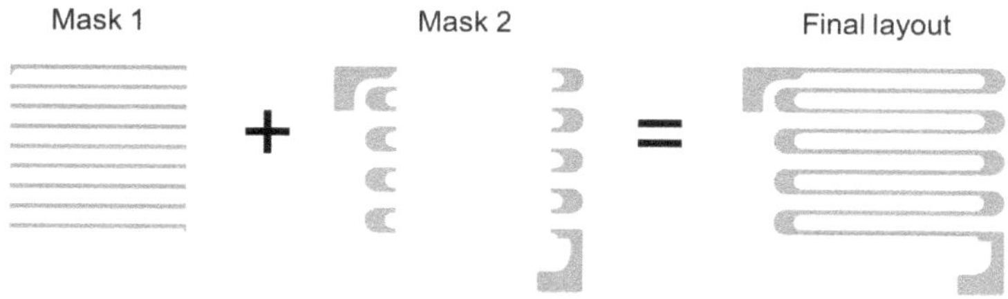

Fig. 2.4.2.20 Schematic example: the sensor structures are divided in different substructures (mask 1 and mask 2) to reduce blurring.

locating pins and pressed onto a surface. Afterward, the lasering starts with the optimized parameters.

The ability of depositing workable sensors directly on technical surfaces has been proved successfully by depositing first test sensors, in this case strain gauges. The technical surface of a 150-mm squared sheet of aluminum (AlMg3) with a thickness of 5 mm has been used as a test substrate. The surface has been polished to a surface roughness of approximately $R_a = 0.2$ μm and $R_z = 1.3$ μm and cleaned with acetone and isopropyl. An approximately 1.3-μm-thick Al_2O_3 layer has been used to insulate the sensor from the aluminum surface. The deposition parameters of such layer equal those described under Section 2.4.2.3. As a layout for the sensor, the structure and dimensions shown in Fig. 2.4.2.18 are used, structured with two masks (compare mask 1 and mask 2 in Fig. 2.4.2.20). As a material for the sensing layer chromium has been used and deposited with the following parameters: power of 200 W, argon flux of 7.5 sccm, sputtering time 16 min, and a pressure of 2.45×10^{-1} Pa before sputtering in process chamber 2.

A deposited sensor is shown in Fig. 2.4.2.21A. A quarter-bridge has been contacted with conductive liquid silver paint. Fig. 2.4.2.21B shows the resistance change from the gained signal for two different deflections.

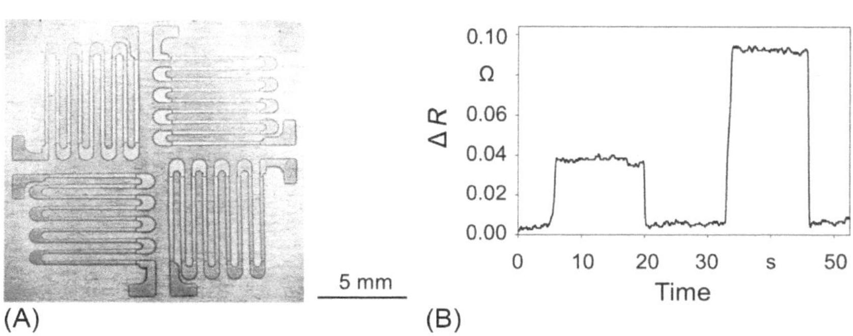

Fig. 2.4.2.21 (A) Sensor deposited on aluminum substrate and (B) gained signal from an evaluated sensor.

The results are proofing the functioning and the high potential of the new coating system. The overall sensor thickness is thinner than former sensors manufactured on carrier substrates. In combination with the arising new application fields, there is a need for further research. The research fields are wildly spread, from characterizing the system to improving various system functions. Depending on the application, it might be useful to investigate the deposition of smaller sensor structures by using, for instance, the lasered shadow masks or by developing another shadow mask manufacturing technology. Moreover, the deposited strain gauges should be further characterized and improved. Especially, the k-factor, that describes the sensitivity of the strain gauge, is of interest.

References

[1] K. Iniewski, Smart Sensors for Industrial Applications, CRC Press, Hoboken, 2013.
[2] W. Menz, O. Paul, J. Mohr, J. Mohr, Mikrosystemtechnik fuer Ingenieure, third ed., Wiley, Somerset, 2012.
[3] H.H. Gatzen, S. Hansen, L. Rissing, A concept for a toolbox of components for modular magnetic microsensors, ECS Trans. 3 (25) (2007) 235–246.
[4] S. Hansen, L. Rissing, H.H. Gatzen, Modularly designed magnetoelastic micro strain gauge, IEEE Trans. Magn. 43 (6) (2007) 2385–2387.
[5] T. Griesbach, L. Rissing, H.H. Gatzen, Concept for the integration of a temperature sensor in the modular, multifunctional micro sensor family, in: Proceedings: 4th Virtual Conference of the EU-funded FP6 I*PROMS Network of Excellence on Innovative Production Machines and Systems, I PROMS 2008, Wales, UK, 2008, pp. 472–476.
[6] T. Griesbach, M.C. Wurz, L. Rissing, Development and fabrication of modular micro sensors on flexible polymer foils, in: Proceedings: 1st Joint International Symposium on System-Integrated Intelligence: New Challenges for Product and Production Engineering, SysInt 2012, Garbsen, Germany, 2012, pp. 66–68.
[7] T. Griesbach, M.C. Wurz, L. Rissing, Modular eddy current micro sensor, IEEE Trans. Magn. 47 (10) (2011) 3760–3763.
[8] T. Griesbach, Technologieentwicklung zur Fertigung von Mikrosensoren auf flexiblen Trägersubstraten, Dissertation Leibniz Universität Hannover, PZH-Verlag, TEWISS - Technik und Wissen GmbH, Garbsen, 2013.
[9] D. Klaas, Slimming down sensors, in: phi - Produktionstechnik Hannover informiert, newsletter no. 9, December 2015. *http://en.phi-hannover.de/research/article/detail/slimming-down-sensors/* (accessed 12.01.16).
[10] Y. Zhang, C. Liu, D. Whalley, Direct-write techniques for maskless production of microelectronics: a review of current state-of-the-art technologies, in: ICEPT-HDP 2009, Proceedings: International Conference on Electronic Packaging Technology & High Density Packaging, Aug 10 - 13 2009, Beijing, China, 2009.
[11] R.F. Bunshah, Handbook of Deposition Technologies for Films and Coatings: Science, Technology and Applications, second ed., Noyes Publications, New Jersey, 1994.
[12] H. Farahani, R. Wagiran, M.N. Hamidon, Humidity sensors principle, mechanism, and fabrication technologies: a comprehensive review, Sensors (Basel, Switzerland) 14 (5) (2014) 7881–7939.

[13] Z.M. Rittersma, Recent achievements in miniaturised humidity sensors—a review of transduction techniques, Sensors Actuators A Phys. 96 (2–3) (2002) 196–210.
[14] D. Weber, Technische Feuchtemessung in Gasen und Festkörpern, first ed., Vulkan-Verl, Essen, 2002.
[15] D. Klaas, M. Rein, M.C. Wurz, L. Rissing, Development of a humidity sensor element based on sputter-deposited thin ZnO-layers, Procedia Technol. 26 (2016) 27–34.
[16] Verein Deutscher Ingenieure - Verband der Elektrotechnik, Elektronik und Informationstechnik, Measurement of gas humidity - Methods of measurement, Beuth Verlag GmbH 07.030, 2013.
[17] N. Fujimura, T. Nishihara, S. Goto, J. Xu, T. Ito, Control of preferred orientation for ZnOx films: control of self-texture, J. Cryst. Growth 130 (1–2) (1993) 269–279.
[18] H. Morkoç, Ü. Özgür, Zinc Oxide: Fundamentals, Materials and Device Technology, Wiley-VCH, Weinheim, 2009.
[19] D. Klaas, P. Taptimthong, L. Jogschies, L. Rissing, Component integrated sensors: deposition of thin insulation layers on functional surfaces, Procedia Technol. 15 (2014) 114–121.
[20] D.M. Mattox (Ed.), Handbook of Physical Vapor Deposition (PVD) Processing, second ed., William Andrew, Oxford, UK, 2010.
[21] A.T. Güntner, V. Koren, K. Chikkadi, M. Righettoni, S.E. Pratsinis, E-Nose sensing of low-ppb formaldehyde in gas mixtures at high relative humidity for breath screening of lung cancer? ACS Sens. 1 (5) (2016) 528–535.
[22] J. Heo, H. Min, M. Lee, Laser micromachining of permalloy for fine metal mask, Int. J. Precis. Eng. Manuf.-Green Tech. 2 (3) (2015) 225–230.
[23] D. Klaas, J.F. Duesing, M.C. Wurz, L. Rissing, Verwendung von Schattenmasken zur Direktstrukturierung individuell adaptierbarer Sensorik auf technischen Oberflächen, in: V. Saile (Ed.), MEMS, Mikroelemente, Systeme; in: Proceedings: MikroSystemTechnik Kongress 2015, October 26–28, 2015, Karlsruhe, Germany, ISBN: 9783800741007, 2015.
[24] S.-J.J. Lee, N. Sundararajan, Microfabrication for microfluidics, Artech House, Boston, 2010.
[25] W.R. Grove, On the electro-chemical polarity of gases, Philos. Trans. R. Soc. Lond. 142 (1852) 87–101.
[26] P.M. Martin (Ed.), Handbook of Deposition Technologies for Films and Coatings: Science, Applications and Technology, third ed., Elsevier, Amsterdam, 2010.
[27] P. Kelly, R. Arnell, Magnetron sputtering: a review of recent developments and applications, Vacuum 56 (3) (2000) 159–172.
[28] U. Helmersson, M. Lattemann, J. Bohlmark, A.P. Ehiasarian, J.T. Gudmundsson, Ionized physical vapor deposition (IPVD): a review of technology and applications, Thin Solid Films 513 (1–2) (2006) 1–24.
[29] J.L. Vossen, W. Kern, Thin Film Processes, Academic Press, New York, 1978.
[30] H.S. Nalwa, Handbook of Thin Film Materials, Academic Press, San Diego, 2002.
[31] D. Depla, S. Mahieu, Reactive Sputter Deposition, Springer, Berlin, 2008.
[32] D. Klaas, Direct deposition of ultrathin sensor systems onto components of arbitrary size, in: B. Denkena (Ed.), Cross Linked Production - Research Results for Industrial Applications, PZH Verlag, Hannover, 2015.
[33] D. Klaas, J. Becker, M.C. Wurz, J. Schlosser, M. Kunze, New coating system for direct-deposition of sensors on components of arbitrary size, in: Proceedings: IEEE SENSORS 2016, October 30–November 2, 2016, Orlando, Florida, USA, ISBN: 9781479982875, 2016.

2.4.3 Laser patterning of thin film strain sensors on technical surfaces

Jan Friedrich Düsing, Jürgen Koch*, Oliver Suttmann*, Ludger Overmeyer†*

*Laser Zentrum Hannover e.V., Hannover, Germany, †Institute of Transport and Automation Technology, Leibniz Universität Hannover, Garbsen; Laser Zentrum Hannover e.V., Hannover, Germany

2.4.3.1 Introduction

The use of sensors and integration of data processing and communication systems in components or machinery is becoming more widespread in many technical branches of industry to offer new functionalities in products or enable more efficient production processes. The use of sensor systems on machine components calls for robust mounting and connection technologies so that reproducible and permanently stable measurements can be guaranteed over the lifetime of the machine or component. Thin film patterning with ultrashort pulsed lasers enables selective metal thin film structuring on stacked film systems, and thus allows for production of strain sensors directly onto the surface of the machine components. Using fast laser scanners, this method even extends the possibilities of conventional photolithographic patterning by the capability of manufacturing electronic structures on complex-shaped components with three-dimensional surfaces.

The layer structure of thin film strain sensors consists of insulation layers and sensor layer which are deposited onto the surface of machine components, as shown in Fig. 2.4.3.1. The insulation layer is produced by depositing oxide films on the

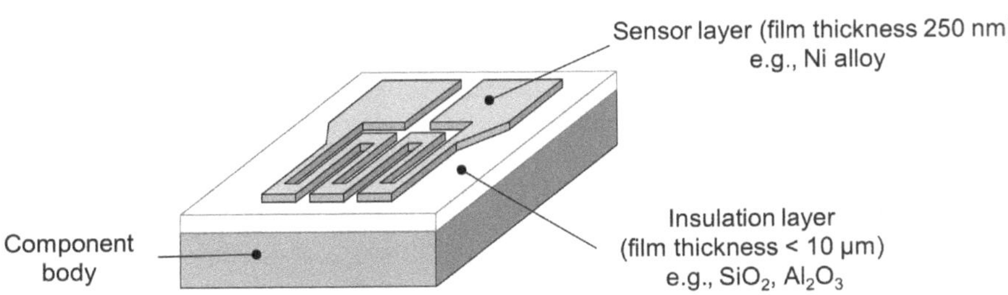

Fig. 2.4.3.1 The principle layer structure of thin film strain sensors.

measurement zone in order to provide a galvanic insulation between the sensor structure and component body. The sensor material used for the strain conductor paths is based on a nickel alloy which offers a small temperature coefficient of electrical resistance.

The function of the sensor element is based on the dependency of the electrical resistance on deformations of the component. The change in the resistance of the metal film is largely attributed to geometrical changes in the geometrical dimensions of the conductor paths. In order to measure small signals the sensor is typically a network of individual strain resistor grids with directional sensitivity. The resistor grids are interconnected to build a Wheatstone bridge configuration which allows for cancelation of common mode errors, such as uniform temperature change, and outputs a differential signal as a function of bridge unbalancing due to component deformation. The signal processing of bridge circuits with large amplification factors requires the individual resistors to be accurately balanced in order to minimize any intrinsic bridge offset.

The manufacturing chain for thin film strain sensors on component surfaces is shown in Fig. 2.4.3.2. The component is mechanically machined and the surface is prepared by cleaning processes. The pretreated surface is subsequently deposited by means of vacuum deposition processes in order to generate insulation and sensing layers on the surface. The sensor layer is then selectively laser processed to generate the electrical conductor paths. The sensor structure may require further laser trimming and is finally connected to the conditioning electronics. In the following, the deposition process and subsequent laser processing is discussed in detail.

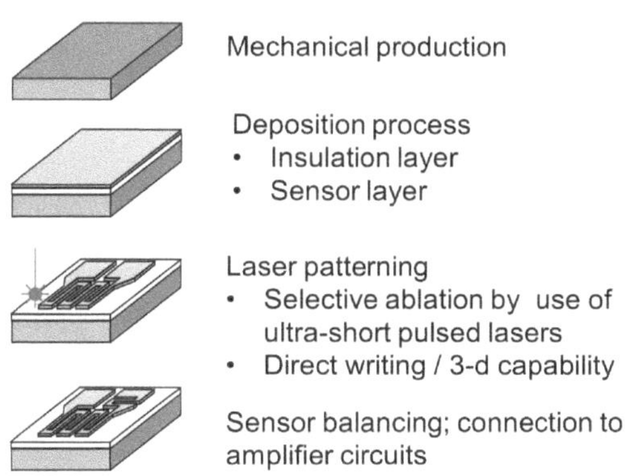

Fig. 2.4.3.2 The manufacturing chain of thin film sensors on component surfaces is based on layer deposition followed by laser patterning process.

2.4.3.2 Surface preparation and thin film deposition

The manufacturing of the thin film system is based on vacuum deposition techniques and aims to generate a thin film stack on the component surface. As an essential requirement for the functionality of the strain sensors the insulation film need to be free of pinholes or electrical contacts to the component body. Even small pinhole defects that would result in electrical bypass connections would render the entire sensor structure nonfunctional, and thus will lead to production rejects. Particles or contamination on the surface during the film process would result in film defects through the insulation layer by preventing a homogenous depositing of the film. Moreover, sharp edges or spikes in the surface can lead to electrical contacts if they are not fully covered by the insulation film. The surface preparation often includes physical or chemical processes, such as surface polishing or etching, respectively.

The components are deposited by vacuum sputter deposition processes. The choice of the deposition process depends on various parameters such as the film material, the tolerable temperatures on the component surface, or required deposition rates. A reliable deposition process for generating an insulation layer composed of Al_2O_3 can be realized by conventional RF sputtering [1,2]. Components with complex geometries, e.g., components of a bearing ring as shown in Fig. 2.4.3.3, require circumferential deposition of the films which is realized, e.g., by attaching the components to rotating spindle substrate holders. In order to generate films that are free from pinholes, the deposition rate needs to be kept at a low level of some 100 nm/h, which results in long process times. During the deposition process temperatures may reach values above 200°C, which can be a limiting factor for temperature-sensitive components.

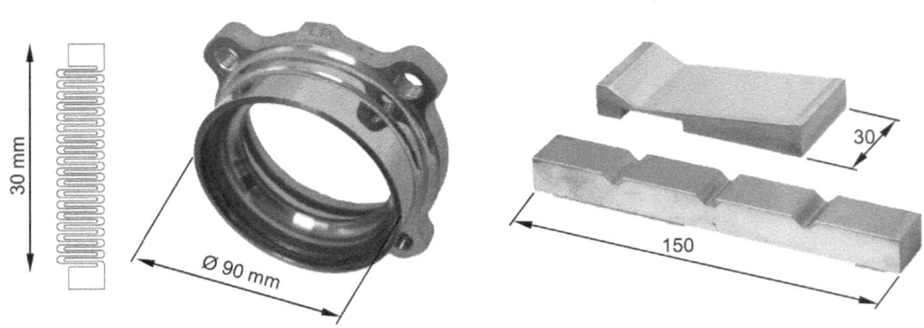

Fig. 2.4.3.3 Components of a bearing ring (left) and test specimen are equipped with thin film strain sensors.

Higher deposition rates are achievable by depositing oxide films via reactive sputter techniques, e.g., by the highly ionized pulse plasma (HIPP) processes [3] or reactive MF sputtering [4]. The deposition rates beyond 10 nm/min are reported. During the process a reactive gas is injected into the processing chamber in order to induce the chemical reaction. To obtain the desired film composition the process requires precise control of critical parameters of the chemical reaction, such as gas flow rate or chamber pressure. Generally, the deposition of defect-free insulation layers with a thickness of several micrometers is a major cost factor in the manufacturing chain of thin film strain sensors.

On complex component surfaces such as that of the wheel bearing ring the deposition rate and hence the resulting film thickness will depend on the motion profile of the component in the deposition chamber. The homogeneity of the film on the surface is influenced by the shape of the component which induces shadowing effects or field disturbance. The optimization procedure of a deposition process can be supported by numerical simulations. For a qualitative analysis the deposition process can be approximated by assuming that particles emitted by the sputtering target are accelerated on a linear trajectory toward the component surface. The particle density on the component surface then depends on the area and emission characteristics of the target as well as the surface geometry of the component including any edges and steps. As shown in Fig. 2.4.3.4, a numerical simulation using the surface geometry of the bearing ring reveals that a considerable lower film thickness can be expected close to the mounting flange of the ring. The flange partially shadows the particle trajectory even if the component is constantly rotated in the vacuum chamber. Thus the deposition rate is locally reduced in this area which results in a reduced film thickness.

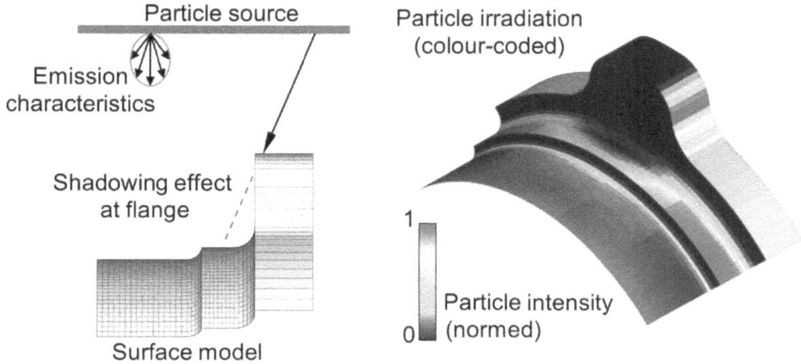

Fig. 2.4.3.4 A numerical simulation of the film deposition process indicates that the film thickness is reduced near the mounting flange of the bearing ring due to shadowing effects.

In order to verify the simulated film thickness distribution on the component the actual film thickness has been measured on test components with high spatial resolution. The insulation layer (Al_2O_3) is optically transparent and can be measured by optical interference analysis. The insulation film is probed with a measuring grid placed circumferentially on the ring. The analysis of the nickel alloy film is based on measuring the sheet resistance (Van Der Pauw method). Cross-like test structures are laser patterned in a measurement grid and probed with electrical contact needles. The resulting sheet resistance of the nickel alloy film allows for conclusion of the film thickness, assuming that the specific resistance of the film is uniform throughout the layer.

In Fig. 2.4.3.5 the results of the film thickness measurement are depicted. Here the measured thickness distribution is shown as a color-shaded image which is mapped onto the corresponding position of the CAD component model. For both deposition processes the film thickness distribution is generally symmetrical on the circumference but shows a significant reduction of the thickness near the mounting flange. This observation qualitatively confirms the findings from numerical simulation.

For the use in the sensor application a film thickness gradient in the insulation layer may not influence the intended insulation function as long as the film is free of defects. However, a thickness gradient in the sensor layer directly results in varying film sheet resistance. Thus the film thickness distribution in the sensorial layer needs to be considered during the manufacturing process, e.g., by adjusting the sensor design or by compensating of these effects during the structuring

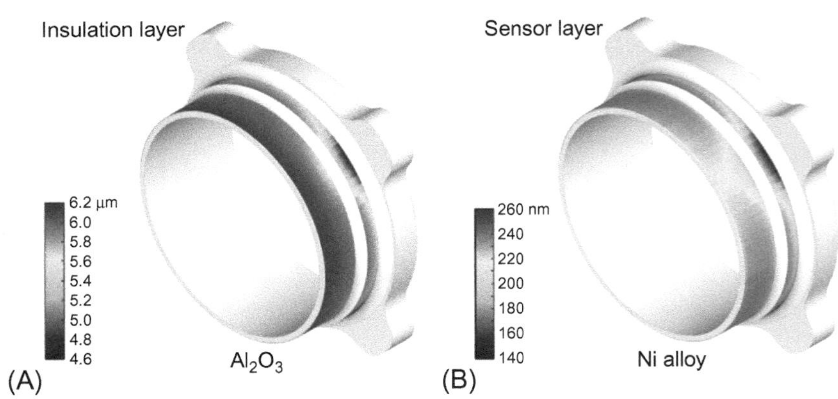

Fig. 2.4.3.5 The measured film thickness distribution on a test bearing ring (A) of the insulation layer measured by optical interference analysis and (B) of the nickel alloy film measured electrically by the Van-Der-Pauw method. Both films show reduced thickness near the mounting flange which is in good agreement with the numerical simulation.

process. Any effects induced by the nonuniform film thickness can be compensated during laser patterning by a laser trimming step. If the application requires manufacturing of sensor bridge networks the sensor trimming can be combined with a bridge balancing step.

As described in the following section, a laser process allows for a combined sensor structuring and bridge balancing process in a single machining setup and is capable to generate high-precision bridge circuits on complex-shaped 3D component surfaces.

2.4.3.3 Laser structuring and bridge balancing

The laser structuring is based on film removal of the sensor layer induced by irradiation of short laser pulses [5–8]. The thin film ablation process exhibits a distinct threshold behavior. The ablation threshold of the film is defined as the critical fluence (optical energy per area per pulse) which results in film removal within the irradiated spot area. If the fluence is too high significant damage can occur in the underlying insulation layer. Thus the process window for homogenous ablation is defined by the ablation threshold of the sensor layer (lower process boundary) and damage threshold of underlying films (upper process boundary). By use of ultrashort laser pulses in the picosecond and femtosecond range, high peak intensities in a locally confined volume of the sensor film are obtained which enables selective film removal without damaging the insulation layer.

Sensor structures are patterned by scanning the laser beam on the surface in order to ablate lines and area segments. High ablation rates, and therefore short processing times are achievable if the laser beam is scanned on the component's surface with high pulse repetition rate and large scanning speed. However, frequent changes in the scanning direction result in a slowdown of the patterning process due to limited acceleration rates of the scanning devices. High scanning dynamics can be achieved by using galvanometer scanner for lateral beam deflection (see Fig. 2.4.3.6). Due to the low mass inertia of the scanner mirrors scanning speed of up to 10 m/s is achievable. On a curved surface the laser focus needs to be adapted to the surface profile. A dynamic focus shifter is included into the machining setup in order to keep the laser focused on the surface on every position in the scanning range. On tilted surface elements the laser intensity is modulated to keep the peak intensity constant even at a large angle of incidence. The experimental setup used for development of the scanning process is shown in Fig. 2.4.3.7.

The beam modulators are controlled by synchronized signals generated by a real-time controller. Additionally to the laser systems the component may be moved by a stage system. This allows for processing of sensors on various positions on the component surface. In this procedure the component surface is positioned into the scan field of the laser setup prior to the laser process and stepped to the next position after

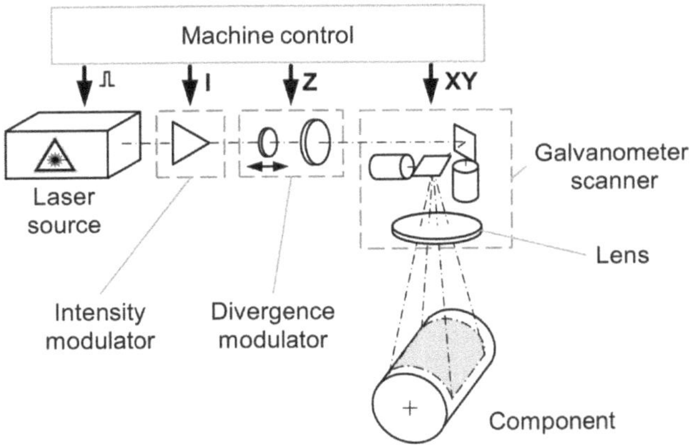

Fig. 2.4.3.6 Machining setup for processing of components with complex 3D surfaces.

Fig. 2.4.3.7 The experimental setup used to test laser scanning and trimming processes on the surface of bearing ring.

the sensor has been created (step-and-scan strategy). The combination of laser scanners and multiple axis stage systems enables short processing times due to the high dynamic of the scanner while maintaining a high degree of positioning freedom depending on the feasibilities of the stage system.

The principle data processing chain [8], as depicted in Fig. 2.4.3.8, processes input data from CAD models of the sensor layout and the component surface contour. The sensor layout can be modeled by a polyline object which is transformed

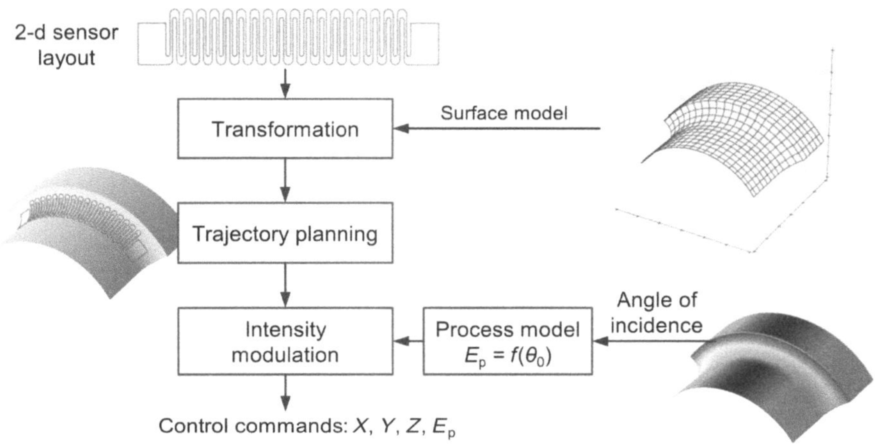

Fig. 2.4.3.8 Data flow graph for generation of machine commands based on CAD sensor and surface models as well as a process model for thin film ablation on tilted surfaces [8].

onto the model of the component surface. The resulting three-dimensional polyline represents the edge of the thin film sensor. The next step involves the generation of the actual focus trajectories in the machine coordinate system. The trajectory is generated by path offsetting of the sensor edge line. Depending on the electrical requirements of the sensor, a series of offset lines is generated to generate a large ablation line width.

The irradiation condition of the laser beam on the surface continuously changes while scanning the beam on the surface. In order to compensate for these nonsteady conditions the laser intensity is modulated synchronously to the scanner motion. The corresponding compensation factor for each surface position is implemented as a surface intensity distribution based on a function of the angle of incidence and laser polarization. The output of the CAM data processor is uploaded to the real-time controller and executed after the component is positioned into the scanner field.

The process model for the intensity compensation requires knowledge of optical parameters of the laser beam and the film material [5,9]. If the laser focus is irradiated on a tilted surface element the circular spot is projected to form an elliptically irradiated area on the surface. Additionally a fraction of the laser energy is reflected by the surface. The reflection coefficient depends on the angle of incidence and the polarization state of the laser beam with respect to the surface orientation as well as the optical properties and of the film system.

The ablation threshold of the Nickel alloy film as a function of angle of incidence is investigated experimentally, as shown in Fig. 2.4.3.9. At incidence

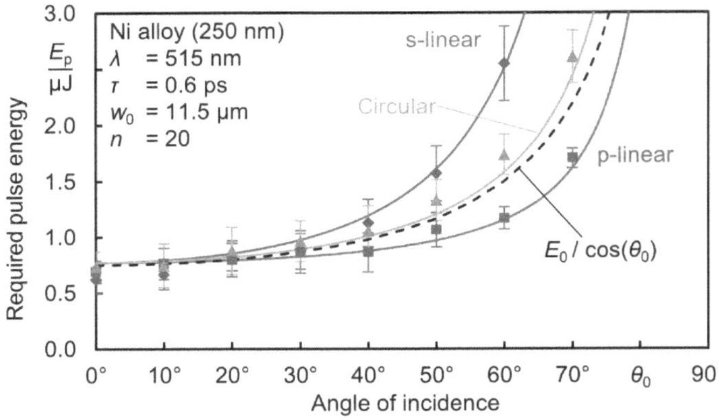

Fig. 2.4.3.9 Ablation threshold of the sensor film as a function of angle of incidence and laser beam polarization. A model based on Fresnel equations is plotted as solid lines [5].

angles less than approx. 30 degrees the angle dependency does vary significantly. However with increasing incidence angles the required pulse energy for film removal significantly increases mainly following a cosine model due to spot projection. This effect is superimposed by polarization dependency which is a result of angle-dependent reflection losses of the interface between thin film and ambient air. The behavior of the ablation process can be approximated by an analytical model (solid lines in Fig. 2.4.3.9) which includes a cosine term to account for spot projection as well as reflectivity coefficients which can be obtained by solving the Fresnel equations for the particular light material properties, e.g., the complex refractive index of the film material for the given laser wavelength and polarization state. The process model can thus be easily implemented into the CAM processor as an analytical function which is parameterized by fundamental laser and material properties.

Fig. 2.4.3.10 shows sensors that have been structured on curved surfaces of the demonstrator components [8]. The sensor structure is generated by scanning parallel to the contour of the layout. The processing time required to complete the structuring thus depends mainly on the length of the contour, the achievable scanning speed, and the number of contour cycles to achieve a required line width. For example, with a laser system delivering 100 kHz and a scanning speed of 100 mm/s a sensor layout as shown in Fig. 2.4.3.10A with a contour length of 330 mm scanned 6 cycles would by processed within 20 s/sensor. The process time can be further optimized by scaling the pulse repetition rate of the laser source and the scanning speed reach, respectively.

After completion of the first contour cut the electrical resistance of the sensor structure is subject to variations which can be attributed to the film properties of

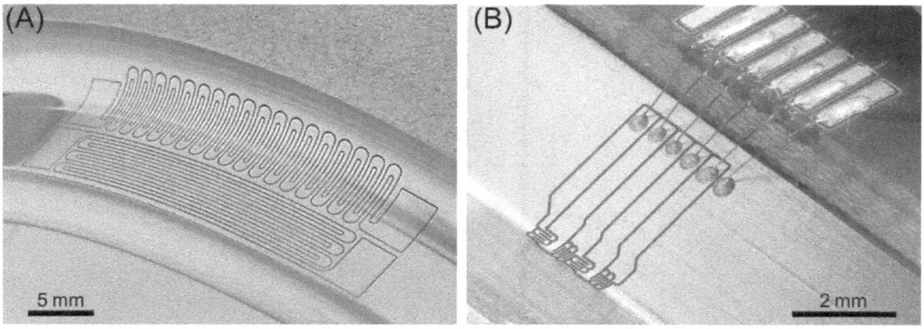

Fig. 2.4.3.10 Laser-patterned thin film sensors placed on the bearing ring (A) and into a v-groove of a component of the tool machine (B).

the Nickel alloy layer and to repeatability of the laser process. While on planar surfaces and uniform film thickness distribution the electrical resistance can be reproduced typically within less than 1% of the nominal sensor resistance, the variance of the electrical resistance on curved surfaces increases with the surface complexity. The key factor for variations on curved surfaces can be attributed to the variation of film thickness of the nickel alloy layer, as shown, e.g., in Fig. 2.4.3.5. The film thickness variations are originated by the variations of the film deposition rate as an effect of the surface shape and by fluctuations of the deposition parameters. Additional, but less significant, are variations induced by positioning errors of the motion and scanning system as well as deviations of the component geometry or clamping device.

Precise measurement of mechanical strain requires accurately balanced sensor bridges which allow signal conditioning with high amplification gain. Thus a laser trimming process is established in order to trim sensor structures or entire sensor bridges on the component surface [10,11]. The laser trim process is based on feeding the results of the laser process (electrical resistance values) back into the machine controller and iteratively process the sensor structure until a target objective is reached.

The feedback loop for sensor trimming [11], as shown in Fig. 2.4.3.11, includes electrical probes for temporarily connecting the thin film structure to a resistance measuring circuit. The circuit converts the signal into a (digital) signal for further data processing in the machine controller. A trimming controller is included into the data flow path of the trajectory generator. The beam path is iteratively adjusted for each sensor structure depending on the reading of the resistance values. The newly generated scan trajectory is then transferred to the real-time controller for execution of the next trim cycle. This trimming procedure is repeated until the objective criteria are reached, e.g., the bridge is balanced and the nominal target bridge resistance is adjusted.

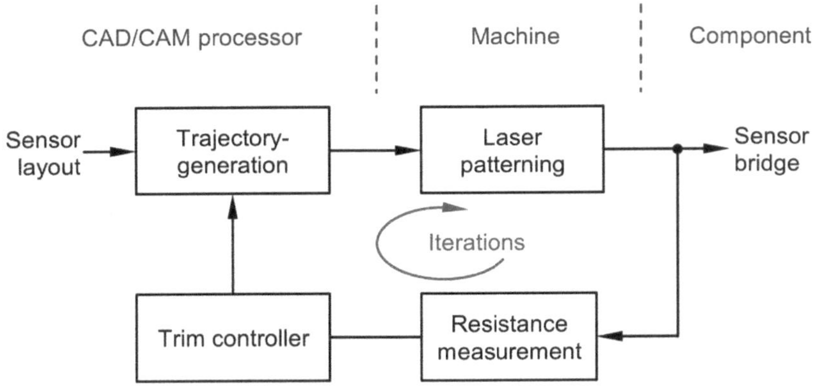

Fig. 2.4.3.11 The process strategy for sensor trimming is based on an iterative approach to process balanced sensor bridges based on a feedback of resistance measurements during the process cycles.

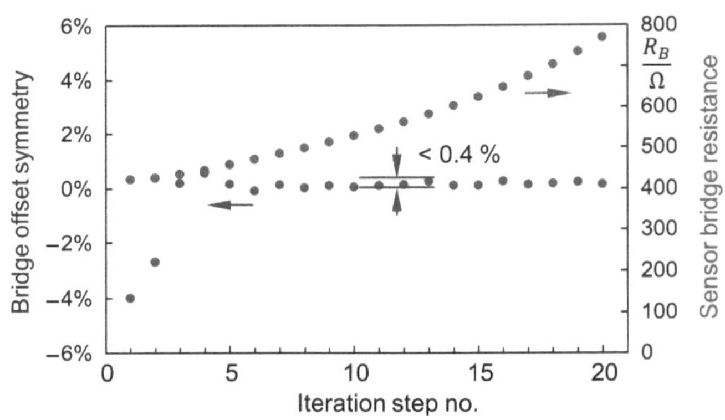

Fig. 2.4.3.12 Trimming of a sensor bridge. During the sequence of trim iterations the initial bridge offset rapidly vanishes within a few trim steps. The total bridge resistance gently increases until the trim process reaches the target objective [11].

The precision of the trimming depends on the trim controller parameters which results in a trade-off situation between remaining resistance variation after trimming and number of trim iterations to reach the final value. However, each trim cycle can be executed within a fraction of a second so that the overall process time mainly depends on the duration of the laser scan step. Fig. 2.4.3.12 displays a recording of a trimming procedure of a Wheatstone bridge network composed of four single resistors. The bridge offset is measured by exciting the bridge with a constant voltage and measuring the diagonal voltage. During this experiment each trim iteration step is scanned within 0.25 s. The bridge offset rapidly vanishes and remains balanced within the fluctuations induced by noise of the feedback loop. During each trim cycle a small amount of material is removed and throughout the course of the process the total bridge resistance iteratively increases. If monitored, this value can be used to define the stop criterion of the trim procedure, e.g., when the nominal bridge resistance is reached.

2.4.3.4 Conclusions

The combination of thin film deposition techniques and laser thin film processing shows potentials for manufacturing of electronic sensors directly attached to the component surface. This approach can be related to the conventional sensor manufacturing techniques on wafer or foil substrates. Wafer- or foil-based manufacturing of strain sensors is very cost effective due to the possible use of mass production technology. However, additional costs arise from the choice of technology to bond the sensor to the component surface. Also, the bonding technology induces further constraints, e.g., the limited temperature operating range of polymer adhesives, long-term stability, and applicability on complex surfaces.

A direct buildup of the sensor system on the component surface aims to overcome these limitations and provides an alternative approach to manufacture strain sensors on component surfaces in order to measure deformation, strain, and related mechanical quantities during the operation of the component. The proposed technology uses robust materials which forms a surface sensor system in very close contact to the measuring zone. By use of scanning laser beams the sensor networks can be accurately patterned und balanced even on complex surfaces over a large range of incidence angles of the laser beam with respect to the surface. The applicability of sensors on 3D surfaces is thus only limited by the accessibility of film deposition process and laser beam to the surface (line of sight).

However, along with the capability to integrate (passive) electronic sensor systems on the component surface new technological challenges arise on how these systems are connected to the higher-level information processing units. For example, new questions arise on where and how to "leave" the component surface, namely, how to connect the sensor to electronic conditioning circuits with the constraints of complex component geometries. Alternatively, one may think of new approaches to integrate data processing units into the surface by generating active electronic circuits on the component surface, and thus increase functionality of the "intelligent" components.

Up to now, the combination of vacuum deposition techniques to generate robust film layers in combination with laser processing for flexible and precise sensor structuring has already proven to be suited for a couple of new measuring applications. The applications may be found in various industry sectors, such as automotive, aerospace, mechatronics, medical engineering, or precision measurement. The technology may be beneficial, in particular, in applications which include at least one of the following aspects

- Small packaging volume of just a few micrometers on the surface.
- Use of highly stable materials in extended temperature ranges (e.g., avoidance of polymer materials).
- Placement of sensors on complex geometries, such as pockets, notches, or other concave/convex surfaces.
- Need for accurately balanced resistor networks in precise measurement applications.
- Ability to fully automate the production line.

References

[1] W.H. Ha, M.H. Choo, S. Im, Electrical properties of Al_2O_3 film deposited at low temperatures, J. Non-Cryst. Solids 303 (1) (2002) 78–82.

[2] M. Voigt, M. Sokolowski, Electrical properties of thin RF sputtered aluminum oxide films, Mater. Sci. Eng. B 109 (1–3) (2004) 99–103.

[3] H. Gerdes, R. Bandorf, D. Loch, G. Bräuer, Reactive sputter deposition of alumina coatings, Mater. Sci. Eng. 39 (2012) 012009.

[4] F.C. Carreri, R. Bandorf, H. Gerdes, M. Vergöhl, G. Bräuer, Highly insulating alumina films by a bipolar reactive MF sputtering process with special arc handling, Surf. Coat. Technol. 290 (2016) 82–86.

[5] L. Overmeyer, J.F. Düsing, S. Suttmann, U. Stute, Laser patterning of thin film sensors on 3-D surfaces, CIRP Ann. Manuf. Technol. 61 (1) (2012) 215–218.

[6] O. Suttmann, J.F. Düsing, J. Koch, L. Overmeyer, Laser patterning of thin film strain gauges, in: Proceedings of the 32th International Congress on Applications of Lasers and Electro-optics, ICALEO, Miami, USA, October 6–10, 2013, pp. 877–881.

[7] O. Suttmann, J.F. Düsing, J. Koch, U. Stute, L. Overmeyer, Patterning of thin film strain gauges on 3D-surfaces, in: Proceedings of the 16th International Conference on Sensors and Measurement Technology, SENSOR, Nurnberg, Germany, May 14–16, 2013, pp. 34–38.

[8] J.F. Düsing, T. Eichele, J. Koch, L. Overmeyer, Laser surface processing of integrated thin film systems on arbitrary shaped components, Procedia Technol. 15 (2014) 122–128.

[9] O. Suttmann, J.F. Düsing, L. Overmeyer, Phenomenological model for prediction of complex ablation geometries in metal films using ultrashort laser pulses, in: Proceedings of the 34th International Congress on Applications of Lasers & Electro-Optics, ICALEO, Altanta, USA, October 18–22, 2015, pp. 23–28.

[10] O. Suttmann, J.F. Düsing, J. Koch, U. Stute, L. Overmeyer, Patterning and trimming of thin film strain sensors, in: Proceedings of the 13th International Conference of the European Society for Precision Engineering & Nanotechnology, EUSPEN, vol. 2, Berlin, Germany, May 27–31, 2013, pp. 340–343.
[11] J.F. Düsing, J. Koch, O. Suttmann, L. Overmeyer, Closed-loop process control for laser patterning of thin film strain sensors, Procedia Technol. 26 (2016) 113–119.

2.5 Integrated communication systems

In this chapter it will be demonstrated how communication systems can be integrated into a component and how energy and information can be transmitted in the component's surface.

The introductory chapter describes the vision of the collaborative research center 653 (see Chapter 1). One basic idea is the fusion of components and sensing elements as well as component-specific information. In order to contribute to this achievement, the use of electromagnetic waves is investigated. Two developments are made to realize wired as well as wireless communication structures within the component (Fig. 2.5.1).

To enable the use of wireless communication, methods are developed which exceed the capabilities of standard radio frequency technologies such as RFID. For this purpose, a highly integrated module with a multilayer composition of optical and electrical components is investigated. The optical component is a highly efficient solar cell combined with an energy storage which guarantees an autonomous functionality. The data processing unit is an ultra-low-power microcontroller, which is also able to store data from integrated sensors. The wireless data exchange is accomplished by an optically transparent radio frequency circuit and antenna positioned on top of the solar cell. The developed wireless communication module ($13 \times 13 \times 4$ mm^3) helps to eliminate the physical separation between the component and the related information (Chapter 2.5.1).

Using optical signals is another alternative for data transfer. For this purpose, optical waveguides are directly integrated into the metallic surface by a selective application of UV-curing transparent polymer. The waveguides integration takes place within the production process of the component. Polymer optical communication structures offer a high bandwidth compared to conventional copper structures. Furthermore, they are resistant to electromagnetic interferences and dynamic stress, which are relevant in case of metallic mechanical parts. A new, automated machine setup is developed in this case in order to achieve the desired integration. Positioning systems with an especially high accuracy and adapted measurement methods are necessary to realize the required solution.

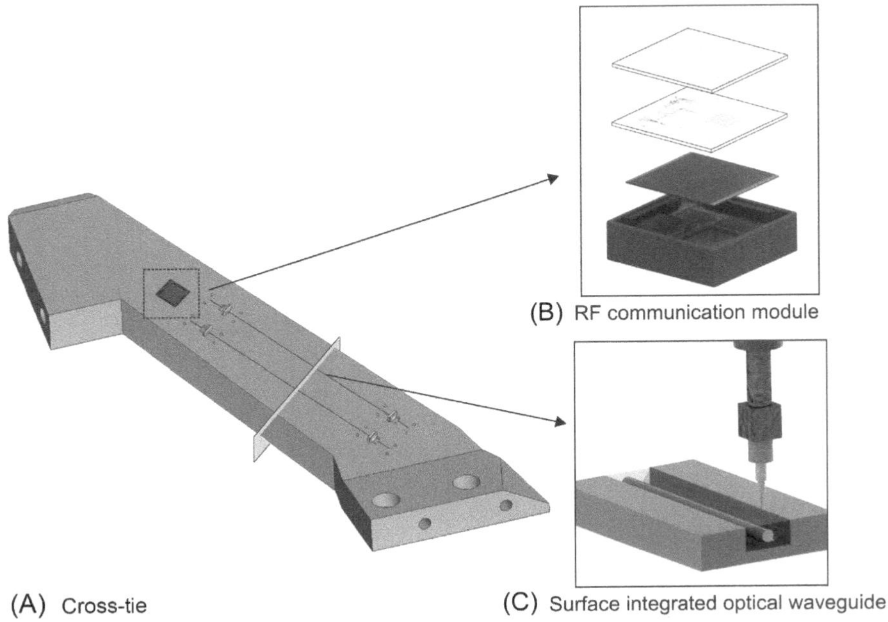

Fig. 2.5.1 Illustration of a cross-tie (A) with surface integrated communication facilities. Subfigure (B) represents a radio frequency communication module. Subfigure (C) represents a surface integrated waveguide for optical communication.

The following subsections present a detailed description of the development steps. In this context, implementation strategies as well as investigation methods leading to the integration of the two mentioned solutions are pinpointed.

See the following chapters for further information

2.5.1 Optoelectronic integration of RF communication systems for gentelligent components
2.5.1 Surface inherent energy and data transmission by optical signal coupling

2.5.1 Optoelectronic integration of radio frequency communication systems in metal components

Quang Huy Dao, Christoph von der Ahe†, Philipp von Witzendorff‡, Ludger Overmeyer†,‡, Bernd Geck**
*Institute of Microwave and Wireless Systems, Leibniz Universität Hannover, Hannover, Germany, †Institute of Transport and Automation Technology, Leibniz Universität Hannover, Garbsen, Germany, ‡Laser Zentrum Hannover e.V., Hannover, Germany

2.5.1.1 Introduction

The size of the communication module (also called as transponder) to be realized is essentially determined by the application scenario limits. Over the past application periods, the research groups of Prof. Overmeyer and Dr. Geck pursued concepts and technologies for a progressive miniaturization. In the first funding period a passive unidirectional communication solution for integration into metallic components was developed, based on a dielectric-filled circular waveguide in the 5.8 GHz Industrial Scientific and Medical (ISM) band, which has a diameter of 20 mm (see [1]). The energy supply of the transponder is carried out by harvesting energy from the electromagnetic field of the reading/writing unit. The communication from the transponder to the reader is based on the backscatter principle.

In the second funding period the working groups achieved a reduction in the diameter of the communication module to 8 mm. Because the carrier frequency determines the size of the antenna, the operating frequency is set within the 24 GHz ISM band. However, the increase in the frequency leads to a significantly higher free space attenuation. Therefore the reading/writing unit was extended with an additional source of energy in the form of a light source and the transponder has been equipped with an optoelectric power converter. In order to ensure a defined guidance of light within the hollow conductor, a hybrid waveguide has been researched [2]. This is formed by a circular waveguide filled with polymer optical fibers. This hybrid waveguide conducts both the light to the solar cells as well as the high frequency data signal. In this way, a bidirectional communication was achieved with a transmission distance of about 0.5 m [3–5]. The structure of the communication module, which is integrated in a screw, is shown on the right-hand side in Fig. 2.5.1.1.

During the third funding period the interaction of optical and electronic components was raised to a higher level of integration through the adoption of new

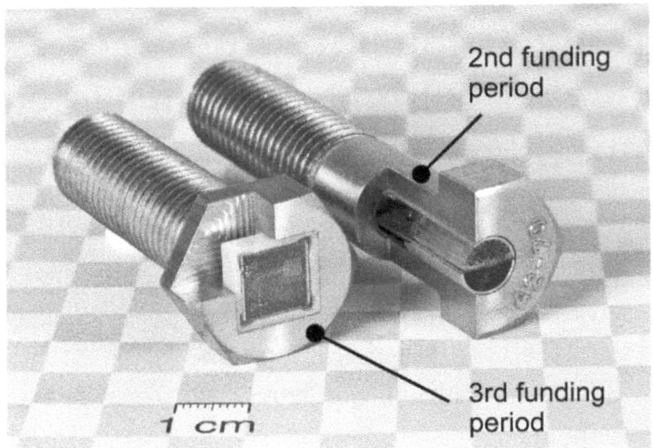

Fig. 2.5.1.1 Application example: Intelligent screws with integrated communication module.

technologies. The suitability of optically transparent and conductive materials for radio frequency (RF) circuit was studied. With this knowledge the optically transparent analog front end of the communication module is structured on a thin quartz glass substrate. This configuration is placed above the solar cell, still realizing a very high level of integration [6]. The reciprocal influences of the solar cell and the RF structure were analyzed which led to further challenges in the design of the antenna as well as the RF circuit. The miniaturization of the communication module leads to new challenges in the construction and connection technology. To connect the different levels of the system to each other and to assemble the integrated components, a new housing made of a three-dimensional molded interconnect device (3D-MID) was used. In doing so, the volume of the communication module could be reduced to $13 \times 13 \times 4\ \text{mm}^3$. For optimizing the entire system for use in a typical production scenario, it was considered how the optical energy supply could be adapted to the application conditions. Optimal utilization of the available light and additional energy harvesting techniques were studied. This knowledge was used for an efficient design of the power supply.

2.5.1.2 Research achievements and outcomes

2.5.1.2.1 System overview

A block diagram of the developed wireless communication system is depicted in Fig. 2.5.1.2. The reading/writing unit is shown on the left-hand side while the communication module, also called as transponder, is illustrated on the right-hand side. The reader/writer consists of two antennas, an RF analog front end, a data processing unit, and a computer to control the whole system. The transponder contains an

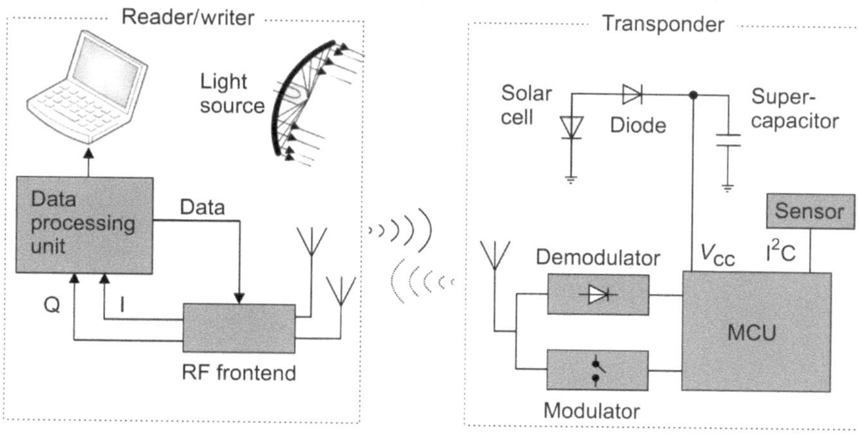

Fig. 2.5.1.2 Simplified block diagram of the 24 GHz wireless communication system.

antenna, a modulator, and demodulator circuit as well as a digital unit which mainly consists of a microcontroller (MCU). Depending on the application suitable sensors can be connected via I^2C bus to the MCU. To power the microcontroller a high efficient solar cell and a supercapacitor acting as energy storage are used. In a low ambient light environment, an additional light source such as a halogen lamp is necessary to provide sufficient energy to the solar cell. The data transmission process from the transponder to the reader/writer, e.g., transmission of the actual temperature value can be described as follows: The reader/writer transmits a request to the transponder via an amplitude-modulated RF signal. Hereby, the carrier frequency is 24 GHz. The request signal is demodulated by an analog front end of the transponder. Subsequently, the microcontroller interprets the command and generates a response by controlling the modulator unit. The backscattered transponder signal is converted into the baseband and this signal is routed through the in-phase (I) and quadrature (Q) channel to the data processing unit of the reader/writer. Afterward, the actual temperature value of the transponder is displayed on a computer. Furthermore, a continuous process monitoring of load parameters can be accomplished by the integration of sensors.

2.5.1.2.2 Design concept of the communication module

The design concept of the RF communication module integrable in the head of the screw is depicted in Fig. 2.5.1.3. The exploded view shows the highly integrated multilayer structure with the fusion of electronical and optical components. This design differs from the original concept, in which vias through the solar cell were

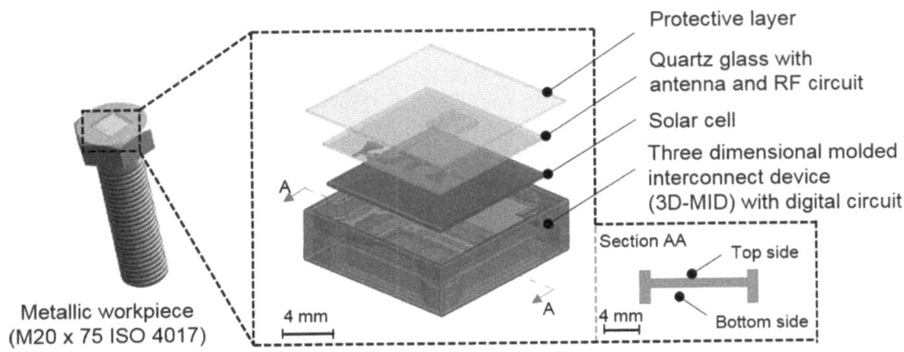

Fig. 2.5.1.3 24 GHz communication module integrated in a metallic workpiece, exploded view of the module and cross section view of the 3D device.

planned, due to the use of the multijunction solar cell. An electrical connection through the solar cell suitable for RF signal is challenging. Therefore a novel design concept is pursued. The whole module consists of four main parts. First of all, a housing is needed to carry the individual layers. A promising approach is the use of three-dimensional molded interconnect devices (3D-MID) acting as circuit carrier [7]. By means of the laser direct structuring (LDS) process, electronic conducting paths can be created directly on the plastic compounds. In this way it is possible to design any 3D circuits by using the third dimension in an efficient way. The bottom and the top side of the housing as well as the inner walls are used to realize the conductor tracks for the electrical connection of the solar cell and the RF circuit. The proposed housing has a volume of $13 \times 13 \times 4\ mm^3$. The length and width of the housing is predefined by the geometry of the solar cell. On the bottom side of this housing, there is room for the integration of a digital circuit consisting of an ultra-low-power microcontroller and an energy storage. To connect the digital circuit with the top layer through hole vias are used. This production step can be achieved completely within the LDS process. Hereby, the laser drills through the plastic compounds and at the subsequent metallization process copper is deposited at the bore hole. In this way, a reliable connection can be achieved.

The main part of the energy harvesting concept is the solar cell which is placed in the 3D-MID housing. Since wire bonding is more prone to damage by vibrations, the electric interconnection of the solar cell with the digital and RF circuit is established by conductive adhesives. Therefore the entire back panel area of the solar cell providing the positive voltage is fully adhered to the top side of the 3D-MID component. Besides, a reliable mechanical connection is realized. The interconnection between the front side (ground connections) of the solar cell, so-called busbars, and the conductor tracks on the sidewalls of the 3D-MID is also realized by conductive adhesives. To avoid short circuits throughout the bonding process an isolation has to be fitted between the two terminals of the solar cell. This isolating layer consists of a non-conductive adhesive.

The RF circuit including an antenna is arranged on the solar cell. A promising approach of the antenna concept is the use of a patch antenna. Moreover, the patch antenna structure can be designed as thin grid lines made of conventional metal in order to achieve optical transparency [8,9]. The RF circuit can be realized in the same way to minimize the shading effects caused by strip lines. Quartz glass is preferred as carrier substrate for these structures since it has a high optical transparency. An additional reason is the low dielectric dissipation factor of this material (see Section 2.5.1.2.3). To protect the communication module against environmental influences such as dust, dirt, lubricants, metal chips, etc., a protective layer made of quartz glass plate is used. This layer should fulfill requirements such as high optical transparency and resistance against environmental impact. Thus it is possible to adapt the surface conditions through specific surface treatments.

2.5.1.2.3 Optically transparent antenna

Since the antenna is arranged on the solar cell, it is necessary to use an optically transparent substrate and the antenna structure itself needs to have a high optical transparency in order not to decrease the performance of the solar cell. Using transparent conducting oxides (TCOs) as the conductive layer is one possible way but these materials have a poor conductivity [10] compared to conventional materials like copper. Our investigations and other research work [11] showed a poor antenna efficiency by using oxide such as indium tin oxide (ITO) which results in a short communication distance. In addition, the reflectance, absorptance, and transmittance of ITO are wavelength dependent [10] in the usable light spectrum of the solar cell. A better way of obtaining a high optical transparency and antenna efficiency simultaneously is designing a microstrip antenna by means of grid lines [12,13]. The antenna is designed in a way to obtain the capability of transmitting circular polarization waves and to ensure a proper communication link between wireless sensor nodes regardless of the orientation of the antenna. There are many possibilities to develop circularly polarized (CP) patch antennas. One way is exciting the patch by two feeds with a phase difference of 90 degrees at adjacent edges of the patch. The other possibility is using a single feed to generate two orthogonal modes. Therefore a perturbation segment such as a slot or other truncated segment methods can be utilized [14]. In this work, single fed truncated corner patch antennas are investigated because of their simple feeding method. Hereby, quartz glass with a thickness of 0.22 mm is used as carrier material to meet the requirement of optical transparency. The dielectric constant is

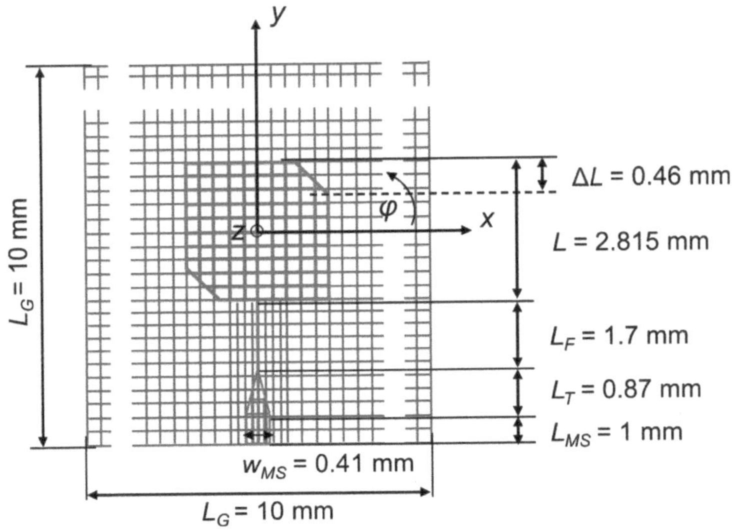

Fig. 2.5.1.4 Layout of the transparent and RHCP antenna.

$\epsilon_r' = 3.81$ and the dissipation factor is $\tan\delta = 0.0004$ at 24 GHz, which is determined by a resonator measurement method. The conductive layers are assigned as copper with a thickness of 1 μm since the intended RF sputtered metal coating layers are in this range of thickness.

Fig. 2.5.1.4 shows the layout of an optically transparent antenna with right-hand circular polarization (RHCP). The square patch consists of eleven horizontal and eleven vertical grid lines with a line width $w_l = 20$ μm and a spacing of $d = \lambda_r/48.1$. Here, λ_r is the resonant wavelength of the antenna. The grid line width of the ground plane and the grid spacing is set to 10 μm and $\lambda_r/46.5$, respectively. Both the patch (colored area) and the ground plane (gray area) are aligned in such a way as to get a high transparency. The geometrical dimensions are a compromise between the antenna performance and optical transparency. The meshed microstrip feed line is designed for an optimal impedance matching. Furthermore, the grid spacing of the microstrip lines and its ground plane is kept small to minimize insertion losses of the incident wave. The whole antenna has a transparency of roughly 92% calculated by the ratio between the nonmetal area and the total area.

A picture of the realized antenna mounted on a metal block for measurement purposes is shown in Fig. 2.5.1.5A. A RO4003C laminate from Rogers Corporation is used to bond the microstrip line (MS) of the printed circuit board (PCB) and the microstrip line of the antenna. The end launch connector is mounted at

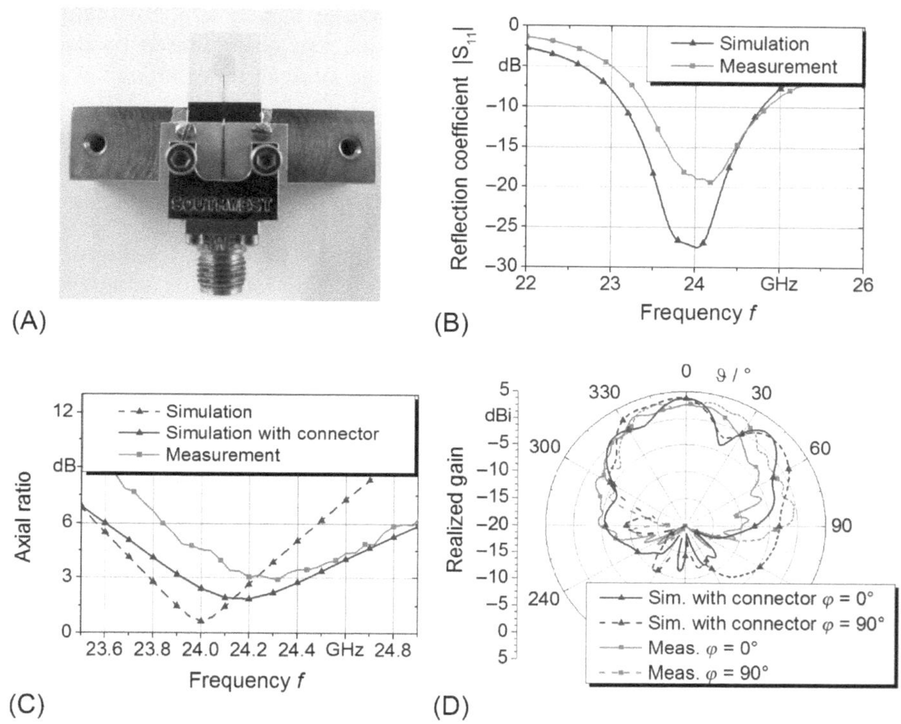

Fig. 2.5.1.5 (A) Picture of the realized antenna. (B)–(D) Simulation and measurement results of the transparent and RHCP antenna.

one end of the PCB. A TRL calibration process [15] is utilized to de-embed the influence of the end launch connector.

The comparison between simulation and measurement results is depicted in Fig. 2.5.1.5B–D. The measured input reflection coefficient $|S_{11}| = -19$ dB at 24 GHz and the bandwidth ($|S_{11}| \leq -10$ dB) is 6%. The axial ratio over the frequency is plotted in Fig. 2.5.1.5C. Since the end launch connector influences the radiation characteristics, it has to be considered in the simulation resulting in the solid line marked with triangles. A good agreement between simulation and measurement is achieved. The radiation pattern of the meshed antenna is shown in Fig. 2.5.1.5D. In the main direction ($\vartheta = 0$ degree), the simulated and measured gain value are 3.8 dBi and 2.6 dBi, respectively. In comparison to this, the simulated gain value for a correspondent nonmeshed patch antenna is 6.2 dBi. Apart from the lower antenna gain, the advantages of using grid lines are that a broader operating bandwidth and axial ratio bandwidth can be achieved compared to the opaque counterpart. The simulated radiation efficiency of the investigated antenna is 43% (opaque counterpart: 85%) which can be increased by using a higher mesh density but in contrast to that, the optical transparency will be decreased.

2.5.1.2.4 Optically transparent RF circuit

In order to process a bidirectional communication between the reading/writing unit and the transponder an RF circuit is needed. This circuit is connected to the antenna and consists of an envelope demodulator and a modulator, which are shown in Fig. 2.5.1.6 [16]. Metal grid lines are used to obtain the optical transparency. All grid line widths are about 0.01 mm. Some pads are not meshed due to the attachment of components on them as well as the realization of vias. The horizontal and vertical spaces between the grid lines are designed in a manner in order to minimize the insertion loss of the meshed microstrip lines (MMS). Thus the meshed ground plane area underneath the MMS has more grid lines than the region further away from the actual structure. According to this design, the overall transparency is about 85% (based on an area of $10 \times 4.4\ mm^2$) which is determined by the simulation software Zemax OpticStudio 15.5.

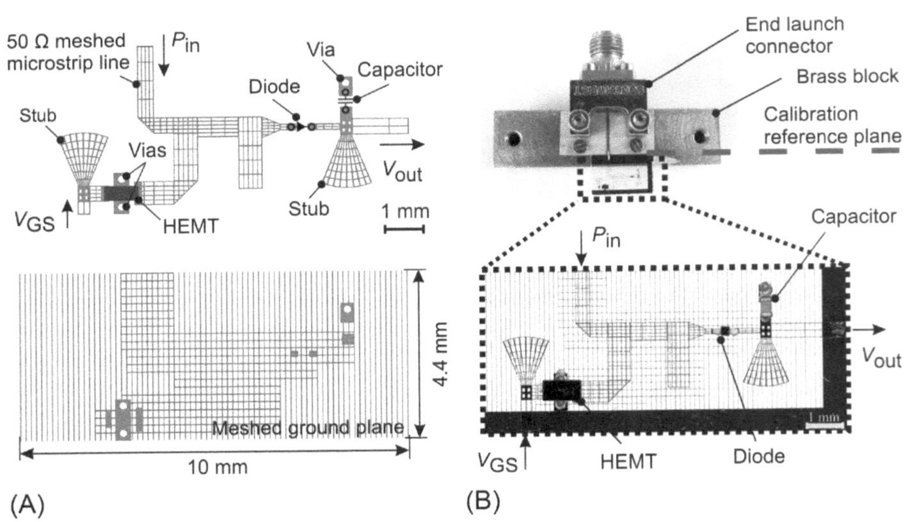

Fig. 2.5.1.6 (A) Layout of the meshed modulator and demodulator circuit, top: component side, bottom: ground plane; (B) picture of the realized circuit mounted on a metal block for measurement purposes and detailed view of the meshed analog frontend with electronic components bonded by two-component adhesives.

The data transfer from the transponder to the reader unit is realized by the backscatter modulation based on amplitude shift keying (ASK). The communication distance d between these units can be calculated as follows [17]:

$$d^4 = \frac{P_T G_R^2 \lambda_0^4 G_{Tag}^2}{P_R (4\pi)^4} |\Gamma_1 - \Gamma_2|^2 \tag{2.5.1.1}$$

Here, P_T is the transmitted power of the reader, P_R is the received input power at the reader, G_R is the gain of the reader antenna, G_{Tag} is the gain of the transponder antenna, and λ_0 is the free space wavelength. The performance of backscattering is mainly determined by two realized reflection states Γ_1 (mismatched case) and Γ_2 (matched case), assuming all other variables are constant. In order to generate these two states an Avago VMMK-1225 high electron mobility transistor (HEMT) is used in a common source configuration acting as a shunt switch. The drain is connected to the antenna feed line while the source is routed to the ground plane. The gate is connected to the microcontroller which controls the transistor by switching the voltage on ($V_{GS} = 0.7$ V) and off ($V_{GS} = 0$ V). In this way, backscattering can be realized. The usage of a HEMT offers low power consumption (6.3 μW @ 0.7 V_{GS}) and a good backscatter modulation performance [18].

For the reception of the reader's commands an envelope detector is realized by a Keysight HSCH-9161 zero bias beam lead Schottky diode and a 0.3 pF capacitor from Johanson Technology. The two stubs (see Fig. 2.5.1.6) are responsible for decoupling the microwave and the baseband signal. The MMS are designed in such a manner as to meet the required impedance matching between the modulator/demodulator circuit and the antenna impedance $Z_A = 50\ \Omega$.

As shown in Fig. 2.5.1.6B, the fabricated glass substrate is fixed on a metal block for measurement purposes. A RO4003C laminate is used to bond the MS of the PCB and the MMS of the analog front end. Applying a TRL calibration process, the influence of the end launch connector can be de-embedded.

Fig. 2.5.1.7A depicts the simulated and measured input reflection coefficients of the analog front end for two different states and an input power of −20 dBm. The curves marked with triangles represent the simulated data and the curves marked with squares are measurement results. There is a slight shift of about 70 MHz (corresponding to 0.29%) concerning the resonance frequency of the matched case. All in all, a good agreement can be obtained for both switching states. At the frequency of 24 GHz the measured reflection coefficients are −2.4 dB and −12 dB.

Since a data rate of 100 kbit/s is intended for the communication between the reader and the transponder, a signal with 100 kHz amplitude shift keying modulation (ASK) is used as test signal. Certainly, a higher data rate can be obtained with the developed circuit. Fig. 2.5.1.7B shows the simulation results of the demodulated envelopes for two different input power values. At an input power level of −30 dBm the measured amplitude of the envelope is 11 mV and the

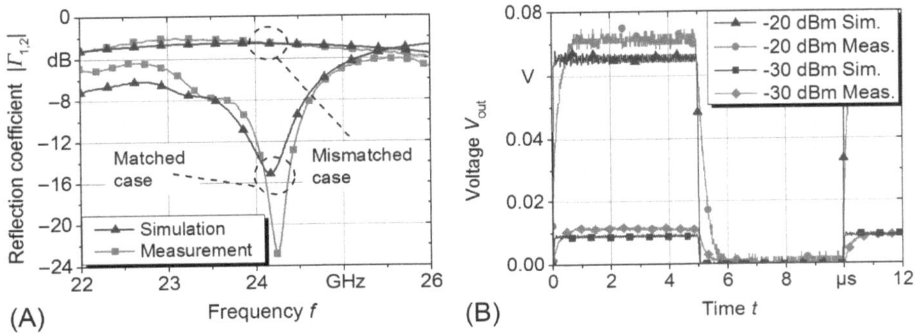

Fig. 2.5.1.7 Simulated and measured results. (A) Input reflection coefficient and (B) Output voltage of the meshed modulator and demodulator circuit.

rectangular waveform still remains unchanged. A good agreement between simulation and measurement is achieved. The mismatch concerning the slope of the envelope is caused by the measurement setup and mainly influenced by the input capacitance of the oscilloscope.

These demodulated signals can be conducted to a comparator that converts the output voltage of the detector to a logic level. Thus a simple decoding with a microcontroller (MCU) is possible. In [18] a low power comparator was used (LMV7271 from National Semiconductor; 16.2 µW @ 1.8 V) which needs at least 30 mV signal level to detect the right data. In order to decrease the threshold voltage and thus extend the communication range, another comparator (TS3021 from STMicroelectronics) is evaluated in the 3rd funding period. The result shows a threshold level of 3 mV corresponding to an incident power of $P_{in} = -37$ dBm, but in contrast to other comparators this component has a higher power consumption (126 µW @ 1.8 V). Further improvement concerning the threshold level, energy demand, and component reduction can be achieved by using the internal comparator of an MCU from Texas Instruments (MSP430FR5738). The investigation shows a minimal sensitivity value of −39 dBm with a simultaneous reduction in power consumption (58 µW @ 2 V). This value is verified by a measurement with the realized prototype. A detailed evaluation concerning the power consumption of the MCU will be presented in Section 2.5.1.2.7.

2.5.1.2.5 Realization of the optically transparent analog front end

The optically transparent antenna as well as the analog circuit is already presented separately in the previous section. The fusion of both components on quartz glass is shown in Fig. 2.5.1.8. The grid lines on the top and bottom side consist of three different metal layers (0.05 µm chromium, 1 µm copper, and

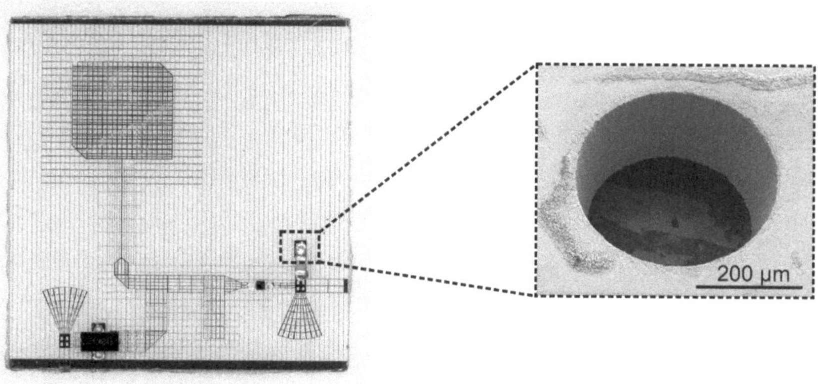

Fig. 2.5.1.8 Picture of the fabricated analog front end with electronic components bonded by two-component adhesives; right: SEM image of a through glass via.

0.3 µm platinum) with an overall thickness of approx. 1.3 µm. The metal grid lines on both sides are realized by the subsequent photolithography process by means of an ion beam etching process. This fabrication process is done in cooperation within the CRC 653. To obtain the ground connections for the transistor and the capacitor, through glass vias are produced by laser ablation [19] with a picosecond laser source with 515 nm wavelength. An example of a via is depicted in Fig. 2.5.1.8. The laser radiation is deflected with a galvanometric scanning system to irradiate the drill hole. The average laser power is set to 3 W at 50 kHz pulse repetition rate. The focal spot size was 20 µm. After the laser ablation process the glass substrate is RF sputtered. As a result, reliable conductive vias can be achieved.

The three electronic components (transistor, diode, and capacitor) are bonded via a two-component silver-filled epoxy system Epo-Tek H20E for testing purposes. For small series or even mass production a vapor phase soldering process can be applied. Therefore platinum is preferred for the top metal layer since it is best suited as a diffusion barrier [20].

2.5.1.2.6 Power supply

As shown in Fig. 2.5.1.3, a solar cell is arranged below the optically transparent analog front end to supply the digital circuit. The power supply concept of the communication module is based on a single metamorphic solar cell in combination with a supercapacitor acting as energy storage (see Fig. 2.5.1.2). The used multijunction III-V concentrator solar cell C4MJ produced by Spectrolab is a

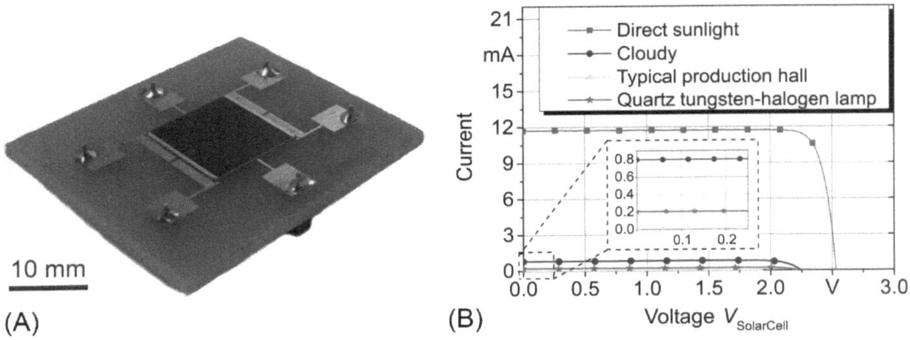

Fig. 2.5.1.9 (A) Picture of the used multijunction solar cell which is bonded on an FR4 substrate for characterization and (B) current-voltage characteristic for different lighting scenarios.

fully qualified triple junction GaInP/GaInAs/Ge solar cell [21,22]. According to the AM1.5D, ASTM G173-03 standard, this solar cell reaches an efficiency of 40% at an irradiance of 500 suns (50 W/cm^2) [23].

The solar cell which is bonded on a FR4 substrate for electrical characterization is depicted in Fig. 2.5.1.9A. The geometric dimensions of the optical aperture are 10×9.9 mm^2. The thickness is 0.19 mm. The busbars on the top side are the ground reference with a width of 0.4 mm while the whole bottom side provides the positive voltage. This mounting configuration is used for a characterization of current-voltage curves for different lighting scenarios. The measurement results obtained by a source measure unit Keysight B2900A are shown in Fig. 2.5.1.9B. By applying direct sunlight to the solar cell, the short-circuit current I_{SC} is 11.7 mA while the open-circuit voltage V_{OC} is 2.52 V. On a cloudy day the solar cell delivers a current I_{SC} of 0.8 mA. In this case, V_{OC} is 2.23 V. The performance of the cell is also evaluated in a typical production hall (Hannover Centre for Production Technology PZH) resulting in the curve marked with triangles. These values are averaged for five different positions within the production hall. The available ambient light provides a current I_{SC} of 9 μA and the voltage V_{OC} is 1.85 V, respectively. This indoor lighting is not sufficient to power the digital circuit since the microcontroller requires at least 1.91 V for proper operation. Thus an external quartz tungsten-halogen lamp can be used as an additional light source. The selected lamp Thorlabs QTH10/M including a condensing lens offers a broadband illumination in the visible and near-IR portions of the light spectrum (400–2400 nm) [24]. The quartz tungsten-halogen bulb has a power consumption of 10 W at a DC voltage of 12 V. To obtain the current-voltage characteristic, this lamp is positioned at a distance of $d=2$ m away from the solar cell since the

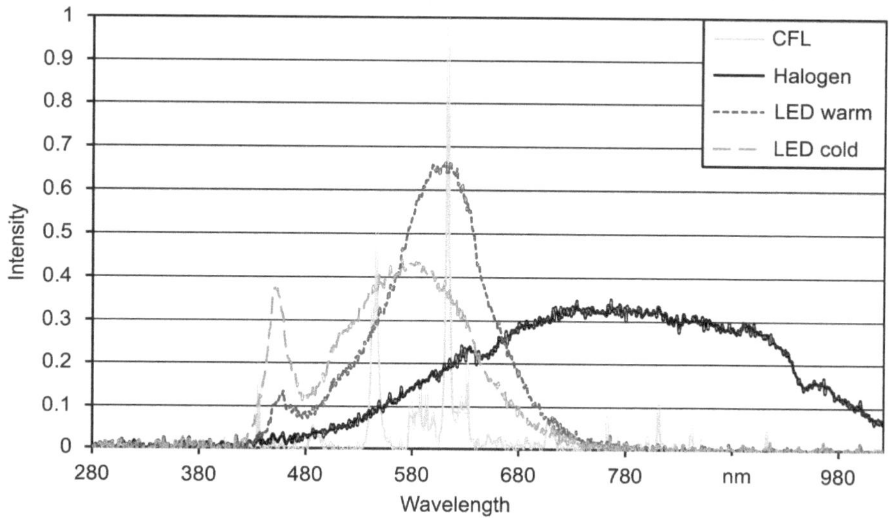

Fig. 2.5.1.10 Spectrum of the used lamps.

maximum communication distance between reader/writer and the transponder is in this range. For this measurement setup the current I_{SC} is 0.2 mA and the voltage V_{OC} is 2.21 V.

According to the standard DIN EN 12464-1 for lighting of indoor working places the illuminance varies between 100 and 1000 Lux, depending on the kind of work. That is why further tests are conducted to characterize the solar cell and its efficiency in indoor lighting conditions. Therefore four different types of lamps, which typically occur in industry facilities, are used: A compact fluorescent lamp (CFL, GE Spherical T2 7 W), a halogen lamp (Osram Halopin ECO G9 20 W), a cold light emitting diode (LED, Osram Parathom Special T26 1.5 W), and a warm light-emitting diode (EGLO LED 4 W). The corresponding light spectrum is depicted in Fig. 2.5.1.10.

For each type of lamp measurements, in the manner as described earlier, with illumination of 100 lx, 200 lx, 500 lx, and 1000 lx are carried out. With the measured voltage and current, it is possible to determine the maximum power point for each lamp type and illuminance condition. The ratio between the incoming radiant flux and the resulting electrical power equals the efficiency of the solar cell for the specific conditions. Fig. 2.5.1.11 depicts the results.

The results can be explained by the functionality of the triple junction solar cell. The solar cell needs a wide spectrum of light, ideally between 350 and 1600 nm, to work at its best performance. Artificial light sources are not able to provide such a wide spectrum. Halogen lamps produce a wider spectrum

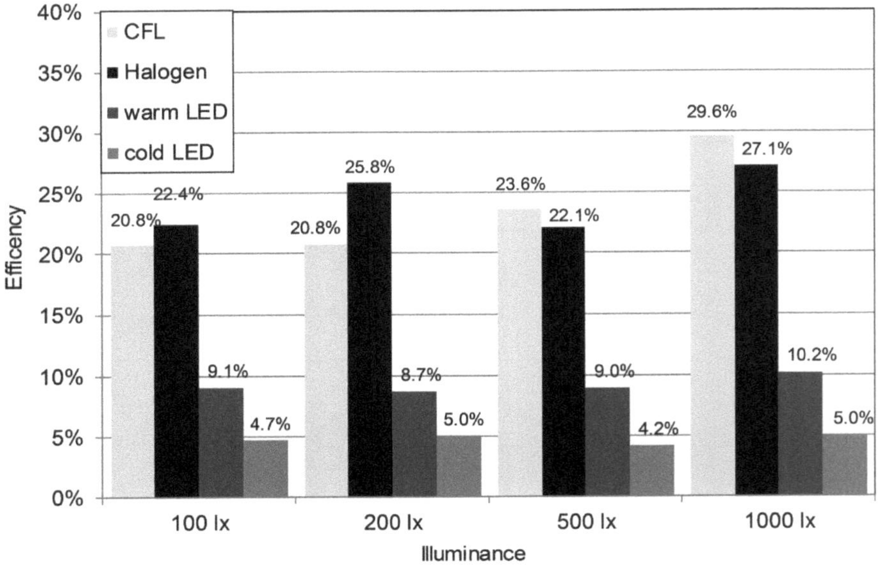

Fig. 2.5.1.11 Solar cell efficiency for different types of lamps and different illumination levels.

than LEDs. The light emitted by CFL lamps is also distributed over a wider range of wavelengths. That is why the efficiency for these two types of light sources is considerably higher. The problem with CFL light sources is that although they produce a good level of illuminance, they emit just a small amount of radiant flux. The amount of radiant flux in relation to the target area is shown as irradiance in Fig. 2.5.1.12.

That means although the solar cell has a good rate of efficiency for the CFL light, but this light contains just too less radiant flux to convert into electrical energy. As found out in the first tests, halogen lamps are the only light sources which supply enough energy for the communication module in combination with the triple junction solar cell. An illuminance level of approximately 300 lx is needed to guarantee continuous operation of the communication module.

The energy harvesting concept is designed in such a manner as that the energy out of the solar cell is buffered in an Elna DCK-3R3E204T614 supercapacitor with a capacitance of 0.2 F. A simple schematic is illustrated in Fig. 2.5.1.13A. An energy harvesting concept based on a power management IC is not applied since the complexity of the power supply should be kept as simple as possible. Therefore a Schottky diode BAT30F4 from STMicroelectronics is used to minimize the discharging process of the capacitor during the low light

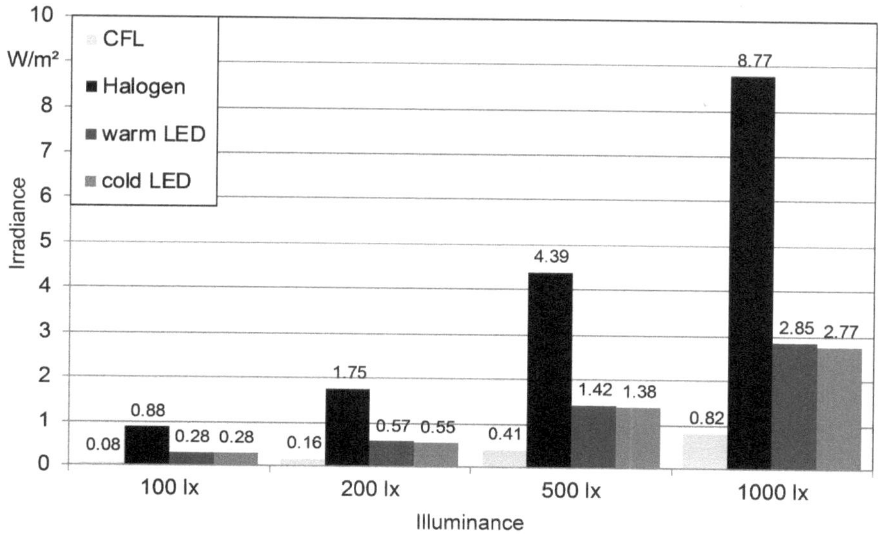

Fig. 2.5.1.12 Irradiance for the light sources under test in relation to the illuminance.

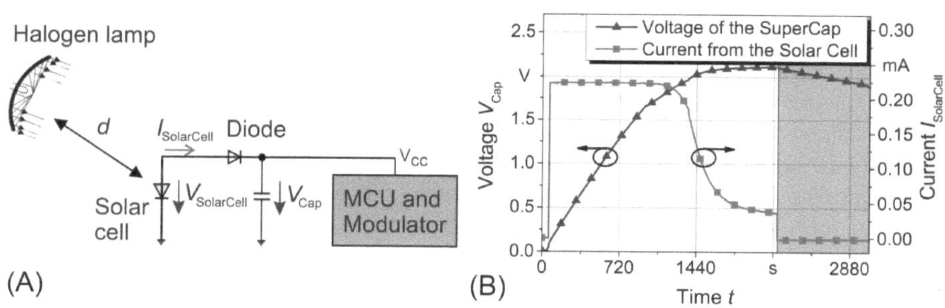

Fig. 2.5.1.13 (A) Schematic of the energy harvesting concept and setup for determining the operating time of the transponder and (B) measured voltage V_{Cap} of the supercapacitor and current $I_{SolarCell}$ of the solar cell during the operating time of the transponder for a defined scenario.

environment. This diode is suitable for this application because of the low forward voltage drop and a reverse leakage current lower than approx. 1 μA at room temperature. The solar cell was favored over other solutions, for example, thermoelectric generators (see [25]), because it provides sufficient power and is best suited for the requirements of the demonstrator scenario.

To evaluate the efficiency of the proposed energy harvesting concept different lighting conditions are tested. In the following, one fixed scenario is defined to find out the operating time of the transponder. In this scenario the halogen lamp QTH10/M is positioned at a distance d of 2 m (see Fig. 2.5.1.13A) which corresponds to an illuminance of 1500 lx. At the beginning, the voltage of the capacitor V_{Cap} is 0.01 V (empty state). A source measure unit Keysight B2900A is utilized to capture the voltage of the capacitor and the current of the solar cell during the charging and discharging process of the energy storage. These curves are depicted in Fig. 2.5.1.13B. First of all, the lamp is switched on after 60 s and the forward current of the diode changes to 0.22 mA. During the whole measurement the reading/writing unit transmits a request for the actual temperature value of the transponder periodically (every second). When V_{Cap} reaches the voltage level of 1.97 V ($t=1364$ s, at bright ambient light such as outdoor environment, the threshold voltage will be reached in approx. 2 s) the microcontroller wakes up, processes the request, transmits a respond to the reader/writer, and goes to sleep mode until the next incoming request. This communication procedure is repeated every second. In the following the current $I_{SolarCell}$ reduces until it reaches a current value of approx. 36 μA ($t=2196$ s). This value is the average current consumption of the microcontroller. This means that within 36.6 min the capacitor is fully charged and V_{Cap} is 2.11 V while the transponder is still transmitting data to the reading/writing unit periodically.

In the next step the halogen lamp is switched off to simulate the operating time of the transponder in a low ambient light. The grayed area in Fig. 2.5.1.13B illustrates the scenario of discharging the supercapacitor. During this period, the reverse leakage current of the diode varies between 1.3 and 0.9 μA. The voltage V_{Cap} decreases steadily. At the time of 3032 s, the voltage level of 1.91 V is not sufficient to power the microcontroller. In this case, a bidirectional communication cannot be maintained. For the investigated scenario, it can be concluded that in a low ambient light the designed transponder can operate continuously for 13.9 min with a fully charged supercapacitor of 0.2 F while transmitting its actual temperature value every second.

2.5.1.2.7 Power consumption of the MCU

To meet the criteria of low power consumption of the transponder, the aforementioned ultra-low-power microcontroller (MSP430FR5738) is used. Since it is a central management unit, its energy requirement should be well known. Therefore the power consumption during a communication procedure (request for the current temperature value) between the reader and the transponder and the corresponding current consumption is shown in Fig. 2.5.1.14. The measured voltage signal V_{out} (curve marked with circles) is recorded at the output of the demodulator while the voltage V_{GS} of the transistor is represented by the curve marked

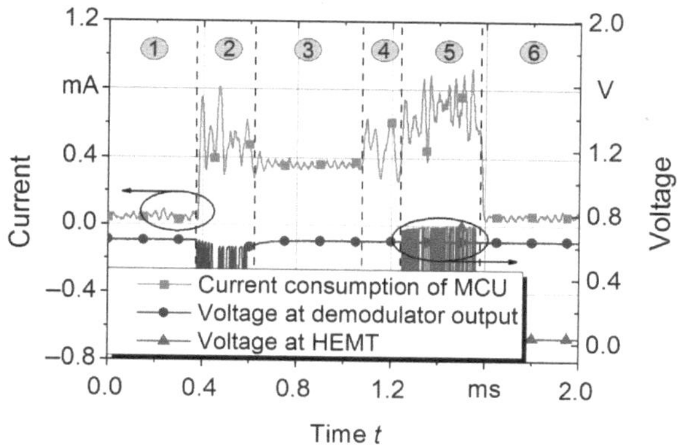

Fig. 2.5.1.14 Communication flow of a reader request for the current temperature value and the current consumption of the MCU.

with triangles. It can be noted that the request from the reader occurs within the first 0.22 ms. Due to the processing time of the MCU, the response of the transponder has a delay of approx. 0.64 ms. The curve marked with squares in Fig. 2.5.1.14 represents the current consumption for a supply voltage of 2 V during a communication sequence. It can be seen that the time varying current consumption can be divided into six states:

- State 1: The MCU is in standby mode (approx. 0.033 mA). This operation mode should be optimized to reduce the power consumption as the transponder is mostly in standby mode.
- State 2: The MCU changes its state of operation to active as it receives a request from the reader. In this case, a request for the current temperature value.
- State 3: In this state the request is processed by the MCU. The current can be measured to approx. 0.8 mA.
- State 4: The A/D converter is activated for measuring the temperature value.
- State 5: The MCU transmits the requested data by switching the gate-source voltage V_{GS} on and off.
- State 6: At the end of the communication procedure the MCU switches back to standby mode.

The presented results show that the power consumption of the used microcontroller varies between 58 μW and 1.85 mW.

2.5.1.2.8 Reading/writing unit

In order to communicate and to exchange data with the transponder, a reading/writing unit is needed. A block diagram of the reader/writer showing the structural concept is depicted in Fig. 2.5.1.15. This unit consists of a FPGA-based

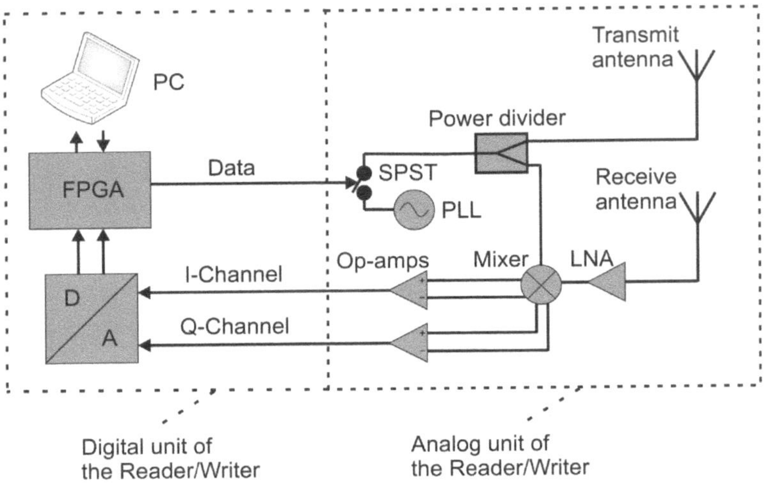

Fig. 2.5.1.15 Block diagram of the reading/writing unit.

data processing unit from National Instrument and commercially available RF modules, which are evaluated in details and connected to each other. The two main parts are the digital unit and the analog unit. The digital unit generates data for the transmitting path and interprets the incoming signals from the transponder. All of these tasks can be controlled via a graphical user interface. The RF components are shown on the right-hand side of the block diagram (see Fig. 2.5.1.15). There are two circularly polarized antennas, one for the transmitting and one for the receiving path. Furthermore, the analog unit consists of a PLL (phase-locked loop) oscillator, a single pole single throw (SPST) switch, a power divider, an IQ receiver IC with an integrated low noise amplifier [26], and an operational amplifier circuit. In order to transmit a request to the transponder the digital unit controls the SPST switch in such a manner as that the 24 GHz signal coming from the oscillator is interrupted by the data cycle. In this way, amplitude shift keying can be realized. The transponder processes the request and transmits a response to the reader/writer. The incoming signal at the receive antenna is amplified to increase the sensitivity of the receiver. Based on the direct conversion receiver concept, the signal is down converted to baseband. The use of the IQ mixer is advantageous since the phase difference between the local oscillator and the input signal is independent from the communication distance between the reader/writer and the transponder [27]. The output signals of the in-phase (I-channel) and quadrature (Q-channel) channel are amplified and routed to the analog digital converter (ADC). The digitized data are forwarded to a field programmable gate array (FPGA) to extract the information from the received data packets. The decoded information is displayed on the graphical user interface.

2.5.1.2.9 Data processing unit and initial system evaluation

The programming of the data processing unit as well as the graphical user interface (GUI) is realized by the development environment LabVIEW. By means of this user interface the communication procedure can be established and controlled. In the event of data exchange, for example, the production plan within the adaptive process planning, these data can be used to process by the "Virtual Planner" (see Chapter 3.1). In this case, the data previously stored on the communication module will be requested and subsequently saved on the computer or a file sharing network for further processing. Conversely, the modified production plan and the process data can be transmitted to the transponder. Since the data is stored within the object the fusion of information and workpiece is achieved. Thus the data can be accessed and updated by any reading/writing station at any time.

Up to now, each unit is evaluated separately. In order to evaluate the performance such as the communication distance of the whole communication system each component is connected together according to Fig. 2.5.1.2. The transmitted power of the reading/writing unit is set to 20 dBm ERP (100 mW effective radiated power, radio regulation). The antenna of the reading/writing unit is aligned with the communication module. Considering the laboratory environment the maximum communication distance is determined up to 2 m depending on the gain of the receiving antenna. This result shows that a reliable data exchange between the reading/writing unit and the workpiece, which are arranged less than 2 m away from each other, can be maintained within the shop floor.

References

[1] E.B. Kaldjob, Design and Analysis of Field-powered Transponders Integrated in Metallic Objects, Dissertation Leibniz Universität Hannover, 2010.

[2] S. Franke, Untersuchung der optischen Übertragungsstruktur eines hybriden Wellenleiters, Dissertation Leibniz Universität Hannover, Berichte aus dem ITA, PZH - Produktionstechnisches Zentrum GmbH, Garbsen, 2011.

[3] J. Meyer, Ein optisch versorgter RFID-Transponder für die Integration in Metall, Dissertation Leibniz Universität Hannover, Berichte aus dem HFT, PZH - Produktionstechnisches Zentrum GmbH, Garbsen, 2015.

[4] J. Meyer, Q.H. Dao, B. Geck, 24 GHz RFID communication system for product lifecycle applications, in: Proc. 2nd International Conference on System-Integrated Intelligence: Challenges for Product and Production Engineering, SysInt 2014, July 2nd–4th, Bremen, Germany, 2014, pp. 369–375.

[5] B. Denkena, T. Mörke, M. Krüger, J. Schmidt, H. Boujnaha, J. Meyer, P. Gottwald, B. Spitschan, M. Winkens, Development and first applications of gentelligent components over their lifecycle, CIRP J. Manuf. Sci. Technol. 7 (2014) S.139–150.
[6] Q.H. Dao, A. Skubacz-Feucht, B. Lüers, P. von Witzendorff, C. von der Ahe, L. Overmeyer, B. Geck, Novel design concept of an optoelectronic integrated RF communication module, in: Proc. 3rd International Conference on System-Integrated Intelligence: New Challenges for Product and Production Engineering, SysInt 2016, June 13th–15th, Paderborn, Germany, 2016, pp. 245–251.
[7] J. Franke, Three-Dimensional Molded Interconnect Devices (3D-MID), Hanser, 2014.
[8] Q.H. Dao, T. Cherogony, B. Geck, Optically transparent and circularly polarized patch antenna for K-band applications, in: 10th German Microwave Conference, GeMiC 2016, Bochum, Germany, 2016.
[9] Q.H. Dao, R. Braun, B. Geck, Design and investigation of meshed patch antennas for applications at 24 GHz, in: European Microwave Conference 2015, Paris, France, 2015.
[10] D.S. Ginley, Handbook of Transparent Conductors, Springer, New York, 2010.
[11] T. Yasin, R. Baktur, C. Furse, A comparative study on two types of transparent patch antennas, in: General Assembly and Scientific Symposium, 2011 XXXth URSI, 2011, pp. 1–4.
[12] K. Ito, M. Wu, See-through microstrip antennas constructed on a transparent substrate, in: Antennas and Propagation, ICAP 91. Seventh International Conference on (IEE), vol. 1, 1991, pp. 133–136.
[13] J. Hautcoeur, L. Talbi, K. Hettak, M. Nedil, 60 GHz optically transparent microstrip antenna made of meshed AuGl material, IEEE Antennas Propag. Mag. 54 (8) (2012) 10–16.
[14] C. Sharma, K. Gupta, Analysis and optimized design of single feed circularly polarized microstrip antennas, IEEE Trans. Antennas Propag. AP-32 (6) (1983) 949–955.
[15] M. Hiebel, Fundamentals of Vector Network Analysis, Rohde & Schwarz, 2011.
[16] Q.H. Dao, B. Lüers, T. Ngo, B. Geck, An optically transparent analog frontend for a solar powered 24 GHz RFID transponder, in: International Microwave Symposium (IMS 2016), San Francisco, California, USA, 2016.
[17] P. Nikitin, K. Rao, R. Martinez, Differential RCS of RFID tag, Electron. Lett. 43 (8) (2007) 431–432.
[18] J. Meyer, Q.H. Dao, B. Geck, Design of a 24 GHz analog frontend for an optically powered RFID transponder for the integration into metallic components, in: European Microwave Conference 2013, EuMC, Nuremberg, Germany, 2013.
[19] P. VonWitzendorff, A. Bordin, O. Suttmann, R. Patel, J. Bovatsek, L. Overmeyer, Laser ablation of borosilicate glass with high power shaped UV nanosecond laser pulses, in: Proc. SPIE 9735, Laser Applications in Microelectronic and Optoelectronic Manufacturing (LAMOM) XXI, San Francisco, California, USA, 2016.
[20] T. Kim, Y. Kim, Sn-Ag-Cu and Sn-Cu solders: interfacial reactions with platinum, JOM Res. Summary Lead-Free Solder 56 (6) (2004) 45–59.
[21] Spectrolab, Datasheet C4MJ, Rev. 7/12/11. http://www.spectrolab.com/DataSheets/PV/CPV/C4MJ_40_Percent_Solar_Cell.pdf, 2016 (accessed 02.05.16).
[22] J.H. Ermer, R.K. Jones, P. Hebert, P. Pien, R.R. King, D. Bhusari, R. Brandt, O. Al-Taher, C. Fetzer, G.S. Kinsey, N. Karam, Status of C3MJ+ and C4Mj production concentrator solar cells at spectrolab, IEEE J. Photovoltaics 2 (2) (2012) 209–213.

[23] R.R. King, D. Bhusari, D. Larrabee, X.-Q. Liu, E. Rehder, K. Edmondson, H. Cotal, R. K. Jones, J.H. Ermer, C.M. Fetzer, D.C. Law, N.H. Karam, Solar cell generations over 40% efficiency, Prog. Photovolt. Res. Appl. 20 (6) (2012) 801–815.
[24] Thorlabs, User guide QTH10(/M) Quartz Tungsten-Halogen Lamp, Rev D, January 4, 2015. http://www.thorlabs.de/newgrouppage9.cfm?objectgroup_id=7541&pn=QTH10/M, 2016 (accessed 28.04.16).
[25] A. Ibragimov, H. Pleteit, C. Pille, W.A. Lang, Thermoelectric energy harvester directly embedded into casted aluminum, IEEE Electron Device Lett. 33 (2) (2012) 233–235.
[26] S. Radar, Datasheet 24 GHz MMIC IQ Mixer RX_024_04, Rev. 0.3., http://www.siliconradar.de/datasheets/Datenblatt_RX_024_04.pdf, 2013 (accessed 12.05.16).
[27] Q. Gu, RF System Design of Transceivers for Wireless Communications, Springer Science+Business Media(2005).

2.5.2 Surface inherent energy and data transmission by optical signal coupling

Bechir Hachicha, Ludger Overmeyer†*
*Institute of Transport- and Automation Technology, Leibniz Universität Hannover, Garbsen, Germany, †Institute of Transport and Automation Technology, Leibniz Universität Hannover, Garbsen; Laser Zentrum Hannover e.V., Hannover, Germany

2.5.2.1 Introduction

The vision of the Collaborative Research Centre 653 is to abolish the separation between mechanical parts and their corresponding information carrier. The use of surface inherent optical communication structures offers a possibility to achieve this goal. In Ref. [1] the selective application of UV-curing, transparent polymers to create communication structures were investigated and validated. Two UV-curing polymers with different index of refraction were applied using pressure-time micro-dispensing system [1].

Fig. 2.5.2.1 shows the necessary steps to create surface inherent optical waveguides based on the internal total refraction. To achieve these steps, the pressure-time micro-dispensing unit was integrated into a 3-axis gantry robot. Two control devices were used for the process. The first control device allows the adjustment of dispensing pressure and dispensing duration. The second device allows a computerized numerical control (CNC) of the gantry robot. This permits the adjustment of the dispensing path and the dispensing feed rate, as well as the distance between dispensing needle and substrate.

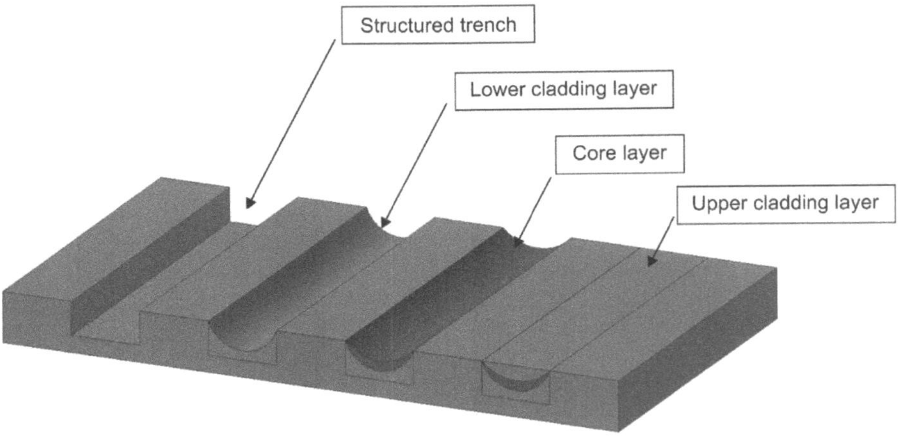

Fig. 2.5.2.1 Four process steps for surface inherent optical waveguides. In the first step a trench structure is prepared on the substrate surface, along the desired path. In the second step the lower cladding layer is achieved by dispensing a UV-curing polymer adhesive. In the third step the core layer is achieved by dispensing a UV-curing polymer with a higher index of refraction than the cladding layer. The fourth step is achieved similar to the second step and the upper cladding layer is dispensed.

Fig. 2.5.2.2 shows the machine setup to dispense surface inherent optical waveguides. The pressure container surrounds the cartridge to prevent its dilatation due to the pressure buildup. The triangulation distance sensor controls the distance between the capillary and the substrate. After production, it is necessary to characterize and evaluate the optical waveguides to make a statement about the process and their properties.

The optical and geometrical characterization of the inherent communication structures was performed in Ref. [2]. Two key figures were defined to evaluate the cross-sectional area and shape of the inherent optical waveguides. The first key figure RS_{cas} is the relative standard deviation of the cross-sectional area over the waveguide length as defined in Eq. (2.1.2.1) and presented in Fig. 2.5.2.3:

$$RS_{csa} = 1 - \left(\frac{\sqrt{\frac{1}{n-1}\sum_{i=1}^{n}\left(A_{\text{Profile}} - \overline{A_{\text{Profile}}}\right)^2}}{\frac{1}{n}\sum_{i=i}^{n} A_{\text{Profile}}} \times 100 \right) \tag{2.5.2.1}$$

A number n of cross-sections is created along the inherent waveguide. The surface area A_{Profile} of each cross-section is calculated with Eq. (2.2.1.2) as presented in Fig. 2.5.2.4:

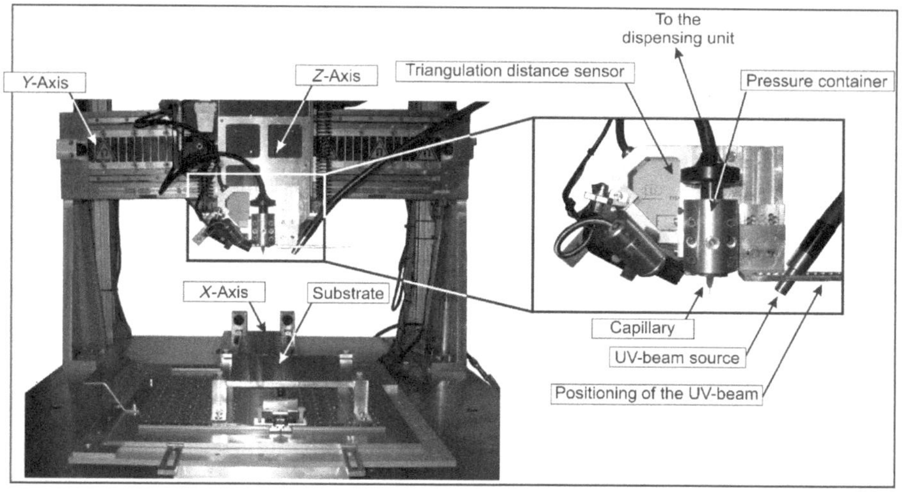

Fig. 2.5.2.2 Machine setup used for dispensing surface inherent waveguides. A gantry robot with three axes is used. On the *Z*-axis a capillary connected with a pressure-time dispensing unit is mounted. A triangulation sensor is used to control the distance between capillary and substrate. A UV beam source is positioned near the capillary for direct curing of the dispensed adhesive.

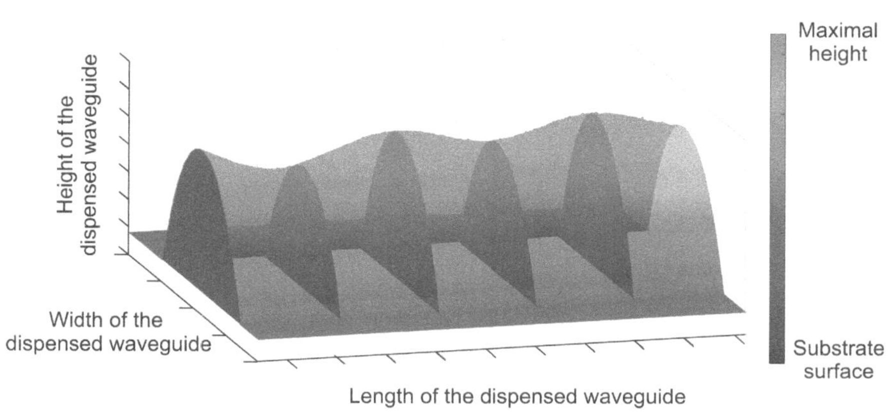

Fig. 2.5.2.3 Examples of realized cross-sections along a dispensed waveguide. A number n of cross-sections is realized along the waveguide and the calculation of the key figure RS_{csa} (Eq. 2.1.2.1) is done using a MATLAB routine.

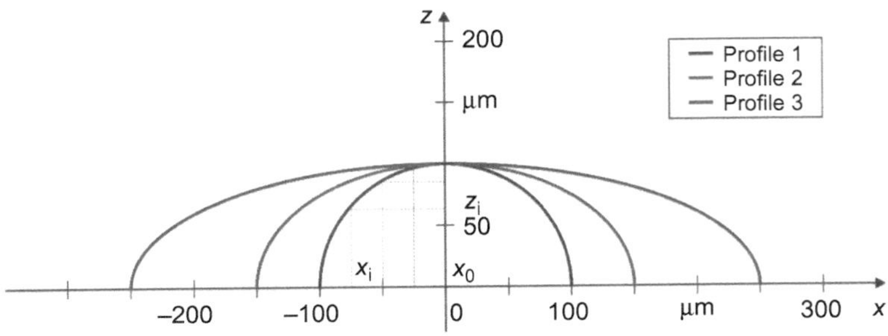

Fig. 2.5.2.4 Schematic presentation of the surface area calculation as well as profile radius deviations to an ideal semicircle (profile 1). The surface area is calculated based on Eq. (2.2.1.2). Profile 2 and profile 3 represent, respectively, 11.41% and 16.64% deviation of the profile radius in comparison to profile 1.

$$A_{\text{Profile}} = \sum_{i=1}^{m} \frac{z_i + z_{i+1}}{2} \cdot |x_{i+1} - x_i| \tag{2.5.2.2}$$

The mean profile $\overline{A_{\text{Profile}}}$ is calculated as:

$$\overline{A_{\text{Profile}}} = \frac{\sum_{i=1}^{n} A_{\text{Profile}}}{n} \tag{2.5.2.3}$$

The second key figure RS_{rad} represents the relative standard deviation of the profile radius in n number of cross-sections and is calculated as presented in Eq. (2.2.1.4):

$$RS_{\text{rad}} = \frac{1}{n} \sum_{i=1}^{n} r_{\text{rad}_i} \tag{2.5.2.4}$$

The relative standard deviation r_{rad} represents the deviation of the profile radius in a specific cross-section and is formulated as:

$$r_{\text{rad}} = \frac{\sqrt{\frac{1}{m-1} \sum_{i=1}^{m} (r_i - \bar{r})^2}}{\bar{r}} \times 100 \tag{2.5.2.5}$$

The mean radius $\bar{r}$ of a cross-section is calculated as:

$$\bar{r} = \frac{1}{m} \sum_{i=1}^{m} r_i = \frac{1}{m} \sum_{i=1}^{m} \sqrt{(x_i - x_0)^2 + z_i^2} \tag{2.5.2.6}$$

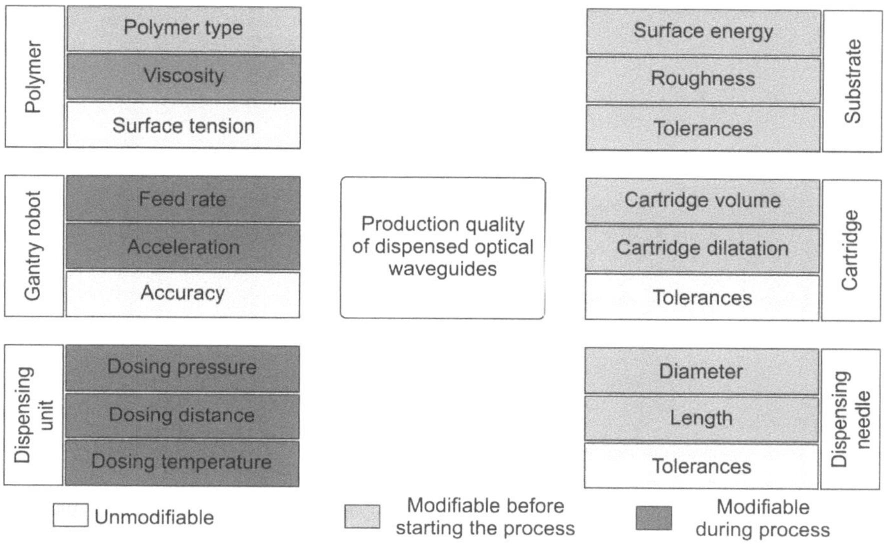

Fig. 2.5.2.5 Relevant process parameters during the dispensing of optical waveguides. Classification of process parameters for the production of surface inherent optical waveguides using time-pressure-dispensing method. The classification is based on the modifiability of these parameters.

This key figure RS_{rad} evaluates the cross-sectional shape of the inherent polymer waveguide. The sensitivity of both key figures to changes in the production parameters was analyzed in Ref. [2]. The relevant process influencing parameters were categorized in three groups as presented in Fig. 2.5.2.5:

1. Unmodifiable parameters
2. Parameters modifiable before starting the process
3. Parameters modifiable during the process

The feed rate, pressure, and dispensing distance were selected as main parameters of the modifiable parameters during the process. Studies in Ref. [2] demonstrate a significant sensitivity of the key figures for changes in feed rate, pressure, and dispensing time at a resolution of, respectively, 20 mm/s, 0.25 bar, and 70 μm. Thus these key figures are suitable for evaluating the quality of the created waveguide's geometry. This evaluation alludes to quantify the geometrical discontinuity, which increases the optical attenuation. Consequently a relative qualitative comparison between produced waveguides regarding the expected optical attenuation is possible using the presented key figures.

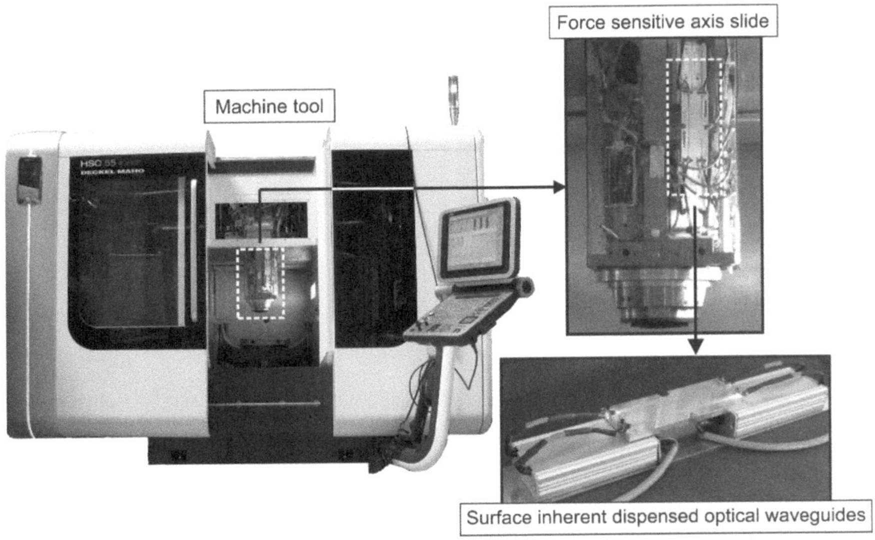

Fig. 2.5.2.6 Inherent optical waveguides integrated in a force sensitive axis slide. In the *z*-slide of the machine tool, strain gauges are integrated to measure applied forces during processing. Inherent waveguides are used to communicate between the strain gauges and a CAN Bus.

Inherent optical waveguides are integrated in a force sensitive axis-slide for a machine tool as presented in Fig. 2.5.2.6. The integrated communication structures are used for data transfer from surface integrated strain sensors into a CAN Bus. The core of applied optical waveguides in this example is about 750 μm wide and 150 μm high. The trench width and therefore the cladding width is 5 mm. This allows a functional density <2 waveguides/cm. Ormocore (as core material) and Ormoclad (as cladding material) polymers from Microresist [3] with, respectively, 1.553 and 1.534 index of refraction at 635 nm were used to create surface inherent waveguides with a 0.242 numerical aperture. The length of the applied waveguide is 100 mm and the measured attenuation is between 0.14 ± 0.02 dB/cm and 0.23 ± 0.02 dB/cm for a coupled beam with a numerical aperture, respectively, between 0.10 and 0.25.

We conclude on the basis of the presented work that it is possible to create surface inherent optical waveguides using the time-pressure dispensing method. However, the developed production environment, as well as the resultant communication structures, shows several weak spots. These weaknesses are summarized in the following Table 2.5.2.1.

Table 2.5.2.1 **The weaknesses of the dispensing of surface inherent optical waveguides**

Weak spots	**Comments**
Production environment	
End-facet preparation	• End-facet preparation is not considered in the production environment • The preparation is achieved separately in a cnc machine using a diamond milling cutter
Coupling	• The properties of beam sender and receiver are not considered during the production process • The correspondence of waveguide geometry to the beam characteristics increases the coupling efficiency
Produced surface inherent optical waveguides	
Low geometrical accuracy	• Minimal achievable $RS_{rad} = 10\%$ • Low geometrical accuracy increases attenuation
High optical attenuation	• Optical attenuation of 0.23 dB/cm by a numerical aperture NA = 0.25 • High optical attenuation excludes the possible use for sensor applications or a data transfer rate $\geq$ 1 Mbit
Low functional density	• Optical waveguide structure has a total width of 5 mm. A functional density < 2 waveguides/cm is achieved

The identified weak spots of the method developed to dispense surface inherent optical waveguides are summarized in this table. They are classified into weak spots of the production environment and weak spots of the produced waveguides.

The points mentioned in Table 2.5.2.1 are the subject of further works. The geometrical as well as optical quality of surface inherent optical waveguides has to be improved. For this reason, the production method has to be adapted to fulfill higher quality requirements. The strategy followed for this purpose is presented in the following section.

2.5.2.2 Surface integration of micro-polymer optical fibers

The lowest optical waveguide core width achieved using time-pressure dispensing is 50 μm. A lower width is not possible without specific conditioning of the substrate surface [4]. Furthermore, the resultant geometrical accuracy decreases with

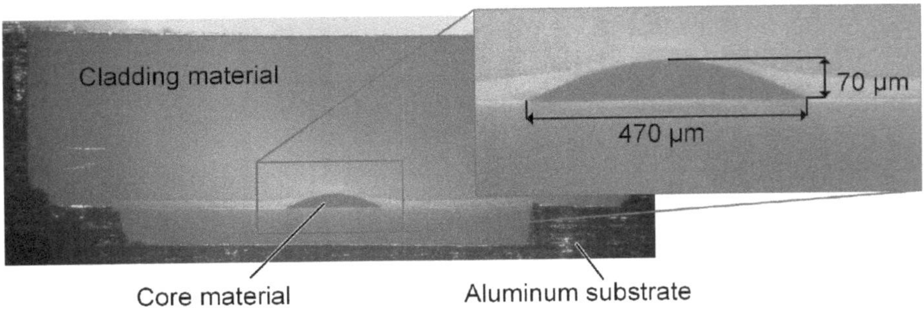

Fig. 2.5.2.7 Dispensed optical waveguide integrated inside a 3 mm wide trench structure [DUM15]. The core has a semielliptical form and is 470 µm wide. The total dimensions of the waveguide including its cladding correspond to the total size of the trench and is 3 mm wide and 1 mm high.

decreasing width [2]. In addition to the core, the cladding increases the total width of the surface inherent waveguide. Analysis based on the key figures and optical characterization previously presented delivers the best results with a 470-µm core width and a 3-mm total width as presented in Fig. 2.5.2.7.

In order to overcome these disadvantages, we investigate the use of polymer optical fibers (POF) as core material. Conventional polymer optical fibers used as optical waveguides have a diameter between 0.5 and 1 mm. POFs are produced with continuous methods and are wound up. Methods such as hot drawing and extrusion are normally used to produce kilometer long fibers [5,6]. The geometrical accuracy of waveguides produced with continuous methods is significantly higher than for those produced with the dispensing process. In this case procedures and guidelines are standardized to control a waveguide's optical, mechanical, as well as geometrical values [7–9].

The goal is to use the available know-how developed for the continuous methods to produce micro-polymer optical fibers (µ-POF). We defined µ-POFs as fibers with a core diameter $\leq$ 50 µm. The µ-POFs are then integrated within the surface to achieve a surface inherent communication structure. Further details about the production and integration method are given in the following subsections.

2.5.2.2.1 Production of µ-POFs

Optical waveguides with a diameter under 50 µm are usually produced with photochemical processes applied to a polymer directly on the substrate surface. Examples are photolocking, photopolymerization, or photobleaching

[10]. These methods need sequential steps such as coating, UV-curing using special masks and tempering [11]. In particular, the preparation of masks reduces flexibility and complicates possible automation, subject to the intended communication path.

The continuous production methods such as extrusion and hot drawing do offer the possibility of core diameters under 50 μm but mostly as multicore fiber [12] or with a significantly high cladding diameter [13]. Single core polymer fibers with diameters under 50 μm are the subject of research [14–16], as well as their diameter monitoring systems for quality control [17]. The main difficulty of achieving a smaller diameter by extrusion or drawing is in realizing feeding mechanisms to ensure a small and controlled melt volume flow.

Combining a miniaturized extruder and drawing coil, we are able to produce polymer optical cores with diameters between 10 and 50 μm. The extruder is driven at a low rotational speed and has a 250-μm nozzle. A gear pump is mounted just in front of the nozzle to allow a controlled slow volume flow. Furthermore, the fiber is drawn by the coil from the melt coming out from the nozzle at a rotation velocity of up to 600 rpm. The result is a thinning of the 250-μm fiber coming out from the extruder. The diameter is monitored while drawing using a newly developed diameter measurement system. The systems available on the market can only measure diameters down to 50 μm. Thus a new development was necessary to monitor diameters below 50 μm. The described setup is presented in Fig. 2.5.2.8.

2.5.2.2.2 Method of surface integration

The produced μ-POFs are single core fibers without cladding. We applied UV-curing adhesives as a cladding for the core and, at the same time, for inherent bonding to the substrate surface. We used a micro-dispensing process to apply the adhesive. The expertise gained from the previous works [1,2] regarding the dispensing of optical waveguides is used for this purpose. Furthermore, we maintained the same integration strategy and adapted it as presented in Fig. 2.5.2.9. In the first step, a trench structure is achieved on the substrate surface along the desired path. The second step consists of dispensing a UV-curing polymer adhesive along the path. This step constructs the lower cladding layer of the waveguide and, at the same time, provides bonding to the substrate surface. In the third step the microfiber core is placed on the first cladding part. The fourth step is achieved analog to the first one and the upper cladding layer is dispensed. The dispensing is followed by curing using UV radiation. With this method a step-index waveguide is integrated into the surface. The adhesive acts as cladding material, in addition to its bonding function.

We tested the production of the trench structure using laser, micro-milling cutters, and saw blades. A pulsed UV laser (375 nm) was tested to construct the

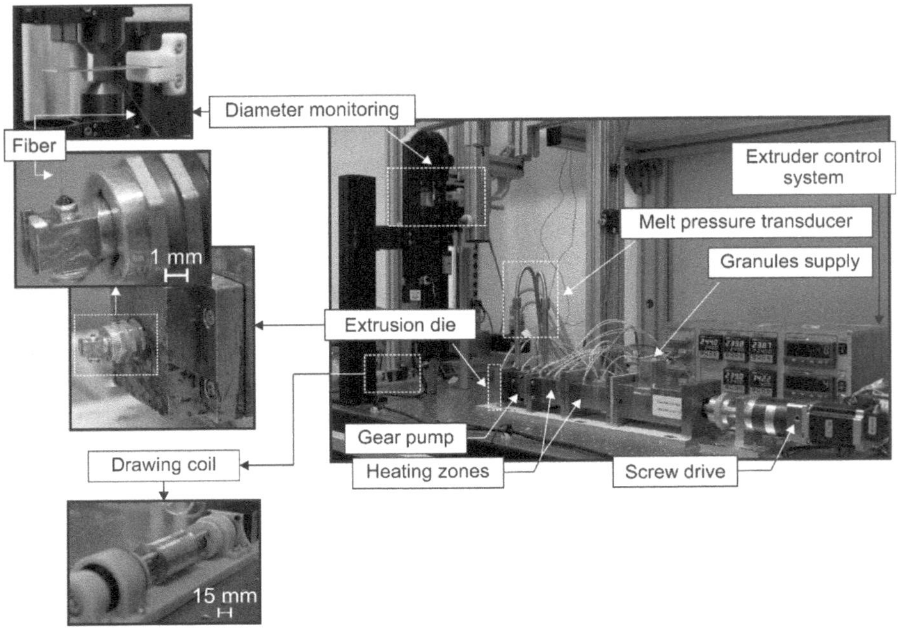

Fig. 2.5.2.8 Achieved laboratory extruder setup. It is possible to produce fiber cores between 10 and 50 μm using a miniaturized laboratory extruder combined with a drawing coil. The diameter of the fiber is monitored using a specially developed optical system.

trench structures. With a focal point diameter of 20 μm, it is possible to vary the trench geometry by adjusting the power *P*, the hatch *h*, and the number of iterations *n*. We are able to construct trenches with widths between 100 and 150 μm and depths between 4 and 300 μm. The laser scanning process to structure the trench is presented in Fig. 2.5.2.10. If the laser spot diameter, hatch, and overlapping values are known, the number of scanning lines can be used to define the width of the trench.

The number of iterations and the material removal per iteration define the depth of the trench. However, the material removal per iteration depends on the adjusted laser power as presented in Fig. 2.5.2.11. Eq. (2.2.1.7) describes the necessary time per length to structure the trench depending on the desired depth *d*, width *w*, hatch *h*, overlap *o* (in percent), pulse frequency *f*, laser spot diameter *D*, and removal rate by iteration *a(P)*, which depends itself on the adjusted laser power:

$$\frac{t}{l} = \frac{w}{h} \cdot \frac{100}{(100 - o) \cdot f \cdot D} \cdot \frac{d}{a(P)} \tag{2.5.2.7}$$

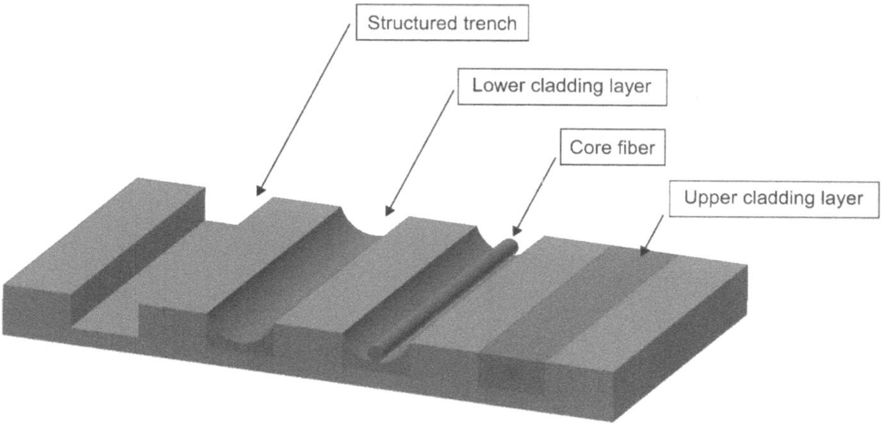

Fig. 2.5.2.9 Process steps for surface inherent optical waveguides. In the first step a trench structure is prepared on the substrate surface along the desired path. In the second step the lower cladding layer is achieved by dispensing a UV-curing polymer adhesive. In the third step the core fiber is placed on the lower cladding layer. The fourth step is achieved analog to the second step and the upper cladding layer is dispensed.

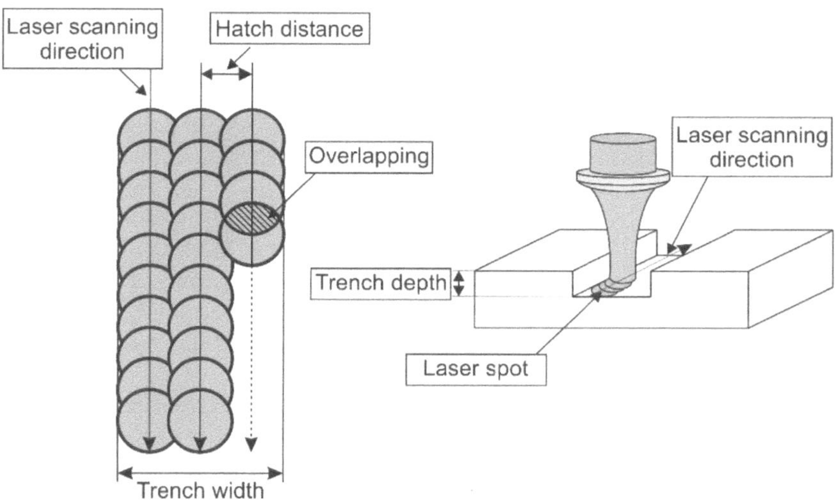

Fig. 2.5.2.10 Functionalizing the substrate surface using laser processing. The trench structure is achieved using laser processing. The width of the trench is defined by the number of scanning lines. The depth of the trench is defined by the number of iterations.

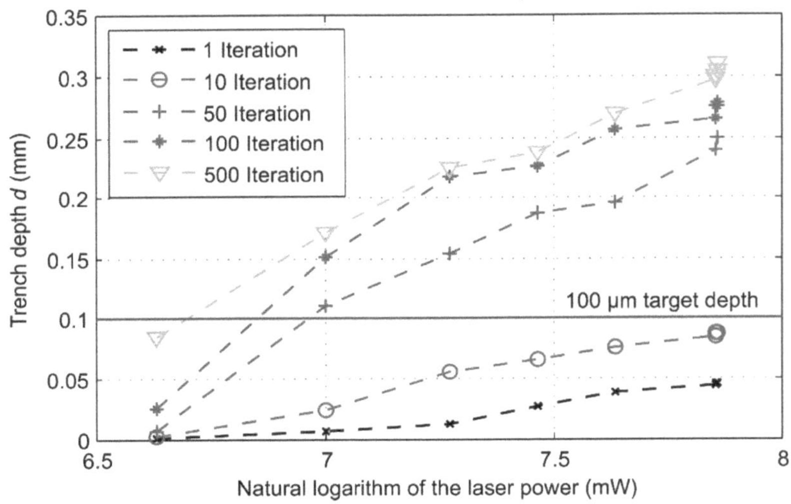

Fig. 2.5.2.11 Realized trench depth with respect to the laser power and the number of iterations. Varying the laser power between 750 and 2580 mW, as well as the number of iterations between 1 and 500, allowed a trench depth of between 0 and 0.3 mm.

Beside the complexity of the parameter sitting demonstrated by Eq. (2.2.1.7), it is a time-consuming process. Table 2.5.2.2 presents examples of structuring results for a 100 μm wide trench and the corresponding theoretical time necessary to achieve a 100 μm depth. The laser is adjusted to a 50-kHz laser pulse frequency. The scanning process has a 0.005 hatch and an overlap of about $\approx$66.67%. In particular, for good surface roughness RMS $<$ 5 μm, the time needed varies between 44 and 11 s/mm with a corresponding laser power of 760 and 2580 mW.

The removal rate by iteration does not increase linearly with the number of iterations, as shown in Fig. 2.5.2.11 and Table 2.5.2.2. With consecutive iterations the depth of the trench increases. Consequently, the surface does not remain at the focal point of the laser, but is still within the range of the Rayleigh length, at about 2.5 mm. This explains the decrease in the removal rate by a higher number of iterations.

Micro-milling cutters and saw blades are used by conventional machining processes. For their part, micro-milling cutters are sensitive to axis oscillations during trench production. High machine rigidity and linear guiding stability are required, otherwise there is a high risk of tool damage. The saw blades are stable during processing and no iterations are necessary to achieve the 100 μm depth. However, feed rate and cooling have to be controlled to avoid high burr formation, on the one hand, and high blade vibration, on the other hand,

Table 2.5.2.2 Necessary time for laser structuring with respect to the adjusted parameters

Laser power (mW) P	Number of iterations n	Adjusted trench width (µm) $w_{desired}$	Actual trench width (µm) w_{actual}	Actual trench depth (µm) d	Removal rate by iteration (µm/iteration) a	Trench roughness (µm) R_{RMS}	Necessary time per path length for a 100 µm depth (s/mm) t
760	20	100	117.8 ± 1.78	9.02 ± 11.96	0.45	3.9	44.44
760	100	100	126 ± 4.41	29.5 ± 27.08	0.29	9.89	68.96
2070	20	100	135 ± 6.04	146 ± 5.09	7.30	11.8	2.73
2070	100	100	131.6 ± 5.12	181.2 ± 3.11	1.81	4.07	11.04
2580	20	100	142.2 ± 9.55	141 ± 9.63	7.05	7.93	2.83
2580	100	100	128.4 ± 2.70	180.4 ± 7.23	1.80	12.8	11.11

Variation of laser power and iteration number to determine the removal rate by iteration. The theoretical time per path length necessary for a depth of 100 µm is determined based on Eq. (2.2.1.7).

Table 2.5.2.3 Selection list of the tested structuring methods based on Ref. [18] as well as their important characteristics

Selection list

Method	Process time per mm path length	Preparation expenditure	Tool wear	Path flexibility	Accuracy	Cost	Selection criteria: (+) Good (−) Poor (?) Unknown Comments	Σ
Laser structuring	−	−	+	+	+	−	Slow and expensive method. Low repetition accuracy. Poor surface roughness	0
Micro-milling tool	+	+	−	+	−	−	Good surface roughness. Machine tool for micro processing with high stiffness is required.	0
Saw blade	+	+	+	−	−	+	Low cost and rapid processing. Good surface roughness. Only straight path is possible.	+ +

The characteristics help to make a statement about the appropriate method of substrate surface structuring.

which leads to wavy trench edges. Furthermore, the final trench width is about 20% higher than the blade thickness and only straight linear paths are possible. Using the saw blade and a forward velocity of 0.6 mm/s, processing is 6 times faster than laser structuring. The following Table 2.5.2.3 summarizes the different surface structuring methods and evaluates their important characteristics based on Ref. [18]. For high process repeatability and a lower production expenditure, we prefer surface functionalization using the saw blade.

The smallest trench width producible with the saw blade is 300 ± 21 μm. The depth is set to 100 ± 14 μm. The structuring of a part of the surface effects its stiffness. We set a 5% maximal allowed stiffness reduction and determined the minimum part dimensions required for this purpose. First, we consider the strain stiffness k_{strain} formulated as:

$$k_{strain} = E \cdot (A_0 - A_{trensh}) > 0.95 \cdot E \cdot A_0, \\ \rightarrow \frac{A_{trench}}{A_0} < 0.05 \tag{2.5.2.8}$$

A_0 corresponds to the cross-sectional area without the trench and A_{trench} corresponds to the cross-sectional area of the trench itself. Second, we consider the flexural stiffness and a maximum allowed reduction of 5%. The bending stiffness $k_{bending}$ depends on the geometrical moment of inertia and can be expressed as:

$$k_{bending} = E \cdot (I_0 - I_{trench}) > 0.95 \cdot E \cdot I_0,$$
$$\rightarrow \frac{I_{trench}}{I_0} < 0.05 \qquad (2.5.2.9)$$

I_0 is the moment of inertia of the part without a structured trench and I_{trench} is the moment of inertia of the trench. Eqs. (2.2.1.8) and (2.2.1.9) determine the minimum dimensions of the part for stiffness greater than 95% of the part stiffness without surface structuring. These limits correspond to forces acting in the direction of the trench. For forces acting orthogonal to the direction of the trench, we use finite the element method to determine the dimension limits. Fig. 2.5.2.12 summarizes these limits and shows the thickness of the part d as a function of its width a. The unhatched area in Fig. 2.5.2.12 represents the dimensions

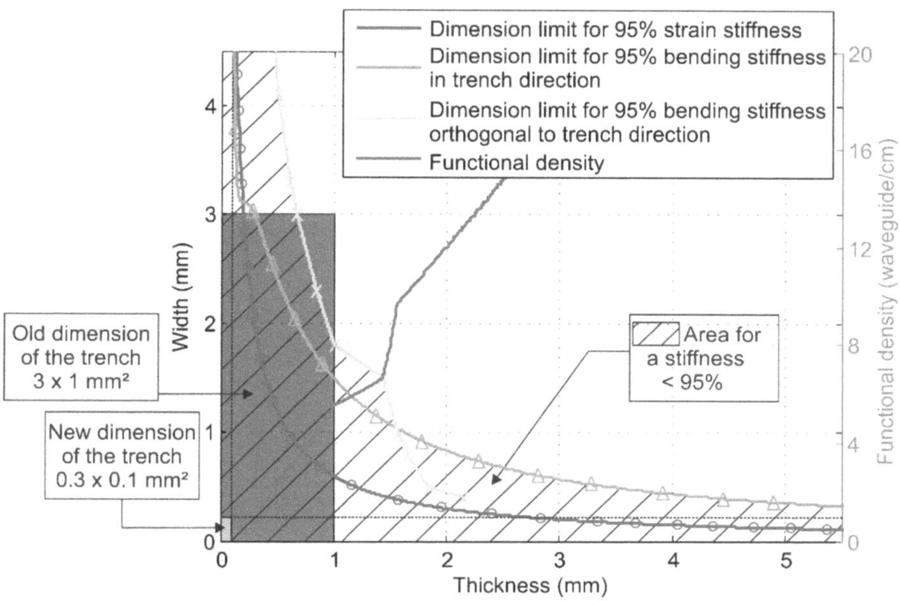

Fig. 2.5.2.12 Limits of the substrate's dimensions needed to achieve a 95% stiffness compared to the unstructured part. Furthermore, the diagram defines the maximum functional density with respect to the necessary substrate dimensions.

necessary for the structured part to have at least 95% stiffness compared to the unstructured original part.

It is possible to determine the functional density of the trenches on a substrate surface based on Fig. 2.5.2.12. Depending on substrate thickness, it varies from 2 to 17 by a thickness of, respectively, 0.5 and 3 mm. In comparison to the previous work, using the presented integration method allows a 99% reduction of the cross-section of the surface integrated waveguide and an increase in functional density of up to 90%.

Beside the production of optical cores and the dispensing of the cladding layers, the positioning and end-facet preparation have to be integrated into the process. The purpose is to achieve an automated process with a ready-for-use surface inherent optical waveguide as a product. A methodical approach is necessary to achieve the process design. We inferred a model of the resultant communication structure. Using this model, it is possible to pinpoint the relevant parameters to be considered for the process design. The model used in our case is presented in the following section.

2.5.2.3 Design of the automated integration process

The intended process has a metallic part with a structured surface as input. The output is a part with a ready-to-use surface inherent optical waveguide. The necessary steps from input to output are automated. We start by modeling the communication path to determine the relevant parameters influencing the optical properties of the optical structures produced. The most important optical property is attenuation, which defines the functionality limits of the waveguide and is caused by intrinsic and extrinsic losses. On the one hand, intrinsic losses are material dependent and cannot be influenced by the surface integration process. On the other hand, extrinsic losses are caused by fluctuations in production quality, such as geometric accuracy, impurities, or scattering. Extrinsic losses can be reduced by regulating and optimizing the production process. The total attenuation of a surface integrated waveguide is modeled in Fig. **2.5.2.13**. Total attenuation is considered as a sum of the coupling attenuation of a transmitter to the waveguide, as well as from a waveguide to the receiver and the attenuation of the waveguide itself. The total attenuation can be expressed using the following equation:

$$a_{\text{total}} = a_c + (a_w \cdot l) + a_c = 10 \cdot \log\left(\frac{P_{\text{in}}}{P_{\text{out}}}\right) \tag{2.5.2.10}$$

Every single instance of attenuation is the result of extrinsic and intrinsic losses as listed in Fig. **2.5.2.13**. This model allows us to pinpoint the fiber core geometric accuracy, the alignment accuracy, and the end-facet quality as attenuation

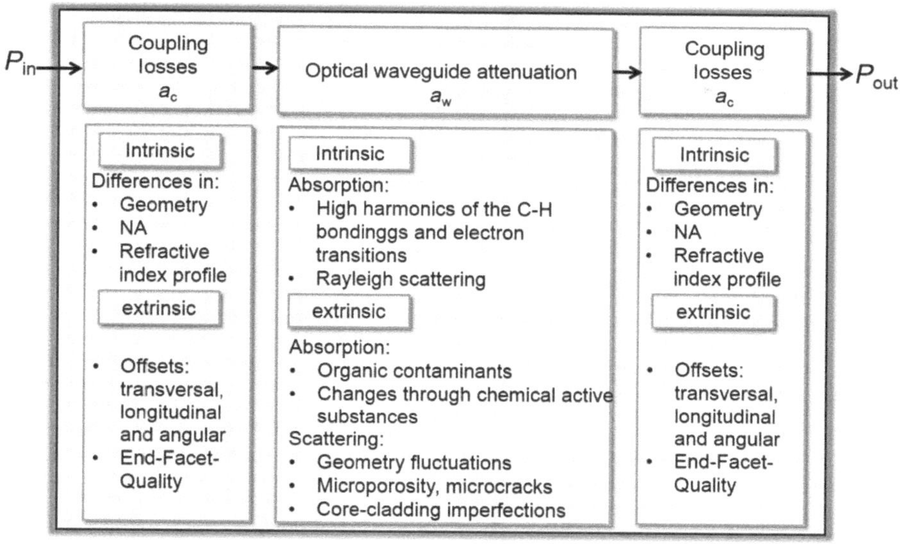

Fig. 2.5.2.13 Modulation of the total attenuation of a surface integrated waveguide. Total attenuation is considered to be the product of the coupling losses and the optical attenuation of the waveguide. They are due to intrinsic and extrinsic losses.

factors, which can be directly influenced by the process and has to be considered during its design.

In line with current technological standards, the alignment and assembly of optoelectronic systems constitute the steps with the lowest degree of automation in actual production chains and represent a relevant issue in scientific research. In waveguide-to-transmitter/receiver alignment two approaches based on the pick, measure, and place method are used [19]. In the first approach, the light intensity coupled into or out of the waveguide is measured. Using high precision positioning systems, the position is actively varied until the maximum beam intensity is reached. In the second approach, precise geometries are achieved whereby the optical components are positioned so that they can directly fit the required positioning tolerances [19]. In the first approach, the focus is placed on the required in-process image-based metrology, as well as on the chain of refined perception needed for a self-optimizing assembly of optical systems [20]. Another important point is the design of micromanipulators and the setup, which can combine active and passive alignment processes, as well as a modular concept [21]. These concepts are used in the packaging of optoelectronic and microelectromechanical systems (MEMS). The purpose here is a system-in-package (SiP), where enabling technologies such as transmitter and receiver chips, optical interconnects, and waveguides are integrated considering optical, electrical, thermal, as well as mechanical interfaces [22]. In this case, the first tolerance estimations can be determined using simulation software such as optical simulation [23] or thermal numerical simulation [24]. Current scientific research about SiP focuses on developing the following aspects [25]:

1. Reduction of the number of parts
2. Reduction of the number of assembly steps
3. Reduction of package size
4. Flexibility concerning the number of optical interfaces
5. Ability to automate the full assembly
6. Reuse of already existing assembly techniques and equipment

In order to contribute to this progress, we investigate a design for automated direct integration of μ-POFs into metallic surfaces. In this case, the number of assembly steps is reduced because of the direct integration of micro-waveguides into the surface and thus avoids intermediate packaging. Consequently, there is a fusion between the data communication route and the end-product surface. This strategy produces a System-in-Surface (SiS) instead of the known SiP. In the next subsections we show how we designed automated integration by considering the aspects listed, as well as the model presented in Fig. **2.5.2.13**.

2.5.2.3.1 Positioning accuracy

We can conclude from the model in Section 2.2.2.4.1 that the positioning accuracy is a source of extrinsic loss. The efficiency of optical coupling between transmitter or receiver and the surface inherent optical waveguide is reduced by the misalignment of the optical axes. We set a coupling loss $< 20\%$ as a target to design the process, which corresponds to a coupling attenuation of about 1 dB. At first it is necessary to determine the relation between position deviation and coupling efficiency to pinpoint the minimum accuracy and tolerances required for the process. For this purpose, we used the optical simulation software ZEMAX to make a statement about the available tolerance scope. This scope also depends on the coupling method used, as well as the optical properties of the waveguide and beam source considered. In the case presented we considered a laser diode with a typical wavelength of 655 nm and a full beam divergence of, respectively, 8° in the parallel direction and 32° in the perpendicular direction. The waveguide considered has a 15-μm core and cladding material with an index of refraction of 1.58 and 1.48, respectively. We considered butt coupling and the use of a GRIN lens to infer the influence on the tolerance range [23]. The setups constructed in ZEMAX for these coupling methods are presented in Fig. 2.5.2.14.

The ideal position of the transmitter to the waveguide is determined based on the beam divergence and the numerical aperture of the surface integrated waveguide.

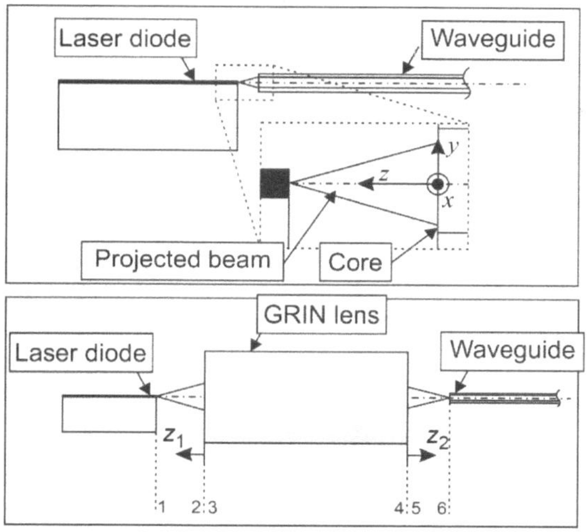

Fig. 2.5.2.14 Constructed setup for the ZEMAX simulation. Starting from the ideal position, the beam source was displaced in the 6 degrees of freedom until a coupling efficiency limit of 80% was reached. Figure a and b represent, respectively, the setup for a butt coupling and a GRIN lens coupling.

Starting from this ideal position, the beam source is moved sequentially in all 6 degrees of freedom until a limit coupling attenuation of 1 dB is reached. Fig. 2.5.2.15 shows the simulation results for the butt coupling. The positioning accuracy should ensure at least 24 μm in the *z*-direction. Depending on this *z*-distance, an accuracy in the *x*- and *y*-direction should be ±7.5 μm at $z = 0$ μm and ±3.5 μm at $z = 19.3$ μm, respectively. Limits for yaw and pitch, determined by the simulation, vary from around ±15° to ±6° depending on the *z*-distance.

Two positions are relevant for the coupling when using a GRIN lens. The first one is the position of the laser diode to the GRIN lens. The second one is the position of the GRIN lens to the waveguide. Starting from the ideal values of these positions the laser diode as well as the GRIN lens is also moved in all 6 degrees of freedom until a limit coupling attenuation of 1 dB is reached. Placing the GRIN lens at a distance of 150 μm from the laser diode allows a 51.2-μm tolerance limit in the z-direction of the waveguide position relative to the GRIN lens. The (*x*,*y*)-positioning tolerances for, respectively, laser diode and GRIN lens achieve approx. ±5 μm and ±3 μm. These results as well as a 3D presentation of the tolerance scope are shown in Fig. 2.5.2.16.

The yaw and pitch limits also depend on the *z*-distance. For the laser diode relative to the grin lens the values vary between 0° and ±6° for yaw and 0° and ±11° for pitch. Considering the position of the GRIN lens relative to the

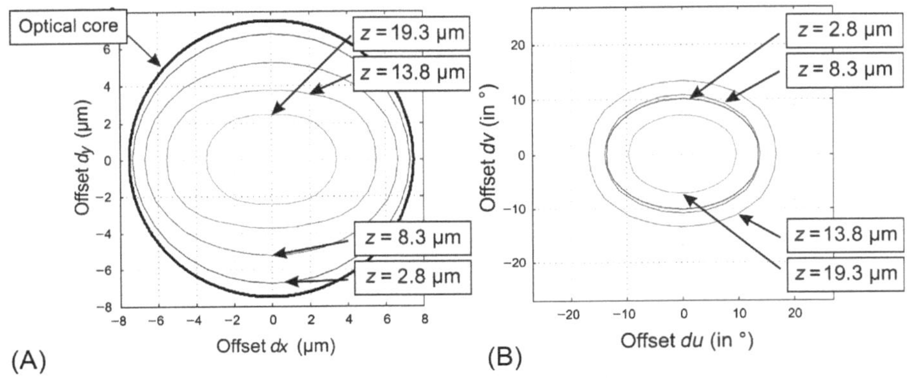

Fig. 2.5.2.15 Simulation results of misalignment limits. Subfigure (A) presents offset limits in *x* and *y*-directions for 1 dB coupling loss. Subfigure (B) presents offset limits in yaw and pitch for a 1-dB coupling loss.

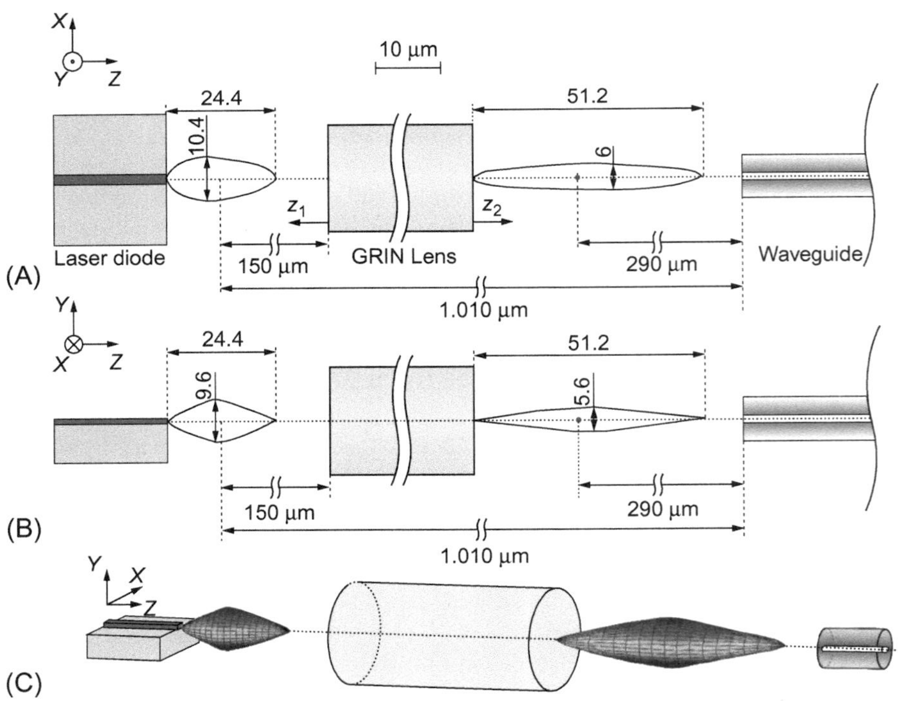

Fig. 2.5.2.16 Simulation results of *x* and *y* misalignment limits. The figure (A) and (B) present the offset limits in *x*- and *y*-direction. Figure (C) presents a 3D illustration of the tolerance scope for a 1-dB coupling loss. The limits in *x*- and *y*-direction depend on the actual *z*-position.

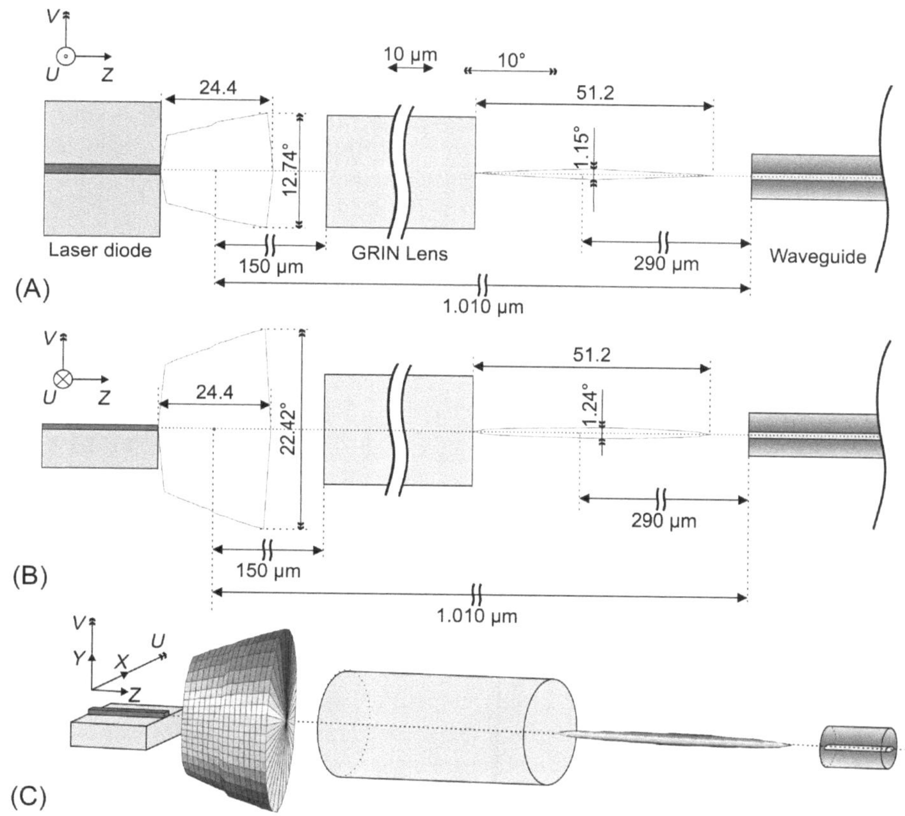

Fig. 2.5.2.17 Simulation results of yaw and pitch limits. The figures (A) and (B) present the yaw and pitch limits. Figure (C) present a 3D illustration of the tolerance scope for a 1-dB coupling loss. The yaw and pitch limits depend on the actual *z*-position.

waveguide, the limits for yaw and pitch are around ±0.6°. These results are presented in Fig. 2.5.2.17. Comparing the tolerance scope of butt coupling and GRIN lens coupling reveals a higher tolerance in the *z*-direction using the GRIN lens. However, in case of the GRIN lens coupling two relative positions have to be considered. The first one is the position of the laser diode relative to the GRIN lens. The second one is the position of the GRIN lens relative to the waveguide. This fact hinders the handling and positioning of laser diode and GRIN lens, since both tolerance scopes of these positions depend on each other. Furthermore, the yaw and pitch tolerance values for the GRIN lens relative to the waveguide are 50% lower than with butt coupling. Consequently, the V-groove needed to place the GRIN lens has to have corresponding geometrical accuracy. In conclusion, the butt coupling method is more suitable for designing automated processes for the surface integration of optical waveguides. For the z-direction, which has the largest tolerances up to 24 μm, we retain use of the linear motor from Etel Motion Technology available in the machine setup with a 3.8-μm resolution and a max. position error of ±0.2 μm. In the case of a maximum error of 4 μm, we infer a required min. accuracy of around 4 μm from the simulation for *x*- and *y*-direction. This accuracy approaches the limit accuracy of the available linear motors. For this reason we complement the Etel linear motor, used for the *x*- and *y*-axes in the machine setup, with piezo-motors from Nanomotion with a 10-nm resolution and a max. 50-nm position error.

2.5.2.3.2 Dispensing process

The dispensing process is used to construct the lower and upper cladding layers as well as enabling bonding to the substrate. Furthermore, the lower cladding defines the height of the core inside the trench. This height is important for positioning and coupling at the end-facet but also for lower bending losses along the path. The dispensing system should fulfill two important requirements. The first requirement is a selective and continuous process with high repeat accuracy. The second requirement is the ability to adjust small volume flow due to the low geometrical dimensions of the structured trench. We selected the ecoPen 300 dispensing unit for the volumetric dosing system. The volume flow can be adjusted from 0.12 mL/min up to 1.48 mL/min. A pressure sensor is integrated in the dispensing unit for process monitoring. We built an experimental setup to characterize the dispensing unit. Its purpose was to select the optimal process parameters, as well as to identify the volume flow deviation using measured pressure values. In this setup, the micro-dispensing unit is placed above a high precision scale. The pressure values are measured online using a Labview data acquisition module. The measured weight using the precision scale is communicated via a serial port to the computer. The measurement of the pressure and weight values is synchronized using a transmitted process start and stop signal from the control unit to the computer. The results are time synchronized measurement data of the dispensing pressure and dispensed weight.

The adhesive Polytec UV 2108 used manifests Newtonian fluid characteristics. Consequently, the volume flow can be deduced based on the Newton rheological model. The following equation is used for the volume flow model [26]:

$$\dot{V}(p) = \frac{\pi \cdot D^4}{128 \cdot \eta \cdot l} \cdot (p - p_0) \cdot k \tag{2.5.2.11}$$

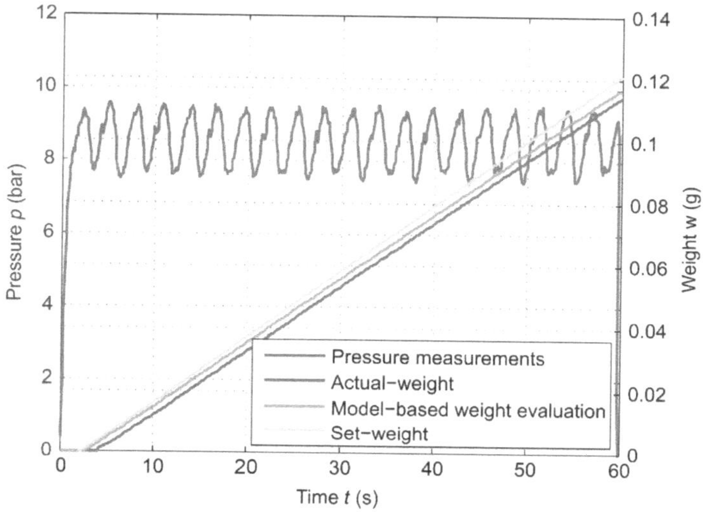

Fig. 2.5.2.18 Measured pressure p and actual dispensed weight w during a 1-minute dispensing process. The set weight and the model-based weight are presented. The set weight deviates with 6.14% from the actual weight, while the model-based weight deviates with only 1.75% from the actual weight.

The model parameter D represents the inner diameter of the dispensing capillary, l the length of the capillary, and η the viscosity of the Newtonian fluid used. The pressure p is the measured pressure in the capillary and p_0 is the pressure at rest. The factor k is a correction factor to adapt the model to the measured volume flow. In the following Fig. **2.5.2.18** an example of the pressure value, as well as the fluid weight measured during the dispensing process, is presented. Furthermore, the set and the model mass flow curve are presented. Using the model allows an estimation of the dispensed volume with an accuracy between ±1.3% and ±2% depending on the dispensing parameters. The deviation between set and actual weight is considerably higher and has values between ±5% and ±10%. The developed model helps to determine the actual dispensed volume with a higher accuracy than the set volume. Furthermore, it is possible to determine the dispensed volume during the process by monitoring the pressure values.

If we consider the trench geometry as well as the interfacial dynamic between substrate surface and the applied adhesive, the wetting behavior can be theoretically determined. The wetting property is characterized by the surface tension σ_S (S: solid) between the substrate and ambient gas, the surface tension σ_L (L: liquid) between the liquid polymer and ambient gas, the surface tension σ_{LS} between the used liquid polymer and substrate, as well as by the contact angle θ between the different surface tensions as defined by the Young equation. If the polar component σ^p and the disperse component σ^d of the surface tension of a liquid and a solid are known, the surface tension σ_{LS} and the contact angle can be determined by the following equations [27]:

$$\sigma_{LS} = \sigma_S + \sigma_L - 2 \cdot \left(\sqrt{\sigma_S^d \cdot \sigma_L^d} + \sqrt{\sigma_S^p \cdot \sigma_L^p} \right) \quad (2.5.2.12)$$

$$\cos\theta = \frac{\sigma_S - \sigma_{LS}}{\sigma_L} \quad (2.5.2.13)$$

Using Eqs. (2.5.2.12) and (2.5.2.13) to determine the angle of contact and considering a finite volume element dV of the dispensed polymer along the trench, its cross-section A can be characterized as presented in Fig. 2.5.2.19. The capillary rise at an infinite vertical board depends on the distance to this board. However, if the considered volume is small enough, the effect of gravity can be neglected and the capillary rise can be considered as circular.

In Fig. 2.5.2.19 the geometrical relation of the dispensed first cladding part is described, where the dimension c represents the sagitta of the presented circle. The height h can be formulated as:

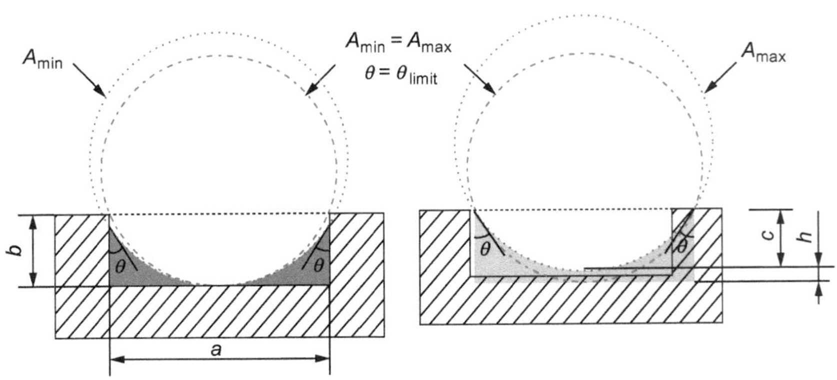

Fig. 2.5.2.19 Geometrical description of the cross-section A of a volume element dV. The figure shows the different geometrical dimensions to describe the cross-section. Thereby, three limits were defined: A_{min}, A_{max}, and θ_{limit}. A_{min} describes the minimal necessary cross-section to just coat the bottom of the trench as presented in subfigure (A). A_{max} is the maximum cross-section before overfilling the trench as presented in subfigure (B). Θ_{limit} corresponds to the limit contact angle, where A_{min} is equal to A_{max}.

$$h = b - c, \tag{2.5.2.14}$$

$$h = b - \frac{a}{2} \cdot \left(\frac{1 - \sin\theta}{\cos\theta}\right), \tag{2.5.2.15}$$

the volume dV is limited by the maximum and minimum values $dV_{\max}$ and $dV_{\min}$ corresponding, respectively, to $h = 0$ (minimum coating of the trench) and $h = h_{\max}$ (maximum coating without overfilling the trench). Based on the geometrical relation in Fig. 2.5.2.19, these limits can be calculated using following equations:

$$\begin{aligned} &dV_{\min} < dV < dV_{\max}, \\ &dA_{\min} \cdot dl < dV < dA_{\max} \cdot dl, \end{aligned} \tag{2.5.2.16}$$

$$\begin{aligned} &\left(a \cdot c - \frac{a^2}{8 \cdot \cos\theta}((\pi - 2\theta) - \sin(\pi - 2\theta))\right) \cdot dl < dV \\ &< \left(a \cdot b - \frac{a^2}{8 \cdot \cos\theta}((\pi - 2\theta) - \sin(\pi - 2\theta))\right) \cdot dl \end{aligned} \tag{2.5.2.17}$$

The volume element dV can be expressed as the product of the volume flow $\dot{V}$ and time dt. The available time for dispensing the lower cladding layer is defined through the forward velocity and the length of the trench. This information allows us to express Eq. (2.5.2.17) as follows:

$$\begin{aligned} &dA_{\min} \cdot dl < \dot{V} \cdot \frac{dl}{v} < dA_{\max} \cdot dl, \\ &\frac{\dot{V}}{dA_{\max}} < v < \frac{\dot{V}}{dA_{\min}} \end{aligned} \tag{2.5.2.18}$$

Depending on the adjusted volume flow, the limits of forward velocity can be determined using Eq. (2.5.2.17). Using Eq. (2.5.2.11) the actual volume flow can be determined from the pressure values. Combining Eqs. (2.5.2.11) and (2.5.2.17), the actual limits of forward velocity are calculated and used to achieve online controlling of the forward velocity.

2.5.2.3.3 End-facet preparation

After production, positioning and bonding of the optical waveguide, end-facet preparation is necessary to establish the optical communication route. In scientific publications, multiple methods have been proven and tested as presented in Ref. [28]. Based on low complexity, we selected and tested the following methods for possible consideration in the process design:

- Milling and polishing
- Hot plate
- Hot cleaving

Table 2.5.2.4 RMS surface roughness of the fiber cutting area from blade cleaving

		Blade temperature		
		22°C	**80°C**	**120°C**
Substrate temperature	22°C	1.738 ± 0.955 μm	0.629 ± 0.131 μm	0.851 ± 0.124 μm
	80°C	–	0.512 ± 0.092 μm	0.772 ± 0.106 μm

The methods and experimental setup we used for these tests are presented in Ref. [29]. At first we tested the hot blade cleaving method as a possibility for end-facet preparation. We created metallic samples with integrated optical waveguides using polycarbonate (PC) fiber as core and UV-curing adhesive polytec UV2108 as cladding [29]. Then we prepared their end-facets by varying the temperature as presented in Table 2.5.2.4. The RMS roughness measurement was performed with a confocal microscope and a 50 times magnification objective lens (Nanofocus, μsurf custom system).

The best values were achieved with a substrate and blade temperature of 80°C. A temperature higher than 80°C exceeded the continuous operating temperature of the applied adhesive. Thereby, the bonding of the fiber while cutting is not sufficiently assured and the cut quality decreases.

Second, we tested milling and polishing as an established preparation method. However, to achieve the best results many process steps are necessary. The surface quality can be increased with more polishing steps, which have less difference in grain size. Better results than the blade cleaving method were possible in this case as summarized in Table 2.5.2.5. Nevertheless, swarf and grinding dust are produced while proceeding and should be extracted by vacuum. This swarf is due to the fact that the optical fibers are integrated into the metal surface. Polymer and aluminum surfaces are simultaneously machined. In particular during polishing, aluminum particles contaminate the polishing tool (see Fig. 2.5.2.20) as well as the fiber end-facet. Consequently, the polishing tool should be replaced often as the end-facet opacity increases. This explains the higher RMS roughness measured in comparison to results presented in Ref. [30].

The third method we tested was the hot plate. Using this method, the integrated fibers are cut and pressed against a hot metallic mirror. The projection length of the fiber is important to achieve optimal results [13]. An example of a projection length influence is presented in Fig. 2.5.2.4. The metallic mirror is heated up to 120°C. At this temperature the core material is close to the glass transition temperature T_g but still below it. At a temperature of around or higher than T_g, the polymer demonstrates a brittle behavior which leads to poor surface quality. With

Table 2.5.2.5 RMS surface roughness of the fiber cutting area from the milling and polishing process

	Translation speed (m/min)	Rotation speed (rpm)	Used tool	Description	RMS surface roughness (μm)
Milling	0.05	120,000	Three-flute router	HSS-Co8 TiAIN	0.683 ± 0.041
	0.25	17,000	Three-flute router	HSS-Co8 TiAIN	0.972 ± 0.157
	0.50	17,000	Three-flute router	HSS-Co8 TiAIN	1.413 ± 0.045
	0.50	34,000	Three-flute router	HSS-Co8 TiAIN	1.014 ± 0.052
	0.05	120,000	Mounted wheel (68 μm)	Winterthur 1A1W	1.154 ± 0.068
Polishing	56	150	Sandpaper (15.3 μm)	Presi Mecatech 334	0.877 ± 0.094
	56	150	Diamond suspension (1 μm)	Presi Mecatech 334	0.041 ± 0.029

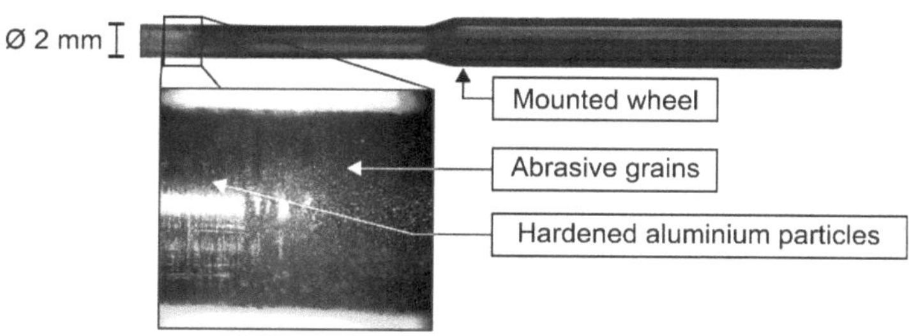

Fig. 2.5.2.20 Damage of the grinding tool due to aluminum particles. The aluminum swarf contaminates the grinding tool during the process and leads to bad quality of the end-facet surface.

this method an RMS roughness value of 0.852 ± 0.087 μm is achieved. A cleaving process is always required for the hot plate, which increases the cost and complexity of this method in comparison to a simple hot blade cleaving. To achieve a hot blade cleaving, a heating cartridge is necessary as well as a temperature sensor to regulate the process. The blade translation during cutting can be

Table 2.5.2.6 **Selection list based on Ref. [29] for the tested end-facet preparation method**

Selection list

Method	RMS surface roughness	Number of steps	Actuator	Sensor	Accuracy	Cost	Selection criteria: (+) Good (−) Poor (?) Unknown Comments	Σ
Blade cleaving	−	+	−	+	+	+	Known and tested method within scientific research with low level of integration complexity	++
Milling and polishing	+	−	−	+	+	−	Good surface roughness, but many machining steps are required. Unsuitable for hybrid materials (polymer and metal)	0
Hot plate	−	−	+	+	−	+	Simple principle but a cleaving process is also required with precise slice	0
UV-curing polymer	+	+	−	−	+	−	Good results depending on the primary surface roughness of the end-facet. In combination with other method possible to decrease coupling loss	+

The selection list helps to select the most adequate method based on the defined criteria as listed.

achieved using the gantry robot axis, which requires no extra actuator. Combined with the index matching polymer, the cleaving method can be considered as a solution for the automated process (see Table 2.5.2.6).

2.5.2.4 Surface integration of μ-POFs in application

In the previous section we pinpoint the causes of extrinsic loss, which can be reduced by controlling the surface integration process. The three main causes are as follows:

- Misalignment of the optical axis between transmitter, receiver, and optical waveguide
- The geometric accuracy of the cladding layer
- End-facet quality

For the position accuracy we determined the tolerance scope for theoretical coupling losses under 1 dB based on optical simulation. The accuracy required in the x- and y-direction is 4 μm. This range is in the limit of accuracy of the axis available in the machine setup. For this purpose we complement these axes with piezo-motors with an accuracy of 0.01 μm for a finer positioning.

Considering the geometric accuracy of the cladding layer, we analyzed the dosing process and its stability during the dispensing. The results show an instability of the volume flow, which can be measured by monitoring the flow pressure in the dosing capillary. Knowing the actual volume flow based on the pressure values, the forward velocity during dispensing can be adapted to ensure constant application of the liquid polymer.

For the last point we chose blade cleaving for end-facet preparation. The choice is due to the simplicity of the cleaving method, the low complexity of integration into the process, as well as the relatively good properties of the resultant end-facet surface as presented in Section 2.2.2.5.

To take advantage of the high accuracy of the piezo-motors it is necessary to locate, on the one hand, the tool position and, on the other hand, the substrate position within the machine coordinate system to a high accuracy. In our case, the dispensing capillary as well as the micro-gripper used to position the optical core fiber is considered as tools. A laser measuring head ODAC xy-18 from Zumbach Electronic is used to determine the (x,y,z)-position of the tools. The head is mounted at a defined position in the machine setup, as presented in Fig. 2.5.2.21, and has a defined measuring field.

Immersing the tool in this field delivers a feedback of its position inside the field with ±0.1 μm accuracy. If the position of the measurement field in the machine coordinate system is known, it is possible to determine the (x,y)-position of the tool itself. Furthermore, it is possible to determine the z-position based on the immersing velocity of the tool into the field as well as the reaction time of the measuring system. With continuous tool immersion, possible changes in its (x,y)-position can be also observed, which reveal a possible inclination, e.g., of the dispensing capillary.

Beside the tools' positions, it is also necessary to determine the substrate surface position in the machine coordinate system. In this case we deploy the confocal sensor IFS2402-0.4 from Micro-Epsilon Messtechnik. In order to achieve these measurements, the confocal sensor is moved toward the surface until this is located within the measurement range. As the measurement range is known to be 0.4 μm, the distance to the surface can be determined and consequently the z-position as well. Then the confocal sensor is moved toward the surface edges. The edge has been reached as soon as the confocal sensor is no longer in its measurement range. Performing this measurement to two different edges allows us to determine the (x,y)-position of the substrate surface. This initialization routine is always executed at the start of the process, following the structure presented in Fig. 2.5.2.22.

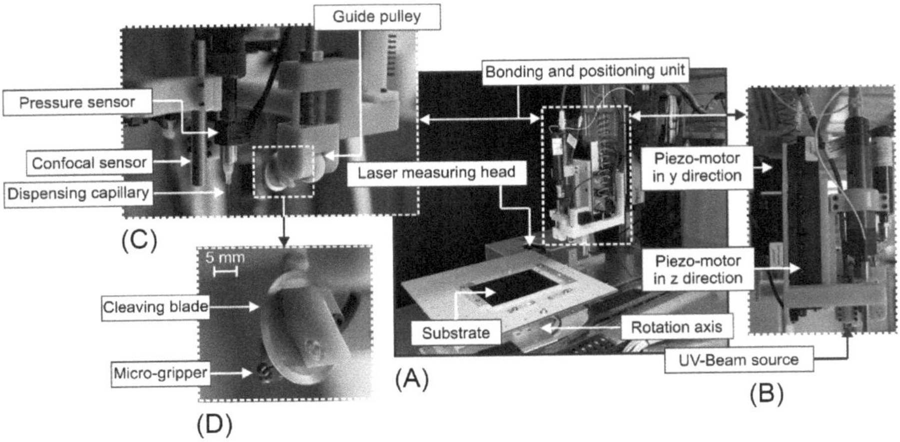

Fig. 2.5.2.21 Newly designed machine setup used to create surface integrated μ-POFs. A perspective view of the new machine setup is presented in subfigure (A). Subfigure (B) presents a side view of the bonding and positioning unit with an indication for the position of the piezo-motors as well as the UV beam source. Subfigure (C) represents a zoom into the unit to show the positions of the confocal sensor, pressure sensor, and dispensing capillary. An enlargement of the micro-gripper and the cleaving blade is presented in subfigure (D).

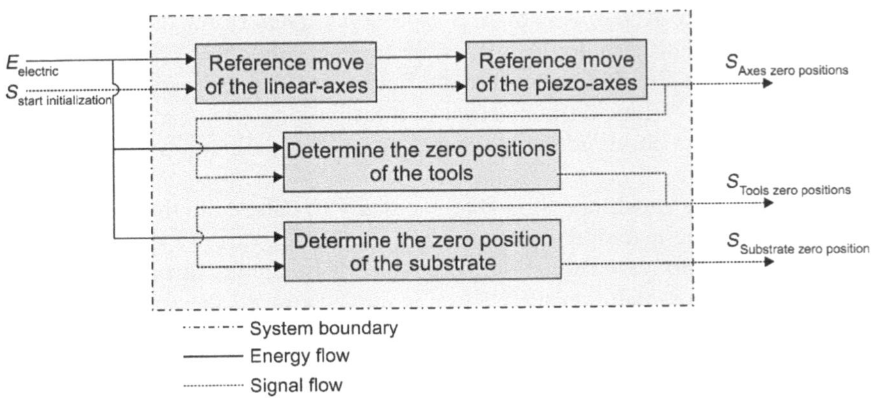

Fig. 2.5.2.22 Initialization routine of the system. With the start signal for initialization, the system begins with a reference displacement of the linear axes, followed by a reference displacement of the additional piezo-axes. Afterward, the zero position of the tools and the substrate is determined successively. Finally the determined positions are communicated to the control system.

Further down the process stream, the confocal sensor is used to measure the trench position and geometry. Based on this information, it is possible to determine the deviation from the set geometry. Consequently, a correction of the dispensing capillary along the trench in (x,y,z) is performed using the additional piezo-motors to maintain the same boundary condition during polymer application. With the measured trench geometry, it is possible to divide the trench into finite volume elements and determine the necessary volume to be dispensed in each volume to guarantee the same filling level and quality. By monitoring the volume flow based on the measured pressure values, the forward dispensing velocity can be determined and regulated based on Section 2.2.2.4.3. This routine is performed for the lower, as well as upper, cladding layer as presented in Fig. 2.5.2.23.

In further work, the regulation of the system presented earlier will be coupled to a graphic user interface. This interface allows the operator to upload the coordinates of the desired path, as well as adapt different settings. Setting examples include the number of finite elements to be considered for the dispensing process and consequently the number of points measured with the confocal sensor along the path. Furthermore, we integrate a rotation axis to the system as presented in

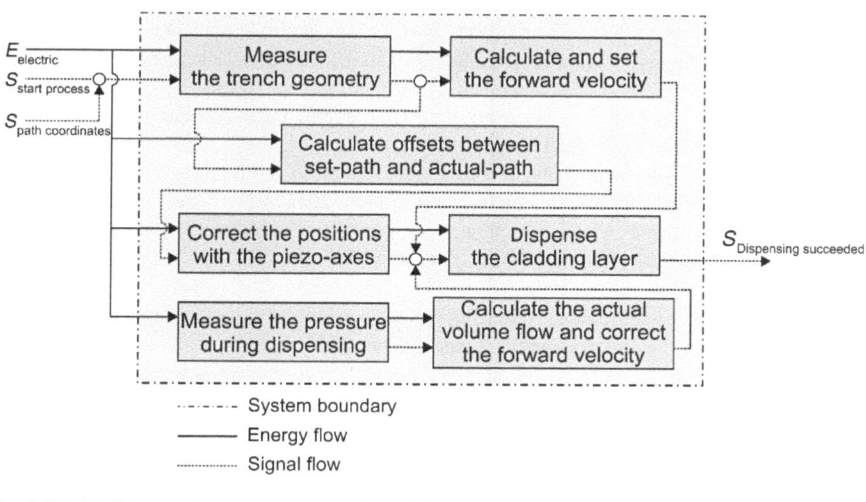

Fig. 2.5.2.23 Dispensing process routine. With the start of the process, the trench is measured using the confocal sensor. The necessary forward velocity during dispensing, as well as the offsets between set path and actual path, is measured based on the measurement data. Afterward, dispensing begins and the pressure is measured online. From the pressure measurement, the actual volume flow is calculated and, accordingly, forward velocity is corrected.

Fig. 2.5.2.21 in order to be able to achieve curved communication paths into substrate surfaces. For this purpose, it is also necessary to consider required transformation between the machine coordinate system and the polar coordinate system of the rotation axis.

References

[1] T. Fahlbusch, Dispensieren polymer Lichtwellenleiter, Dissertation Leibniz Universität Hannover, Berichte aus dem ITAPZH-Verlag, Garbsen, 2007.

[2] M. Dumke, Untersuchungen zur optischen Dämpfung mikrodispensierter Wellenleiter, Diessertation Leibniz Universität Hannover, Berichte aus dem ITA, PZH-Verlag, Garbsen, 2015.

[3] Micro resist technology GmbH, OrmoCore and OrmoClad datasheet. *http:/www.micro resist.de/sites/default/files/download/PI_OrmoCore_OrmoClad_2015_0.pdf*, 2015 (accessed 03.10.16).

[4] G. Hoffmann, T. Reitberger, J. Franke, L. Overmeyer, Conditioning of surface energy and spray application of optical waveguides for integrated intelligent systems. in: Proc. 3rd Joint Symposium on System-integrated Intelligence: New Challenges for Product and Production Engineering, SysInt 2016, June 13th–15th, Paderborn, Germany, 2016, pp. 169–176, http://dx.doi.org/10.1016/j.protcy.2016.08.023.

[5] E. Hering, R. Martin, Photonik: Grundlagen, Technologien und Anwendung, Springer Verlag, Berlin, Heidelberg, Deutschland, 2006.

[6] W. Scheel, Optische Aufbau- und Verbindungstechnik in der Elektronischen Baugruppenfertigung, Verlag Dr. Markus A. Detert, Templin/Uckermark, Deutschland, 2002.

[7] DIN 58141-1, Messung von faseroptischen Elementen, Deutsche Fassung DIN 58141-1:2015, Beuth, Berlin, Deutschland, 2015.

[8] DIN EN ISO 12180, Geometrische Produktspezifikation (GPS), Deutsche Fassung DIN EN ISO 12180-1:2011, Beuth, Berlin, Deutschland, 2011.

[9] VDI/VDE Richtlinie 5570, Prüfung von konfektionierten und unkonfektionierten Kunststoff-Lichtwellenleitern (POF), Deutsche Fassung ICS 33.180.01, Düsseldorf, Deutschland, 2006.

[10] O. Ziemann, POF-Handbuch: optische Kurzstreckenübertragungssysteme. 2. Auflage, Springer Verlag, Berlin/Heidelberg, Deutschland, 2007.

[11] P. Henzi, UV-induzierte Herstellung monomodiger Wellenleiter in Polymeren, Dissertation Forschungszentrum Karlsruhe in der Helmholtz-GemeinschaftWissenschaftliche Berichte Forschungszentrum FZKA 6978, Karlsruhe, Deutschland, 2004.

[12] G. Barton, M. Van Eijkelenborg, G. Henry, M. Large, J. Zagari, Fabrication of microstructured polymer optical fibres, J. Opt. Fiber Technol. 10 (2004) 325–335.

[13] D. Bosc, C. Toinen, Full polymer single-mode optical fiber, IEEE Photon. Technol. Lett. 4 (7) (1992) 749–750.

[14] A. Nain, C.A. Fellow, M. Sitti, Proximal probes based nanorobotic drawing of polymer micro/nanofibers, IEEE Trans. Nanotechnol. 5 (5) (2006) 499–510.

[15] X. Huo, S. Pan, S. Wu, Fabrication of optical fiber probe nano-tips by heated micro-pulling combined with static chemical etching, in: Proc. 1st IEEE Conference on Nano/Micro

Engineering and Molecular Systems, NEMS 2006, January 18th–21th, Zhuhai, China, 2006.
[16] B. Li, X. Xing, N. Lupu (Ed.), Polymer Nanowires, Nanowires Science and Technology, INTECH Press, Croatia, 2010.
[17] F. Warken, H. Giessen, Fast profile measurement of micrometer-sized tapered fibers with better than 50-nm accuracy, Opt. Soc. Am., Opt. Lett. 29 (15) (2004) 1727–1729.
[18] G. Pahl, W. Beitz, J. Feldhusen, K. Grote, Konstruktionslehre: Grundlagen erfolgreiche Produktentwicklung, Methoden und Anwendung, Springer Verlag, Berlin/Heidelberg, Germany, 2007, pp. 162–182.
[19] G. Boettger, H. Schroeder, R. Jordan, Active or passive fiber-chip-alignment: approaches to efficient solutions. in: Proc. SPIE Optoelectronic Interconnects XIII, Vol 8630 Integration Technologies, February 2nd, San Francisco, USA, 2013http://dx.doi.org/10.1117/12.2014176.
[20] S. Haag, D. Zontar, J. Schleupen, T. Müller, C. Brecher, Chain of refined perception in self-optimizing assembly of micro-optical systems. J. Sens. Sens. Syst. 3 (1) (2014) 87–95, http://dx.doi.org/10.5194/jsss-3-87-2014.
[21] C. Brecher, N. Pyschny, S. Haag, L.V. Guerrero, Automated alignment of optical components for high-power diode lasers. in: Proc. SPIE High-Power Diode Laser Technology and Applications X, vol. 8241 High-Power Packaging and Cooling, January 21st, San Francisco, USA, 2012http://dx.doi.org/10.1117/12.910173.
[22] T. Tekin, Review of packaging of optoelectronic, photonic, and MEMS components. IEEE J. Sel. Top. Quantum Electron. 17 (3) (2011) 704–719, http://dx.doi.org/10.1109/JSTQE.2011.2113171.
[23] B. Hachicha, J. Kuklik, L. Overmeyer, Simulation of coupling efficiency for surface integration of optical waveguides. in: IEEE Proc. 6th International Conference on Modeling, Simulation, and Applied Optimization, ICMSAO 2015, May 27th–29th, Istanbul, Turkey, 2015http://dx.doi.org/10.1109/ICMSAO.2015.7152259.
[24] G. Gao, H. Wang, G. Yang, H. Zhu, Thermal numerical simulation for advanced package development. in: Proc. IEEE International Conference on Electronic Packaging Technology & High Density Packaging, ICEPT-HDP 2009, August 10th–13th, Beijing, China, 2009, 2009http://dx.doi.org/10.1109/ICEPT.2009.5270755.
[25] B. Velsher, Application-specific optoelectronic packaging. in: Proc. 52nd Electronic Components and Technology Conference, ECTC 2002, May 28th–31st, San Diego, USA, 2002http://dx.doi.org/10.1109/ECTC.2002.1008190.
[26] F. Roess, Charakterisierung des Fließverhaltens der Dosiereinheit ecoPEN 300 und Ermittlung der Korrelation zwischen Volumenstrom und Druckmesswerten, Bachelorarbeit Leibniz Universität Hannover, Institut für Transport- und Automatisierungstechnik, Garbsen, 2016.
[27] C. Miller, P. Neogi, Interfacial phenomena: equilibrium and dynamic effects, second ed., Surfactant Science Series, vol. 139, CRC Press/Taylor & Francis Group, Boca Raton, FL, 2008.
[28] D. Moll, H. Poisel, Polymer optical fiber termination—a never ending story, in: Proc. 9th International Conference on Plastic Optical Fibers, POF 2000, September 5th–8th, Boston, USA, 2000.

[29] B. Hachicha, L. Overmeyer, Metal surface inherent micro-POFs, in: Proc. 24th International Conference on Plastic Optical Fibers, POF 2015, September 22nd–24th, Nuremberg, Germany, 2015, pp. 295–301.

[30] H. Baghsiahi, K. Wang, W. Kandulski, R. Pitwon, D. Selviah, Optical waveguide end-facet roughness and optical coupling loss. J. Lightwave Technol. 31 (16) (2013) 2659–2668, http://dx.doi.org/10.1109/JLT.2013.2271952.

Applications in manufacturing

3

Increasing competitive pressure and rising demands on small-lot production result in high requirements for today's manufacturing technologies. Consequently, companies need to combine the flexibility of a workshop production and the efficiency of a high volume production. Moreover, agile production planning and scheduling becomes increasingly important. Digitalization and the use of gentelligent machine tools and workpieces may offer a solution to meet these challenges. The vision of a gentelligent production site includes parts and machines that are independent entities and capable of sensing their own condition. Thus production processes can be monitored with detailed information measured by the machine tool or the workpiece itself and a comprehensive picture of every manufacturing process arises. Within this vision, production machines and workpieces communicate and share information, perceptions, and knowledge among each other.

This chapter presents applications that pave the way toward a gentelligent production by using the basic technologies described in Chapter 1. An essential part are gentelligent components with the ability to sense, store, communicate, and even decide. Examples for such components are spindle slides (see Chapter 3.2.1), clamping systems (see Chapter 3.2.3), forging dies (see Chapter 3.2.6), or the workpiece itself (see Chapter 3.2.2). Based on the information gathered by these components, novel approaches for adaptive process planning and control as well as teachless process monitoring and self-optimizing machine systems can be established. In contrast to current approaches, process planning and control in a gentelligent production is characterized by more decentralized decision-making. Consequently, cascaded control loops result in the production phase as shown in Fig. 3.1.

In the following, approaches for the planning and control of a gentelligent production by means of a *Virtual Planner* (Chapter 3.1), intelligent machines and their

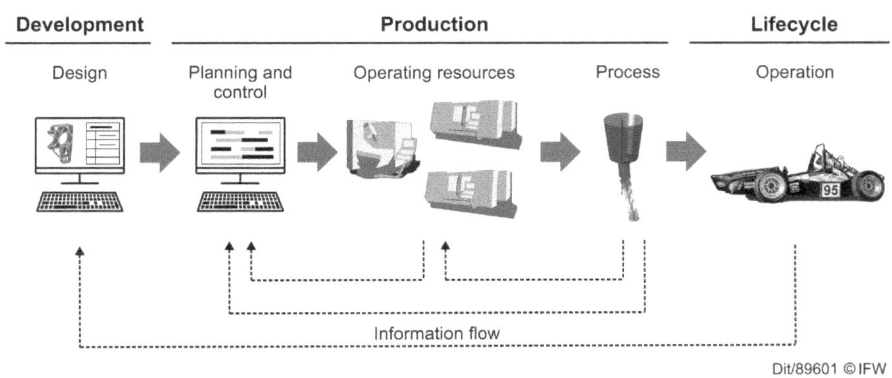

Fig. 3.1 Cascaded control loops.

Cyber-Physical and Gentelligent Systems in Manufacturing and Life Cycle. http://dx.doi.org/10.1016/B978-0-12-811939-6.00003-0

components (Chapter 3.2), and additional use cases for the newly developed piezo-actuatory milling process (Chapter 3.3) are presented.

3.1 Process planning and control in a networked production

In this chapter it will be demonstrated how cross-linked production processes can be planned and controlled, how quality monitoring is realized in a process-integrated way and how self-learning process models can be applied for a process planning.

High productivity and product quality are essential to succeed in international competition. Companies are especially challenged by the increasing trend of individualized manufacturing and a greater importance of flexibility. This tendency leads to a vast variety of products and a large number of possible processes and process chains. Particularly in single-item and small-lot production, decisions concerning the production are often based on unstructured practical knowledge. In order to find the optimal solution, this knowledge can be extended by an effective data analysis and process monitoring in real time. The vision of a gentelligent production is to control and adjust the manufacturing process based on real-time data communication of the machine and the workpiece. This information is processed by a so-called *Virtual Planner*.

The *Virtual Planner* is an integrated platform for adaptive process planning, control, monitoring, and modeling (see Fig. 3.1.1). Process and workpiece information as well as knowledge about already applied process parameters are the groundwork for process planning. In case of disturbances in the production, e.g., due to machine failures and subsequent downtimes or waiting queues in front of single processes, the process control unit updates the process plan and controls its execution (Chapter 3.1.1). The role of the process planning and the process control module can be compared to a navigation system in individual transportation. The process plan obtained from the process planning

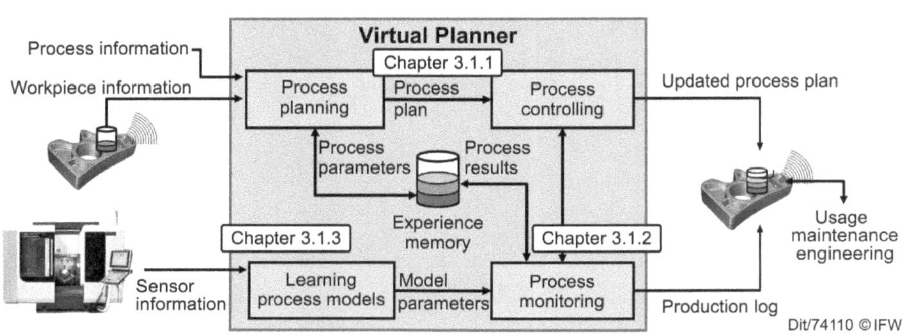

Fig. 3.1.1 Elements of the *Virtual Planner*.

module corresponds to the initial planning of the route at the beginning of the trip. The process control module corresponds to the process of controlling the actual position and speed as well as adapting the route according to the current traffic situation. In order to allow a consistent retraceability, the individual process plan is stored on each part. Based on learning process models (Chapter 3.1.3), the process monitoring module of the *Virtual Planner* monitors single process steps (Chapter 3.1.2). The obtained process results are stored in the experience memory of the *Virtual Planner* for upcoming process planning and directly on the part as a production log.
See the following chapters for further information

3.1.1 Adaptive process planning and control

Berend Denkena, Marc-André Dittrich, Florian Winter
Institute of Production Engineering and Machine Tools, Leibniz Universität Hannover, Garbsen, Germany

3.1.1.1 Introduction

Technological developments and customer's expectations have led to an increased complexity of production systems. Customers and companies interact directly via the internet, nowadays. Moreover, the internet let customers compare products more easily which results in a higher demand for innovations and individualized products. However, individualized products require complex production processes because of a vast variety of assorted characteristics (e.g., different features, colors, materials, etc.). In addition to these challenges, companies face cyclic demands, stochastic disturbances, and increased quality awareness [1]. Thus efficient process planning becomes increasingly important in order to achieve economic success.

Within the *Virtual Planner* adaptive process planning is a key factor for truly agile manufacturing, since conventional concepts of production process planning are not suited to cope with the mentioned challenges. In case of unexpected events and downtimes conventional and nondynamic planning offer a limited number of alternatives which endangers the economic success of the production process. However, the concept of adaptive process planning allows a quick reaction to changes in the production system. Nonlinear process plans, which contain alternative sequences of process steps, create the basis for the adaptive process planning. The best sequence of process steps is determined according to the preferences (e.g., fastest route, cost-efficient route) and according to the status of the production system by an integrated valuation method. In this chapter, an overview about existing planning approaches, adaptive process planning, and its integration into the *Virtual Planner* are given. Subsequently, the performance of the novel planning approach is evaluated in an application scenario.

3.1.1.2 Approaches for process planning and control

Process planning in production systems is a part of the operations scheduling. Operations scheduling encompasses all nonrecurring planning activities, which are conducted in order to achieve an efficient production [2]. Furthermore, operations scheduling has long-term perspectives as well as middle- and short-term perspectives. Design and organization of production means are the main objectives in the long-term perspective whereas the process planning is the focus of the middle- and the short-term perspective. At the process planning stage of the production process the workpiece is described, the production technologies and strategies as well as the sequence of process steps are defined. Moreover, machines, tools, and process parameters are selected. In general, process planning is based on practical knowledge of the employees, in particular in small and medium enterprises. As a result, process planning may lead to nonoptimal solutions despite high efforts due to manual data processing and a low reuse of production data [1].

After developing the sequence of process steps during process planning specific information about the production process are required. *Process control* contains four main aspects: order release, capacity control, detailed sequencing, and process monitoring. During order release the starting time of a job is defined, either concerning the start at just one workstation or the start in an entire production system. Company-specific conditions like number of workstations, capacity, and availability of workstations are setting the framework for dividing the different jobs to the different workstations. Regarding this framework, the capacity control is another instance to control and to execute the detailed production process. Subsequently, the detailed sequencing determines the final process information for production. As a result, process planning and control are temporally separated. Actual state information concerning the workpiece and the production system are not considered in the phase of process planning. Without any influence of the actual situation there is a risk to design process plans that do not match the capacity or the availability of the production system. Moreover, in terms of reducing the planning efforts not all resources of the production system are assigned to the possible process steps during process planning. Consequently, the integration of process planning and production control offers high potential for further improvement [1,3,4]. An increased flexibility is one of the main goals of integrating process planning and production control. Areas of flexibility are the flexibility of alternative processing steps for producing a design feature, alternative sequences of process steps, and finding alternative production resources [1,5,6]. Fig. 3.1.1.1 depicts three common approaches for an integrated process planning and control.

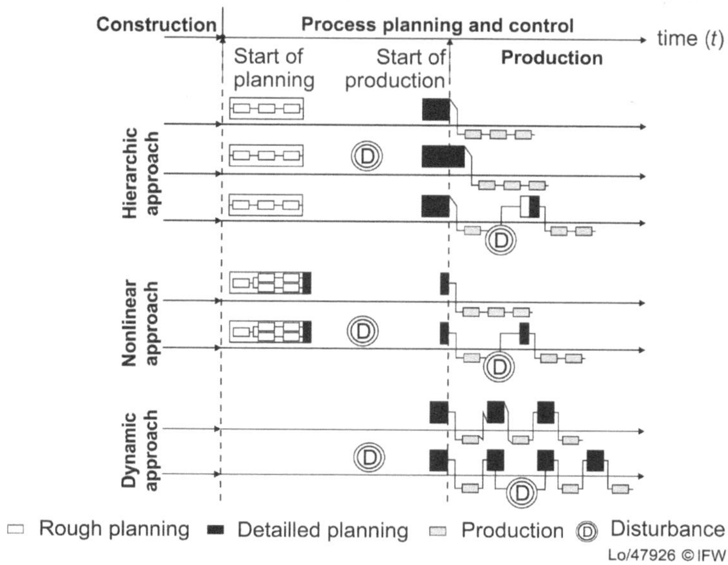

Fig. 3.1.1.1 Existing approaches for process planning and control [7].

The hierarchic process planning is the approach with the lowest flexibility. Main parts are the rough planning and detailed planning stage. Within the rough planning stage, the sequence of production operations is determined whereas the operations are assigned to the production resources during the detailed planning. No alternative operations are planned before the start of the production process, consequently the flexibility of this method is low. If an interruption in the process appears the complete planning process will be executed again. As a result, there is a low effort at the beginning of the production process but, on the other hand, a low flexibility, respectively, high efforts in case of disturbances [3,7].

In contrast to the hierarchic approach, nonlinear process planning considers alternative operations and process sequences via a process plan network. The allocation of the resources to the process step is executed shortly before the process is about to start. A valuation method is analyzing which sequence of process steps is the "best" regarding to the set preferences (e.g., quality, time, and cost). As this is executed right before the production starts, a reaction to unavailability and maximum utilization through rescheduling the process based on this information is ensured [7,8].

Dynamic process planning offers an even higher flexibility. No planning activities are executed before the start of the production process. The resources are allocated directly before the process step starts regarding availability and utilization. For this reason, the ability to react is very high. However, the planning effort is also very high [7].

All three approaches display different preferences between planning effort and flexibility. The hierarchic approach which divides rough and detailed planning enables a low

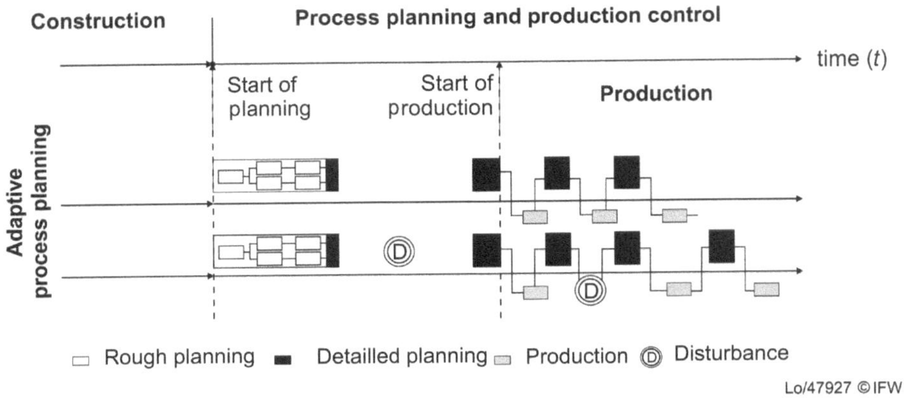

Fig. 3.1.1.2 Novel approach for adaptive process planning [7].

planning effort along with low flexibility. For standardized processes with mass goods, high availability, and low utilization of production resources the hierarchic planning is well suited. Higher flexibility and higher efforts compared to the hierarchic planning are characteristics of the nonlinear process planning. Moreover, the dynamic planning ensures maximum flexibility with high efforts. Combining the beneficial characteristics of these three approaches, namely, the distinction between rough and detailed planning, the use of alternative process plans, and the use of dynamic and actual state information, results in the concept of *adaptive process planning* (Fig. 3.1.1.2).

Before start of production the rough planning including generating all possible process plans and production step sequences is executed. Based on defined and prioritized criteria (e.g., processing time, cost, quality) all sequences of process steps are evaluated by the analytical hierarchy process (AHP). The main aspect to be highlighted is the continuous detailed planning parallel to manufacturing. As a consequence, the actual state of the production system regarding machine availability and actual process parameters is considered [6]. This integrated planning approach has been developed in the CRC 653 and is an essential part of the *Virtual Planner*.

3.1.1.3 Development of a concept for gentelligent process planning

Nonlinear process plans are the basis for a flexible reaction to changes of the current state in production systems. Adaptive processes planning requires nonlinear process plans. In order to identify these nonlinear process plans alternative processing steps have to be defined in a first step. Lorenzen proposed a method to derive the necessary process steps from design features, like the ones defined in the ISO 14649-10. The method is summarized in Fig. 3.1.1.3. Based on the extracted design features

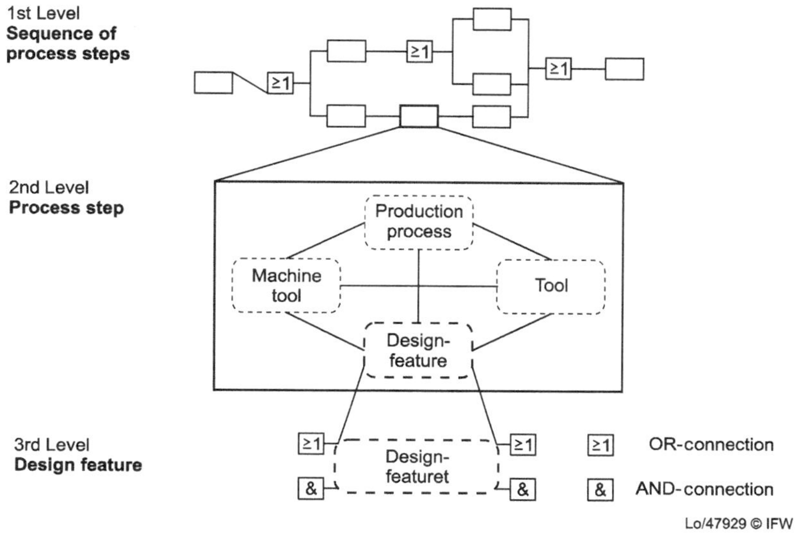

Fig. 3.1.1.3 Structure of nonlinear processing sequences [6,9,10].

(3rd Level) an algorithm allocates the recommended tool, machine, and process parameters for processing of the respective design feature (2nd Level).

Using simple logical operators the identified process steps are connected to alternative processing routes (1st Level). An AND-Connection symbolizes a fixed sequence of necessary process sequence by technological reasons. Changing the sequences is not possible, whereas an OR-Connection allows for changes.

In order to implement the developed method in different production scenarios a systematic structure for data storage and evaluation needs to be defined. The proposed structure is called *onthology concept* and consists of static (*core onthology*) and dynamic (*domain ontology*) parts (Fig. 3.1.1.4). The *core onthology* stores standards like manufacturing processes, design features, tools and machines, and their relation to each other. The core ontology is a standardized framework for all possible design features, their required resources, and their relationships, e.g., design feature A is processed by production process B. However, there is no specific relation to the company and the current situation in the production system. In contrast to this the *domain onthology* contains company-specific information, like currently available machines and tools.

As a result of this merging of both onthologies, the application-specific ontology is created. At this point the specific workpiece is taken into account for the first time. Comparable to level 3 in Fig. 3.1.1.3 the specific design elements of the workpiece are sequenced for the production processes. Consequentially, all possible sequences of processing steps are associated with the available tools and available machines

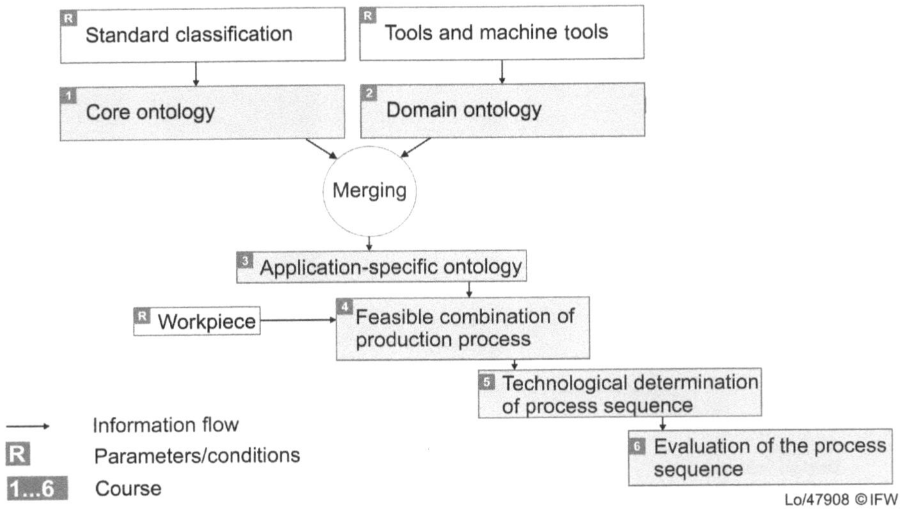

Fig. 3.1.1.4 Ontology concept [11].

and a nonlinear processing plan is created [6,11]. At the last step an evaluation of the process sequence is executed.

The adaptive process planning is used for the following production control considering current state information of the production system. At each time (t_e) tasks of process planning as well as production control are executed. A decision about the situation-specific optimal process sequence at time t_e requires the availability of up-to-date state information of all involved planning objects (e.g., machine, workpiece), see Fig. 3.1.1.5. The workpiece, which stores the information about possible and preferred routes inherently, communicates with the production control in order to receive information about the current state of the production system. Route 1 as the

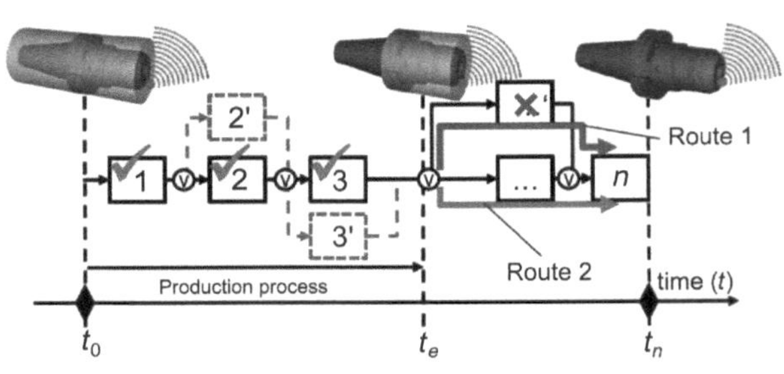

Fig. 3.1.1.5 Adaptive process planning based on current state information [1].

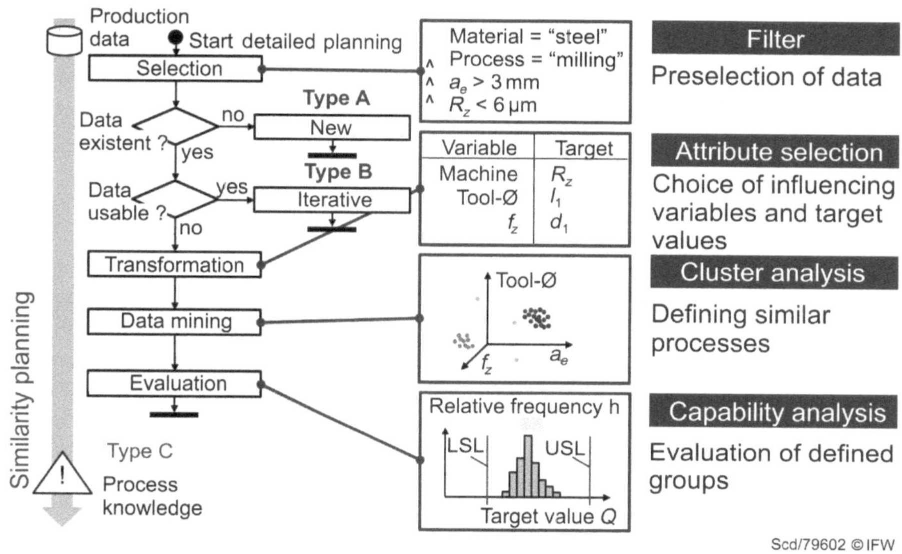

Fig. 3.1.1.6 Procedure of the data evaluation [1].

preferred route is unavailable. As a consequence, Route 2 is chosen to circumvent the bottleneck in Route 1.

The used process parameters (e.g., cutting speed, feed rate) and resulting targets (e.g., surface roughness) are saved directly in a structured database in order to create and extend the planning basis for future evaluations and detailed process planning. A transformation from former process results to the actual planning activity is enabled by four steps (data selection, data transformation, data mining, and data interpretation), see Fig. 3.1.1.6 [1,12].

Within the *data selection* parameters like the material of the workpiece, the activities to carry out the process step and target figures to be reached are defined. Considering this input, comparable data sets are preselected. In *data transformation* crucial influencing factors and target figures are defined. An example for a target figure is the surface roughness of the workpiece. Machine, tool diameter, tool length, feed rate, and cutting rate are the most important influence factors for the resulting surface roughness. Regarding the following *data mining* process, it is recommended to limit the amount of influence variables in order to guarantee good results [1,13].

Analysis of the preselected data results in finding process parameters with high reliability and repeatability. This is done by cluster analysis. Continuously and parallel to the process, the cluster analysis is executed directly before the next production process is starting. Depending on the scattering of influence variables in the former processes, the cluster has different qualities and reliabilities. The k-means algorithm which is arranging data sets in cluster with the maximum difference to the next cluster is used for grouping data sets into clusters. In the last step the quality and reliability of the clustering is measured. Therefore the ratio between range of target values and range of input variables is calculated and the operating points can be evaluated concerning their process reliability by means of a capability analysis.

3.1.1.4 Application of the adaptive process planning

To evaluate the adaptive process planning the method was implemented in a software-based demonstrator called *Virtual Planner*. In order to adapt the process sequence based on current state information input interfaces to the design, Enterprise Resource Planning (ERP) and machines are developed. For the purposes of gentelligent process planning the GIML (Gentelligent Markup Language) as a specified definition of the XML format is used. Output interfaces of the Virtual Planner contain the current process plan including process sequence and parameters, tool, and the process results including surface roughness and dimensions, see Fig. 3.1.1.7.

An application scenario was chosen for testing the demonstrator. Within this scenario two different workpieces are manufactured. A pin with eight design features and a more complex wheel carrier requiring 19 design features are evaluated. The material flow of the wheel carrier and the pin as well as machines and their parameters are

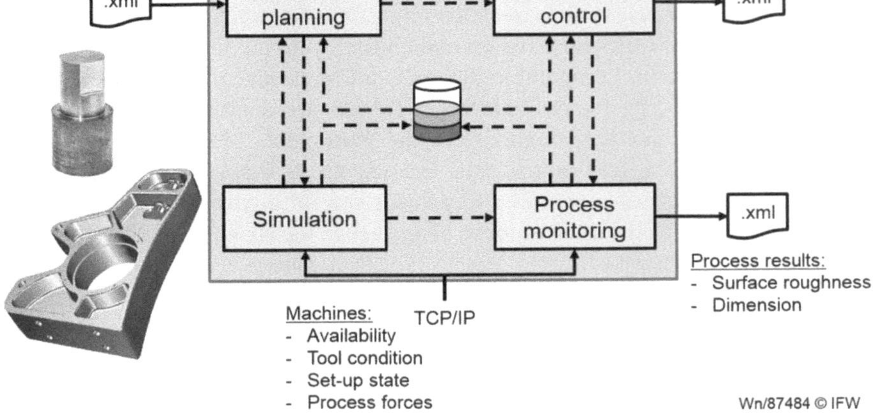

Fig. 3.1.1.7 Interfaces of the Virtual Planner.

modeled with the software Siemens PLM Plant Simulation. The nonlinear process plans are integrated in the model and the production system is evaluated under various circumstances. The process monitoring is done by integrating a sensory spindle in one machine of the application scenario. This allows the online measurement of process forces and the surface roughness of the workpiece in order to get information about the workpiece quality.

The evaluation of the adaptive process planning approach in the described testing environment consists of three different examinations. Fundamental for the preparation and feedback of production data is the ability to draw conclusions from the process parameters regarding the production targets, e.g., surface roughness. As an example four milling processes for producing one design feature of the wheel carrier are executed with different operating points and repeated five times. Target value is the average of the surface roughness, while the influence variables are given by the feed rate, the machine, and the tool. After preselection and analysis of data the influence of the feed rate on the surface roughness is confirmed, which is also proofed in the literature [1,14]. Furthermore, there are also smaller influences by the tool and the machine, however the tool has a stronger influence than the machine. By this examination it is proofed that the feed rate can be used as an indicator for the surface roughness of the workpiece. Further examinations are concentrated on the database quality of this interrelation.

Based on this knowledge, the data mining method depicted in Fig. 3.1.1.6 is evaluated. The basic concept is to use the data mining for finding similar processes to the current process and adapt the process parameters. The quality of the clustering depends on the spreading of the influence variables in the underlying database. In case of higher spreading, a higher amount of data is needed for identifying clear clusters and different operating points. Fig. 3.1.1.8 depicts that fact.

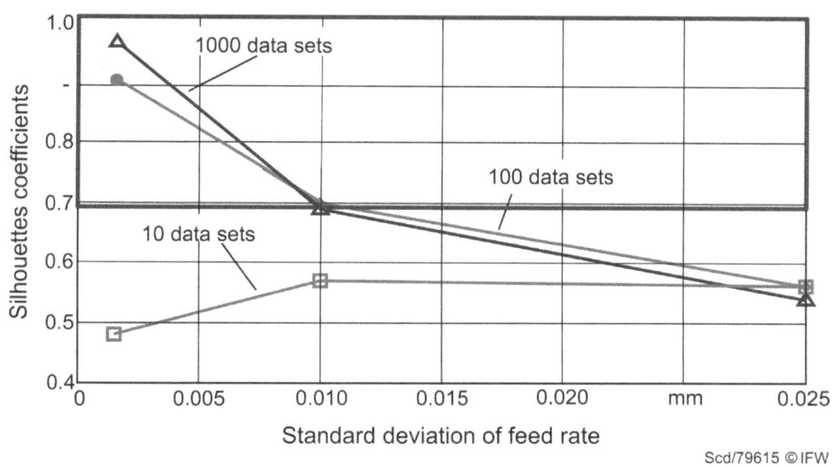

Fig. 3.1.1.8 Quality of clustering measured by the silhouettes coefficient [1].

The silhouettes coefficient is an analytical measurement for the quality of the clustering. A silhouettes coefficient above 0.7 is regarded as an indicator for a "strong" structure of the clustering [1,15]. A clustering consistent of 10 data sets is not reliable for finding groups in the data sets, regardless of the standard deviation of the input parameters. A higher spreading of the input variables causes a lower silhouettes coefficient. Even with a higher amount of data sets, the standard deviation is the limiting condition for the applicability of the clustering method in order to find similar process parameters. The increase of the amount of data sets is not sufficient in case of a wide spreading of input parameters. As a conclusion, the decision, if the data mining method is applicable for finding good process parameters, depends on the quality of the database and has to be proofed by evaluating the silhouettes coefficient.

As the final part of the evaluation the adaptive process planning based on current state information is evaluated. For this purpose the lead time and the production costs are considered. To simulate a real production system, the model of the production system in Siemens PLM Plant generates randomly controlled downtimes of the machines, while the production process with alternative process planning is executed. As a reference process, a linear process plan without any alternative routes is executed simultaneously, see Fig. 3.1.1.9.

The average lead time is increasing in both cases of linear and nonlinear process planning, when the utilization of the production system increases. Waiting times for the processing are the reason for that effect. For small batch sizes, there is no significant difference between linear and nonlinear process planning. This can be explained by the target function in the AHP. When there are just short waiting times, the effect of

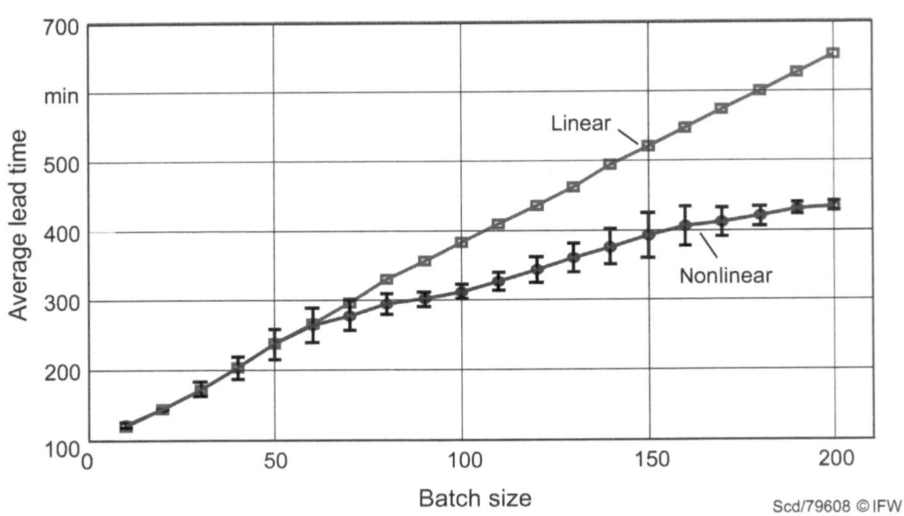

Fig. 3.1.1.9 Average lead time of linear and nonlinear process planning [1].

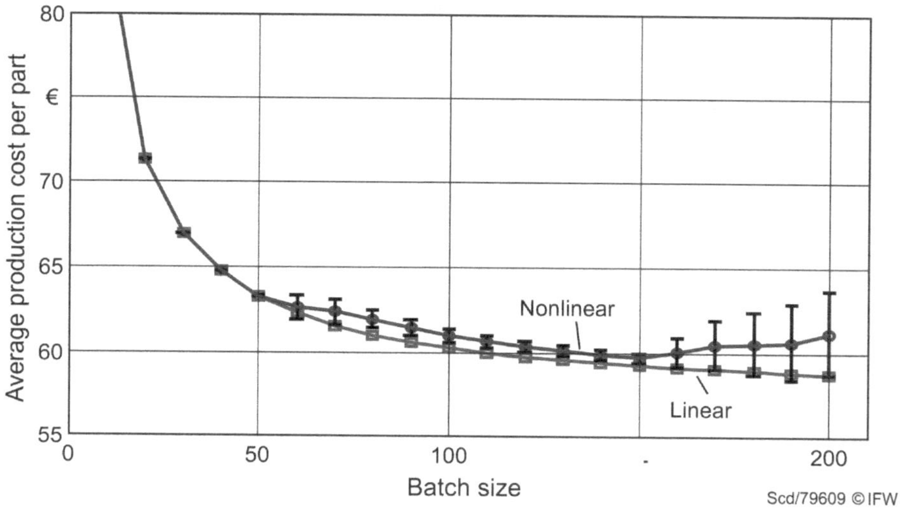

Fig. 3.1.1.10 Production costs of linear and nonlinear processing planning [1].

the time variable on the target value is not essential for initiating an adaption. In case of more parts in the production system, the adaptive process planning realizes considerable improvements concerning the process time because of the bypassing of bottlenecks.

Regarding the production costs, the linear and nonlinear process planning are similar, see Fig. 3.1.1.10. Especially with high utilization of the production system, the costs of linear process plans are lower compared to the nonlinear plans. In this case, the adaptive process planning divides the batches into smaller amounts in order to reduce the lead time. Some of these time-saving alternative production steps cause higher costs than the conventional ones. At this point, a consideration of the cost evaluation method is necessary. Several aspects such as inventory costs, capital costs, and penalties in case of delayed deliveries are not integrated, which would be positively influenced by a reduced lead time. As a result, the total costs of production per part cannot be estimated exactly by this simulation model.

Furthermore, the adaptive process planning is evaluated regarding the complexity and their resultant amount of necessary production steps. As expected, producing more complex parts requires higher lead times and higher production costs, see Fig. 3.1.1.11.

The potential for a high utilization using nonlinear process plans concerning lead time is higher for the simple product. In general, this could be unexpected because more process steps enable a higher potential for bypassing by the adaptive planning. At this point the conditions of the production system have to be taken into account. The nonlinear plans offer more possible routes through the production system, nevertheless the capacity of the production resources is the same. This can lead to unavoidable queues and waiting times in case of complex products with a high amount

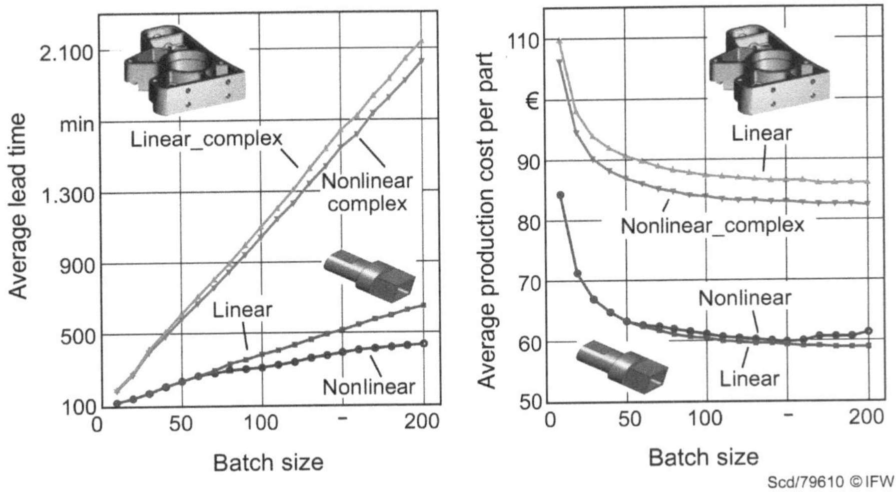

Fig. 3.1.1.11 Lead time and production costs for complex and simple products [1].

of required process steps. However, a significant reduction of production costs for complex workpieces is realized by using nonlinear process plans. In conclusion, the potential of the approach reached in terms of improving the process planning is shown. Connecting the process planning and the production control through evaluation and feedback of structured data and knowledge enables a decrease of lead times and production costs in most applications.

References

[1] J. Schmidt, Integrierte Arbeitsplanung und Fertigungssteuerung auf Basis von Zustandsinformationen, Dr.-Ing. Dissertation, Leibniz Universität Hannover, 2015.

[2] W. Eversheim, Organisation in der Produktionstechnik 3: Arbeitsvorbereitung, fourth ed., Springer, Berlin, Heidelberg, 2002.

[3] J. Kreutzfeldt, Planen mit Bearbeitungsalternativen in der Teilefertigung, Dr.-Ing. Dissertation, Universität Hannover, 1995.

[4] D.T. Pham, C. Gologlu, A computer aided process planning system for concurrent engineering, Proc. Inst. Mech. Eng. B J. Eng. Manuf. 215 (8) (2001) 1117–1131.

[5] H.A. ElMaraghy, Flexible and reconfigurable manufacturing systems paradigms, Int. J. Flex. Manuf. Syst. 17 (4) (2005) 261–276.

[6] L.-E. Lorenzen, Entwicklung einer Methode zur gentelligenten Arbeitsplanung, Dr.-Ing. Dissertation, Leibniz Universität Hannover, 2011.

[7] B. Denkena, J. Schmidt, M. Krüger, A quality-oriented approach to product-driven production, in: B. Denkena, J. Gausemeiser, B. Scholz-Reiter (Eds.), System-Integrated Intelligence: New Challenges for Product and Production Engineering. Proceedings of the 1st Joint International Symposium on System-Integrated Intelligence 2012 (SYSINT2012), Garbsen, 27th–29th June 2012, 2012, pp. 6–8.

[8] S. Kim, J. Woo, S. Park, B. Jung, H. Cho, Integrated development of nonlinear process planning and simulation-based shop floor control, in: E. Yücesan, C.-H. Chen, J.L. Snowden, J.M. Charnes (Eds.), Proceedings of the Winter Simulation Conference, San Diego, 2002, pp. 1465–1468.
[9] A. Brandes, Positionierung technologischer Schnittstellen – Beitrag zur ganzheitlichen Auslegung fertigungstechnischer Prozessketten, Dr.-Ing. Dissertation, Leibniz Universität Hannover, 2008.
[10] B. Denkena, P.-O. Woelk, A. Brandes, Flexible Process Chains by Template Based Configuration. Production Engineering 12/2, Springer, Berlin, Heidelberg, 2005, pp. 81–84.
[11] B. Denkena, J. Henjes, L.-E. Lorenzen, An ontology aided process planning system for gentelligent production, in: Proceedings of the 6th CIRP International Conference on Intelligent Computation in Manufacturing Engineering, Naples, 23rd–25th July 2008, 2008, pp. 87–94.
[12] U. Fayyad, G. Piatetsky-Shapiro, P. Smyth, From data mining to knowledge discovery in databases, AI Mag. 17 (3) (1996) 37–54.
[13] M. Ester, J. Sander, Knowledge Discovery in Databases—Techniken und Anwendungen, Springer, Berlin, Heidelberg, New York, 2000.
[14] T.L. Saaty, L.G. Vargas, Models, Methods, Concepts & Applications of the Analytic Hierarchy Process, vol. 2. International Series in Operations Research & Management Science, vol. 175, Springer, New York, 2012.
[15] L. Kaufman, P.J. Rousseeuw, Finding Groups in Data: An Introduction to Cluster Analysis, Wiley, Hoboken, 2005.

Further Reading

B. Denkena, H. Henning, L.-E. Lorenzen, Genetics and intelligence: new approaches in production engineering, Production Engineering—Research and Development 4:1, Springer, Berlin, Heidelberg, 2010, pp. 65–73.
B. Denkena, H.-K. Tönshoff, Spanen. Grundlagen., vol. 3, Springer, Heidelberg, Dordrecht, London, New York, 2011.
E. Westkämper, Einführung in die Organisation der Produktion, Springer, Berlin, Heidelberg, New York, 2006.

3.1.2 Process-integrated quality monitoring

Berend Denkena, Florian Uhlich
Institute of Production Engineering and Machine Tools, Leibniz Universität Hannover, Garbsen, Germany

3.1.2.1 Introduction

Currently existing process monitoring approaches observe process signals and trigger an alarm or process stop, when certain signals exceed assigned threshold values [1]. Applied methods for signal evaluation can be classified into three categories [2]:

- Signal-based approaches
- Artificial intelligence-based approaches
- Model-based approaches

Signal-based approaches allow to identify process disorders due to machine collisions and tool breakage [3] or to monitor tool wear based on cutting forces [4]. However, either these systems need to be trained from several processes to obtain suitable signal limits or the threshold values have to be set according to statistical analysis from historical processes. Artificial intelligence-based approaches are capable to interpret signals from one or more sensors and classify the process state. These approaches use neural networks [5] or support vector machines [6] to monitor tool wear or estimate surface roughness. These approaches are based on machine learning methods that have to be trained with historical data. Approaches that are based on models use explicit models of the process or the involved dynamic components to reconstruct derived measures from process signals [7].

However, there is still a missing link between available process signals and technical quality measures, such as surface roughness or shape and form errors. Therefore, the model-based transformation of process signals will play a major role for the observations of process and quality conditions based on process signals. The following section describes a model-based approach for monitoring of shape deviations of machined workpieces, based on actual cutting forces measured with "feeling" machine tool components (see Chapter 1).

3.1.2.2 Method development

To study the fundamental formation of shape deviations in peripheral milling, a milling process is designed that exhibits different widths of cut to provoke varying cutting forces and occurring tool deflection. The process conditions performed during the milling processes are summarized in Table 3.1.2.1.

The shape deviations in peripheral milling are provoked by using a slender end mill with a diameter of 10 mm and an overall length of 60 mm and a large depth of cut $a_p = 30$ mm. In order to mill the workpiece in a single process, that exhibits the desired variations of the width of cut, the workpiece is prepared according to

Table 3.1.2.1 Process parameters for the experimental milling processes

Experiment Nr.	1	2	3	Unit
Engagement depth a_p		30		mm
Engagement width a_e	0.5	1.5	2.5	mm
Feed per tooth f_z		0.05		mm
Spindle speed n		500		1/min

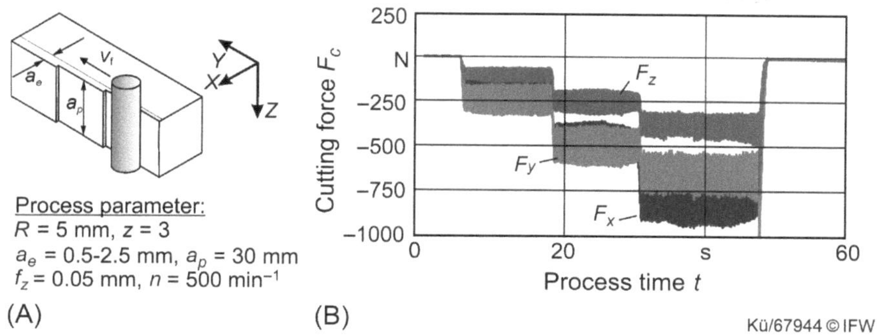

Fig. 3.1.2.1 (A) Prepared workpiece and (B) recorded process forces.

Fig. 3.1.2.1A. The resulting cutting force components F_x, F_y, and F_y are shown in Fig. 3.1.2.1B for a selected process. The different stages of the process, which are characterized by variable widths of cut, lead to distinct stages in the course of the process forces.

To examine the effect of the process forces on the surface quality, a digital model of the workpiece is built up using a 3D laser scanner. The difference between the measured surface and reference plain represents the shape deviation. In fact, the surfaces display different shape errors and surface properties with respect to varying cutting conditions and cutting forces as presented in Fig. 3.1.2.2. According to DIN 4760 the shape of the form error can be classified into dominant form errors of first error

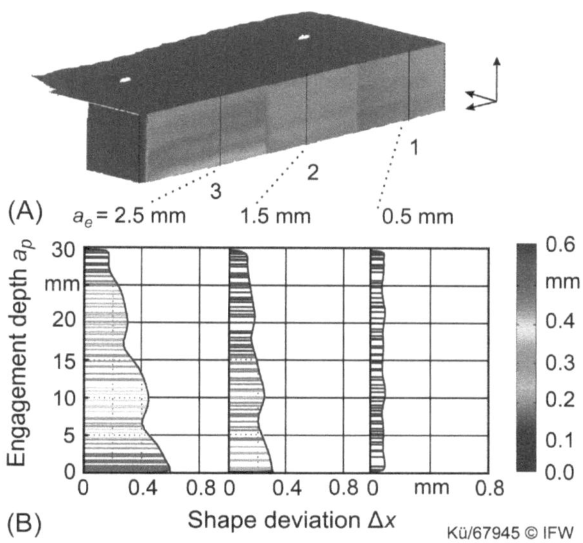

Fig. 3.1.2.2 (A) Measured workpiece shape and (B) resulting shape deviations of the machined surface for different widths of engagement.

(tilt error) and second order (waviness). The tilt error is typically caused by the static deflection of the machine tool or the end mill due to process forces. The waviness of a machined surface typically results from dynamic milling effects. Analyses of the recorded workpiece reveal the influence of the cutting forces on the shape error. Generally, higher process forces lead to higher shape errors. The maximum shape deviation of 0.6 mm occurs at the engagement width $a_e = 2.5$ mm.

The shape error and the surface roughness are important properties of technical components and require costly quality control. If the resulting shape could be predicted in the process based on process signals, the subsequent end-of-line quality control could be replaced by a process-integrated approach. Consequently, quality issues could be identified and appropriate actions be taken earlier, leading to less defective parts. Ultimately, the manufacturing system could adapt its process parameters even before the tolerance limits are exceeded, leading to a zero defect production.

Based on the reasoned assumption that the process forces define crucial surface properties and a model-based monitoring system considering static and dynamic effects has been developed and is described in the following section.

3.1.2.3 Design of the model-based monitoring system

The fundamental functional principle of the online quality monitoring system is based on the observed relationship between process force and shape error that can be explained with the laws of elasticity as illustrated in Fig. 3.1.2.3. The resulting deflection of the tool results from the process force as well from the rigidity of the tool.

The outlined model-based quality method consists of four subsystems. First, a process force measurement has to be integrated into the machine tool. In this case, a

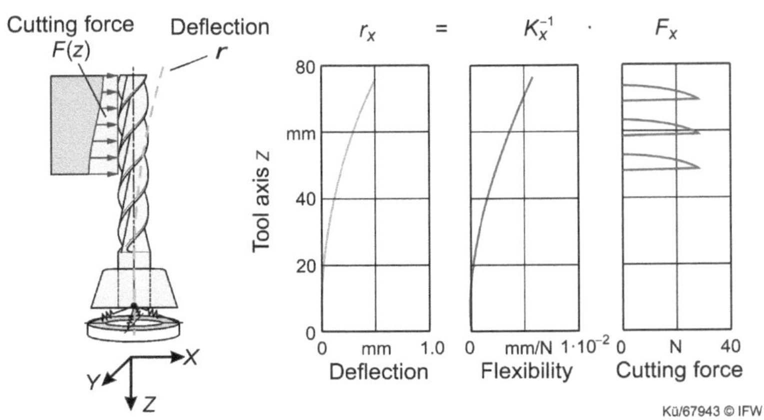

Fig. 3.1.2.3 Underlying principle of the surface prediction method.

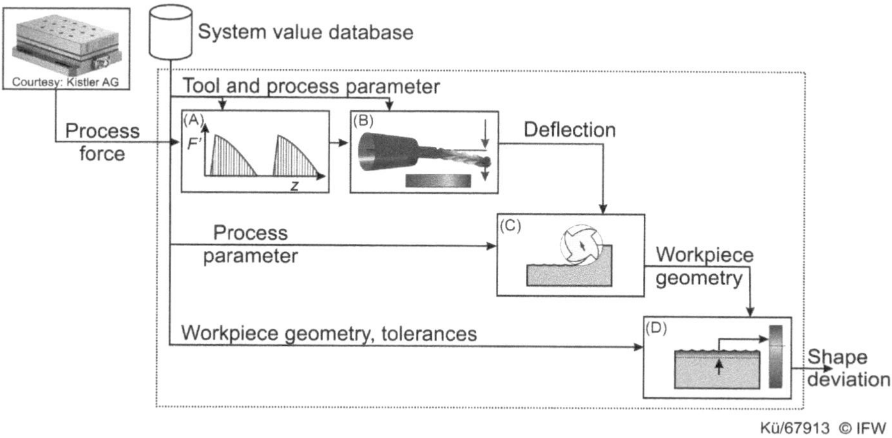

Fig. 3.1.2.4 Concept of the model-based monitoring system and its subsystems: (A) force distribution model, (B) tool deflection model, (C) material removal model, and (D) quality inspection module.

conventional piezoelectric dynamometer is implemented. However, the process force can also be measured using machine-integrated sensors as presented in Chapter 3.2.1.

Next, the obtained process forces are distributed along the tool length according to the tools' contact areas (Fig. 3.1.2.4B). Subsequently, a material removal simulation with the bended tool is conducted in order to calculate the resulting surface with respect to shape deviations and surface roughness (Fig. 3.1.2.4C). Finally, an evaluation of the reconstructed surface reveals whether the surface properties such as the roughness or the shape error comply with the specified requirements (Fig. 3.1.2.4D).

3.1.2.4 Evaluation

After the model-based surface monitoring system has been developed, a series of experiments are conducted to analyze the capability and reliability of the method. Exemplary results are presented in Fig. 3.1.2.5. The input force, which has been recorded during the process, is depicted in the upper part of the figure. Following, the calculated surface structure can be seen. Finally, a sectional view of the surface with both the reconstructed and the measured profile is shown. Due to the experiments, which are based on typical cutting conditions for finishing using common tool dimensions, it is demonstrated that the developed method is capable of calculating a shape error >0.1 mm with an error below 10%. Furthermore, the approach is able to calculate the profile waviness of a magnitude >20 μm with an error below 15%. Additionally, it becomes evident that the magnitude of the shape error depends on the magnitude of the force and on its distribution along the tool. It should be noted

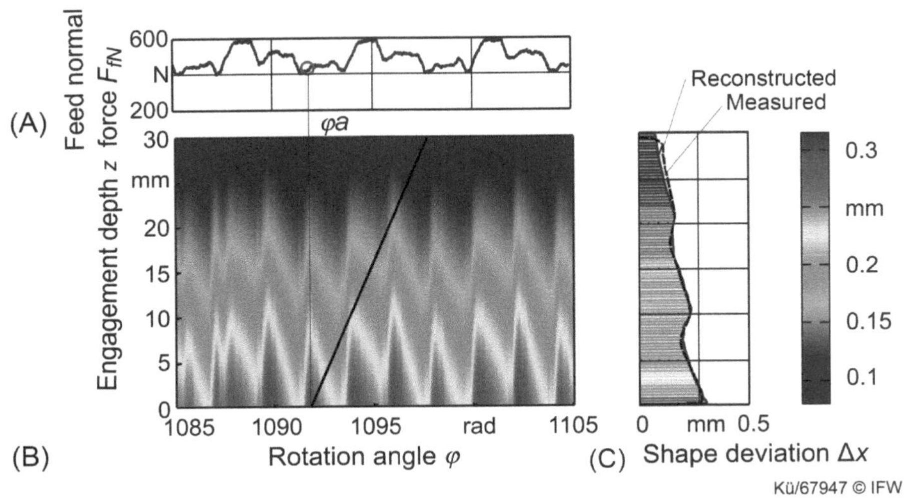

Fig. 3.1.2.5 Cutting forces (A), resulting tool deflection over tool rotation (B) and resulting shape deviation of the workpiece (C) of experiment 2.

that the presented results are obtained with a tool showing a low rigidity. However, the validation reveals that the type and magnitude of effects were mapped on the basis of a model accurately. It can be assumed that the method can also be applied on processes with lower forces and tools with a higher rigidity. The validation of a shape error smaller than 0.1 mm has not been accomplished in this work due to the increasing ratio of noise during the measurements.

The presented results show that the developed method for predicting the surface and shape quality is capable to provide quality information under common industrial conditions. Following the idea of a gentelligent production, current state information can be calculated and evaluated for every individual workpiece at low expense directly from the process. Because of the online calculation and evaluation of the shape error, shortcomings of the quality can be detected while they arise. Moreover, the gained information can be provided to the Virtual Planner in order to adapt crucial process parameters and avoid defect workpieces.

References

[1] P.W. Prickett, C. Johns, An overview of approaches to end milling tool monitoring, Int. J. Mach. Tools Manuf. 39 (1) (1999) 105–122.

[2] B. Denkena, M. Krüger, A model-based approach for monitoring of shape deviations in peripheral milling, Int. J. Adv. Manuf. Technol. 67 (9) (2013) 2537–2550.

[3] F. Klocke, M. Reuber, Process monitoring in sculptured surface milling, Prod. Eng. Res. Dev. 7 (1) (2000) 1–4.

[4] B. Denkena, K. Litwinski, D. Dahlmann, H. Boujnah, J. Kiesner, M. Krüger, J. Schmidt, T. Neff, Industrie 4.0 in der Zerspanung, ZWF 7 (8) (2014) 537–541.
[5] D.A. Dornfeld, M. DeVries, Neural network sensor fusion for tool condition monitoring, CIRP Ann. Manuf. Technol. 39 (1) (1990) 101–105.
[6] D. Brezak, D. Majetic, T. Udiljak, J. Kasac, Tool wear estimation using an analytic fuzzy classifier and support vector machines, J. Intell. Manuf. 23 (3) (2012) 797–809.
[7] A. Larue, B. Anselmetti, A prediction of the machining defects in flank milling, Int. J. Adv. Manuf. Technol. 24 (1–2) (2004) 102–111.

3.1.3 Process planning with self-learning process models

Berend Denkena, Florian Uhlich
Institute of Production Engineering and Machine Tools, Leibniz Universität Hannover, Garbsen, Germany

3.1.3.1 Introduction

Process planning defines the means of production that are deployed for a production process. In the case of machining, process planning also defines the cutting parameters [1]. As the cutting parameters have an influence on the resulting costs and obtained quality, process planning is responsible for the efficiency of the production.

Parameter optimization is part of the process planning and has a long history in production research [2,3]. Recent research work focuses on evolutionary algorithms [4], especially Genetic Algorithms, Simulated Annealing, and Particle Swarm Optimization. Planning methods based on analytical models are available to predict cutting forces and dimensional errors [5]. However, the models have to be parametrized for every cutting process and require several trials to find optimal values.

Simulations of cutting processes are widely used to model mechanical and thermal phenomena and to reduce the effort of finding appropriate machining conditions [6,7]. Cutting simulations are used to calculate internal process states and the resulting shape of the workpiece. However, the used technological models also need to be parametrized for every material tool combination in order to generate reliable results.

Hence, there is still a demand for a method that allows to gain information automatically from every single process and to obtain additional knowledge continuously and with low effort.

3.1.3.2 Method development

The key for achieving a constantly growing knowledge base is the automatization of the knowledge generation in the sense of a gentelligent production. The developed approach pursues the idea to aggregate simulated process information, machine data, and measurement values to allow for spatial-temporal allocation of all information. On this basis, the information are put into the right context and technological models are parametrized and correlation is modeled.

For that purpose, a framework for the developed planning method is built up as presented in [8]. This setup consists of a machine tool for milling operations that is connected to a virtual planner. The virtual planner is the component of the setup that is responsible for aggregating, storing, and deploying all obtained information to generate machining knowledge to plan and optimize production processes. The machine control provides drive information, which was originally used for internal control tasks but it is to be used to gain supplement insights of machining processes. The information of interest are the axis positions, feeds, and currents as well as the spindle speed and the spindle load factor. Drive currents and the spindle load are monitored continuously and transmitted to the virtual planner, see Table 3.1.3.1. The system setup and all information flows are depicted in Fig. 3.1.3.1.

The process parameters and the shape error, which is measured with an integrated machine probe, are transferred with an NC-based communication routine. The communication is based on memory segments, which can be read and written by the NC program as well as by applications running on the machine. To ensure a safe

Table 3.1.3.1 Summary of measured signals

Data type	Symbol	Unit	Sampling frequency
Tool position	–	mm	25 Hz
Spindle current	I_S	A	25 Hz
Feed force	F_f	N	100 Hz
Normal feed Force	F_{fN}	N	100 Hz
Shape error	Δd_S	mm	4 per process

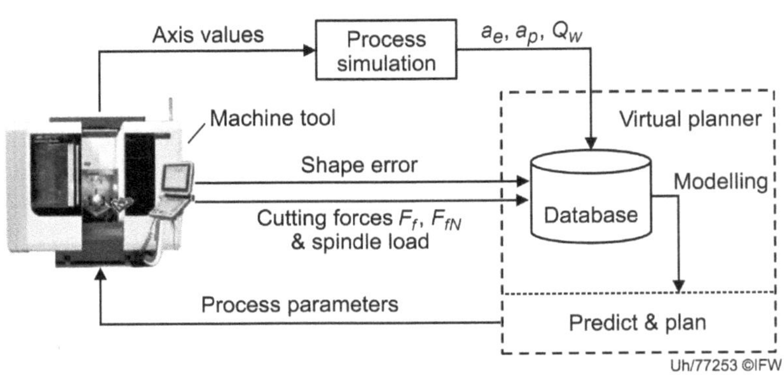

Fig. 3.1.3.1 Information flow between machine tool, process simulation, and Virtual Planner.

transmission of critical data, like process parameters, whose faulty transmission may result in a machine damage, a communication protocol is developed. The protocol makes use of status indicators and read/write restrictions for variables as well as an error detection to securely transfer data from the machine tool to the Virtual Planner and the other way around.

The data types described so far include process forces and machine internal states. However, the system is lacking geometrical information about the actual process in terms of the geometric process conditions that describe the interaction of the tool and the workpiece in the course of the process. This information is indispensable to build up correlations between sensor information and the cutting process.

To gain this information, a simultaneous process simulation is developed that mirrors the machine tool and the workpiece with the actual shape including the removed material. For that purpose, the material removal simulation CutS, which is developed at the Institute of Production Engineering and Machine Tools is extended in such a way that the axis positions of the machine tool are directly assigned to the axis of the virtual machine model. The virtual machine model including the workpiece is highlighted in Fig. 3.1.3.2. By this means, the process is available in a digital form, which allows to perform calculations and to analyze the process. The simulation describes the tool and workpiece interaction in terms of width of cut, height of cut, and the material removal rate and also mirrors the actual feed rate. This simplifies the signal interpretation and avoids error from filtering and segmenting of sensor signals. An overview of the data that is calculated parallel to the machining process is presented in Table 3.1.3.2.

Exemplarily, the collected and augmented data from one process is described in the following. The first diagram (see Fig. 3.1.3.3) shows the history of the axis values from a slot milling process. These values on their own do not describe the process

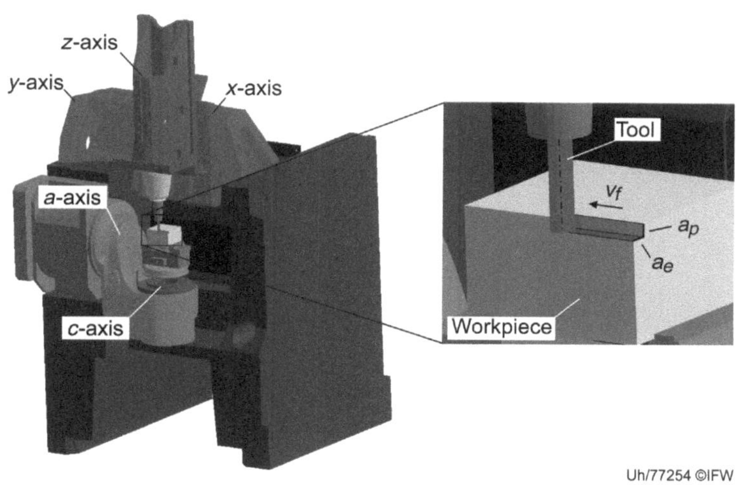

Fig. 3.1.3.2 Machine and process simulation.

Table 3.1.3.2 Summary of calculated values

Data type	Symbol	Unit	Sampling frequency
Width of cut	a_e	mm	25 Hz
Depth of cut	a_p	mm	25 Hz
Material removal rate	Q_w	cm^3/min	25 Hz

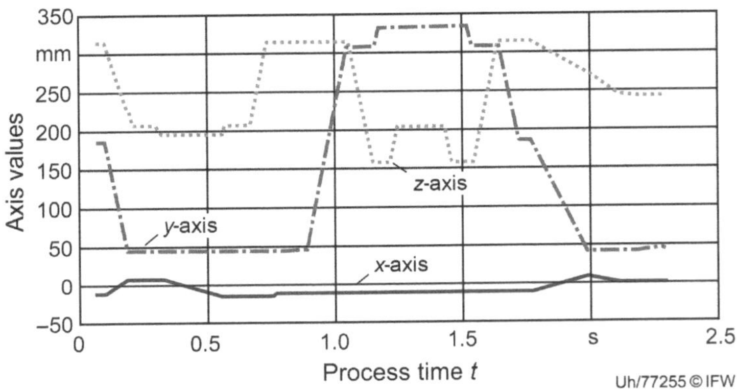

Fig. 3.1.3.3 Course of axis values.

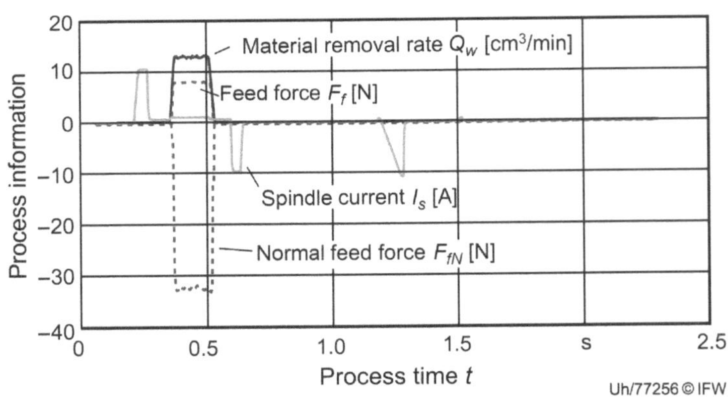

Fig. 3.1.3.4 Aggregation of machine control data, measured values, and calculated values.

completely, as they do not directly give an indication on the material removal process and the cutting conditions. To obtain the relative position, orientation, and speed between the tool and workpiece, the axis positions of the real machine tool are assigned to the virtual twin. The next diagram in Fig. 3.1.3.4 shows sensor data, machine control data, and the data obtained from the process simulation.

It can be noticed that the spindle current rises before the tool is engaged due to the acceleration of the spindle. However, using the data from the cutting simulation, the beginning of the material removal can be determined precisely.

Measured forces can also be used to identify the cutting phase. Still, forces are the result from various cutting parameters, such as depth of cut, width of cut, and feedrate and also varies due to tool wear. A measured force though cannot unambiguously be correlated to the cutting parameters. As process planning needs to find appropriate cutting parameters and not process forces, considering the force exclusively is not sufficient in the context of process planning. Additionally, a spatial allocation of sensor information and shape error is only possible if the cutting process is modeled geometrically.

The shape error of a machined workpiece is part of the requirements that has to lie within the specified tolerances. Choosing the right machining parameters to comply with the required specifications is an elementary part of planning cuttings processes. This decision can be made with reliable prognosis models based on machine learning.

3.1.3.3 Experiments and modeling

To choose an appropriate machine learning approach and analyze the achieved prediction accuracy of the shape error, a series of cutting processes with varying process parameters is conducted. The design of experiments is presented in Table 3.1.3.3. As each parameter combination is performed three times, 108 cuttings processes are conducted in total.

As all resulting machine data is stored in a common database, numerical analyzes can be executed by performing database queries and applying suitable aggregations. The shape error of a machined surface can be attributed mainly to the tool deflection that is caused by the process forces. Therefore an examination of the influence between normal feed force and shape error is conducted, see Fig. 3.1.3.5. According to laws of elasticity, an unambiguous relationship between the shape error caused by the deflected tool and the normal feed force was expected. Though, the analysis does not show a distinct relationship. This observation could be caused by dynamic effects and interdependencies within the cutting process. The low correlation is also indicated by a low correlation coefficient of $R = 0.2321$.

Table 3.1.3.3 Summary of experimental design

Parameter	Symbol	Levels	Unit
Cutting speed	v_c	400, 500, 600	m/min
Feed per tooth	f_z	0.01, 0.02, 0.03, 0.04	mm
Depth of cut	a_p	2	mm
Width of cut	a_e	0.5, 1.0, 2.0	mm

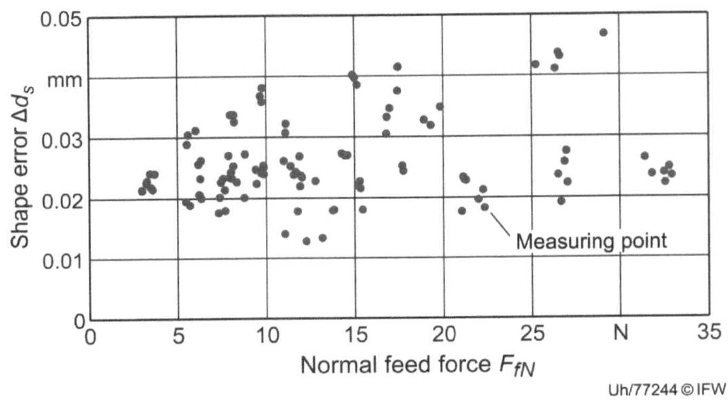

Fig. 3.1.3.5 Plot of measured shape error and corresponding normal feed force.

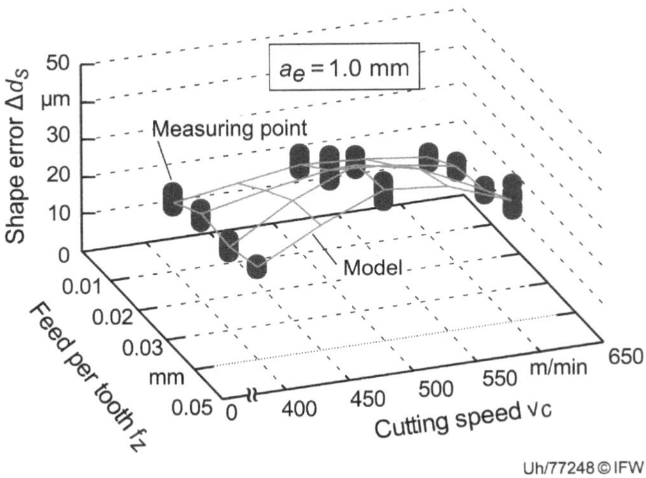

Fig. 3.1.3.6 Measurements and trained model.

For process planning, the effects of parameters have to be known as accurate as possible to choose appropriate parameters that lead to results that satisfy the requirements. Especially, when individual parts are produced, defect parts have a disproportional influence on the economic success of a production order. A Support Vector Machine is used to create a predictive model (see Fig. 3.1.3.6) that forecasts the effect of cutting parameters. A major advantage of machine learning approaches is that the overall shape of the model has not to be known in advance. Also, new measurements can be regarded by retraining the model and thereby increasing the accuracy of the prediction.

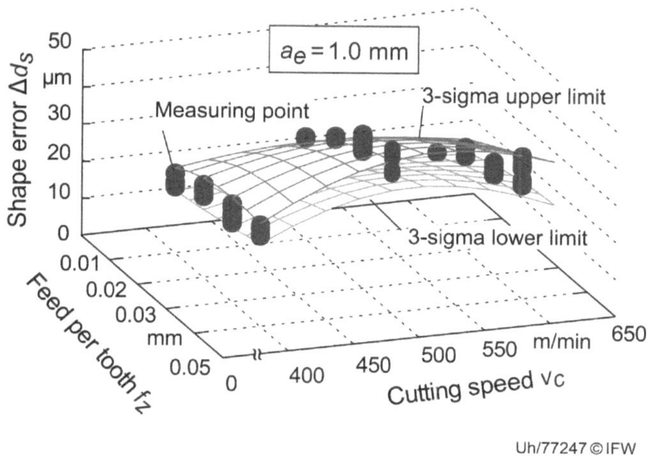

Fig. 3.1.3.7 Enhanced models with variance estimation.

To evaluate the predicting accuracy of the model the experimental results are shuffled in advance and split up in to separate data sets with a ratio of 75% and 25%. The first data set, the training set is used to build the model, the test data set is used to compare the prediction to the measured results. This evaluation yields a mean square error of 2 μm^2, which is low compared to the observed shape error. The difference between the prediction and the actual measured value is less than 12% in all cases. Using a Support Vector Machine as machine learning approach, a high accuracy of forecasting is achieved.

To increase the process reliability the variance of the measurements is taken into account. While the model in Fig. 3.1.3.6 approaches the mean value of the measurements, the model in Fig. 3.1.3.7 estimates the upper and lower three sigma process limits. By this means the model is used to choose parameters that predict the process outcome within a freely selectable confidence interval.

The obtained models can be sampled with a low processing effort. Therefore an optimization is applied in short amount of time and a suitable combination of process parameters is chosen that is both efficient and reliable.

As the process of machine data recording, aggregating, knowledge generation, and process parameter selection was completely automatized, a self-learning, self-optimizing cutting process is achieved.

References

[1] H.B. Marri, A. Gunasekaran, R.J. Grieve, Computer-aided process planning—a state of art, Int. J. Adv. Manuf. Technol. 14 (1998) 261–268.

[2] F.W. Taylor, On the Art of Cutting Metal, ASME, New York, 1907.

[3] Y. Yusof, K. Latif, Survey on computer-aided process planning, Int. J. Adv. Manuf. Technol. 75 (1) (2014) 77–89.
[4] I. Mukherjee, P. Ray, A review of optimization techniques in metal cutting processes, Comput. Ind. Eng. 50 (1–2) (2006) 15–34.
[5] N. Yusup, A.M. Zain, S.Z. Hashim, Evolutionary techniques in optimizing machining parameters: review and recent applications (2007–2011), Expert Syst. Appl. 39 (10) (2012) 9909–9927.
[6] A.D. Spence, Y. Altintas, A solid modeller based milling process simulation and planning system, J. Eng. Ind. 116 (1) (1994) 61–69.
[7] Y. Altintas, P. Kersting, D. Biermann, E. Budak, B. Denkena, I. Lazoglu, Virtual process systems for part machining operations, CIRP Ann. Manuf. Technol. 63 (2) (2014) 585–605.
[8] V. Böß, D. Nespor, A. Samp, B. Denkena, Numerical simulation of process forces during re-contouring of welded parts considering different material properties, CIRP J. Manuf. Sci. Technol. 6 (3) (2013) 167–174.

Further Reading

B. Denkena, M.-A. Dittrich, F. Uhlich, Augmenting milling process data for shape error prediction, Procedia CIRP 57 (2016) 487–491.

3.2 Intelligent manufacturing systems

In this chapter it will be demonstrated how components of a production process (machine tool, workpiece, clamping system) are enabled to sense and how this feature can be used effectively in production processes.

The CRC 653's vision of manufacturing in the future stands out notably from the present state of technology. In contrast to the current situation, workpieces and machines will be independent entities capable of sensing their own condition to provide online information about the machining process and even to control and optimize the machining steps.

Innovative approaches for novel sensors (Chapter 2.4) and signal communication (Chapter 2.5) enable the realization of "feeling" machine tools (Fig. 3.2.1). This new kind of machines has the specific functionality of sensing vibration and heat which occur during machining and of controlling parts of the process in case of detected failures. The "feeling" machine tool is realized by integrating sensory machine components, which are directly involved into the flux of force and closely located to the process. In milling, the spindle slide, which carries the working spindle (Chapter 3.2.1), the clamping system, which holds the workpiece (Chapter 3.2.3), and even the workpiece, which represents the focus of the process (Chapter 3.2.2), are suitable for sensor application. A big challenge, thereby, is to achieve a high sensitivity of the component without affecting its mechanical properties. Within the CRC

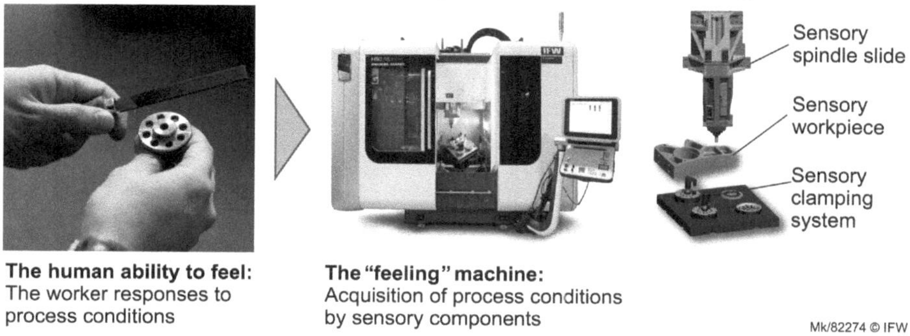

Fig. 3.2.1 Sensory components enable machine tools to "feel."

653, strain gauges are integrated into spindle slides and clamping systems of milling centers to detect the process forces in real time. Furthermore, temperature sensors are applied into workpieces to get online signal of the occurring heats. The provided process signals represent a basic information source for process and condition monitoring systems. Additionally, these signals can be used in many applications, such as the optimization of the machine components themselves and the design of work tools.

Based on the provided process signals, teachless online process monitoring and control systems are developed. Thereby, the main goal has been an increased process security of production facilities for single item production. This allows the machining of individualized products resulting from adaptive technologies. The investigated approaches offer robust and reliable monitoring for production starting with the first manufactured item. To substitute ramp-up processes, which have been necessary so far, a reference for the parameterization of monitoring limits is simulated during the process planning phase. Relevant information of process planning is used to simulate monitoring limits. Hence, teaching of the monitoring system during ramp-up is completely substituted by a simulation during process planning. Thus an NC-based simulation is not only used to validate and optimize the planned process but provides process conditions to calibrate monitoring limits (Chapter 3.2.4).

With respect to milling processes, the force sensing by machine components allows the identification of machine properties like the axis or tool stiffness and enables the online detection of failures during machining, such as the tool deflection (Fig. 3.2.2). A measurement of the tool stiffness is realized by a controlled soft collision of the tool and the workpiece. Knowing the traveling distance of the axis and the collision force, the stiffness of the tool is calculated and used for deflection estimation. Based on a communication between the monitoring system and the machine control, the tool deflection is compensated online in the milling center by adjusting the axis feed or the axis positions. The approach shows considerable improvement regarding the geometric tolerance (Chapter 3.2.5).

With respect to bulk metal forging processes, "feeling" forging dies are developed. Based on integrated cooling channels, the process can be monitored and controlled.

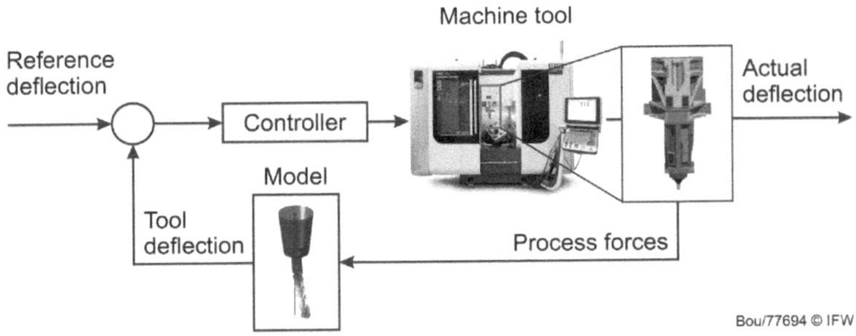

Fig. 3.2.2 Sensory spindle slide enables process control based on process forces.

In hot bulk metal forming processes, a regulated temperature control for dies permits an effective prevention of defects as well as a reduction in process interruptions by means of temperature stabilization. Furthermore, a reduction of thermally induced wear leads to a favorable decrease in tool costs. Such a die temperature control is of particular importance for processes like precision forging, in which geometrical integrity of the produced parts is the main concern (Chapter 3.2.6).
See the following chapters for further information

3.2.1 Feeling machine
3.2.2 Sensory workpieces
3.2.3 Feeling clamping system
3.2.4 Teachless process monitoring for single item production
3.2.5 Online monitoring and control of tool deflection in milling
3.2.6 Tempering control of forging processes by integration of cavities

3.2.1 Feeling machine

Berend Denkena, Haythem Boujnah
Institute of Production Engineering and Machine Tools, Leibniz Universität Hannover, Garbsen, Germany

3.2.1.1 Introduction

Nowadays, the trend in the production sector is edging toward more individual products. As a consequence, the producers have to adapt to a further increase in the variance of the product spectrum with decreasing its lot size. This development results in generally higher demands on intelligent systems for process monitoring and control that may help to reduce incurred additional costs in comparison to mostly cost-optimized series productions.

In order to enable systems for process monitoring and control, process signals are indispensable. The process force represents the most important physical size to describe a process during manufacturing. The first difficulty lies in the generation of signals, which correlate well with that force. To get such signals, sensors can be integrated in machine tool components. For instance, the incorporating of sensors in tool holders is achieved in [1], whereby the integration of the force sensor system led to a loss of stiffness and a sophisticated radio transmission of signals was needed. In the works [2,3], a sensing clamping system with piezo actuators was developed in order to reduce the vibration of workpieces during manufacturing. In the works [4,5], strain gauges and acceleration sensors were combined into a sensor network and integrated in a sensory clamping system. In a further step, this clamping system was combined with an adaptronic spindle and allows a robust force-related process monitoring with high sensibility [6].

Within the collaborative research center CRC 653 [7], "feeling" machine tools are being developed. This new kind of machines has the specific functionality to sense the vibrations and forces which occur during machining and to control the process in case of detected failures. The "feeling" machine tool is realized by the integration of sensory machine components, which are closely located to the process and directly guide the flux of force. In milling, the spindle slide, which carries the working spindle, or the clamping system, which holds the workpiece, are suitable for sensor application. The load detection is realized by strain measurements on the component structure. Such machine components are generally designed with respect to their stiffness in order to achieve the requirements for high machining accuracy and process stability. Therefore the strains in the structure are very low and thus can hardly be measured. A big challenge lies on achieving a component with a high sensitivity without lowering its stiffness.

3.2.1.2 Strain state in selected machine components

Finite element analyses were conducted in order to estimate the occurring strains in the spindle slide of a selected milling center. The resulting effective strains are illustrated in Fig. 3.2.1.1 for a force of 1 kN which is applied at the TCP in x- and y-directions. As the figure demonstrates, the achievable strain amplitudes in the structures are generally very low. Most of the structure shows strain values smaller than 10 μm/m. It can be concluded that an optimal placement for strain sensors is crucial to achieve reasonable sensor signal amplitudes.

3.2.1.3 Strain signal improvement in machine components

The strain in the structure of the machine component is measured by strain gauges. To raise the strain signal amplitudes, a concept to increase the measured strains is necessary. This can be achieved by locally adapting the force flux in the component

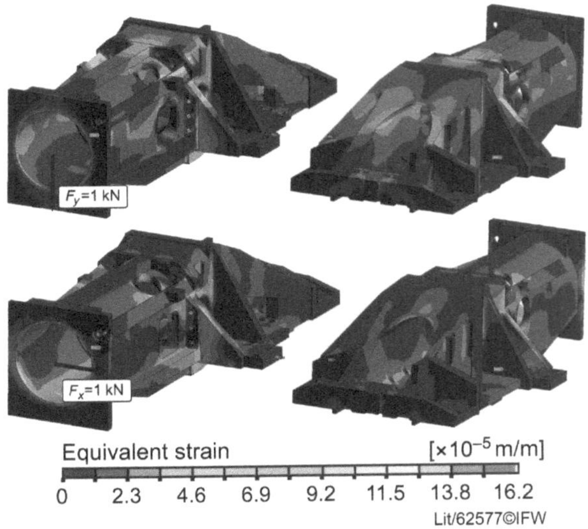

Fig. 3.2.1.1 Strain state in spindle slide structure.

structure [8]. For the local force flux adaption, notches in the structure of the spindle slide [9,10] are investigated. The principle of the notching effect is depicted exemplarily in Fig. 3.2.1.2A. When a load is applied on a notched structure, local strain intensification occurs in the notch ground. By integrating strain gauges into the notch ground, higher signal amplitudes can be measured compared to a sensor application onto a nonnotched flat surface.

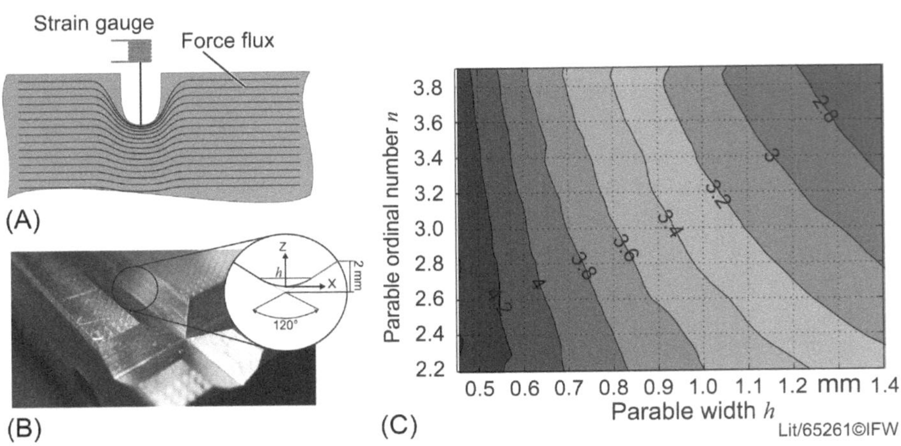

Fig. 3.2.1.2 Signal amplification by notching effect.

The investigated notching approach suits very well with the extremely small sensors, which are researched and used within the framework of the CRC 653 [11,12]. These sensors permit the use of small notch geometries and consequently lead to less influence on the structural rigidity. The design of one of the notch geometries is illustrated in Fig. 3.2.1.2B. As accessibility is of great importance for the application of the sensors, a chamfer with an angle of 90–120 degrees is chosen. Additionally, the depth of the notch is restricted to 2–4 mm due to the material thickness of the slide. The notch has a parabolic curved ground to increase strain maxima. The curved ground transits to a linear slope from the respective tangents. To achieve high strain amplitudes in the notch ground, parametrical FE simulations were conducted to determine the optimal profile parameters. Within the simulations, the parable ordinal number n and the parable width h are varied to compute the maximal amplification factors with the given geometrical boundary conditions. As the result of these simulations, the amplification factor of the measurable strains compared to nonnotched surface measurement is shown in Fig. 3.2.1.2C as a function of h and n. As can be seen, decreasing h and n leads to higher strain amplification.

3.2.1.4 Developed machine components

A sensory spindle slide is realized prototypically and integrated into a milling center DMG HSC55 linear [9,10] (Fig. 3.2.1.3). For the strain detection, miniature strain gauges are integrated into the manufactured small notches in the spindle slide. Two different kinds of miniature strain gauges, which were developed within the CRC 653, are applied: the first kind includes laser structured strain gauges (L-SG) which are directly sputtered on the 3D surface and subsequently structured using a laser beam [11]. The second kind consists of micro strain gauges which are based on a flexible polymer substrate and can only be applied to flat surfaces (μ-SG) [12]. These strain gauges are applied in 12 positions across the slide structure. These positions correspond to locations in the spindle slide which exhibit the highest sensitivity to mechanical strain as calculated from FE simulations. All used strain gauges are connected to electronic signal processing devices to compute and communicate the signals to the unit control. Because of the negligible dimensions of the notches compared to the slide, the change of the slide stiffness due to the notches lies in the range of 0.1% (Fig. 3.2.1.4A). Their influence can therefore be ignored.

Fig. 3.2.1.4B shows the sensing sensitivities of the strain gauges in the positions P1 to P4 according to static forces at the slide TCP in the directions x, y, and z. In general, the strain gauges integrated in the notch (L-SG, μ-SG) clearly show a big increase in the sensitivity to load compared to conventional strain gauges (Ref.) which are located very closely to the positions but are not integrated in notches. In comparison to the x- and y-directions, the spindle slide is not sensitive to load in the z-direction. This is due to the higher stiffness of the slide in this direction.

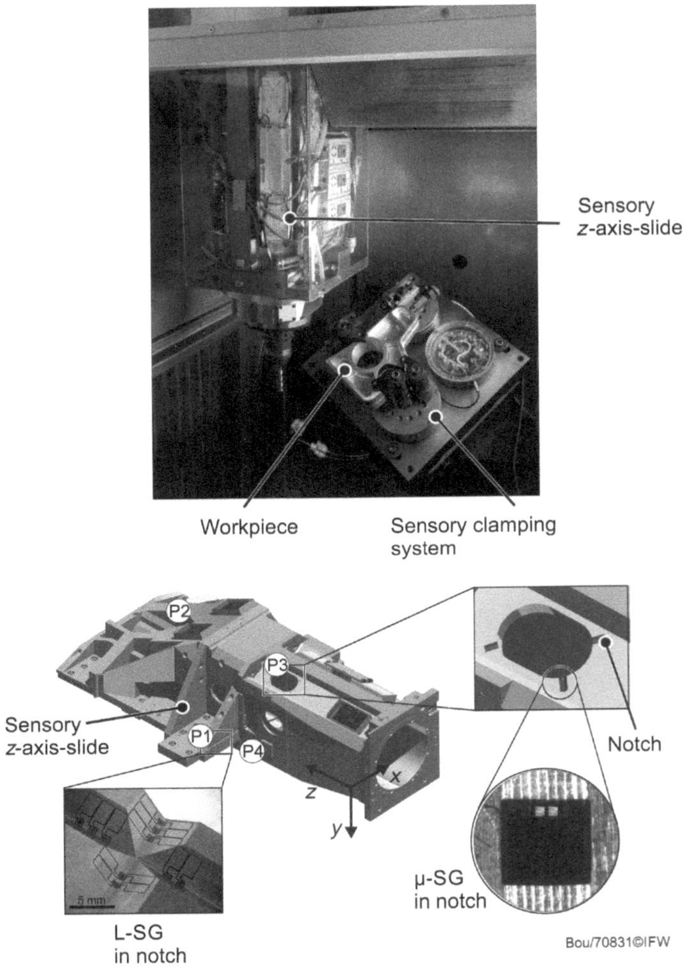

Fig. 3.2.1.3 Integrated sensory components in the test machine tool.

3.2.1.5 Force calibration by machine components

The aim of the force calibration is to determine mathematical matrices which are able to convert the signals of the strain gauges into force signals. The calibration matrices are computed using a linear regression analysis using reference force signals and all strain signals of the spindle slide.

With respect to the spindle slide, the low sensitivity in the z-direction impedes the force measurement in this direction. This is why the calibration neglects this force. The forces in the x- and y-directions can be detected accurately due to the high sensitivity to load of the strain gauges. Because each integrated strain gauges is sensitive

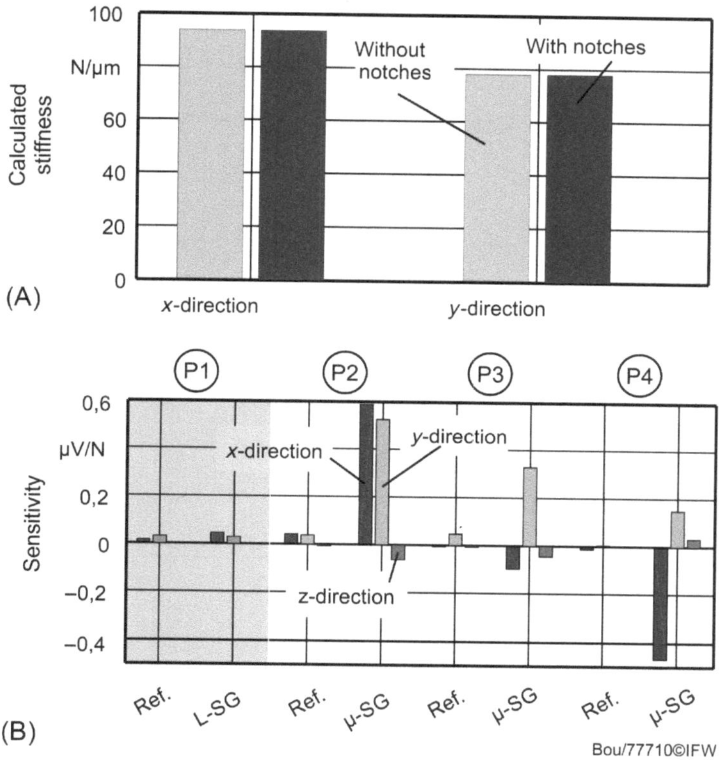

Fig. 3.2.1.4 (A) Influence of the notches on the slide stiffness and (B) influence of the notches on the sensitivity to load.

in both directions, the force calibration depends on how the load force is applied at the TCP. Two situations can be distinguished: the first situation arises during the milling process. During this process, process forces are applied simultaneously in the x-, y-, and z-directions. Each strain gauge provides a signal as a response to the combined loads from both directions. In this case, the calibrations of the x- and y-forces should be computed in parallel using the process forces. The second situation occurs in offline applications where forces are used to statically excite the slide in only one direction, either the x- or y-direction, such as the stiffness measurement of the spindle slide or the tool. Each strain gauge delivers a signal as a response to a simple load direction. In this case, the slide calibration in x- and y-direction should be carried out separately using statically applied forces at the TCP.

In the offline force calibration, the reference force signals are measured by a conventional force sensor. Fig. 3.2.1.5A shows the reference forces and the forces of the spindle slide measured after static calibration. In the in-process calibration, the signals are measured during up and down milling, where only the cutting depth is changing.

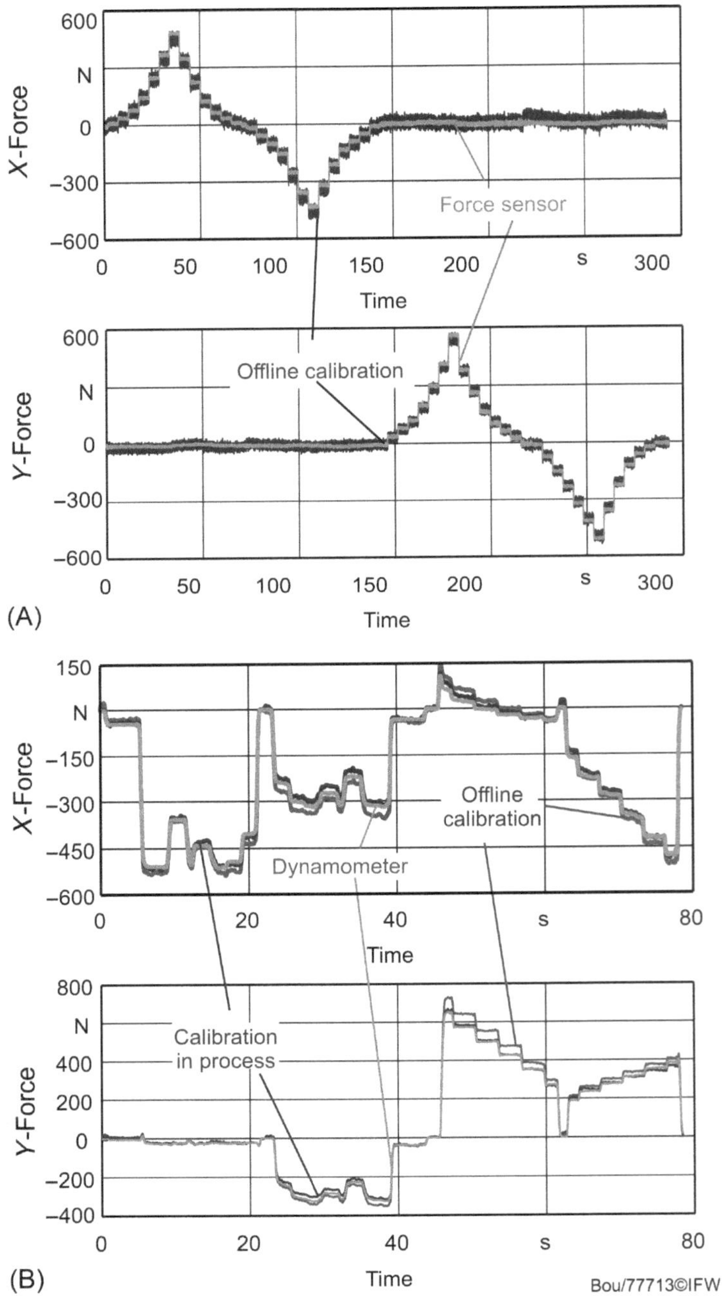

Fig. 3.2.1.5 (A) Offline force calibration of the spindle slide and (B) calibration of the spindle slide forces using process signals.

The reference forces are provided by a dynamometer. The filtered reference forces and the filtered forces of the spindle slide after calibration are mapped in Fig. 3.2.1.5B. This figure shows that the forces of the spindle slide can be approximated well with standard linear regression for both calibration methods. The calculated deviations of the forces to the reference grow to maximal 10% in the range of 500 N, where the force resolution is about 10 N. The calculated forces of the spindle slide based on the offline calibration are also depicted in Fig. 3.2.1.5B. A comparison of these forces with the reference forces shows a deviation up to 40 N that can occur when an unsuitable force calibration method is used.

3.2.1.6 Conclusion and outlook

This chapter describes the development of "feeling" machine tools by the integration of strain gauges into selected machine components. The use of the notching effect shows a significant improvement of the component sensitivity and the measuring quality of forces without lowering the component stiffness. Chapters 3.2.2 and 3.2.3 describe the development of further "feeling" machine components and even "feeling" workpieces. Based on the provided force signals from the "feeling" machine tool, process monitoring and process control systems can be enabled. Chapter 3.2.5 presents an approach for the monitoring and control of the tool deflection in milling.

References

[1] B. Denkena, J.W. Brinkhaus, D. Lange, Possibilities of sensor transponders in machine tools, in: 10th International Scientific Conference on Production Engineering, Computer Integrated Manufacturing and High Speed Machining, 2005.

[2] J. Leopold, D. Clauss, M. Klrner, A. Poppitz, M. Baldoli, A. Merlo, Investigations to new fixturing principles for aerospace structures, in: International Conference on Applied Production Technology, 2007, pp. 173–189.

[3] J. Hesselbach, H. Hoffmeister, B. Schuller, K. Loeis, Development of an active clamping system for noise and vibration reduction, CIRP Ann. Manuf. Technol. 59 (2010) 395–398.

[4] B. Denkena, H.C. Moehring, K.M. Litwinski, Design of dynamic multi sensor systems, Prod. Eng. 2 (2008) 327–331.

[5] K.M. Litwinski, Sensorisches Spannsystem zur Überwachung von Zerspanprozessen in der Einzelteilfertigung, Dissertation Leibniz Universität Hannover, Berichte aus dem IFW, PZHVerlag, Garbsen, 2011.

[6] H.C. Möhring, K.M. Litwinski, O. Gümmer, Process monitoring with sensory machine tool components, CIRP Ann. Manuf. Technol. 59 (2010) 383–386.

[7] B. Denkena, T. Mörke, M. Krüger, J. Schmidt, H. Boujnah, J. Meyer, P. Gottwald, B. Spitschan, M. Winkens, Development and first applications of gentelligent components over their lifecycle, CIRP J. Manuf. Sci. Technol. 7 (2) (2014) 139–150.

[8] H.A. Richard, M. Sander, Technische Mechanik – Festigkeitslehre, Vieweg & Sohn Verlag, 2006.

[9] B. Denkena, K.M. Litwinski, D. Brouwer, H. Boujnah, Design and analysis of a prototypical sensory Z-slide for machine tools, Prod. Eng. Res. Dev. (WGP) 7 (1) (2013) 9–14.

[10] B. Denkena, K.M. Litwinski, H. Boujnah, Detection of tool deflection in milling by a sensory axis slide for machine tools, Mechatronics 34 (2016) 95–99.
[11] L. Overmeyer, J.F. Duesing, O. Stuttmann, U. Stute, Laser patterning of thin film sensors on 3-D surfaces, CIRP Ann. Manuf. Technol. 61 (2012) 215–218.
[12] T. Griesbach, M.C. Wurz, L. Rissing, Application of sacrificial layers for the modular micro sensor fabrication on a flexible polymer substrate, in: Sensor+Test Conference, 2011, pp. 355–360.

3.2.2 Sensory workpieces

Markus Mücke, Dominik Dahlmann, Berend Denkena
Institute of Production Engineering and Machine Tools, Leibniz Universität Hannover, Garbsen, Germany

3.2.2.1 Introduction

Workpieces are the focus of each manufacturing process and the origin of acting loads. Consequently, the signals measured at the workpiece can show a high sensitivity to mechanical and thermal loads. So far, workpieces have not been used to detect process loads in machining processes. Components such as spindle slides [1,2] and clamping systems [3], which are located near to the process zone and directly involved in it, are preferred for sensor integration in milling machines. In general, the signal quality decreases with the increasing distance from the sensor position to the origin of the process loads. Therefore the workpiece is predestined as an information source for process loads.

With sensory workpieces it will be possible to monitor and compensate the effects of mechanically and especially thermal process loads. In addition to the measurement of acting forces, the sensor integration into the workpiece allows the capture of its temperature. Machine components have a low sensitivity to the workpiece thermal because of a complex distribution of the generated heat into the tool, chip, and workpiece. By monitoring mechanical and thermal loads, process failures can be detected. Additionally, by determining the resulting mechanical and thermal displacements, and controlling the machining process, shape deviations in the final workpiece shape can be reduced. Although the sensors must be integrated into each workpiece, this approach enables monitoring and control in any machine tool without sensing capabilities.

Process load measurements with workpieces have been subject to little research. Commonly, dynamometers are used to measure process forces for investigating and developing milling processes. Disadvantages of dynamometers are high costs and restricted flexibility. In case of temperature measurement, sensors are only used temporarily in workpieces for process developments and experimental investigations [4–6]. Mostly, the thermal effects are simulated. However, simulating the generated heat and the distribution into tool, chip, and workpiece are challenging. Heat generation and distribution are dependent on the workpiece material, the process parameters, the tool, and the tool wear and can change dramatically with different process conditions. To describe these dependency, various analytical models [7–9] are existing. In these models, every combination of tool, workpiece material, and processing parameters requires an identification of the model parameters. With numerical removal simulation [10,11], it is possible to reduce the effort for the parameter identification by simulating the cutting conditions of different processing parameters. In addition, the thermally induced distortions are determined with workpiece models and taken into account when creating the tool paths [12]. These methods are used in the work preparation and are, in general, not transferable to process parallel monitoring or control tasks because of long computing times. In Ref. [13] a process parallel modeling of the head conduction in machine tools is presented. The machine tool model is generated with model order reduction (MOR). With this method the amount of equations of a Finite Element Modell (FEM) is reduced dramatically and thus the computing time. The aim of the model is to estimate and compensate thermal deformations of the machine tool in real time. An approach to determine the generated heat during turning is an inverse heat conduction procedure [14]. Therefore an analytical model of the heat flow from the tool into the tool holder is developed. The heat flow is then calculated based on temperature changes measured on the tool holder. Consequently, the direct measurement at the workpiece has advantages for detecting process loads during machining.

For the development of sensory workpieces, as the third developed sensory component within the CRC 653, simulation-based methods for sensor placement are established. The FE simulations are used to identify load-sensitive sensor positions. The simulations have been carried out for spindle slides and clamping systems for different load conditions. However, integrating sensors on workpieces leads to novel challenges concerning sensor placement, energy, and data transmission [15].

3.2.2.2 Technical challenges of sensory workpieces

The most important challenge in the development of sensory workpieces lies in finding optimal positions in the workpieces for the sensor integration. These positions should provide sufficient sensitivities to correctly detect the mechanical or thermal loads. Generally, two factors may influence such sensor positions. On the one hand, the sensor positions depend on the changing workpiece geometry during processing. In case of milling of solid blocks, the volume of the workpiece is dramatically decreasing, while geometry differs slightly from the initial between process steps by semifinished parts like forged or casted parts. On the other hand, considering that each process step often requires different clamping configurations in the machine tool, the process steps also influence the sensor positions. The changes in the workpiece structure and the clamping situation generally lead to variations in the workpiece stiffness and the thermal boundary conditions and are therefore the reason for variable sensitivities. Furthermore, the tool movement continuously shifts the point of load in one process step. The sensor sensitivity has to be independent from the tool movement or the changes have to be considered in signal processing.

The cost-effectiveness, the practicability, and the robustness of the sensing system represent challenging aspects to be taken into account when designing sensory workpieces. With regard to cost-effectiveness, the sensor integration in workpieces should demonstrate a functional benefit for process monitoring and optimization. Here, as few sensors as possible should allow the accurate detection of the caused forces and thermal loads. Concerning the practicability, the sensor system should not prevent the makeready or the cutting within the machine tool. This leads to challenges, especially for the power supply of the signal devices and the data communication. Considering that sensors and signal devices are exposed to several external influences during cutting such as coolants, metal chips, and electromagnetic disturbances, the sensing system finally requires a robust protecting design.

3.2.2.3 Concept for sensory workpieces

The sensing system for workpieces can consist of a single sensor or a network of sensors in combination with miniature electronic devices that allow the signal processing and the communication with an industrial PC. Depending on the use case, the sensors and the electronic devices can be fully embedded in the workpiece, either temporarily mounted to it or as a variation in-between. Reuse of temporarily mounted sensors reduces the cost. For example, in large workpieces sensors can be moved to the section of the workpiece that is currently machined. Permanently integrated sensors can be economical if they are used in live cycle. As on a wheel carrier, the mechanical stress on the component and the bearing temperature can be monitored. For this research purpose, in which several machining experiments carried out, a low-cost concept is created (Fig. 3.2.2.1). In this concept, strain gauges for mechanical and resistance temperature detectors (RTD) for thermal load detection are used.

As shown in Fig. 3.2.2.1, a strain gauge or an RTD and a socket are applied and connected on the workpiece. An aluminum housing with the integrated electronic device can be plugged onto the socket and allows hereby the temporarily connection with tension relief between the strain gauge and the device. In addition, the housing allows sealing and protection of the whole sensor system against humidity, chips, and electromagnetic disturbances while cutting. The power supply is realized by rechargeable batteries. With an isolator between the electronics and the temperature sensor, the two components are separated to prevent interference in the temperature measurement. The data communication works wirelessly. The individual sensors are combined to a network and the data is received by one industrial PC. To allow process control the industrial PC is linked to the control of the machine tool.

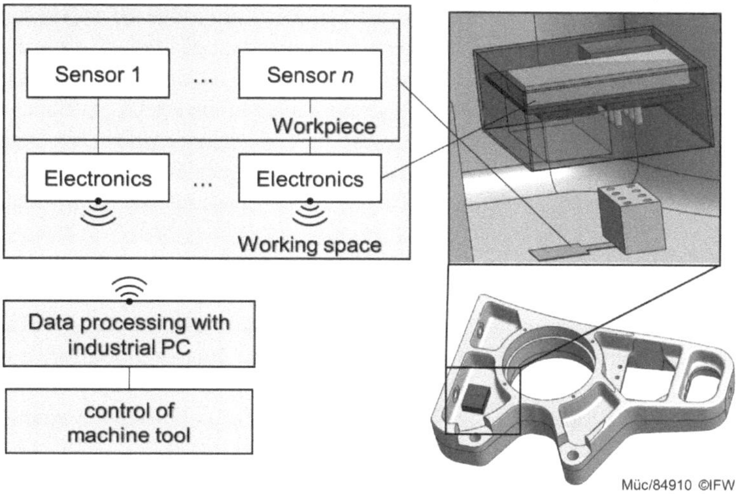

Fig. 3.2.2.1 Concept for sensor integration in a wheel carrier as a casted workpiece.

3.2.2.4 Sensor placement for force detection

To estimate the occurring strains and to determine optimal sensor positions in workpieces, static structural finite element analyses are conducted with ANSYS Workbench. First tests with the developed simulation-based approach are performed using an easily understandable analogy part of an L-shaped aluminum workpiece. It is assumed that the workpiece is clamped on selected clamping surfaces on its one side and that it is milled on the machining surface on its other side. The remaining surfaces on which sensors can be safely integrated are selected as sensing surfaces for sensor application (Fig. 3.2.2.2).

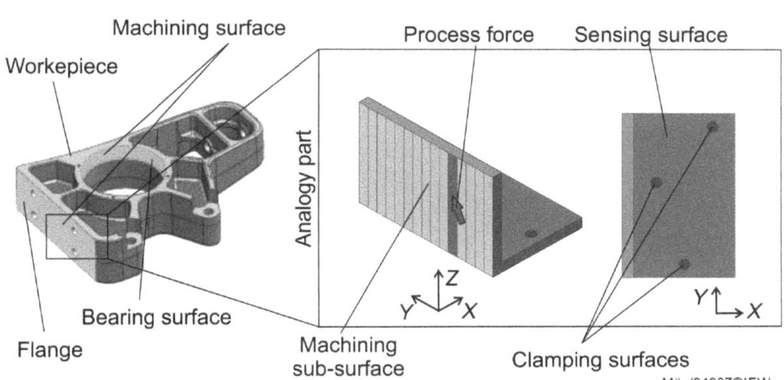

Fig. 3.2.2.2 Setups for simulation.

After meshing the workpiece the clamping surfaces are modeled as fixed supports. During simulation, the forces and the corresponding point of load are shifted equidistantly along a predefined path in the machining surface. The surface is split into subsurfaces to apply the load. The computed strain values of the sensing surfaces for each subsurface are exported. A script to generate the subsurfaces offers big benefits for case studies especially in the simulation of various workpieces with more complex shapes, many machining surfaces, and by modeling various load and engagement conditions.

The optimal positions and even the load sensitivity of the sensors depend, in general, on the resulting force flux, which is mainly affected by the point of load of the applied forces and the clamping situation of the structure. The movement of the point of load along the tool path causes a variation in the force flux and therefore a variation in the strain state. Fig. 3.2.2.3 shows the resulting von Mises strains in the sensing surface for different tool positions. To find the optimal positions the change of the strain in the sensing surface is analyzed.

Based on the exported strain values of the sensing surfaces, each mesh node of these surfaces is statistically evaluated by building the mean value and the deviation of its strain for the simulated load steps (Fig. 3.2.2.4). The strain mean value provides information about the sensitivity to a load of the mesh node. The deviation of the strain

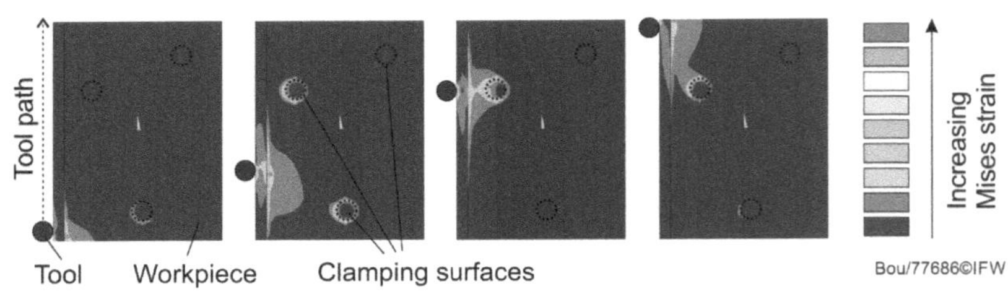

Fig. 3.2.2.3 Simulated strain state in workpiece by a moving tool.

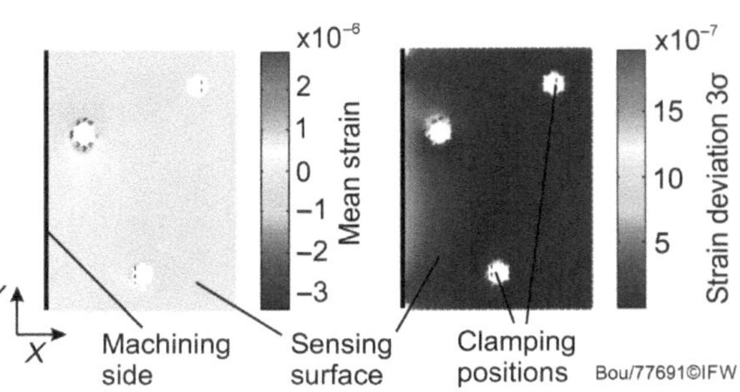

Fig. 3.2.2.4 Statistical evaluation of the resulted strain in *Y*-direction for an applied force of 100 N.

of a mesh node is a measure of the sensitivity variation on that node while the tool is moving along its path. Accumulations of adjoining mesh nodes showing similar sensitivity behavior represent optimal positions for sensor integration.

Two kinds of sensor positions can be distinguished (Fig. 3.2.2.5): the first kind comprises path-dependent positions. Such positions show high mean values of strain and strain deviations. In these positions strain sensors would generate intense signals showing strong deviation along the tool path. In this case, a sensor calibration with respect to tool movement is indispensable for an accurate measurement of the process forces in the investigated workpiece. The second kind of positions are path-independent sensor positions. They generally show lower mean values and significantly lower deviations of strain in comparison to path-dependent positions. However, in such positions, sensor signals with sufficient amplitudes, but mostly independent from the actual tool position would be provided.

Because the force flux is generally concentrated near the machining surface and around the clamping positions, the sensor positions mostly appear near to these positions or in-between. Simulations of the workpiece show that the identified sensor positions depend on the direction of the applied force, the considered direction of the resulting strain (Fig. 3.2.2.6), and the clamping situation of the workpiece (Fig. 3.2.2.7). The size of the identified sensor position depends on the applied restriction by node filtering such as the minimum of required strain and the maximal permissible strain deviation. Positions on the outer side of the workpiece represent noisy nodes and cannot be used as sensor positions. This approach is also transferable to workpieces with more complex shapes and many processing steps. Here, the statistical evaluation of the strain should be considered for all tool positions over all processing steps. It is expected that the number of path-dependent sensor positions (high strain, high deviation) increases with the number of process steps, while the number of path-independent sensor positions (low strain, low deviation) decreases.

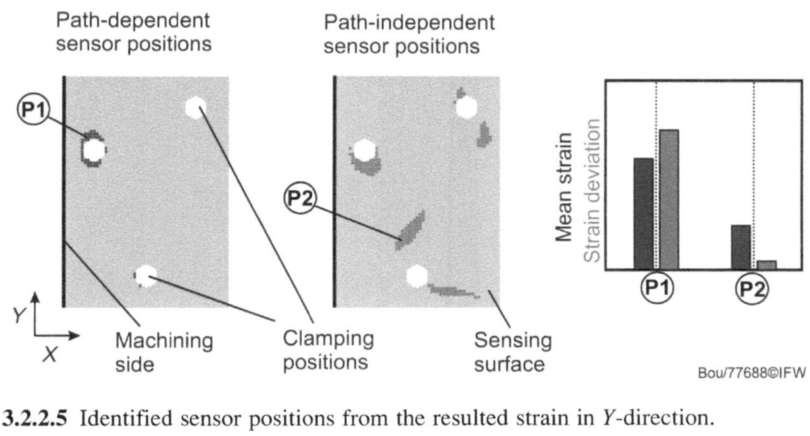

Fig. 3.2.2.5 Identified sensor positions from the resulted strain in *Y*-direction.

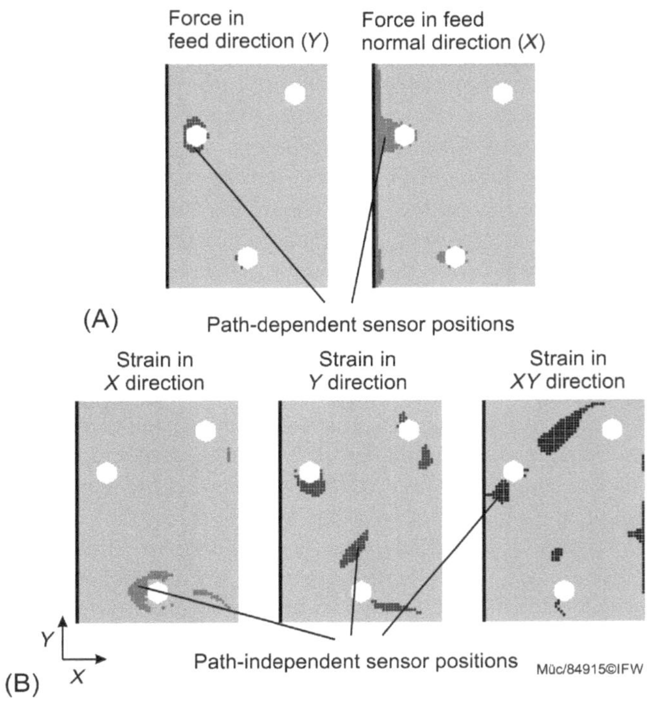

Fig. 3.2.2.6 Identified sensor positions: (A) path dependent for different force directions and (B) path independent for different strain directions.

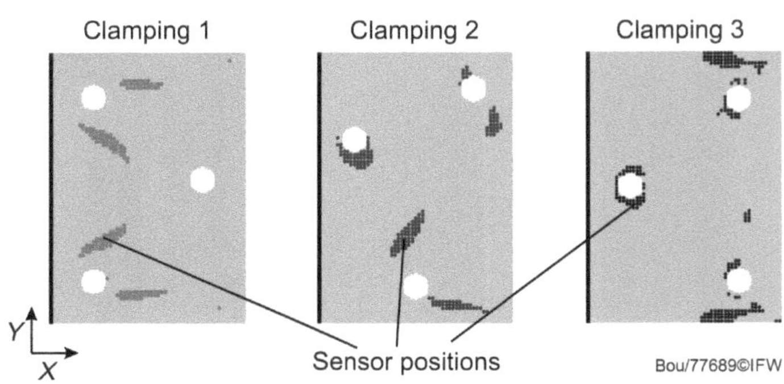

Fig. 3.2.2.7 Identified sensor positions (path independent) for different clamping situations.

3.2.2.5 Reconstruction of the workpiece temperature distribution

During cutting process, heat is generated and one part from this heat is induced into the workpiece. The long lasting heat conduction inside of the workpiece leads to an inhomogeneous and over the time changing temperature distribution. This temperature distribution cannot be measured directly and economical with locally measuring RTD. To cover the whole workpiece surface with RTD in an appropriate sampling distance, a huge number of sensors is needed. Furthermore, the internal temperature distribution remains unknown. In addition, the temperature change depends on boundary conditions such as the workpiece material, the tool, the processing parameters, and the ambient temperature. To overcome this problem by reconstruction of the temperature distribution in parallel to machining, a model-based approach was developed. Therefore the heat conduction is simulated. By correcting the simulation model behavior with as few as possible local temperature measurements changing boundary conditions are considered. Fig. 3.2.2.8 shows the developed model for reconstruction of the workpiece temperature distribution with local temperature measurements.

To simulate the heat conduction in the workpiece during machining, a FE model was created. The machining surfaces on the workpiece model are defined by the processing steps. The clamping is simulated by constant heat dissipation on the surface of the workpiece, which is in contact with the machine table. All other clamps are ignored to simplify the model. On the remaining surfaces a free convection boundary

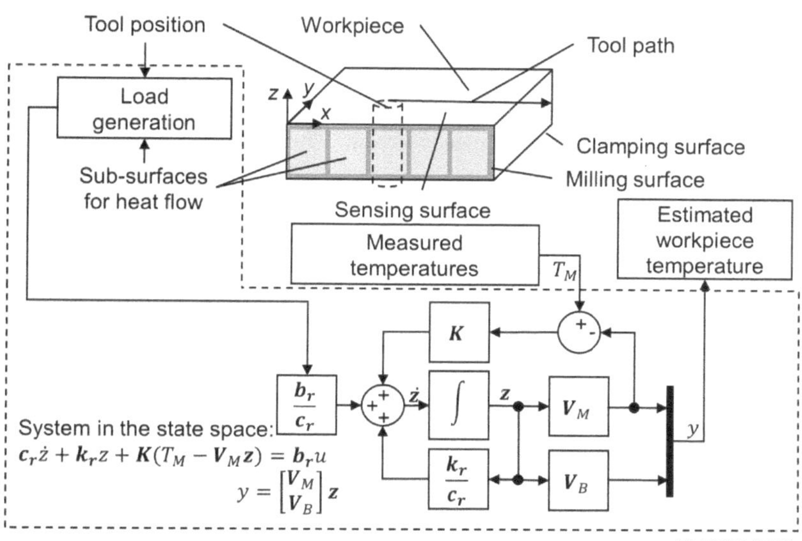

Fig. 3.2.2.8 Model-based temperature distribution reconstruction.

condition is applied to take the ambient temperature into account. The processing of the workpiece is simulated by a moving heat source [16]. Therefore the machining surface is divided into subsurfaces. Depending on the tool position, a load is applied to the subsurface, which is in contact with the tool at this moment. The applied load is a heat flux density $\dot{q}$ which depends on the heat flow of the process q, the process time t_P, and the number of subsurfaces n:

$$\dot{q} = \frac{q \cdot n}{t_P} \tag{3.2.2.1}$$

To allow process parallel calculation, the number of equations of the model, which are correlated with the computing time, was reduced by the method of MOR. The order reduction is done with the software mor4ansys [17] by moment matching via the Arnoldi algorithm, leading to a description of the workpiece model in the state space:

$$\begin{aligned} \boldsymbol{c_r}\dot{z} + \boldsymbol{k_r}z &= \boldsymbol{b_r}u \\ y &= \boldsymbol{V}z \end{aligned} \tag{3.2.2.2}$$

The matrixes $\boldsymbol{c_r}$ and $\boldsymbol{k_r}$ describe the heat capacity and the thermal conductivity of the system. The moving heat source for the reduced model is defined by a vector of time stamps and applied heat flux densities $\dot{q}$ for each machining subsurface. This load vector is transformed to the state space via the matrix $\boldsymbol{b_r}$. Vice versa $\boldsymbol{V}$ transforms the state variables into the output values, in this case the temperatures on selected positions on the workpiece. The complexity and thus the computing time of the reduced model is given by the number of the used nodes. The required number of nodes is approximately three times the number of machining subsurfaces. To avoid the time-consuming identification or modeling of exact boundary conditions (process heat, clamping, and ambient conditions) estimated values were used. Imperfections in the boundary conditions and the influence of changing environmental conditions will be considered by the temperature measurement. The temperatures are measured in sensitive positions on the real workpiece during machining. A state observer is used to estimate the temperature at any position on the workpiece based on these measurements. By splitting the matrix $\boldsymbol{V}$ into $\boldsymbol{V_M}$ and $\boldsymbol{V_B}$, the process parallel simulated temperatures are used as inputs for the observer and as estimated workpiece temperature outputs. The observer compares the simulated and the measured workpiece temperature and fed back the difference to the system through the gain matrix $\boldsymbol{K}$. With this feedback the simulation system as shown in Fig. 3.1.1.8 follows the behavior of the real workpiece and the real workpiece temperatures are estimated on the system output. To generate the load matrix for the simulation in parallel to the machining, the input to the system is generated based on the actual tool position. The tool position is read out from the machine tool control. Big changes of shape during milling are conceivable in connection with a material removal simulation. Therefore if an allowable shape deviation is exceeded by the material removal, a new MOR model is created and during the processing, the model has to be chosen based on the machining progress.

3.2.2.6 Sensor placement for temperature measurement

The occurring temperature differences, the response time, and the process parallel simulation model characteristics have an influence on the sensor positioning. By defining a discrete number of machining subsurfaces, differences between the simulated and the real temperature distribution occur. As shown, the number of equations in the reduced models depends on the number of the subsurfaces used. In a relatively small example part with the dimensions of $100 \times 160 \times 50\ mm^3$, this fact has no great influences on the computation time. But with bigger workpieces in mind, the sensor integration is investigated for one and two subsurfaces with the example workpiece as a part of a bigger workpiece. With this enlargement of the subsurface area and thus an abstraction and simplification of the machining process the number of subsurfaces in bigger workpieces decreases. To investigate the influences on the sensor positioning a FEM simulation of the temperature distribution of a process (represented by 20 subsurfaces) is compared to a simulation with one or two subsurfaces. The sensors will be placed on top of the cuboid example workpiece in the sensor surface as marked in Fig. 3.2.2.8. With the increasing area of a subsurface local peaks in the occurring temperature distribution are averaged (Fig. 3.2.2.9). The peaks arise in the real process because of the high heat generation at the small contact surface between tool and workpiece. These higher temperatures fade with the depth of the part and transform into a temperature distribution similar to that of the simplified model. This leads to areas of the sensing surface with differing temperature distributions between the real process and the simulation. The use of the state observer prohibits these areas for sensor integration. The aim of the observer is to consider varying workpiece temperatures caused by changing boundary conditions. However, the occurring temperature differences in these areas are additionally caused by the model simplifications. By avoiding these areas, influences on the observer from these systemic differences are eliminated. A further selection criterion for sensor placement is the sensor sensitivity, in this case the time between the start of the machining and an increase of the temperature at the

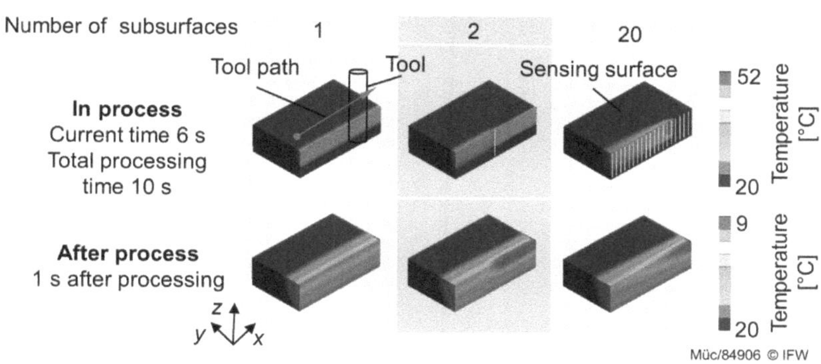

Fig. 3.2.2.9 Influence of the number of load surfaces on the temperature distribution for two time points.

sensor positions. This response time reaches a minimum when the sensors are located in the permissible area close to the cutting edge and at the beginning of the tool path. It is important to ensure that temperature changes at the sensor positions are measurable before the next operation starts. Thus the time the heat needs from the cutting edge to the sensor must be less than the process time for the corresponding machining operation. Otherwise the workpiece temperature is not reconstructed by the observer when the next machining operation starts.

Fig. 3.2.2.10 shows suitable surfaces for sensor placement resulting from the comparison of the simulation with 20 subsurfaces and a simulation with one and two subsurfaces. In the marked surface for sensor placement, the temperature difference between the compared temperature distributions is less than ±0.1 K. The difference was observed over the duration of the process (10 s) and the homogenization of the workpiece temperature (240 s). After the homogenization the workpiece temperature is equal and the heat conduction is finished. The permissible area for sensor placement for sensors as input for the observer rises from 47% of the entire surface with one subsurface to 62% with two subsurfaces. This low correlation in the thermal behavior represented by the proportion of the suitable surface from 47% or 62% arises from the observation over all of the brief extreme differences. Actually 1 s after the end of the process the overall temperature difference falls to less than 1 K. The requirement of minimal response times leads to sensor position 6 (response time 8.1 s) for one subsurface and positions 3/2 (response time 3.8 s) for two subsurfaces. Additional sensors in position 1 or 4/5 are added to capture environmental influences. Three sensors for each setup were chosen to make them comparable. In both sensor arrangements at least one sensor responds quickly enough to determine a temperature change before the operation (machining duration 10 s) finishes. Investigations on different feed rates and heat flux densities show no significant differences in the surface for sensor

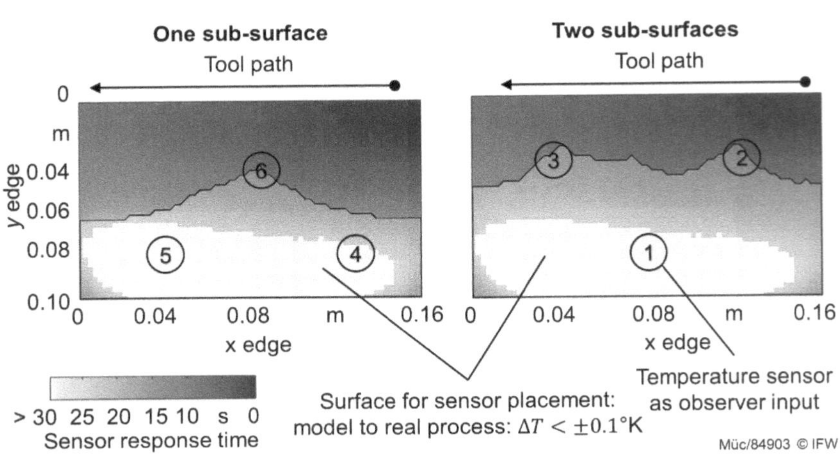

Fig. 3.2.2.10 Sensor placement for temperature measurements.

placement. With increasing feed rate the response time decreases. The faster tool movement brings the heat source faster near to the sensor and their response time decreases. With this method it is possible to determinate sensor positions on a workpiece surface for the processing parallel model-based reconstruction of the temperature distribution in the workpiece.

3.2.2.7 Process force monitoring with sensory workpieces

The presented concept of a sensory workpiece is evaluated in milling experiments. The measurement with strain gauges on the workpiece will be compared to signals of a dynamometer. The L-shaped workpiece is mounted to the dynamometer in this experiment (Fig. 3.2.2.11). To reuse the analogy workpiece, a block for machining was mounted to the workpiece. The simulation results for the changed shape show no big differences in the sensor characteristic.

The chosen path-independent sensor positions on the sensing surface are selected according to Section 3.2.2.4 and shown in Fig. 3.2.2.12. To analyze the influence of the force direction, the simulation of the strain is done separately for the cases of force in the feed and the feed-normal directions. The process force is a combination of the feed and normal force. Fig. 3.2.2.12 shows the resulting optimal sensor positions in the sensing surface with respect to strain occurring in the feed and the feed-normal directions. In addition to the force-induced strain, thermal strain is caused by the process heat. To compensate this effect and measure only the force-induced strain, strain

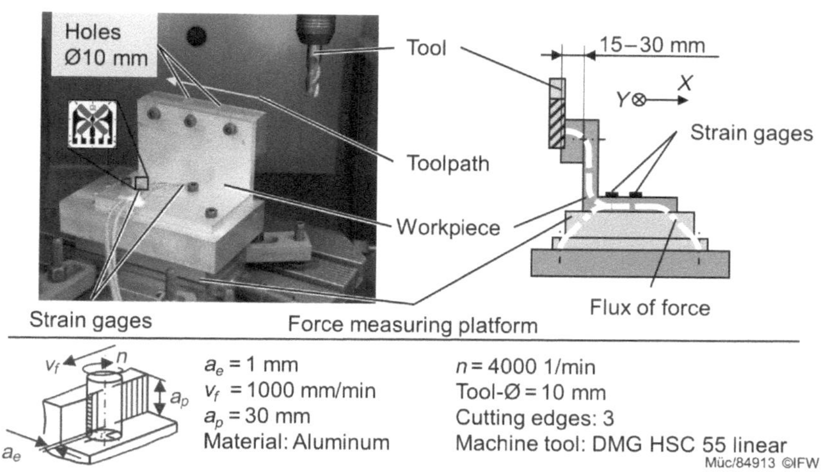

Fig. 3.2.2.11 Setup for comparing the measurement of a dynamometer and the sensory workpiece.

gauges in a full bridge with four measuring grids (HBM 1-VY43-3/350) are used. This arrangement compensates thermally induced strain, but therefore mechanically induced strain is measured in two directions perpendicular to each other. The two path-independent signals (S2-Y and S1-X at feed force) have a low amplitude and therefore no significant influences on the overall measurement. Only the signals S1-Y and S2-X show a path significant dependency due to load in the case of feed normal direction with a change of −20% and −80%. By combining the signal in the two directions, it leads in total to a path dependency of 6% for sensor two and 20% for sensor one. Under consideration of a temperature measurement of the workpiece, it is possible to measure the strain in each direction individually. The thermal strain can be calculated from known temperature changes and can be compensated. In consequence, the signal S1-X is only sensitive to the feed-normal force. By using this

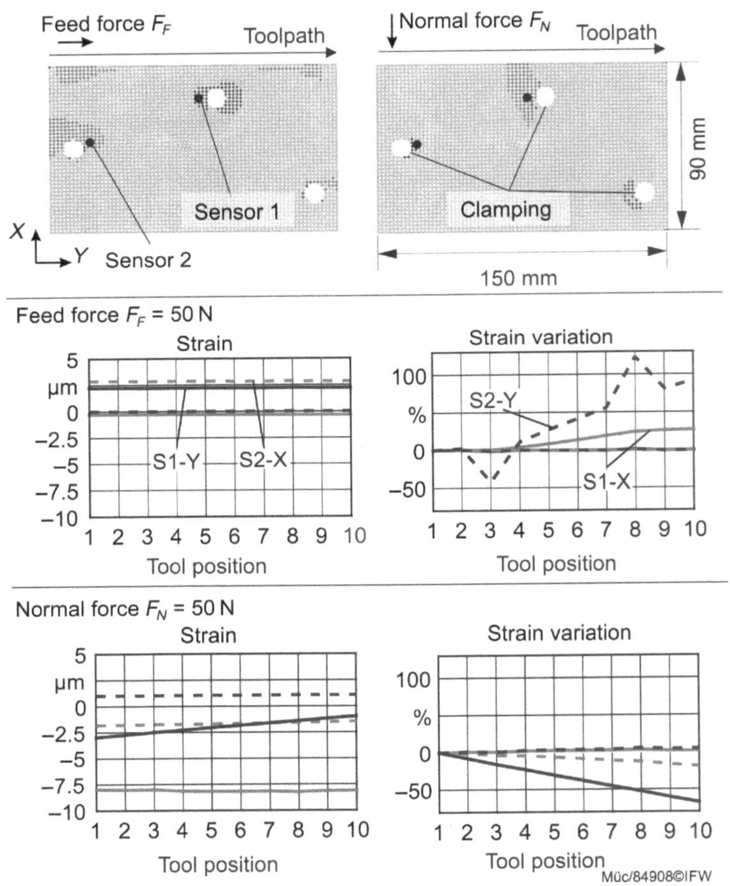

Fig. 3.2.2.12 Sensor placement on the L-shaped workpiece.

fact, even the force direction can be determined. Therefore the feed force can be calculated from the signal S2-X with the known feed-normal force.

The milling experiment was performed on a DMG HSC 55 with a tool with three cutting edges. Fig. 3.2.2.13 shows the results for two cutting processes with the same process parameters The correlated strain signals show that the measuring is reproducible under similar conditions. The trend of the measured values meets the expectations. The signal of sensor one changes by 52% and the signal of sensor

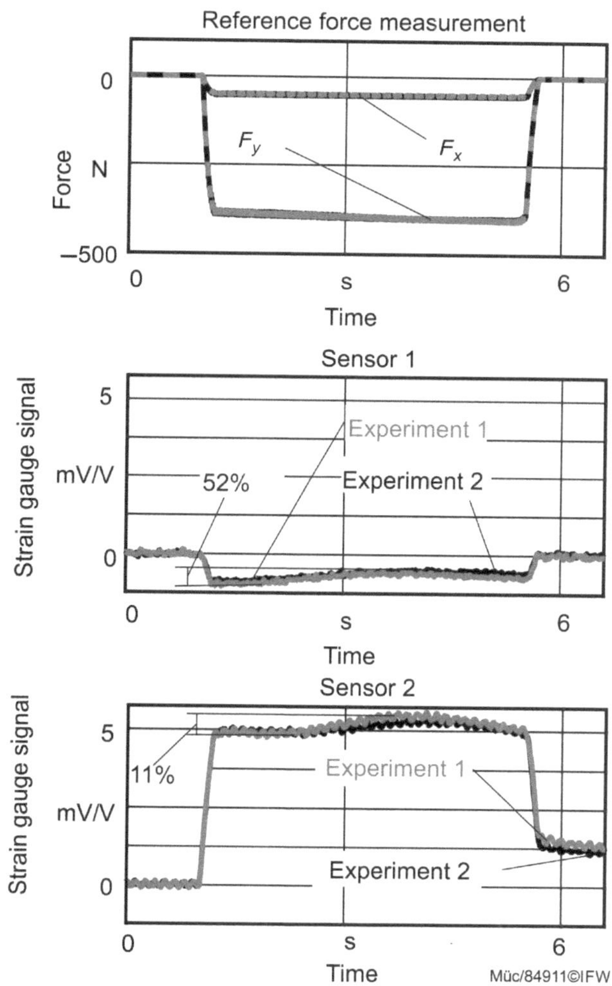

Fig. 3.2.2.13 Force and strain gauge signals for two cutting experiments with equal processing parameters.

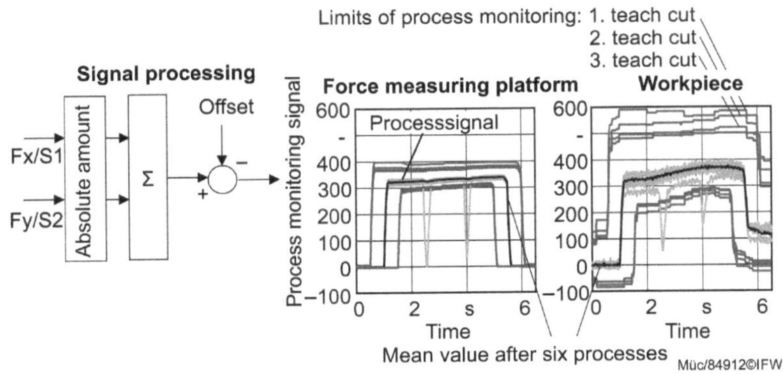

Fig. 3.2.2.14 Results of process monitoring.

two by 11%. The higher changes in comparison to the simulation are caused by the strain gauge surface of $10 \times 10\ mm^2$. In this surface, signals with other characteristics than the signal at the simulated point are detected. Additional experiments are conducted with different cutting speed and cutting width (a_e). The results show that the strain signals are depending on the mechanical load. In case of sensor two, hysteresis occurs. These are caused by tensions in the component and thermal effects.

In order to investigate a process monitoring approach, the workpiece is prepared by two holes. The used monitoring method computes statistical envelope curves [18,19]. The system is trained with the first six processing operations with a confidence limit of 98%. After the third processing the monitoring is activated. Errors can be detected if the process signal exceeds the trained limits. In the seventh processing step the error occurs. The results in Fig. 3.2.2.14 show a better signal-to-noise ratio of the dynamometer compared to the sensory workpiece. The differences are caused by the force flux. In the case of the dynamometer the total force flux is measured. In the workpiece only an amount of the force flux causes the measured deformations. This leads to tighter process monitoring limits for the dynamometer. However, the process monitoring based upon the sensory workpiece detects the workpiece errors. Consequently, the sensory workpiece is suitable for process monitoring and there is no need for additional dynamometers.

References

[1] B. Denkena, K.M. Litwinski, D. Brouwer, H. Boujnah, Design and analysis of a prototypical sensory Z-slide for machine tools, Prod. Eng. Res. Dev. 7 (1) (2013) 9–14.

[2] B. Denkena, K.M. Litwinski, H. Boujnah, Process monitoring with a force sensitive axis-slide for machine tools, Procedia Technol. 15 (2014) 416–423.
[3] B. Denkena, D. Dahlmann, J. Kiesner, Production monitoring based on sensing clamping elements, Procedia Technol. 26 (2016) 235–244.
[4] B. Davoodi, H. Hosseinzadeh, A new method for heat measurement during high speed machining, Measurement 45 (8) (2012) 2135–2140.
[5] P.S. Sivasakthivel, R. Sudhakaran, Optimization of machining parameters on temperature rise in end milling of Al 6063 using response surface methodology and genetic algorithm, Int. J. Adv. Manuf. Technol. 67 (9–12) (2012) 2313–2323.
[6] C.H. Lauro, L.C. Brandão, S.L.M. Ribeiro Filho, Monitoring the temperature of the milling process using infrared camera, Sci. Res. Essays 8 (23) (2013) 1112–1120.
[7] R. Joliet, A. Byfut, P. Kersting, A. Schröder, A. Zabel, Validation of a heat input model for the prediction of thermomechanical deformations during NC milling, in: 14th CIRP Conference on Modeling of Machining Operations, vol. 8, 2013, pp. 403–408.
[8] M. Schweinoch, R. Joliet, P. Kersting, A. Zabel, Heat input modeling and calibration in dry NC-milling, Prod. Eng. Res. Dev. 9 (4) (2015) 495–504.
[9] D.J. Richardson, M.A. Keavey, F. Dailami, Modelling of cutting induced workpiece temperatures for dry milling, Int. J. Mach. Tools Manuf. 46 (2006) 1139–1145.
[10] D. Niederwestberg, B. Denkena, Simulation of thermal and mechanical workpiece load, CIRP J. Manuf. Sci. Technol. 7 (2014) 315–323.
[11] D. Biermann, I. Iovkov, Modelling, simulation and compensation of thermal effects for complex machining processes—the priority programme 1480 "CutSim" Prod. Eng. Res. Dev. 9 (4) (2015) 433–435.
[12] B. Denkena, A. Schmidt, P. Maaß, D. Niederwestberg, C. Niebuhr, J. Vehmeyer, Prediction of temperature induced shape deviations in dry milling, in: 15th CIRP Conference on Modelling of Machining Operations (15th CMMO), vol. 31, 2015, pp. 340–345.
[13] K. Großmann, C. Städel, A. Galant, M. Andreas, Berechnung von Temperaturfeldern an Werkzeugmaschinen, ZWF 06 (2012) 452–456.
[14] L. Liang, H. Xu, Z. Ke, An improved three-dimensional inverse heat conduction procedure to determine the tool-chip, Int. J. Therm. Sci. 64 (2013) 152–161.
[15] B. Denkena, D. Dahlmann, H. Boujnah, Sensory workpieces for process monitoring—an approach, in: Proc. 3rd International Conference on System-Integrated Intelligence: New Challenges for Product and Production Engineering, SysInt 2016, June 13th–15th, Paderborn, Germany, 2016, pp. 129–135.
[16] R. Pabst, Mathematische Modellierung der Wärmestromdichte zur Simulation des thermischen Bauteilverhaltens bei der Trockenbearbeitung, Dissertation Universität Karlsruhe (TH), Forschungsberichte aus dem wbk, Institut für Produktionstechnik, Shaker, 2008.
[17] E.B. Rudnyi, J. Lienemann, A. Greiner, J.G. Korvink, mor4ansys: generating compact models directly from ANSYS models, in: Technical Proceedings of the 2004 Nanotechnology Conference and Trade Show, Nanotech, 2004, pp. 279–282.
[18] J.W. Brinkhaus, Statistische Verfahren zur selbstlernenden Überwachung spanender Bearbeitungen in Werkzeugmaschinen, Dissertation Leibniz Universität Hannover, Berichte aus dem IFW. PZHVerlag, Garbsen, 2008.
[19] H.C. Möhring, K.M. Litwinski, O. Gümmer, Process monitoring with sensory machine tool components, CIRP Ann. Manuf. Technol. 59 (2010) 383–386.

3.2.3 Feeling clamping system

Berend Denkena, Johann Kiesner
Institute of Production Engineering and Machine Tools, Leibniz Universität Hannover, Garbsen, Germany

3.2.3.1 Introduction

Clamping systems are the interface between the machine tool and the workpiece. They define the workpiece position, guide the flux of force, and thus determine the quality of the finished workpiece. Because of their close proximity to the workpiece and thus the machining process, clamping systems with integrated sensors offer a promising possibility for monitoring of machining processes.

Due to its position in the working space, clamping systems operate under harsh environmental conditions, with contact to chips and coolants. In particular, hydraulic actuated clamping systems can be subjects to wear phenomena at the piston surfaces and seals when overloaded. Therefore in addition to the process control, monitoring of the system state is given an important role. The warranty of an intact system state and an automated check of the clamping situation improve the profitability of a clamping system by preventing failures, e.g., loosening of components during processing, and the production of rejects, considerably.

The integration of sensors in clamping systems, however, results in several challenges: The sufficient sensor sensitivity, the harsh environmental conditions including electromagnetic interference, and the power and data transfer at applications with rotating machine tables.

3.2.3.2 Objective

The integration of sensors close to the process and an application-specific signal conditioning enable the detection of small errors, e.g., imperfections in the workpiece material. Beyond the detection and monitoring of the process behavior, the monitoring of the functional state of the clamping system itself represents a significantly added value to sensory components. In particular, automatically operating devices need features to enable the detection of malfunctions and thus the avoidance of serious process failures. The main objective is to realize a "feeling" clamping system that enables a robust monitoring of its own functionalities and the process behavior under industrial conditions in cooperation with two companies.

3.2.3.3 Methods

In cooperation with two companies, selected results from the fields of sensor integration and monitoring described in Chapter 3.2.1 were developed toward an industrial application. In order to resolve the existing conflict between the maximum sensitivity of strain gauge signals and maximum stiffness of the clamping device, a method for sensor placement has been explored by the CRC 653 [1]. This method compares the signals of virtual sensors within a FEM-based simulation of a clamping system at different signal inputs, taking into account interactions between sensors. A modification of this procedure extends the application for universal use under industrial conditions even with hydraulic clamping elements. Furthermore, the results of the CRC point out that a robust implementation of the data and energy transmission to the sensor system contributes significantly to the applicability in an industrial environment.

Requirements for an exemplary clamping system are derived together with industry partner and from failures that result from practice-relevant common causes. Based on these requirements, methods for construction and conceptual design were applied to realize and evaluate variants for the integrated sensor system. These include, in particular, the use of the morphologies, computer-aided designs, and stiffness analysis using the finite elements method. Promising approaches have been verified in preliminary studies with experimental models. The variant for realization was selected on the base of specified criteria, for example, space, costs, and signal quality.

3.2.3.4 Application scenario and system specification

Increasing product variants and thus a higher number of different workpieces lead to the need of high flexibility and availability of clamping solutions. To ensure this essential requirement, the decision, to equip only the individual clamping elements with sensory properties, instead of the overall system, was taken. This conceptual design enables a flexible solution even for widely varying clamping tasks.

The development of the "feeling" clamping system bases on a representative application scenario from the automotive industry from the portfolio of the company ReiKam. It deals with the hydraulic clamping of cast casing covers for multiaxis machining. The exemplary hydraulic clamping fixture, see Fig. 3.2.3.1A, is equipped with clamping elements of the manufacturer Roemheld. The assembly consists of a fixture plate, three hydraulic swing clamps (type 1844-L090-L24) with appropriate supports, and one hydraulic work support (type 1953-320), compare also Fig. 3.2.3.2. The hydraulic clamping elements are supplied with oil by drilled hydraulic lines inside the fixture plate [2].

Fig. 3.2.3.1B summarizes the concept of the "feeling" clamping system, see also [3]. To provide the original hydraulic clamping fixture with sensing abilities, strain gauges and temperature sensors are applied to the clamping elements. Furthermore, the sensors are connected to electronic preprocessing devices that are integrated into the base plate. The direct proximity of analog sensors and digitizing units shortens the susceptible analog transmission paths, e.g., to EMC and offers the beneficial use of fieldbus communication with an industrial PC via CAN at an early stage. By providing every clamping element with its own microcontroller device it is easily possible to expand the "feeling" clamping system by additional elements. A communication via a radio link and an inductive energy transmission substitute wired connections at applications with endless rotating tables.

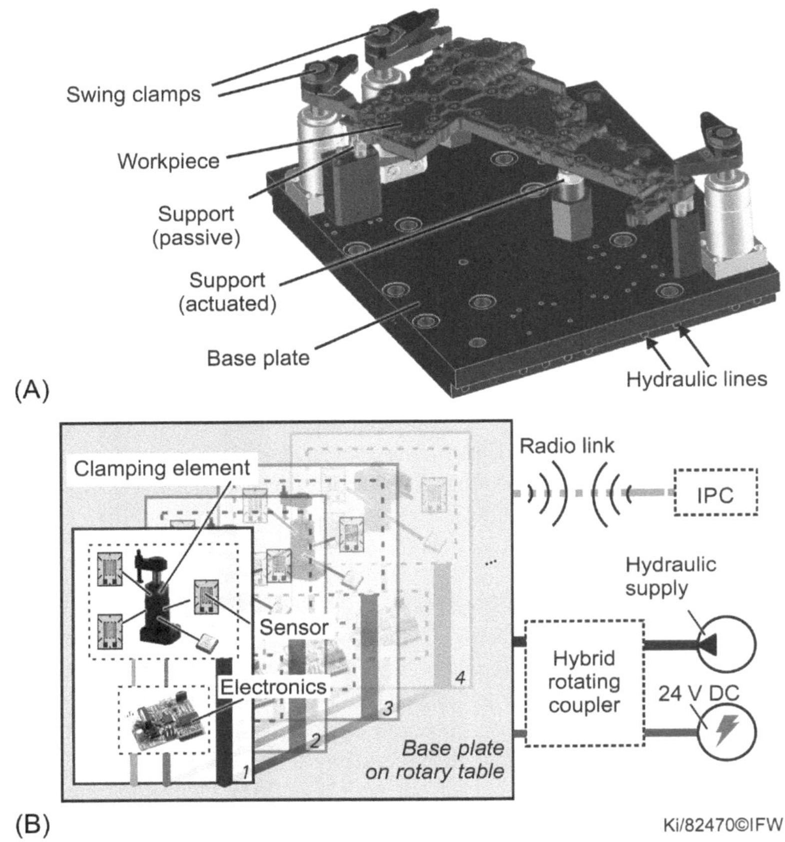

Fig. 3.2.3.1 (A) Hydraulic clamping fixture and (B) overall concept of the "feeling" clamping system.

The estimated energy consumption of the remote electronics amounts 5 W. But, to provide a legitimated value, the electrohydraulic rotating coupler must be able to supply components with higher power consumption, for example, hydraulic valves. For this reason the transmission of electrical power of 100 W was set as a target. An industry-standard 24 V DC power source should serve as the electrical power supply.

To establish the monitoring task, different causes of failure from the perspective of end users, the clamping technology manufacturer, and the fixture construction service provider were discussed and listed in Table 3.2.3.1. In most cases, the multiple causes of failures, errors, and problems with the workpiece clamping can be detected and quantified by measuring the values of the hydraulic pressure, the force, and the piston position. Table 3.2.3.1 summarizes selected examples with associated measurands. For example, based on the pressure measurement it is possible to detect impermissible

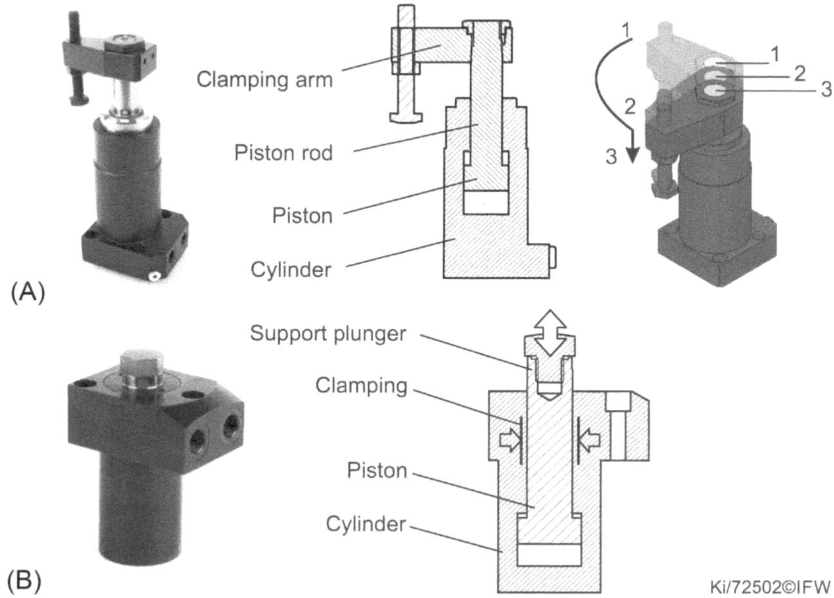

Fig. 3.2.3.2 Selected hydraulic clamping elements: (A) swing clamp and (B) support.

Table 3.2.3.1 Failure causes and measurands in hydraulic workholding

Failure causes/errors/problems	Measurands
Exceeding flow rates/pressure peaks Oil pollution/destroyed strippers by chips Changes of the oil temperature Loss of pressure/leakages/corrosion/wear	Hydraulic pressure
Exceeding clamping arm lengths/wrong support Exceeding transverse forces/collision/overload Too high machining forces on support elements No adaption of the clamping force to the machining Chatter/tool breakage/tool wear Collision with loader or spindle	Clamping force and process force
External chips (strippers, chip accumulation, lever clamp) Pollution/chips inside the oil circuit/stagnation pressure Elasticity and deformation of the workpiece Wrong location of the workpiece/incorrect clamping	Piston position
Aggressive cooling medium Water infiltration	Other

flow rates in the form of peaks in the pressure signal. On the other hand, contaminated oil lines throttle the volume flow and thus cause a too slow pressure increase. Even without any volume flow, differences of the hydraulic pressure arise due to temperature-related expansion of the oil. Any other causes of failure that are not attributable to the three measurands in Table 3.2.3.1 do not correspond to the given objectives. They are included only for completeness.

To capture such measurands, different measuring principles, e.g., piezoelectric, strain based, inductive and optical, were considered. Their usability to realize sensory clamping elements was evaluated. Conventional, housed sensors including their electronic devices are unsuitable for integration because of high space requirements. Due to the good integration capabilities and low degree of complexity for digitalization, conventional strain gauges (SG) are preferred. For example, in contrast to optical sensors, strain gauges can be protected well against environmental influences and in contrast to piezoelectric sensors they allow long-term measurements. Strain sensors that are researched in CRC653 are not available at the mass market yet, but may be interesting for a future market launch.

An analysis of the use cases with regard to the responsiveness has shown that the detection of faults and the generation of additional information are more important than the achievement of challenging real-time capabilities. Therefore latencies that occur due to the application of radio links in the communication of sensor data are considered as not critical.

3.2.3.5 Sensor integration

According to the application scenario, three hydraulic swing clamps of the same type and one hydraulic support, see Fig. 3.2.3.2, are chosen for the equipment with sensors.

The application of the sensors is done in two steps. In the following, an explanation of this process is given in reference to the application of strain gauges at the hydraulic swing clamp. The first step is a preselection of components and surfaces for the SG application. Therefore unsuitable sensor locations are identified with the exclusion principle. This excludes, e.g., locations, where the hydraulic oil prevents strain gauge application. The moving parts are also not appropriate, because a robust energy and data transmission in the harsh environment in proximity of the machining process is not realizable at reasonable costs. Thus the outer surface of the swing clamp housing has been chosen as application location.

In the second step, strain distributions at the swing clamp housing are calculated in reference to different, desired measurands utilizing the finite element method (FEM). Here, normal strains that occur in axial and tangential directions on the cylinder's surface are of main interest, because they are directly measureable with SG. The most important parameters that were varied are, according to Table 3.2.3.1, the hydraulic pressure, the piston position, externally applied forces, and combinations of these parameters. For this purpose, Fig. 3.2.3.3 depicts the simulation model with an exemplary strain distribution. All subsequent strain distributions on the cylindrical surface

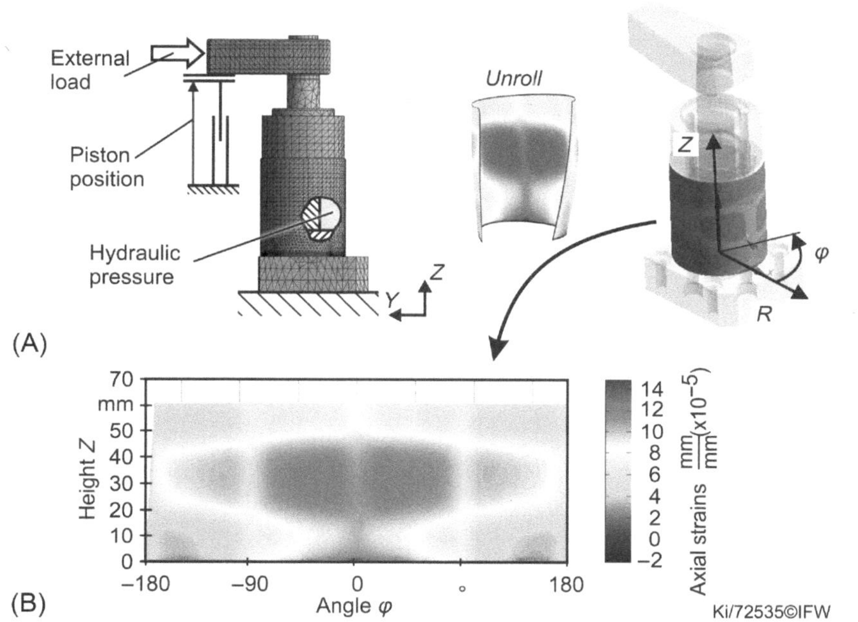

Fig. 3.2.3.3 (A) FEM setup and (B) visualization of an exemplary strain distribution.

of the swing clamp housing are shown in the unrolled shape, see Fig. 3.2.3.3B, due to a clearer presentability.

The simulation results are transmitted from the simulation software ANSYS to the software MATLAB via a programmed interface to analyze the strain distribution. The aim of the analysis is the identification of most suitable positions and alignments for the strain gauge application that best meet the following two objective criteria:

Criterion 1: Position with highest strains that are caused by the desired measurands
→ high sensitivity for desired measurands

Criterion 2: Position with lowest strains that are caused by other influences
→ low unwanted interactions

In the following, this method is explained on an exemplary determination of strain gauge locations that are suited for pressure measurement:

Fig. 3.2.3.4A depicts the simulated distribution of the tangential strains on the cylindrical surface of the swing clamp that results from the increase of the hydraulic pressure from ambient pressure to 15 MPa. At the set clamping situation, the piston is located at highest position.

In this case, criterion 1, the maximum strains due to the increase of the pressure, is met at eight positions at a height of $Z = 30$ mm. Hereby, the symmetrical distribution results from the specific geometric arrangement of the mounting screws inside the clamping element, which can differ completely from other components. The strain distribution in Fig. 3.2.3.4B is established at a lower piston position and

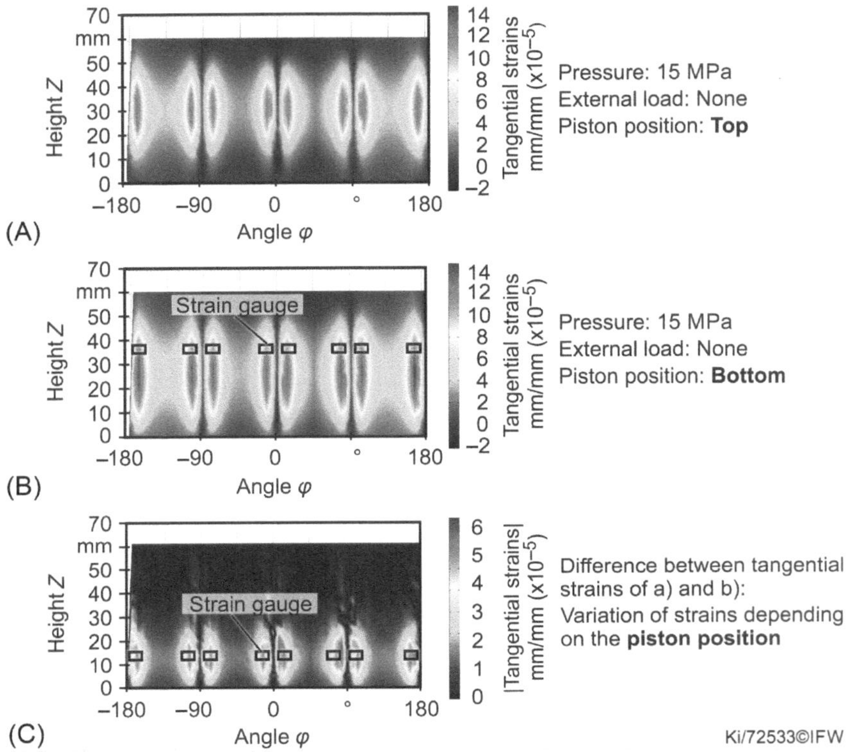

Fig. 3.2.3.4 (A) Tangential strain distribution at the highest piston position and (B) tangential strain distribution at the lowest piston position and (C) variation of tangential strains that depends on the piston position.

otherwise unvaried conditions. It is observable that the area of maximum strains also expands to a lower level. The comparison of both strain distributions by calculating the difference shows the influence of different piston position, see Fig. 3.2.3.4C. Hereby, areas with low strain differences meet criterion 2. In this way, the two criteria were evaluated for further strain distributions, e.g., axial strain distribution and strains, that are caused by external acting forces. The superimposition of the criteria exposes positions for the strain gauges application, which are particularly sensitive to changes of the hydraulic pressure and simultaneously interact hardly with other causes, like different piston positions and external forces. The sensitivity is increasable by additional introduction of notches to adapt the flux of force. Thus at the expense of 1.3% of stiffness an increase of sensitivity of 71% is achievable, see [2]. However, to avoid the stress concentration at notches as a possible initiator of fatigue, this possibility of increasing the sensitivity will not be considered in the following case.

The simulation-based approach enables the determination of the ideal SG positions and appropriate alignment of the measuring grid for the measurement of the hydraulic

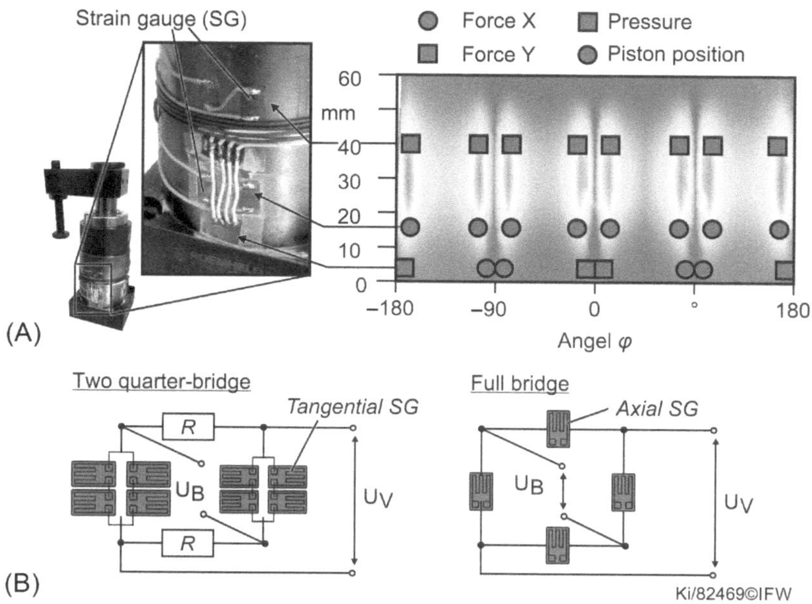

Fig. 3.2.3.5 (A) Determined positions for strain gauges and (B) strain gauge interconnections.

pressure, the piston position, and the force that acts from the *X*- and *Y*-directions. For validation, a functional model was build up using the determined positions and axial and tangential orientations that are depicted in Fig. 3.2.3.5A schematically and with exemplarily applied SG at the swing clamp cylinder.

It is desirable to interconnect four strain gauges to form a Wheatstone Bridge in order to maximize the measurement signal [4]. The axial strain gauges are connected to such full bridges, which compensate temperature drifts due to the measuring principle. The order of the measurement grid interconnections of the bridge determines whether bending or axial forces are measurable. The tangential SG are interconnected to a special form of a two quarter-bridge circuit, see Fig. 3.2.3.5B. Thereby, the influence of bending forces on the pressure measurement is compensated, see also [5]. For the compensation of temperature effect, for example, from heat that is introduced by the hydraulic oil, a Pt100 temperature sensor is mounted nearby the measuring grids.

The applied SG must be protected against environmental influences, in particular, against thermal, chemical, and mechanical stress. Roemheld developed a protection concept that provides a protective tube, which surrounds the SG and so forecloses them against external influences. Furthermore, the protection concept implies that all wires have to be integrated and guided through the base plate to avoid vulnerable cables at the top of the fixture. A more detailed explanation of the realized protection concept follows in Section 3.2.3.8.

3.2.3.6 Energy transmission

According to requirements in Section 3.2.3.4 the targeted value for the contactless energy transmission was set to the electrical power of 100 W. Hereby, an industry-standard 24 V DC serves as electrical power source. The challenging task in this case is the miniaturization of the necessary electronic power transmission components. This is necessary to reduce the additional space requirement for the integration into the selected hydraulic rotating coupler (type 9284-205) from Roemheld's product portfolio.

The realized contactless transmission of electrical energy is carried out by an inductive energy transmission principle. The circuit principle for generating the AC voltage corresponds to a resonant converter according to the Royer converter [6]. In comparison to other circuit principles, experimental investigations and optimizations with the Royer converter revealed best results in terms of the four objective criteria: the maximum power transfer, the efficiency, the requirements of space, and the robustness, for example, by varying air gaps and inductivity deviations, in order to compensate production and assembly tolerances.

The prototypical primary and secondary electronics in Fig. 3.2.3.6A with coils that are enclosed by ferrite cores meet the aim of contactless power transmission up

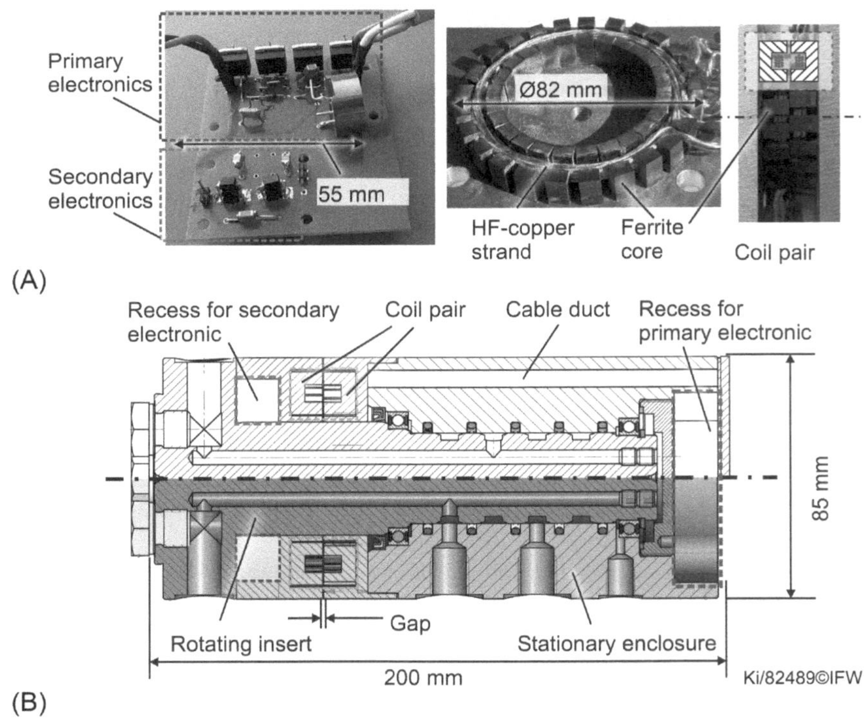

Fig. 3.2.3.6 (A) Power transmission electronics and (B) CAD of hybrid rotary coupler.

to 100 W. In this case, an efficiency of 73% was measured. The power loss is observable mainly in form of increasing temperature of the power electronics, which constitutes the limiting factor of the small electronic components. Therefore without active cooling only short-term operations are permitted. As active cooling is to be avoided, due to space savings, a continuous operation is possible up to 65 W. In this case, the achieved efficiency of the entire energy transmission chain, with an air gap of 300 microns between the ferrite halves, measures 84% and is acceptable.

Based on a feasibility study in CAD with a constructive adaptation of an existing hydraulic rotary feedthrough by the cooperation partner Roemheld, see Fig. 3.2.3.6B, the integration of all necessary power transmission components could be proofed. The new design extends the rotating unit in comparison to the original dimensions by 60–200 mm.

The prototypical production of the electric-hydraulic rotary coupler, to verify the inductive energy transmission as a composite, could not be completed due to capacity reasons. Instead, the verification was carried out with analogous experiments. Hereby, it was proofed that both a metallic environment and the rotation of the secondary coil hardly influence the energy transfer. In addition, a contactless energy transfer is ensured even for the electrical case of hardship of an impulsive inductive load when switching electromagnetic valves.

3.2.3.7 Data transmission

The communication structure of the sensory system from the clamping elements to a central processing unit consists of the following paths: First, the analog signals of the four strain gauge bridges and one temperature sensor are transmitted to nearby signal preprocessing device via cable. The direct proximity of analog sensors and digitizing unit shortens the susceptible analog transmission paths and offers the beneficial use of fieldbus communication at an early stage. For the digital transmission the CAN bus was selected among others, due to the small number of necessary cables and the high availability of components for the realization of miniaturized solutions.

3.2.3.7.1 Distributed preprocessing device

The digitizing and preprocessing of the sensor signals are performed by a preprocessing device that was designed especially for this purpose, because similar electronics that meet the demands of cost and space requirement do not exist. The realized printed circuit board (PCB), see Fig. 3.2.3.7A, processes four strain gauge bridges and one PT100 temperature sensor.

The space requirements of the PCB measure $30 \times 40 \times 10\ mm^3$. The required electrical voltage of 9–36 V DC is provided together with the two CAN bus lines via a four-wire cable. Using the flexible I^2C bus interface, the sensor system is extendable by further digital sensors.

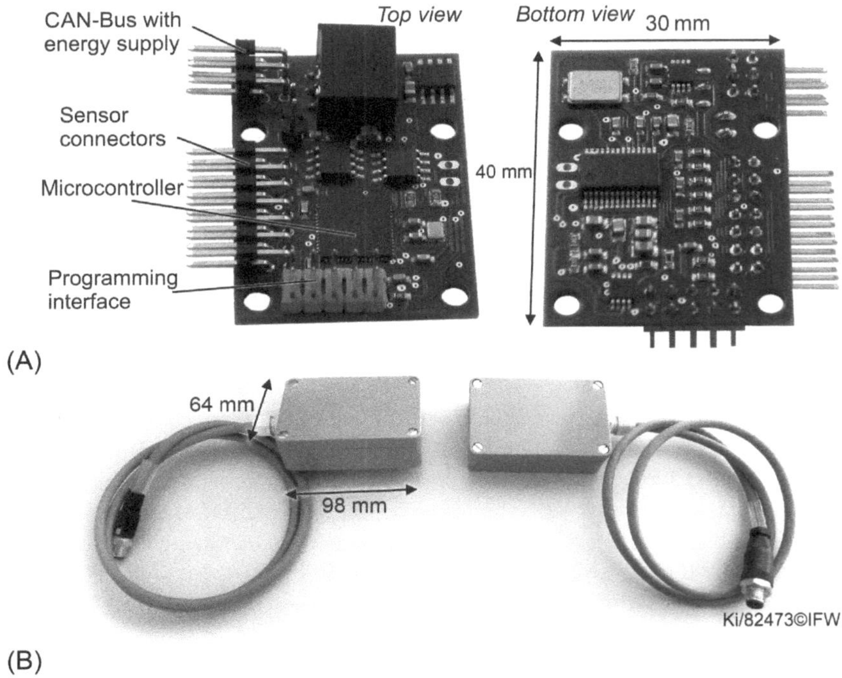

Fig. 3.2.3.7 (A) PCB for the signal preprocessing and (B) CAN bus radio link.

The loading of the CAN bus system depends on the number of messages that are sent from the bus station in a defined period. In the considered application scenario every clamping element has its own PCB. Therefore with four clamping elements at the maximum CAN bus baud rate of 1000 kbps and a bus load of 80% the measured data are transmittable with a rate of 300 Hz. An optional, more efficient message encoding doubles the transfer rate by neglecting the time stamps. By restricting the number of simultaneously sending sensors, potential data rates up to 7 kHz are realizable.

3.2.3.7.2 Contactless data transmission

In accordance to the defined requirements the need for real-time capability is not given. Therefore industry-appropriate solutions exist for contactless radio transmission of CAN bus data. From the comparison of different solutions, in particular, based on the requirement of robustness for the use in the working space of a machine tool, the wireless CAN bridge of the company Avisaro was selected, see Fig. 3.2.3.7B. The tested wireless connection with the maximum baud rate of 1000 kbps was successful.

3.2.3.7.3 Control interface

The control interface consists of an industrial PC that is expendable with various terminals. The graphical user interface for the visualization and the control of the monitoring functionalities was designed using the software TwinCAT 3. The function blocks for the online processing of the measured data were developed using Matlab/Simulink and then exported to structured text according to the international standard IEC 61131, which deals with the basics of the programmable controller. Thus the realized function blocks are applicable independently of the Beckhoff system at any other PLC control system.

3.2.3.8 Realization of the prototype

Based on the simulation-supported sensor integration and the performed functional tests of the individual components, a prototype of the "feeling" clamping system for the use in a milling center has been set up, see Fig. 3.2.3.8A and B [7–10]. Hereby,

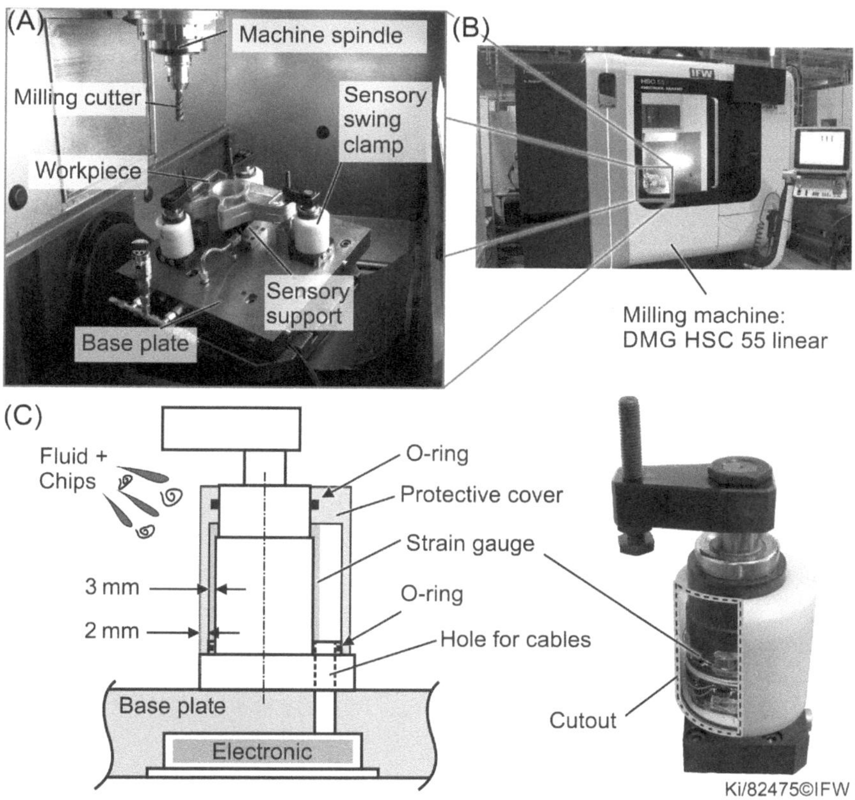

Fig. 3.2.3.8 "Feeling" clamping system for use in the machining center.

Roemheld designed three hydraulic swing clamps that ensure the protection of the integrated sensors by means of a protective plastic tube. Fig. 3.2.3.8C depicts the protection concept and a realized swing clamp. The protective tube seals the integrated sensors with O-ring sealings against the environment.

According to the protection concept, the base plate includes introduced pockets for the integration of the PCBs under every "feeling" clamping element. Hereby, the PCBs are attached to the covers that close and seal the pockets. The sensor cables are fed through holes in the base plate. In this way, any cables on the top of the fixture were avoided. For the connection to the central control interface, standardized four-pole M8 plug connectors are used. Furthermore, the base plate serves the purpose of supplying the hydraulic clamping elements with oil via integrated hydraulic lines.

To investigate the dynamic response characteristic of the integrated force sensors, dynamic forces up to 600 Hz have been applied to the clamping components using a shaker with a sweep excitation over a period of 120 s. The first three seconds of the force signal are depicted in Fig. 3.2.3.9A and B. For this purpose, the shaker is coupled with a rod and an adaptation to the clamping arm of the swing clamp. An additional force sensor in the flux of force measures the input signal of the system that is being compared to the integrated force sensor as the output signal. The resulting amplitude and phase frequency responses in Fig. 3.2.3.9C show that dynamic forces up to 600 Hz are measurable with the integrated sensors without any significant damping. Hereby, the phase shift is negligible and has its extreme value of −25 degrees at an excitation frequency of 300 Hz. The measurements were repeated at different clamping pressures. As a result, the hydraulic preload only influences the static offset but hardly influences the measurement of dynamic forces, which is proofed by the two curves in Fig. 3.2.3.9C that are similar despite the different hydraulic preloads.

3.2.3.9 Signal processing

The simultaneous use of multiple sensors involves both the disadvantage of higher costs and the benefits in terms of better signal quality and higher contents of information. The sensor signals are analyzed and fused to extract monitoring features for the evaluation.

3.2.3.9.1 Sensor evaluation and fusion

The fusion of the sensor signals is carried out with analog and digital signals. The analog sensor fusion takes places in terms of interconnecting the single strain gauges to measuring bridges. For example, based on the sensor design tangentially directed measuring grids at eight different positions are suitable to measure the hydraulic pressure, see Fig. 3.2.3.5. Every single measuring grid at each of these eight positions responds to changes of the hydraulic pressure. The interconnection of these redundant measuring grids to a two quarter-bridge offers three advantages:

1. The amplitude of the useful signal is doubled, in comparison to a single measuring grid.
2. The combination of the measuring grids by the circuity performs an averaging that smooths the measured signal and reduces the influence of interfering signals at every measuring grid.

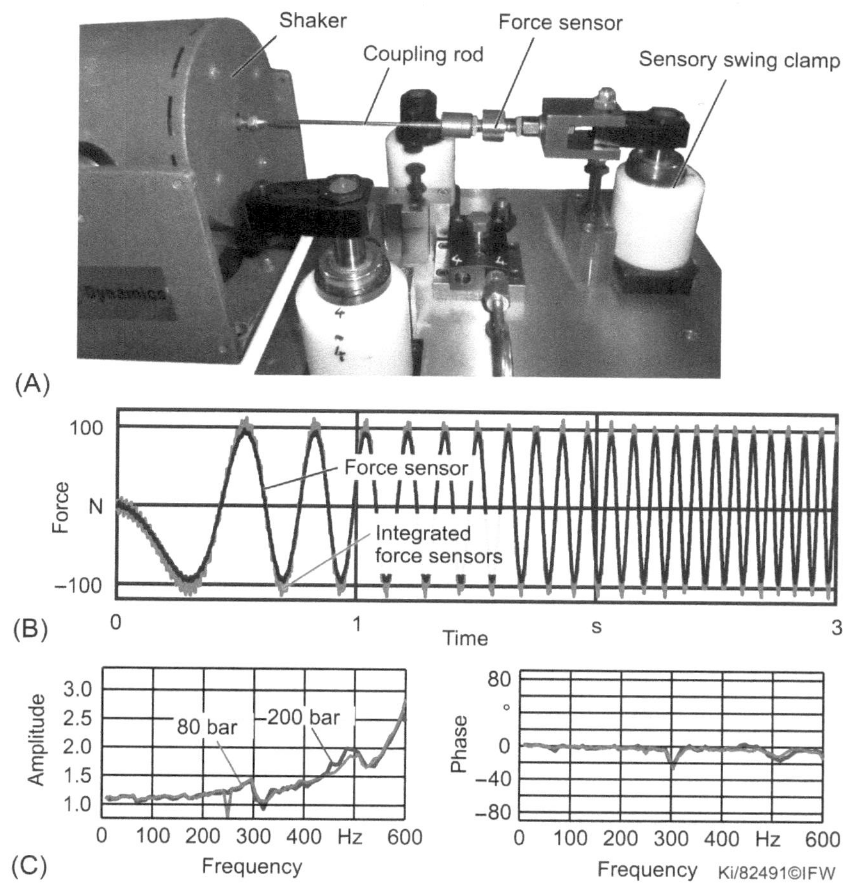

Fig. 3.2.3.9 (A) Experimental setup for frequency response measurement, (B) exemplary time series of the measured forces, and (C) frequency responses at different clamping pressures.

3. The interconnection reduces the unwanted influence of mechanical bending loads on the measured signal significantly. Bending causes strain changes with inverted signs but same amplitude on the opposite side of a cylinder. The averaging cancels this influence on the signal.

When interconnecting the axial measuring grids to full bridges in order to measure forces, the two generally known advantages result:

1. The useful signal is quadrupled, in comparison to a single measuring grid.
2. The influence of temperature on the measured signal is compensated [4].

After the analog signal acquisition and digitization the digital fusion of the measured data of the clamping elements follows to improve the signal-to-noise ratio of the common measurands. The calculation of an average value out of the data of multiple sensors represents a simple variant of sensor fusion. This includes, for example, the determination of the hydraulic pressure of the system as the mean value of the hydraulic pressures of every "feeling" clamping element as an additional value. Here, it is assumed that the pressure gradients between the clamping elements are negligible, which is a permissible assumption for normal operation.

The process force is the variable of interest for process monitoring. Usually, it is determined by calculating the sum of forces that are measured with force sensors in the flux of force. For the present application, however, this is not a solution, because only a part of the flux of force passes the "feeling" clamping elements. An unknown amount of the flux of force passes the passive support elements without being measured. A determination of the unknown force components is possible with model-based approaches, for example, by utilizing a Kalman filter, like in the sensory clamping system developed by the CRC 653 [1]. However, the detailed modeling of the sensory clamping fixture requires a system identification that is associated with high effort and expense and that has to be repeated for new, individual clamping tasks. For practical use, in favor to enable flexible usability and a significantly reduced expenses at setup, model-based approaches are not used. Thus the fusion of the sensors of the clamping elements focuses not on the measurement of the absolute cutting force but on their effect. For this purpose, a combined monitoring signal that depends on the process force and that is not constrained to any specific clamping task is generated by means of sensor fusion.

First, a band-pass filter limits the bandwidth to a small range along the tooth engagement frequency to avoid static interferences and dynamic interfering influences on the monitoring value, as long as they occur in a different frequency band. Then, the root mean squares (RMS) are calculated and smoothed by low-pass filtering. This signal processing has the advantage that even for dynamic signals boundaries for the upper and lower amplitudes can be determined and monitored. After that the fusion of the individual signals is achieved by a summation. On the one hand, the merged signals offer the opportunity of a better signal-to-noise ratio. On the other hand, the sum reduces multiple force-sensitive sensor signals to one single signal and thereby reduces the effort for the teaching of limits.

3.2.3.9.2 Monitoring features and evaluation

The hydraulic pressure, the clamping force, and the piston position of the swing clamp are being extracted as features to enable the monitoring of the state of the clamping system in accordance with the sensor concept. The hydraulic pressure is derived from the measured signal of the upper SG two quarter-bridge. It is directly proportional to the hydraulic pressure according to the simulation-based design. For the calibration, a reference signal of a pressure sensor is necessary.

To measure the forces that act at the clamping arm, e.g., the clamping force, a calibration is performed on a reference force signal. Furthermore, with a known relationship between the clamping force and the hydraulic pressure the calibration can be performed using only the pressure sensor. Then, the application of a force sensor is not necessary for the calibration.

For extraction of the piston position, the measured signals of both SG two quarter bridges must be evaluated simultaneously during a stationary clamping state. Both measurement signals change their values as a function of the hydraulic pressure, but only the lower one depends additionally on the piston position. The difference between the two measuring signals is the desired position-dependent value. Because of a nonlinear relationship, the determination of a characteristic curve at different piston positions is necessary.

To enable process monitoring a dimensionless quantity of the process force is extracted as a feature of the digital sensor fusion. This feature can be monitored but allows only limited physical interpretation.

To evaluate the monitoring features different methods are applicable. This includes the monitoring of the violation of limits that are fixed, changing over time, or taught empirically.

3.2.3.10 Verification

The application of a "feeling" clamping system inside a production system offers many benefits, see [11]. Also, many different potential application scenarios exist. The followings sections show the use for the condition and process monitoring.

3.2.3.10.1 Condition monitoring

The main task of hydraulic clamping elements is the conversion of hydraulic pressure into mechanical clamping force. During the period of use, different error conditions, e.g., leakages at hydraulic connections and blocked-up hydraulic lines, arise and cause pressure differences in the hydraulics. Further error conditions occur by operating the hydraulic pump with incorrect parameters, e.g., excessive flow rates, often without even knowing it. The integrated pressure-sensitive sensors purpose the monitoring of the hydraulic supply at every "feeling" clamping element. Fig. 3.2.3.10A summarizes different scenarios that lead to clamping failures. If the flow rate of the oil exceeds the permitted setting (red curve) impermissible pressure peaks occur. These lead to faster motions of the swing clamp and increase the load on the bearing of the piston and the swing mechanism. Insufficient pressure buildup must be avoided as well, otherwise a sufficient clamping force at the start of the machining is not guaranteed. By monitoring the hydraulic pressure, faults can be detected and removed before serious consequences occur.

The crucial advantage of the integrated pressure measurement, in comparison to a common hydraulic pressure sensor, is the proximity to the place, where the supplied hydraulic pressure is converted to clamping force. Because of the acquisition directly at the clamping element, the integrated sensors measure the local pressure variations in the piston chamber with a higher sensitivity. Even a momentary pressure drop during the stroke movement of the piston, see Fig. 3.2.3.10B, is detectable. The analysis of such effects, e.g., the duration of the drop, provides additional information of the wear

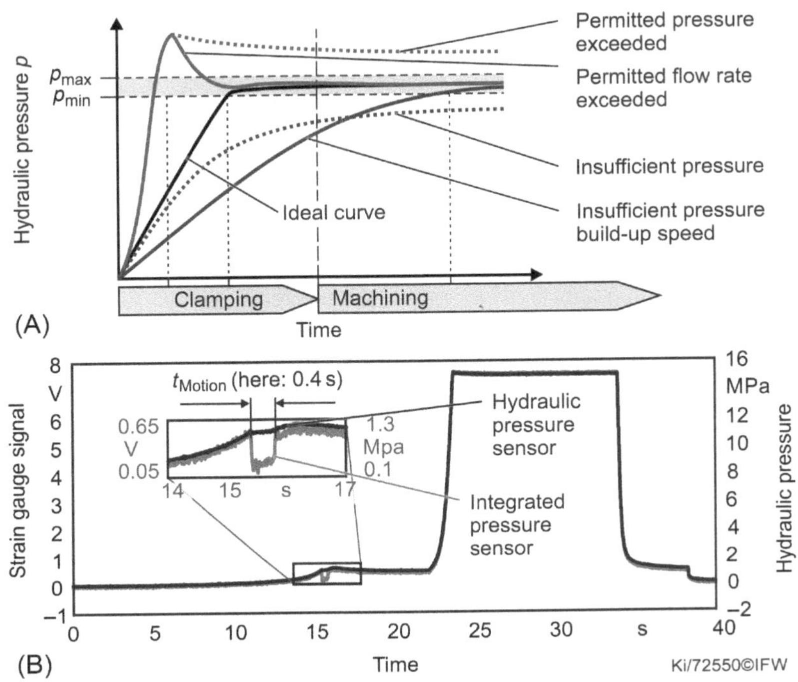

Fig. 3.2.3.10 (A) Hydraulic pressure buildup scenarios before machining and (B) integrated pressure sensor of the sensory swing clamp versus conventional hydraulic sensor.

condition of the piston guide. By use of a conventional pressure sensor near the hydraulic pump this information gets lost. Although, in the illustrated example, the hydraulic line amounts only 1.5 m between the pressure sensor and the clamping element.

Forces that act at the clamping arm, e.g., clamping forces or process forces, apply bending loads to the swing clamp. The task of the strain gauges at the lowest level of the cylinder, see Fig. 3.2.3.11A, is the measurement of strains that are induced by the bending.

With the known lever arm L they are convertible into absolute forces. In the shown example, a horizontal force is applied sequentially in positive *X*- and *Y*-directions on the clamping arm. A conventional load cell measures the absolute values of the applied force of approx. 60 N, which corresponds to a simple press with the thumb. The diagrams in Fig. 3.2.3.11B present the measured force of the load cell (F_{FS}) and the integrated sensors (F_{SGX}, F_{SGY}) that measure the forces in independent directions. It is obvious that the integrated strain gauge-based force sensors are able to detect the applied forces that act from different directions.

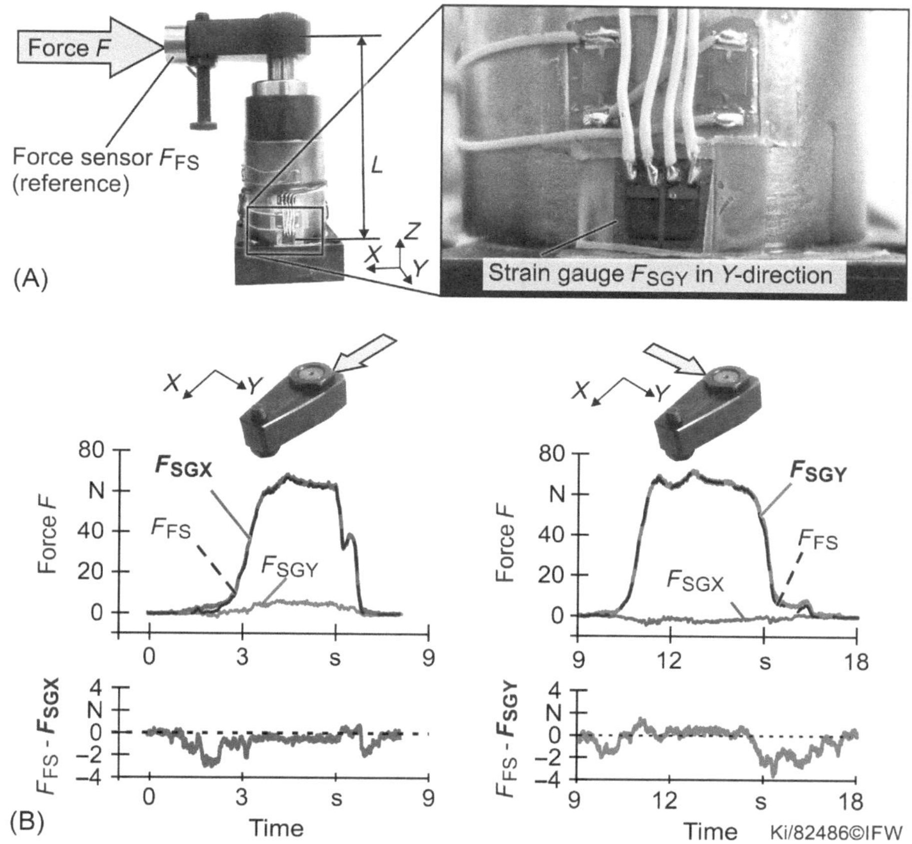

Fig. 3.2.3.11 (A) Setup for force application and (B) comparison between a conventional force sensor and the integrated sensors for force measurement.

In addition, it is shown that even the variation of forces less than ±4 N can be detected with this measuring principle, despite the very robust and rigid construction of the clamping cylinder. Application potentials exist where the use of conventional force sensors is too expensive, not practicable, for instance, because of the increase of the mechanical compliance, or only feasible with great setup effort, for example, inside the working space of machine tools.

The position-dependent strain gauge signal, that results from the difference of the two two-quarter bridges, see Section 3.2.3.5 and Fig. 3.2.3.5, has a quadratic relation to the piston position. This relation has been approximated to a second-order polynomial function, see Fig. 3.2.3.12A. Using the mathematical equation, the integrated

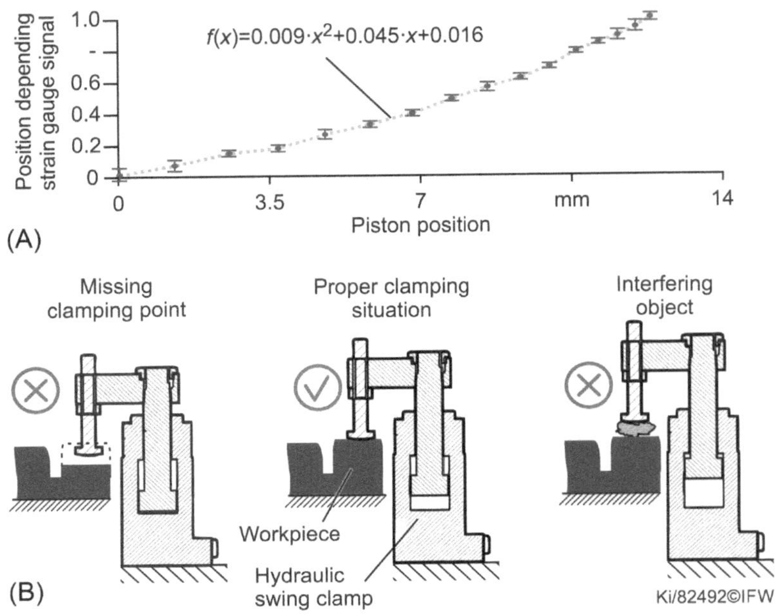

Fig. 3.2.3.12 Integrated position measurement: (A) characteristic curve of the SG and (B) application potentials.

sensors allow the measurement of the position. With this, numerous applications for industrial practice are realizable, for example, the detection of missing clamping points and interfering object or the notification of a properly executed clamping of the workpiece, compare Fig. 3.2.3.12B.

For the estimation of the polynomial coefficients, the experimental setup in Fig. 3.2.3.13A has been used. It consists of a hydraulic swing clamp with applied strain gauges for the position measurement and a laser triangulation sensor, which has an accuracy below 5 μm. It measures additionally the absolute stroke as reference. By implementation of the characteristic curve, the strain gauge-based measurement of the piston position accomplishes a precision of approx. ± 0.3 mm. The comparison of the integrated position sensor and the laser triangulation sensor is shown in Fig. 3.2.3.13B. Hereby, each of the standard deviations at different piston positions, which were varied with spacers at the clamping point, results from twelve clamping cycles.

Measuring the stroke of the clamping elements in a production system, e.g., inside a machine tool, is rarely done. Therefore the space requirements of common measuring systems are often too challenging. Instead, small limit switches are used more frequently to test whether the piston has reached the desired end position or not. In those cases, the integrated position sensor system (strain gauges) has the following benefits: It consumes less space on the fixture than conventional position sensors and provides more information than limit switches, because it acquires intermediate stages as well.

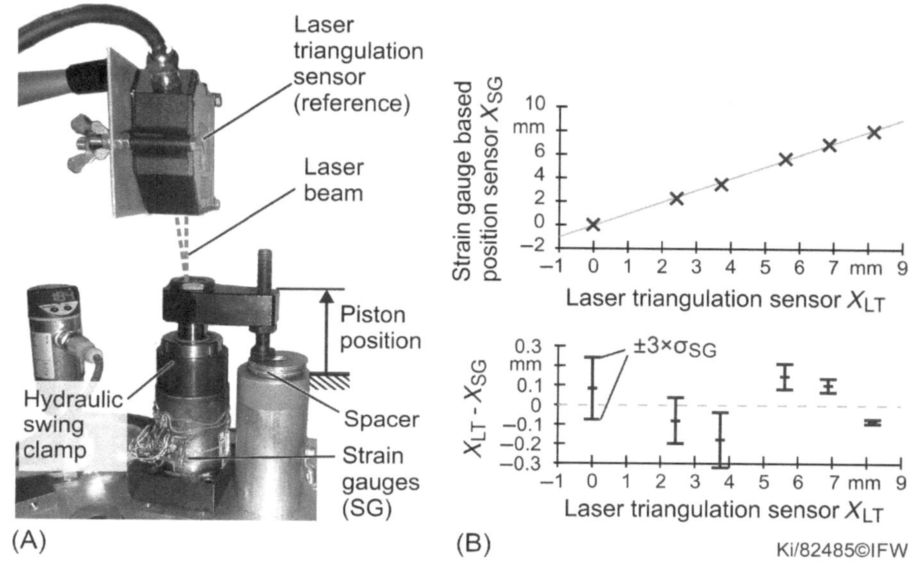

Fig. 3.2.3.13 (A) Experimental setup to measure piston position at different heights and (B) comparison of the measured piston positions.

3.2.3.10.2 Process monitoring

The "feeling" clamping system was used for the machining of aluminum and steel, which cause different cutting forces. The amplitudes of the measurable, running RMS values that are generated for process monitoring according to the feature extraction in Section 3.2.3.9 strongly depend on the clamping situation: the higher the compliance of the workpiece support, the higher the measured amplitude because of the higher forces that acts at the sensory swing clamps, see also [12]. Fig. 3.2.3.14 superimposes the resulting monitoring signals from machining 24 similar milling paths. Hereby, the monitoring limits were taught in accordance to the monitoring method using statistical limit curves [13]. With each additional data set the monitoring limits, which equal the 98% confidence interval, tighten more around the expected value. In the case of a process failure, here provoked by two intentionally introduced defects in the aluminum workpiece, the signal deviates significantly from the expected value and indicates a process failure by violating the monitoring limits. Actually, for the illustrated process error only four learning steps were necessary. That means, the occurrence of this error could be detected starting from the fifth milling path.

To sum up the demonstrated results, the "feeling" clamping system is useable for both: for the condition as well as for the process monitoring. Hereby, for the use in process monitoring the sensitivity of the measurement system strongly depends on the clamping situation.

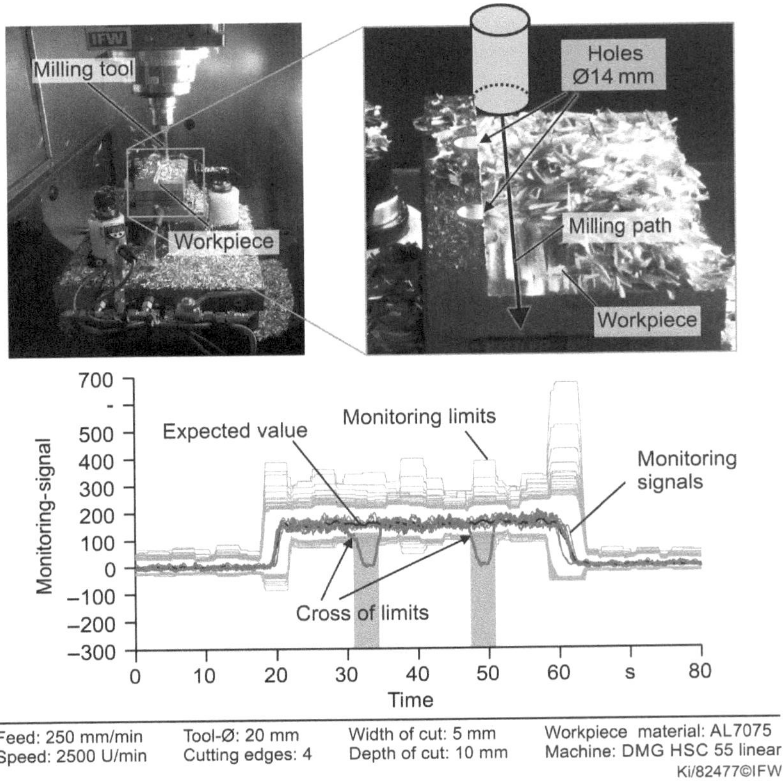

Fig. 3.2.3.14 Process monitoring.

References

[1] K.M. Litwinski, Sensorisches Spannsystem zur Überwachung von Zerspanprozessen in der Einzelteilfertigung, Dissertation Leibniz Universität Hannover, Berichte aus dem IFW, PZH Verlag, Garbsen, 2011.

[2] B. Denkena, D. Dahlmann, J. Kiesner, Sensor integration for a hydraulic clamping system, in: Proc. 2nd International Conference on System-Integrated Intelligence: Challenges for Product and Production Engineering, SysInt 2014, July 2nd–4th, Bremen, Germany, 2014, pp. 465–473.

[3] B. Denkena, K.M. Litwinski, D. Dahlmann, H. Boujnah, J. Kiesner, M. Krüger, J. Schmidt, T. Neff, Industrie 4.0 in der Zerspanung, ZWF - Zeitschrift für wirtschaftlichen Fabrikbetrieb 109 (7–8) (2014) 537–541.

[4] K. Hoffmann, Eine Einführung in die Technik des Messens mit Dehnungsmeßstreifen, DRACH Print Media GmbH, Darmstadt, 1987.

[5] B. Denkena, J. Kiesner, Strain gauge based sensing hydraulic fixtures. Mechatronics 34 (2016) 111–118, http://dx.doi.org/10.1016/j.mechatronics.2015.05.008.

[6] J. Rehrmann, Das Netzteil- und Konverter-Handbuch, edition trifolium, Kassel, 2003.
[7] B. Denkena, J. Kiesner, Sensorisch überwachen, NC-Fertigung 12 (2015) 46–47.
[8] B. Denkena, J. Kiesner, Sensorisches Spannsystem zur Überwachung des System-und Prozesszustands, SFB 653: Die vernetzte Produktion - Forschungsergebnisse für die. Praxis, 2015, p. 11.
[9] B. Denkena, J. Kiesner, Monitoring of system and process state by application of sensory clamping elements, CRC 653: Cross Linked Produktion—Research Results for Industrial Application, 2016, p. 11.
[10] B. Denkena, J. Kiesner, Fühlende Spannelemente überwachen die Werkstückeinspannung, ZWF - Zeitschrift für wirtschaftlichen Fabrikbetrieb 110 (11) (2015) 698–700.
[11] B. Denkena, D. Dahlmann, M. Shanib, J. Damm, J. Kiesner, T. Neff, Prozessüberwachung in der Zerspanung, ZWF - Zeitschrift für wirtschaftlichen Fabrikbetrieb 111 (4) (2016) 174–177.
[12] B. Denkena, J. Kiesner, Production monitoring based on sensing clamping elements, in: Proc. 3rd International Conference on System-Integrated Intelligence: New Challenges for Product and Production Engineering, SysInt 2016, June 13th–15th, Paderborn, Germany, 2016, pp. 235–244.
[13] J.W. Brinkhaus, Statistische Verfahren zur selbstlernenden Überwachung spanender Bearbeitungen in Werkzeugmaschinen, Dissertation Leibniz Universität Hannover, Berichte aus dem IFW, PZH Verlag, Garbsen, 2009.

3.2.4 Teachless process monitoring for single item production

Berend Denkena, Thomas Neff
Institute of Production Engineering and Machine Tools, Leibniz Universität Hannover, Garbsen, Germany

3.2.4.1 Process monitoring for machining production processes

Process monitoring systems are used in machining production to detect process errors and to ensure process safety. Critical situations such as upcoming collisions, tool breakage, tool wear, and process instabilities must be detected quickly and reliably in order to react correctly [1,2]. Monitoring systems help to avoid expensive consequential damages and the resulting machine downtimes as well as to ensure the availability and production quality [3].

One or several monitoring variables have to be gathered to monitor the production process on a machine tool. This can be done, on the one hand, by machine internal sensors and, on the other hand, by external, subsequently installed sensors. The construction of monitoring and diagnostic systems is divided into signal acquisition, signal processing, and signal interpretation or rather decision-making. In manufacturing processes, the cutting force causes most effects that are detected by the process monitoring systems. Direct force measurement for process monitoring is not common in industry. This is partly due to the high cost of the relevant sensors and the considerable expense for sensor mounting and lack of overload protection. Indirect force measurement provides an alternative. This is done, for example, by strain measurements in force bypass [4]. Due to high stiffness of machine components, a significantly lower sensitivity has to be tolerated compared to direct measurement methods [2]. Alternatively, machine's internal signal sources, which correlate with the process forces, can be used for monitoring [5]. This method offers several advantages, such as relatively low costs and robustness. Furthermore, less time and effort is required for installation and maintenance. Especially the measurement of motor current and active power at main spindles and screw spindles [4,5] can be used for the reconstruction of process forces.

Different approaches for the reconstruction of process forces from drive signals are described scientifically [4,6,7]. For example, signal correction methods use experimentally parameterized filters. They are very inflexible to changes in the drive configuration. In the model-based process force reconstruction, a path model of process force transfer behavior from the place of application of force to the drive sensors displays physical drive properties. The process requires an elaborate identification and parameterization of the transfer behavior. A transfer to other drives or machines is therefore not guaranteed. A third possible option is the so-called artificial neural networks. It is a very efficient approach that can even illustrate nonlinear effects. However, the parameterization of such a network requires an extensive experimental program that does not necessarily draw any conclusions on the drive structure or machine structure.

Based on measured signals, conclusions about the process status have to be drawn in order to monitor a machining process. For that, there are signal-based methods, feature-based methods, and model-based methods. Signal-based methods detect defective system states based on measured sensor data without knowing the analytic correlation between process status and sensor information. Feature-based methods extract significant characteristics of the signal for monitoring. Model-based methods approximate the probable signal characteristic, which is being compared to the actual signal characteristic [8]. Errors are detected by comparing the monitoring variable to the predefined limit values. These limiting values can be set statically or adjusted dynamically.

In order to monitor a production process reliably, reference data must be available, which enables a comparison of target process signals and actual process signals. Today's process monitoring methods are initially being taught-in in series production during a reference process. In single item production, the teaching is associated with economically not justifiable expenses. Therefore the main goal of the presented work is the teachless process monitoring of production facilities for single item production in order to ensure process security. Therefore part of the parameterization effort is done during the work preparation. Here, a monitoring reference is created by a multilevel simulation system in form of the calculated tool engagement conditions. Semiempirical cutting force models allow a parallel to the process determination of monitoring limits.

3.2.4.1.1 *Preliminary work and required fundamentals*

Selected results from the areas of simulation (see Chapter 3.1.3) and monitoring (see Chapter 3.1.2) were transferred into industrial applications. In particular, research results for the simulation-based limit value generation as well as for acquiring and evaluating monitoring variables of machine tools were transferred into industrial application.

The groundwork for an envelope curve-based error identification has been done based on the machine's internal signals (see Chapter 3.2.1). Furthermore, the CRC 653 elaborated basic principles concerning simulation-based limit value generation for process monitoring and learning process simulation (see Chapter 3.1.2). Thereby, essential conditions for a simulation-based method for monitoring machining processes with a parameterization parallel to the process are supplied (compare Fig. 3.2.4.1).

3.2.4.1.2 *Simulation-based parameterization of process monitoring*

Based on the specification of the reference processes, an essential work content was the simulation-based parameterization of process monitoring. For that purpose, an NC-based segmentation of a production process for future monitoring was made possible. This was done by means of the simulation system CutS, with help of which the program run for the defined reference component was investigated. Afterward, segments with constant or rapidly changing intervention variables during machining were derived. The goal was that these segments can always be assigned based on the calculated data for further procedure. The ARTIS monitoring system CTMv5 offers the possibility of automated recognition of different process stages, so that the referenced segments can be synchronized during monitoring. For all further simulation tests, both the kinematics of an INDEX R300 turn mill center, as well as of a GILDEMEISTER CTX 420 linear universal turning center, used for analogy tests, were modeled. Fig. 3.2.4.2 illustrates simulation results of the reference process. For workpiece discretization, a dexel-based method, based on Cartesian coordinates, was used, which is availabel in CutS. For this purpose, the workpiece is depicted by

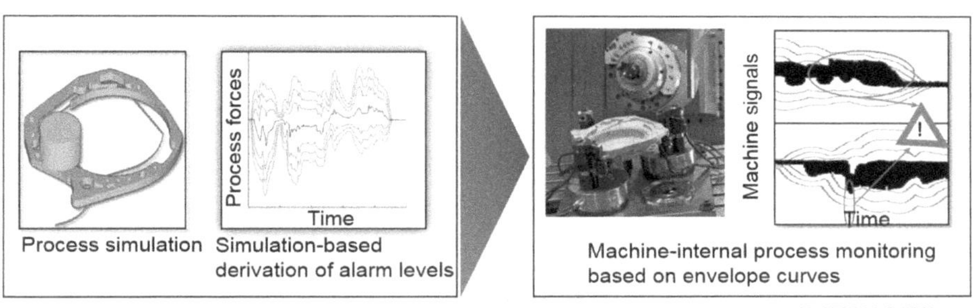

Fig. 3.2.4.1 Preparation work—simulation-based process monitoring from CRC 653.

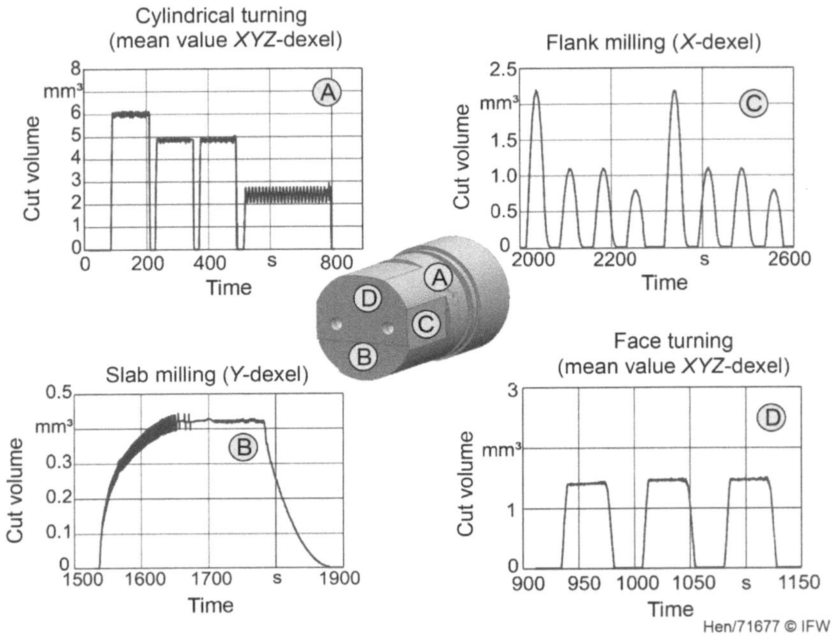

Fig. 3.2.4.2 Simulated chipping volume per simulation step in four selected workpiece segments.

a total of three dexel grids (*x*-, *y*-, and *z*-direction). A dexel is a bar-shaped discretization element with a defined start and end.

Depending on the used dexel grids, discretization errors occur while calculating the material removal rate. The determining factor here is the orientation of the dexel relative to each cutting movement. The discretization error is lowest when the dexel grid with orientation orthogonal to the cutting direction is used. For machining operations with rotating workpiece and thus rotating dexel model, the average value of all three dexel grids is used and allows the calculation of an interference-free process variable sequence. It was also established that the approach is not suitable for calculating machining variables during thread cutting. A detailed tool model with individual teeth and a very high dexel resolution is required here. Therefore it is not possible to execute a simulation run in an acceptable computing time. For this reason, the threading operations was classified as not monitorable with the pursued approach.

The ARTIS GmbH carried out internal preliminary studies for the integration of segmentation in the monitoring system. INDEX made extensive adjustments to the INDEX-own simulation system "Virtual Machine" in order to create the necessary external interfaces.

The experiments on the reference component continued to form the database for the segmentation model. For segmentation of the NC code, it is necessary to distinguish the tool engagement from noncontact movements of the machine, tool changes, and other nonmachining movements. By using the ARTIS monitoring system, which can detect automated process phases, a temporal comparison is ensured between the reference for process monitoring and the real process. A common way here is to interrupt the process monitoring when no cutting takes place. Thus expensive false alarms can be avoided.

The approach of segmentation is, in general, effective in order to ensure a robust referencing for processing within a segment. Furthermore, the segmentation is used in combination with the process monitoring system to enable the time synchronization between the reference for monitoring and the real process between segments. The approach of connecting process simulation CutS to the Virtual Machine of INDEX (VM) is effective in order to enable just-in-time references within a segment. This was clearly demonstrated by relevant preliminary investigations. Therefore a concept for connecting the existing simulation solution VM with the Simulation System CutS was established. For this purpose INDEX provided possible interfaces. Furthermore, advantages and disadvantages of various options of data communication were discussed, thus a decision for the method of connection could be made, which is shown in Fig. 3.2.4.3. This allows achieving the reconstruction of the actual axis movements, due to the use of real control and set parameters, which have been virtualized in the machine simulation by INDEX.

The interpolated axis positions and information on tool change and spindle commands are passed on to the process simulation in CutS within the NC cycle. Current tool engagement in the cutting process and intervention variables can be simulated with the resulting tool paths. Based on this workpiece model, an experiment was

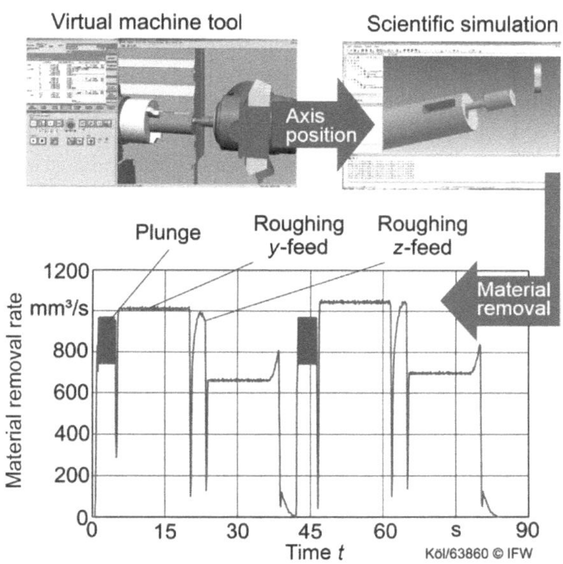

Fig. 3.2.4.3 Axis position data from INDEX "Virtual Machine," used in CutS.

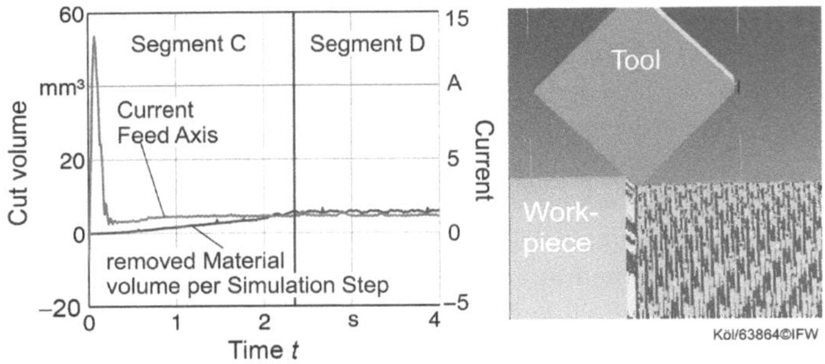

Fig. 3.2.4.4 Comparison by segment of simulated intervention variables and measured driving current.

performed on the testing machine CTX 420 Linear. Different calculated intervention variables have been compared to the measured signals. Fig. 3.2.4.4 shows the current cut volume and the measured driving current of the feed axis for longitudinal turning (on the outside) on the test component. By combining different intervention variables, which change depending on the operation, such as current cutting width or projected contact surfaces, an approximation toward the measured signal can be simulated. This signal can be passed on to the ARTIS monitoring system by an INDEX turn-mill center during production as a segment-specific reference.

3.2.4.1.3 Reconstruction of process variables of machine-internal signal sources

For the reconstruction of process variables, a qualification of signal sources was necessary. The goal here was to identify the signals that have a high sensitivity toward process errors such as tool breakage and correlate with the simulated process variables. To ensure the economic benefits of the application, great emphasis was put on the usage of control-internal existing sensors. The digital drive data was recorded by the SIEMENS compile cycle and the SinuCom Trace function (Fig. 3.2.4.5).

The sensitivity analysis of all sensor signals was initially carried out on the testing machine CTX 420 linear. For the tests, NC control signals and external sensor signals were recorded. Force measuring platforms for turning, milling, and drilling operations were used. All signals from additional sensors were combined in the dSPACE system DS1103 for recording with a scan time of 120 μs. Unprepared experimental workpieces as well as workpieces with additional boreholes were used for the tests. The goal was to provoke a change in the signal amplitudes due to discontinuous cutting. Furthermore, worn tools were specifically inserted and workpieces clamped faulty. Based on the described experimental setup, the recorded signals were analyzed with regard to their sensitivity to process errors in order to identify and qualify them as sources for process monitoring. Therefore correlation analyses of time curves were

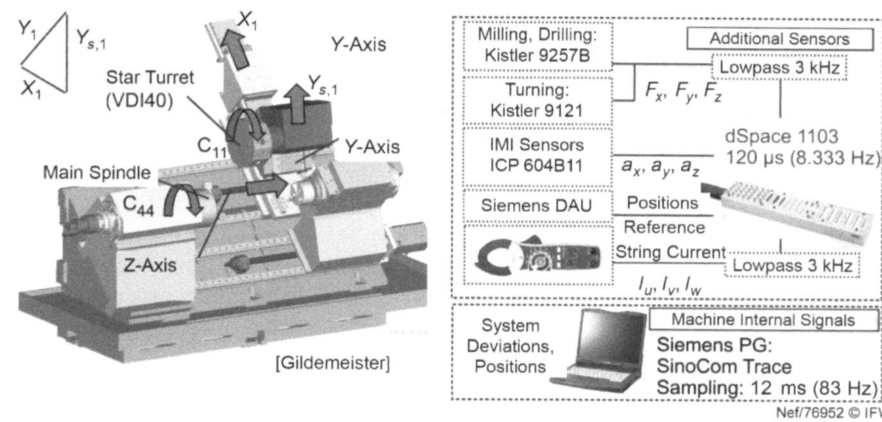

Fig. 3.2.4.5 Testing machine CTX 420 linear and applied measuring technology.

performed. For this purpose, the force signals were analyzed and correlated with a conceptual model of the respective process, which was expressed by particular correlation coefficients. Subsequently, additional sensor signals (currents, accelerations) were examined in order to validate the findings regarding the sensitivity. Fig. 3.2.4.6 shows examples of process signals during longitudinal turning (on the outside) for a workpiece with defects and for the corresponding process without workpiece (idle). The boreholes and the resulting cut interruptions were recorded

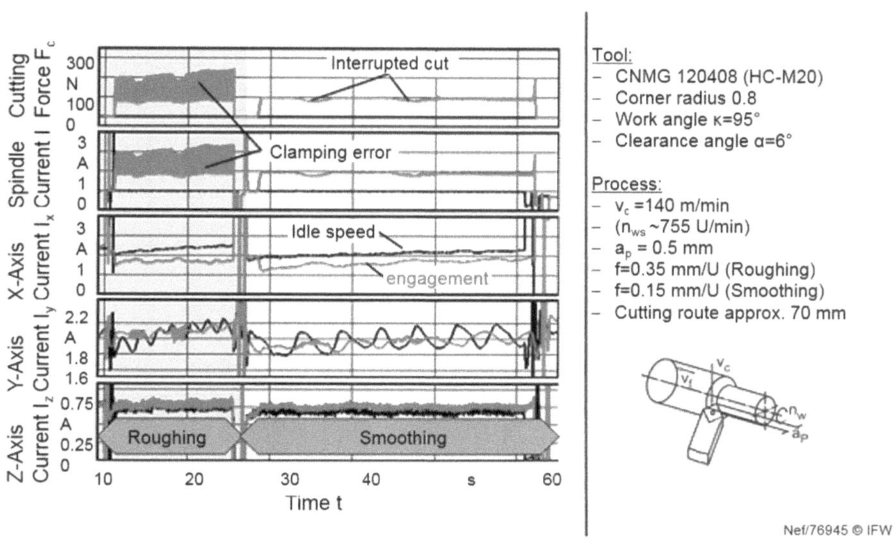

Fig. 3.2.4.6 Analysis during longitudinal turning.

clearly. The suitability of control-internal recorded data could be proved by low-pass filtering of forces with 12 Hz cutoff frequency and by correlation coefficients from a cross-correlation analysis between the power signals and other investigated signals. During the described process, it became apparent that the spindle power had a high sensitivity.

Table 3.2.4.1 presents examples of the results for longitudinal turning operations (on the outside).

Similar results were achieved in the same way for other relevant process types. Consequently, the motor current of the milling spindle or turret spindle for milling processes and the motor current of the main spindle for turning processes were defined as a monitor signal. Both have excellent correlation properties with the process forces.

Cutting tests were also carried out at INDEX. The results were used for qualification of the suitable signal sources and for the sensitivity analysis in relation to the process. As an example, current signals of various axes and spindles during a milling process are shown in Fig. 3.2.4.7.

Table 3.2.4.1 Correlation coefficients of different longitudinal turning processes

	F_c	F_f	F_p	I_C	I_X	I_Y	I_Z
Cutting force F_c	1.00	0.99	0.99	0.99	−0.96	0.36	0.97
Feed force F_f		1.00	0.99	0.99	−0.96	0.36	0.97
Passive force F_p			1.00	0.99	−0.98	0.39	0.96
Cutting force F_c	1.00	0.98	0.99	0.99	−0.98	−0.53	0.62
Feed force F_f		1.00	0.95	0.97	−0.99	−0.59	0.62
Passive force F_p			1.00	0.99	−0.96	−0.49	0.60
Cutting force F_c	1.00	0.96	0.97	0.99	−0.06	0.00	0.38
Feed force F_f		1.00	0.96	0.96	−0.38	−0.21	0.51
Passive force F_p			1.00	0.99	−0.39	−0.14	0.55
	RD S	**RD X**	**RD Z**	**Process**			
Cutting force F_c	0.91	−0.17	−0.99	Rising feed rate			
Feed force F_f	0.91	−0.17	−0.99	a_p = 0.1–1 mm			
Passive force F_p	0.87	−0.17	−0.99	f = 0.15 mm/U			
Cutting force F_c	0.97	0.00	0.97	Rising depth of cut			
Feed force F_f	0.96	0.00	0.99	a_p = 0.15 mm			
Passive force F_p	0.96	0.00	0.94	f = 0.05–0.3 mm/U			
Cutting force F_c	0.57	−0.08	−0.20	Both processes			
Feed force F_f	0.45	0.00	−0.05				
Passive force F_p	0.39	0.00	−0.26				

RD = system deviation; I = current; v_c = 140 m/min.

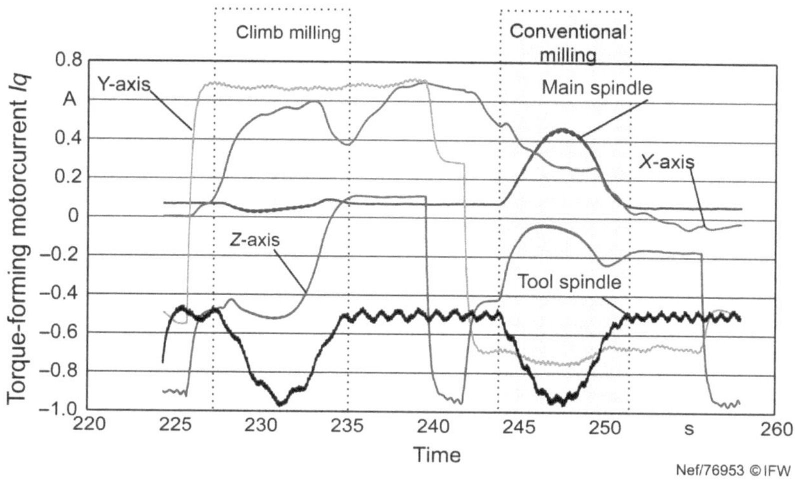

Fig. 3.2.4.7 Milling in half-section with variable cutting depth.

3.2.4.1.4 Development of robust monitoring algorithms based on reconstructed process variables

3.2.4.1.4.1 Development of a robust error identification method and a monitoring methodology

To ensure robust error identification, it is necessary to investigate the occurrence of process errors in the machine signals, which have to be monitored. For this purpose, in a number of reference tests tool breakages, wrong clamping positions, and defective component contours were caused, in order to have a reference for the development of robust error identification. Monitoring signals are always compared with set quantities for process monitoring, in order to identify process errors due to corresponding significant deviations. This implies predictions for the monitored spindle currents, which are reconstructed from the simulation. In order to derive target variables, a model of monitoring signals (digital driving data, spindle currents, and axis currents) is parameterized. Based on this model, a prediction for monitoring signals and monitoring limit values can be derived from the process simulation. This parameterization can be carried out online (parallel to the process) and as a result can be used for process monitoring. This approach was taken in implementing teachless monitoring on the CTX 420 Linear. In addition to this approach, which was parameterized parallel to the process, a complete offline parameterization was examined by means of simulation data and measured data from a parameterization cut (see Section 3.2.3.7.1).

The motor current of the main spindle, respectively, the turret spindle was chosen as a monitoring signal. The strong correlation between the spindle current (main spindle or tool spindle, respectively) I and the material removal rate Q_W justifies the assumption of a linear model $I = k \cdot Q_W + b$. This model can be configured online for each individual process step. The parameter b describes the so-called idle current of the rotational speed at that moment, this means without material intervention. This parameter can be accessed just before the actual machining segment. The parameter k can be determined by linear regression and be updated each time in the NC cycle.

Both parameters b and k are also referred to as process parameters. The prediction quality is becoming more accurate with each additionally recorded cycle. In regard to the functional relationship between the simulated material removal rate and the spindle currents, the underlying decision logic of the monitoring can be kept simple. Based on the data, which were recorded at the beginning of each machining segment, and the associated simulation reference, a statistical confidence interval can also be derived at any given level of confidence concerning the model prediction. A confidence level of $P = 0.999$ was defined and used here. The context is described in more detail in [9,10]. It is assumed that the particular deviations follow the t-distribution. If the currently measured signal leaves this confidence interval, a process error is detected. Such a process error is shown as an example in Fig. 3.2.4.8 on the right. For monitoring, the entire process is already segmented during work preparation based on simulation data. For this purpose, phases with material removal (material removal rate $Q_W > 0$) are identified. An initial classification (milling, turning, or drilling) within these sections is made via the segment-specific tool ID. For a more precise specification, the simulated tool paths, which are generated by the virtual NC core, are analyzed. The currently

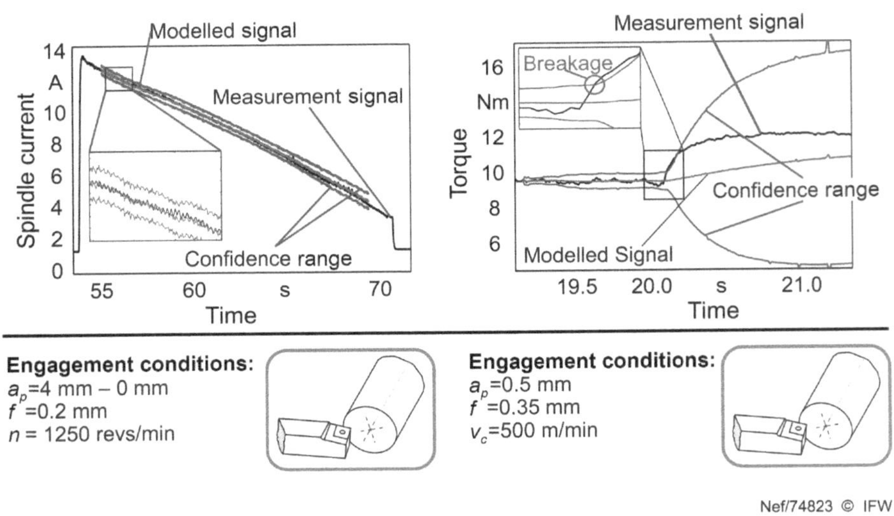

Fig. 3.2.4.8 Error-free process with properly acquired monitoring limits on the left, defective process (detected tool usage) on the right.

active axes and the current feed direction are determined, making an accurate classification of individual elementary processes possible (horizontal surfacing, longitudinal turning (on the outside), etc.). Due to the fact that the machine kinematics were modeled in the simulation with CutS, the feed directions of the axes movements can be transformed. The segments are synchronized for monitoring during machining with the currently measured process signals (see Section 3.2.2.6). Monitoring is performed in relation to specific segments, based on the presented decision logic.

A concrete catalog of errors and characteristics has been reviewed as not effective according to the adapted monitoring concept, because this monitoring concept does not search for predefined error characteristics in the signal, but generates monitoring limits parallel to the process based on machine signals and process simulation data. Spindle and axes currents, as well as position data of all active axes were recorded. Furthermore, spectral analyzes on the process signals were made (in analogy to Chapter 3.2.1), in order to investigate frequency-dependent characteristics of process errors. However, the implemented simulation-based monitoring approach with parameterization parallel to the process soon proved to be more stable and efficient. This approach does not need a fixed definition of characteristic error patterns in the signal or in the frequency spectrum. The obtained results from the spectral analysis have therefore not been included in the implemented system. For the same reason the formation of crest factors in machine signal was unnecessary for working out the signal-inherent error characteristics. Thus the corresponding analysis was left out.

Synchronization of process and monitoring

To ensure an error-free monitoring by the presented model, a real-time synchronization between monitoring segments and corresponding simulation segments is necessary. For this purpose, machine signals recorded during machining must be temporally synchronized with stored simulation data. The SIEMENS compile cycle was used as an interface for supplying machine-internal information, which provides machine data in real time via Profibus. The process-specific simulation data were saved on the monitoring platform in use, and thus were directly available for synchronization.

Two different procedures were tested as a synchronization method. One method intended to be defined directly in the NC code by user-specific H-commands at the beginning and the end of a processing segment. A stored simulation segment was used in the corresponding section. This procedure was used in the implementation of process monitoring on the BENZINGER TNC-S turning machine of the ARTIS GmbH. A synchronization within the NC cycle and thus the guarantee of robustness against deviations during the process (e.g., by varying the feed rate override) was not possible with this method. In order to consider these deviations, a synchronization based on axis position data was used, which are available as simulation data and can also be measured during processing, that means a path-based synchronization.

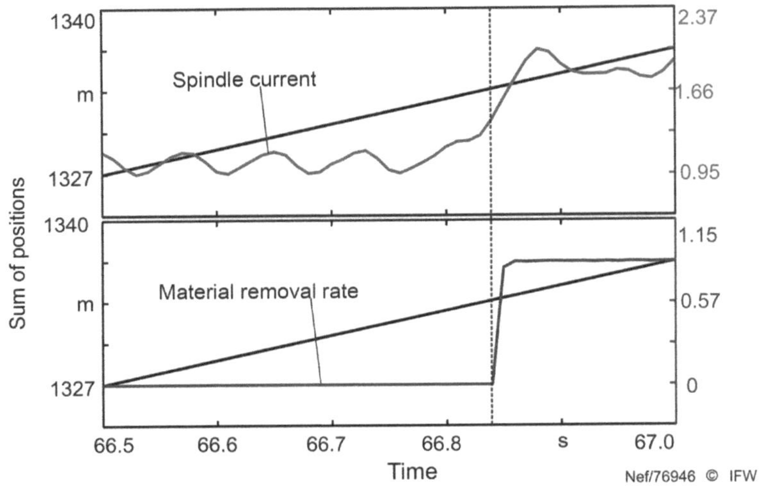

Fig. 3.2.4.9 Position-based synchronization of simulation and measurement.

The simulated data were in compliance with the measured data due to the use of a virtual NC core inside the Virtual Machine. The method is therefore very stable and also invariant to parameter modifications by the operator. A separate identification of such modifications, for example, due to the use of residual generators, was not necessary. In order to exclude errors because of repeatedly reached positions, the sum of all axis positions, summed up within the interpolation cycle, is calculated. This path information is used as a synchronizing signal, see Fig. 3.2.4.9. This method was used in the implementation of process monitoring on the CTX 420 turning center. It also enabled the accurate detection of deviations between simulation and process parameters (e.g., spindle rotation speed). As a result, a synchronization method was available that was able to use reliably the simulation-based process segmentation, thus enabling a reliable process monitoring.

3.2.4.1.4.2 Prototypical implementation of developed process monitoring in the machine tool

Unfortunately, the process chain (implementation of the entire monitoring concept from simulation up to monitoring by monitoring hardware) could not be implemented as planned at Premium AEROTEC due to restructuring measures and associated scheduling difficulties. Because of that, different components of the monitoring system were individually developed and examined. Therefore different prototypical commissioning procedures have been developed. The monitoring of a turning process on the BENZINGER TNC-S lathe was implemented. A typical test component was used, provided by the cooperation partner ARTIS. This test component provides, on the one hand, a relatively wide range of variation at different process segments and widely varying intervention variables and, on the other hand, it offers good comparability with other monitoring approaches.

Monitoring was carried out by the ARTIS DP master system (Fig. 3.2.4.10). This allows transmitting simulated waveforms via the PROFIBUS interface directly from the PC to the CTM card, where they are treated as acquired reference curves from a series production. The segment-based CTM process monitoring can be parameterized based on these teach-in cuts. A process monitoring could therefore be implemented based on simulated material removal rates.

The parameterization can then be carried out for components without making new reference cuts in work preparation based on the simulated removal rate. According to the commissioning process, characteristic process errors were provoked, which were recognized by the system. After commissioning on the BENZINGER TNC-S lathe, the ARTIS GmbH is able to use teachless process monitoring also for other applications.

In another approach, the GILDEMEISTER CTX 420 linear has been used for implementing the online parameterization of the monitoring system. The system was implemented as Simulink model on a Beckhoff Industrial PC.

The relevant machine data were provided by a PROFIBUS interface in real time. The test component was manufactured by a longitudinal turning process (on the outside) and two milling processes on the lateral surface. Fig. 3.2.4.11 shows an example of the implemented monitoring. A process error is detected in the first turning segment. This results from an incorrect tube geometry of the part to be processed.

Process parameters, which were derived by the monitoring system in the process, were recorded and saved in order to be available to the system later. Due to the automatic parameterization of the monitoring model, the decision logic adapted itself specifically to the corresponding machine. In order to ensure the desireda

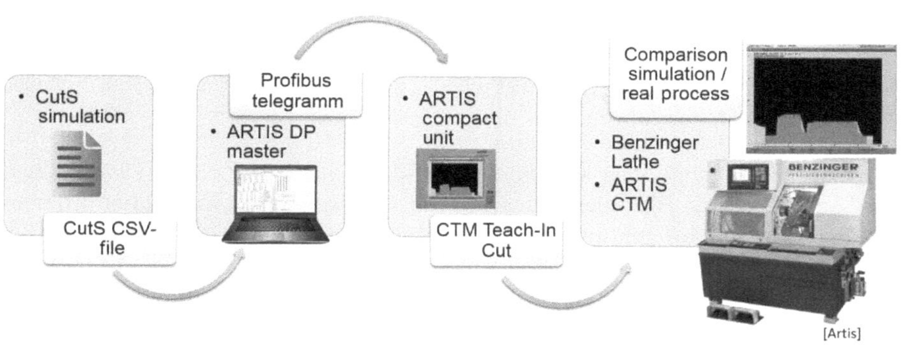

Fig. 3.2.4.10 ARTIS DP master system.

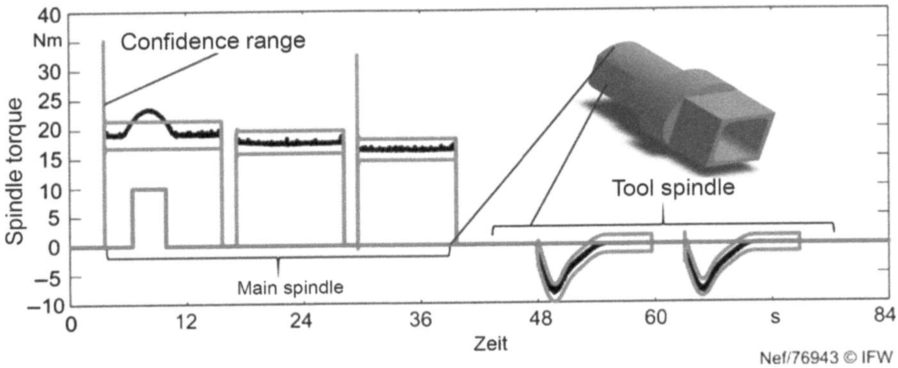

Fig. 3.2.4.11 Online monitoring on the GILDEMEISTER CTX 420 turning center.

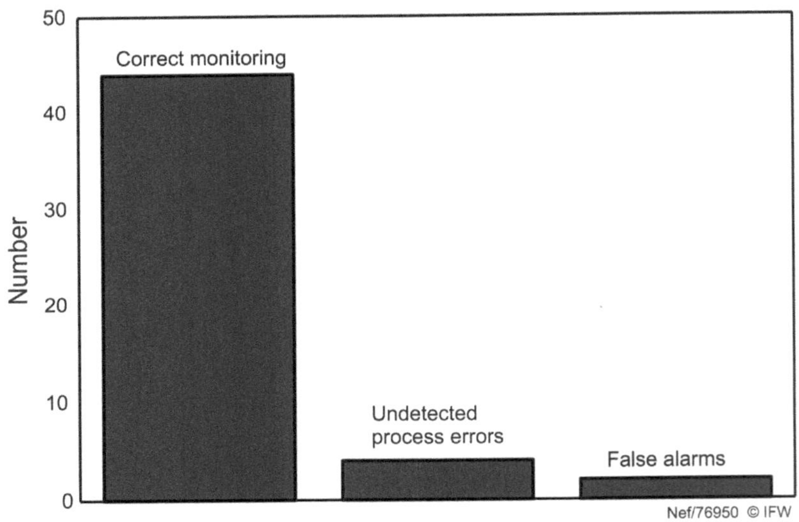

Fig. 3.2.4.12 Representation of undetected process errors and false alarms in relation to the quantity of all validation tests.

sensitivity of the monitoring system, the confidence level of $P = 0.999$ was defined (see Section 3.2.2.6). The monitoring method was positively evaluated at different machine tools with different spindles. This enabled a verification of the transferability of the approach. Tests on the main and the opposed spindle for ensuring the transferability were not necessary in this context.

It has been possible to implement all the single parts of the process chain on two different machine tools (see Section 3.2.3.7.1), from simulation up to online parameterization and monitoring. The functionality of all interfaces was as well demonstrated. The results of the validation processes were documented and analyzed (Fig. 3.2.4.12). In 44 out of 50 performed machining processes all prepared errors were detected. In four cases errors remained undetected. There were two false alarms. The considered approach can therefore be described as very successful.

3.2.4.2 Conclusion

In conclusion, high effort for parameterization of previous monitoring models can be stated based on reference cuts (cutting processes needed to parameterize a monitoring system). The applied methods can therefore only be used economically in large-scale production. So far, there are no methods to parameterize process monitoring in single item production with economically justifiable expense.

References

[1] G. Byrne, D. Dornfeld, B. Denkena, Advancing cutting technology, CIRP Ann. Manuf. Technol. 52 (2) (2003) 483–507.
[2] G. Byrne, D. Dornfeld, I. Insaki, G. Kettleler, W. König, R. Teti, Tool condition monitoring (TCM)—the status of research and industrial application, CIRP Ann. Manuf. Technol. 44 (2) (1995) 541–567.
[3] L. Kühne, Entwicklung eines universellen Überwachungs-und Diagnosesystems für Fertigungseinrichtungen, Dissertation, (1985).
[4] H.-C. Möhring, K.M. Litwinski, O. Gümmer, Process monitoring with sensory machine tool components, CIRP Ann. Manuf. Technol. 59 (1) (2010) 383–386.
[5] A. Albrecht, Wärmedehnungskompensierte Rekonstruktion von Prozesskräften an Vorschubantrieben, Vulkan-Verlag, 2009.
[6] E. Fries, Anwendung neuronaler Netze zur Werkzeugverschleißerkennung beim Fräsen, Dissertation, (1999).
[7] C. Lapp, Indirekte Erfassung und Regelung von Bearbeitungskräften an direkt getriebenen Vorschubeinheiten, Dissertation, (2002).
[8] M. Rehse, Flexible Prozessüberwachung bei der Bohr-und Fräsbearbeitung in einer autonomen Produktionszelle, RWTH Aachen, 1999.
[9] B. Denkena, K. Litwinski, D. Dahlmann, H. Boujnah, J. Kiesner, M. Krüger, J. Schmidt, T. Neff, Industrie 4.0 in der Zerspanung, ZWF 7 (8) (2014) 537–541.
[10] B. Denkena, R. Fischer, D. Euhus, T. Neff, Simulation based process monitoring for single item production without machine external sensors, Procedia Technol. 15 (2014) 341–348.

3.2.5 Online monitoring and control of tool deflection in milling

Berend Denkena, Haythem Boujnah
Institute of Production Engineering and Machine Tools, Leibniz Universität Hannover, Garbsen, Germany

3.2.5.1 Introduction

Systems for online process monitoring and control in milling are widely used. They allow early detection of process failures such as chatter [1], tool wear, tool collision, tool breakage [2,3], and clamping failures [4]. Furthermore, they support the manufacturer in case of detected failure by stopping immediately the machine axis or try even to stabilize the process [5]. Resulting failure costs can be ultimately reduced. However, previously developed systems are generally restricted to optimizations regarding process load [6] or process stability [5]. Approaches for process quality monitoring and control have been subject to only little research.

The tool deflection represents one of the most important quality-degrading effects in milling. It occurs generally in any cutting process due to the compliance of the used tool and process forces and causes a deviation between the real and the reference path of the tool. The results are shape and dimension failures on the workpiece side. Required manufacturing tolerances can no longer be maintained. Furthermore, especially in the processing of complex free-form geometries and in finishing processes, the tool deflection has a decisive influence on the productivity.

For the online monitoring of the tool deflection during manufacturing, a process force signal and the knowledge of the stiffness of the used tool are indispensable. The developed "feeling" machine tool in CRC 653 [7] is already able to measure the process forces on its tool center point (TCP). Based on a machine-integrated force sensing system, a measuring cycle is developed and implemented into the machine control, allowing the identification of the tool stiffness, within the machine tool [8,9]. Since the tool deflection can be monitored by known process forces and tool stiffness, it can be even controlled [10].

This chapter shows the development of a monitoring and control system for tool deflection as an assistance system for milling operations. First, the machine-integrated identification of the tool stiffness is described. Based on the provided force signals from the "feeling" machine and the identified tool stiffness, a model for detection of the tool deflection is presented. Finally, a feed-based tool deflection control is realized by adjusting the axis feed using a data communication between the monitoring system and the machine control.

3.2.5.2 Detection of tool deflection

The stiffness of the tool represents the ratio of the applied force on that body to the caused deformation. To measure the stiffness, both deformation and force have to be measurable. The stiffness measurement can be carried out in the "feeling" machine tool through a soft collision of the tool with the workpiece. Here, the spindle slide of the machine is driven slowly into the workpiece until a contact force is detected. From that position, the slide continues the movement for a predefined distance of some micrometers. With this known distance and the measured force deviation, the static stiffness of the spindle slide can be calculated. Fig. 3.2.5.1 shows the deviation of the force and the tool path during a collision along the feed direction with a predefined collision distance of 150 μm using a tool with a diameter of 12 and 70 mm length. The change in the collision force is detected by the developed spindle slide of the "feeling" machine tool and the collision distance is assigned by the machine tool control. The stiffness of the spindle slide can be determined with all relevant components for the process such as the linear guide, the spindle, the tool holder, and even the tool.

During a soft collision, the displacement of the spindle slide was also measured externally with laser triangulation sensors at the locations L1 to L7 (Fig. 3.2.5.2), where L1 was located on a linear guiding carriage of the spindle slide. This measurement enables the determination of the relative deformation of the spindle slide going from the reference location L1. L8 is measured on the workpiece side to detect its deformation and slipping during the collision. It shows that the tool (L5 to L7) presents the part on the spindle slide with the highest deformation (Fig. 3.2.5.2). In the collision, most of the set collision distance is converted into tool deflection.

In milling, the tool is statically deflected due to the process forces in the same way as during the collision. To estimate the static tool deflection in the process, the clamped tool in the tool holder can be modeled as a cantilever beam. Due to

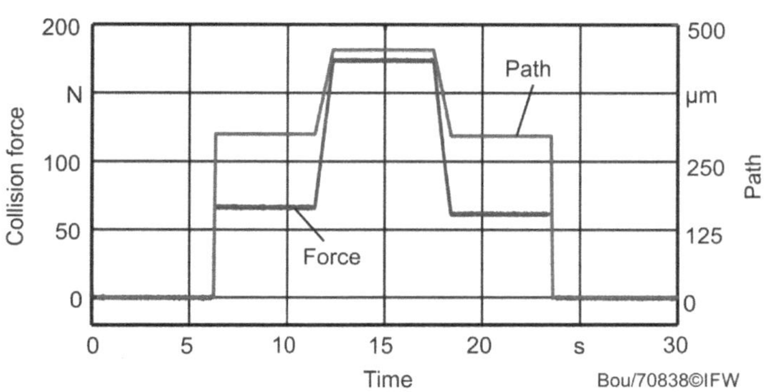

Fig. 3.2.5.1 Measured force and tool path by a soft collision with the workpiece.

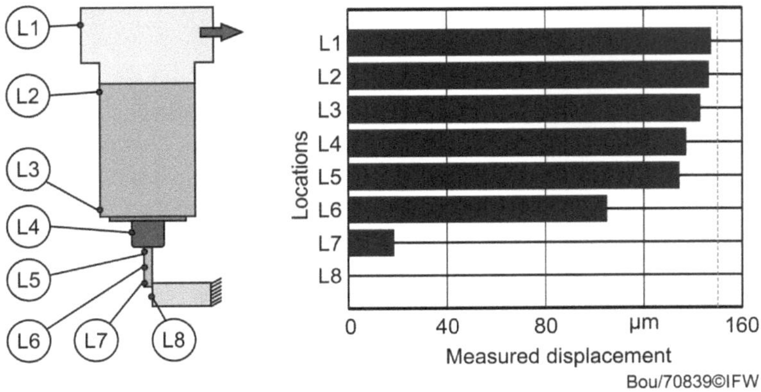

Fig. 3.2.5.2 Measurement of the spindle slide deformation.

the concentrated force F, the tool deflection w at position x along the tool axis can be determined by:

$$w = Fx^2(3a - x)/6EI \qquad (3.2.5.1)$$

for $0 < x < a$, otherwise by:

$$w = Fa^2(3x - a)/6EI \qquad (3.2.5.2)$$

where a is the position of the concentrated force, E is the modulus of elasticity, and I is the area moment of inertia of the tool. The bending stiffness (EI) is estimated from the soft collision, while a is set equal to the tool length L. The distance a in milling can be formed as a function of the depth of cut a_p, assuming the force transmission point at the half of a_p:

$$a = L - a_p/2 \qquad (3.2.5.3)$$

As F is measurable by the spindle slide and L and a_p are set parameters, the tool deflection during milling can be estimated based on Eqs. (3.2.5.1), (3.2.5.2).

In order to verify the measurement of the tool deflection, the deflection has to be provoked during milling. This is achieved by milling of premachined workpieces which show alternating ridges of about 0.25 mm depth and 10 mm width. The milling process is carried out with a tool of 12 mm diameter and 70 mm length. The depth of cut is set to 40 mm. In order to provide reference surface curves, the machined surfaces of the workpieces are measured additionally after milling using a perthometer at three different positions A, B, and C along the depth of cut: A at the top, B in the middle, and C at the bottom of the surface. Fig. 3.2.5.3 shows the surface

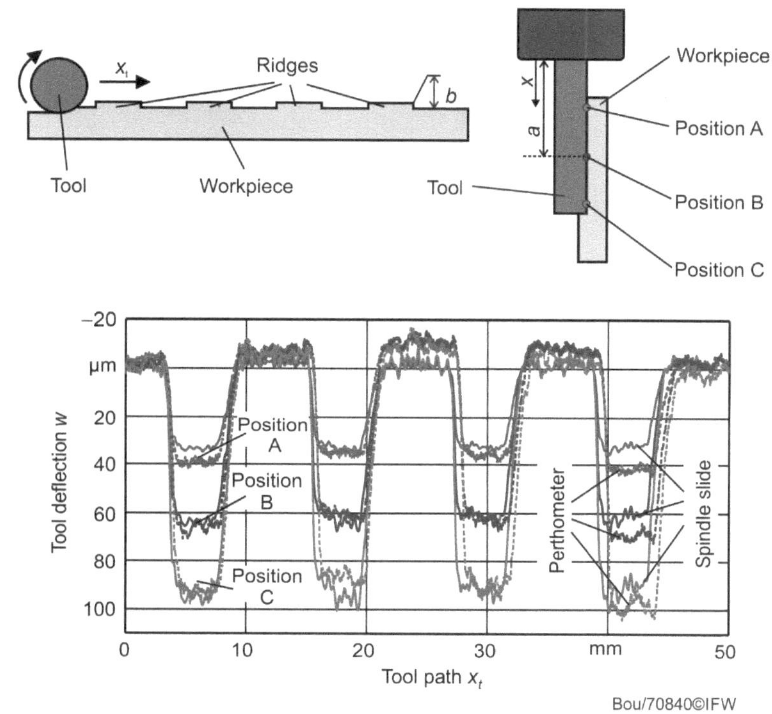

Fig. 3.2.5.3 Measurement of the tool deflection.

curves measured with the perthometer and the spindle slide. Because of the alternating ridges on the surface, the process forces on the tool are changing, which means that the tool deflection is changing as well. From the top and the bottom curves of the perthometer the deflection of the tool along the depth of cut can be approximated to about 55 µm. The curves measured by the spindle slide at positions A, B, and C correlate very well with the reference curves on these positions. The deviation of the slide curves to the reference is smaller than 10 µm and can be explained by the used simplified deflection model which supposes a perfect clamped beam with ordinary geometry.

3.2.5.3 Feed-based control of the tool deflection

In order to reduce the tool deflection and to achieve the desired manufacturing tolerances, the tool deflection has to be controlled while machining. The main objective of the control loop (Fig. 3.2.5.4) is to maintain the actual tool deflection d_{Act} while milling constantly at a desired reference deflection d_{Ref} despite of disturbances in process

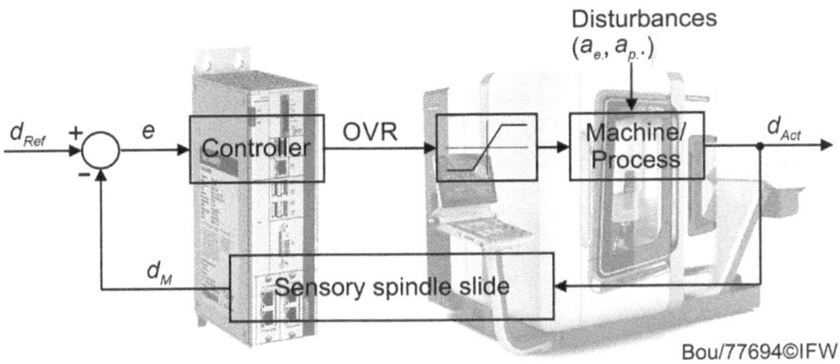

Fig. 3.2.5.4 Control loop for tool deflection.

by adjusting the axis feed of the machine. Such disturbances may be, for example, unexpected fluctuations in allowances in casting parts that cause changes in the cutting parameters like the cutting width a_e or the cutting depth a_p within a straight path. The actual deflection is measured by the sensory spindle slide and is compared to a given reference, which can be calculated from the required shape and dimension tolerances of the workpiece. The occurring difference e, called also the residual error, is applied to the control system, which provides the required feed override *OVR* for the machine control. For safety reasons, a signal limitation is integrated at the input of the machine control, so that the feed override can be varied only between 0% and 120%. For the implementation of the control loop, an industrial PC with real-time controller and programming environment TwinCAT3 from Beckhoff is used. The used "feeling" machine tool is equipped with a machine control of type Siemens SINUMERIK 840d [10]. The override signal of the control system and further command signals are transferred between the industrial PC and the machine control via PROFIBUS communication. The signal transfer on the Siemens control is performed by programming synchronous actions, which are implemented within the NC program. The signals can be read out in the interpolation cycle rate of the controller achieving a sampling rate of maximum 250 Hz.

The required control for tool deflection should be fast enough to react quickly to changes in the measured deflection signals and accurate to eliminate completely the detected residual error. Furthermore, the control should show a quiet behavior by providing the override signal for the machine control. Disturbances in the override can cause damage to the machined surface and the control application has no more benefits. PID controllers are probably the most used controller structures in industrial applications. However, for the deflection control, the PID controller may be unsuitable because of the unrest of its derivative component in combination with noisy signals. Instead, a PI controller is more suitable for this control task. It combines the advantage of the P controller, namely, the rapid response to error, with the exact settling of an I-controller. The PI controller is fast and accurate. Its manipulation algorithm is in Eq. (3.2.5.4), where k is the number of sample, K_p the gain of the

P-controller, the K_i the gain of the I-controller, T_a the sample time, and OVR_{mid} the initial override value:

$$OVR_k = K_p \cdot e_k + K_i \cdot T_a \cdot \sum\nolimits_0^k e_k + OVR_{mid} \tag{3.2.5.4}$$

The initial override value OVR_{mid} is set to 70% to allow the controller settling in both directions. To get an effective starting point for the controller tuning, the second method of Ziegler-Nichols is used [11]. This method is practicable for systems, whose transmission behavior is unknown or not easily identifiable, and which can be driven to their stability limits without causing damages. Under this method, only the P-controller is activated in the control loop. Fig. 3.2.5.5 shows the deflection signal at the stability limits by applying the Ziegler-Nichols method in several milling processes with different values of proportional gain K_p. The test is performed for a reference deflection d_{Ref} equal to 80 μm and constant cutting parameters with exception of the axis feed. The initial override OVR_{mid} of 70% corresponds in this test to an initial feed F_{mid} of 2000 mm/min. The deflection control is started within the air cutting, 15 mm before reaching the workpiece. The gain K_p is increased step by step up to 1.75. At some critical value $K_{cri}=1.25$, sustained oscillation in the signal of the measured deflection with corresponding period $T_{cri}=0.145$ s occurs for the first time. Based on

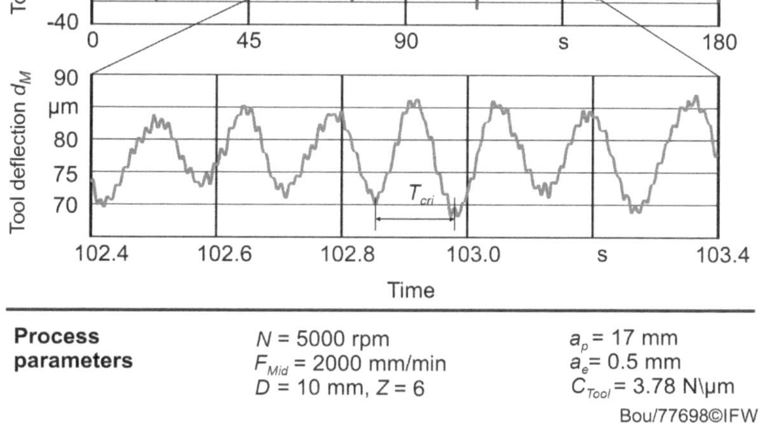

Fig. 3.2.5.5 Deflection signal at stability limit while controller while tuning.

the notified critical values, the controller gains K_p and K_i can be set with respect to the tuning method as following:

$$K_p = 0.45 \cdot K_{cri} \tag{3.2.5.5}$$

$$K_i = K_p / (0.85 \cdot T_{cri}) \tag{3.2.5.6}$$

Fig. 3.2.5.6 shows the deflection signals by using the PI controller with its determined parameters for varied references between 70 and 100 μm. The process parameters and the control starting conditions are the same as in the tuning test. Generally, it shows that the measured deflection signal follows well the set references. The settling behavior of the controller with respect to determined controller parameter appears accurate and quiet. However, high overshoots appear in the measured signal. They show up also significantly over the first 20 mm of the workpiece surfaces.

In Fig. 3.2.5.7, the signals of the controller test by a reference of 80 μm are plotted. As previously mentioned, the control is already switched on within the air cutting before the tool reached the workpiece. So no deflection is occurring and a positive residual error is established. Consequently, the control begins increasing the feed override in order to increase the tool deflection and to reduce the existing error. However, the tool deflection stays unchanged and the feed override exceeds immediately and significantly the upper limitation of the machine control. As the tool attends the workpiece and the material cutting begins, it comes to a change of sign of the residual error and the I-controller begins reducing its output. That takes time until the output reaches again the upper limit of the override and until the error is settled to zero. In this way, an overshooting is produced at each beginning of material cutting. Such overshooting is a typical problem for control loops with I-controllers in combination with limited system input and is known as the windup effect. Technically, many antiwindup solutions and algorithms exist and allow maintaining the controller output at the upper limit. They may reduce the settling time of the residual error. However, they are not able to decrease the overshooting amplitude, because the tool is still hardly entering the workpiece with the maximum axis feed.

A promising solution to damp the overshoots is to shift the starting time of the deflection control. As shown in Fig. 3.2.5.8, starting the control as a material cutting

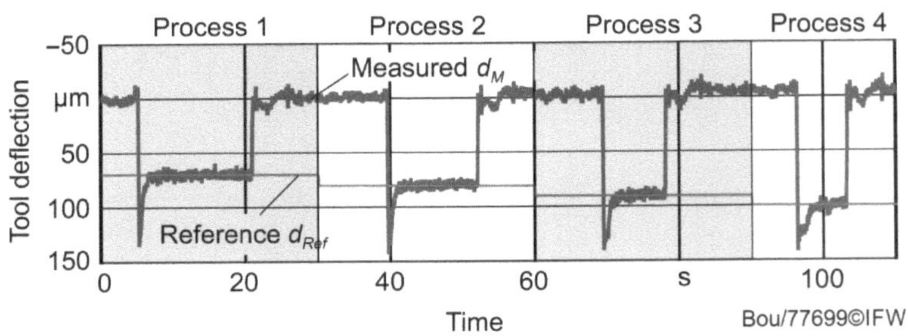

Fig. 3.2.5.6 Deflection control for references between 70 and 100 μm.

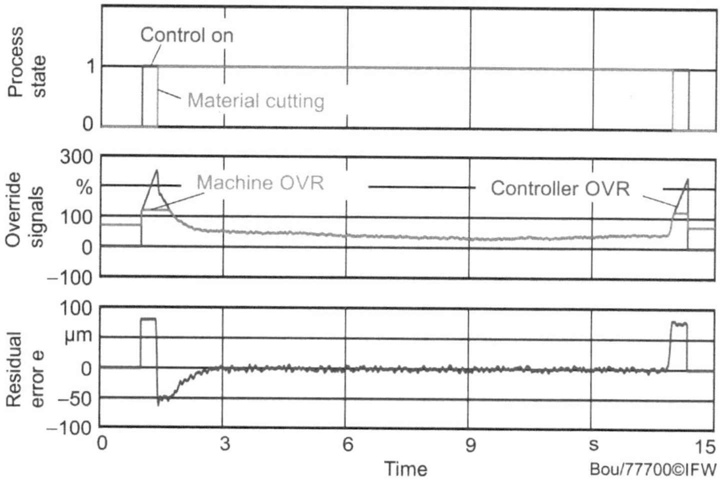

Fig. 3.2.5.7 Control signals while deflection control by 80 μm.

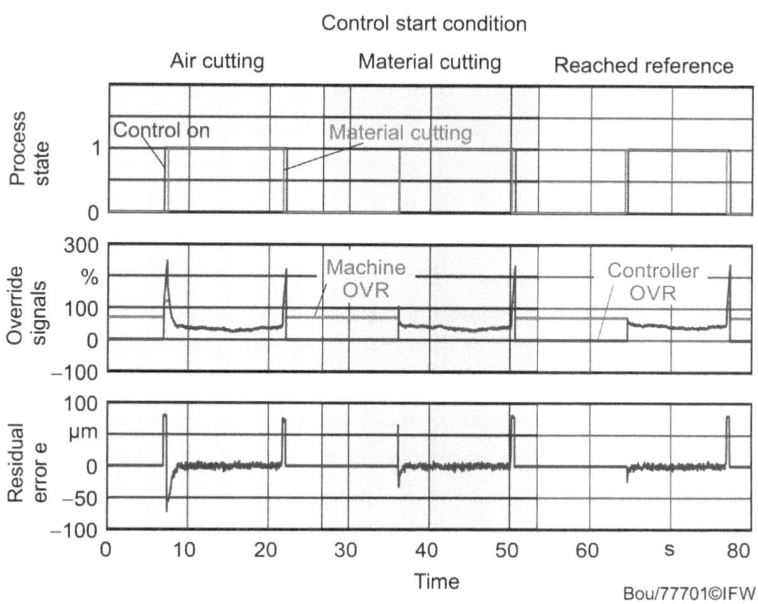

Fig. 3.2.5.8 Different starting strategies for the deflection control.

is detected allows a significant decrease of the overshooting amplitude and even the settling time in comparison with previous starting while air cutting. That is because the tool deflection is occurring during material cutting and is responding to the override changes provided by the controller. In this way, the controller output stays mostly within the limitation area of the machine control and the tool is no more reaching the workpiece with the maximum axis feed. Furthermore, starting the control first as the tool deflection reaches the reference shows smoother entering of the tool into the workpiece with only small overshoots.

To assess the presented control approach regarding cutting accuracy and cutting time, milling tests are performed with varying cutting widths and constant cutting depth. The test steps are shown in Fig. 3.2.5.9A. In the first step the workpiece is prepared by machining a ramped shape. In the second step the milling of a straight path is performed. During this step, the ramp simulates a variation of the cutting width between 0.5 and 1 mm and causes a changing in tool deflection. In the final third step, the marginal positions of the workpiece are measured by a touch probe allowing the determination of the resulting shape offset. The process is repeated 10 times, with and without deflection control. In Fig. 3.2.5.9B the results of the process tests are depicted. It shows that processes with deflection control produce considerably smaller shape offsets compared to uncontrolled processes. Consequently, the milling is performed

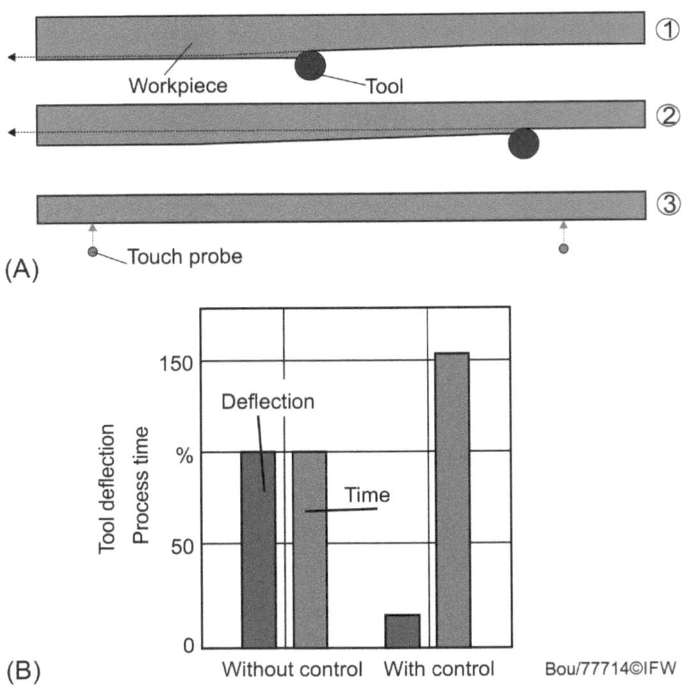

Fig. 3.2.5.9 Setup for tool deflection evaluation tests (A). Results of milling tests (B).

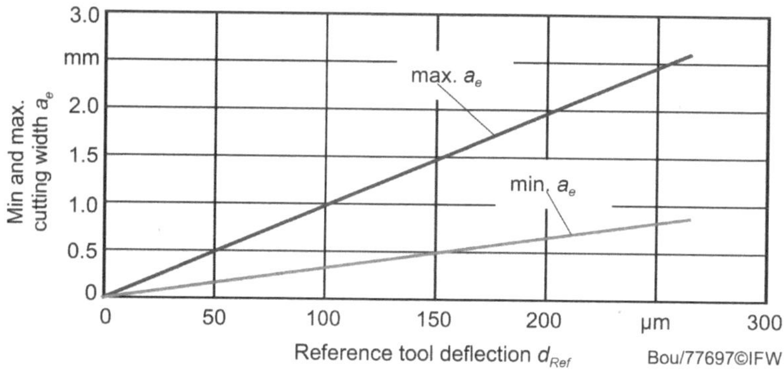

Fig. 3.2.5.10 Boundary cutting width for deflection control.

more precisely by using the deflection control despite of changing cutting width. However, controlled processes take significantly more time due to permanently changing axis feed by the control.

A further control restriction factor for the deflection control is the override limitation on the machine control side. The override limitation bounded the area of the axis feed and therefore limits the permissible cutting width in process (Fig. 3.2.5.10). If the actual cutting width lies outside of the permissive area of the cutting width, the deflection control cannot be done entirely without error. The permissible area of the cutting width grows with increasing reference deflection.

3.2.5.4 Conclusion

As the products are becoming more and more individual, intelligent manufacturing systems are required to reduce additional costs in comparison to conventional series productions. This chapter presents an approach for a monitoring and control system for the tool deflection, one of the most frequent process failures in milling. The monitoring system is realized by using a "feeling" machine tool, which is able, on the one hand, to sense online the occurring process forces by integrated strain gauges in its sensory spindle slide, and on the other hand, to measure autonomously the tool stiffness. The control system allows to maintain the measured tool deflection constantly at a set reference value despite of disturbances in process like unexpected changing in the cutting parameters. The control loop is based on a PI controller with the benefits of fast and accurate settling. The controller parameters are determined experimentally with respect to the method of Ziegler-Nichols. It shows that the control starting condition has a big influence on the settling behavior of the controller and on the overshooting amplitudes. Starting the control within the material cutting allows smoother entering of the tool into the workpiece and damps therefore overshoots. Further process tests show that the deflection control increases significantly the milling accuracy but also the cutting time.

References

[1] Y. Altintas, M. Weck, Chatter stability of metal sutting and grinding, CIRP Ann. Manuf. Technol. 53 (2) (2004) 619–642.
[2] M. Nouri, B.K. Fussell, B.L. Ziniti, E. Linder, Real-time tool wear monitoring in milling using a cutting condition independent method, Int. J. Mach. Tools Manuf. 89 (2015) 1–13.
[3] R. Koike, Y. Kakinuma, T. Aoyama, K. Ohnishi, Tool collision detection in high-speed feeding based on disturbance observer, in: 6th CIRP International Conference on High Performance Cutting HPC2014, 2014, pp. 478–483.
[4] B. Denkena, J. Kiesner, Strain gauge based sensing hydraulic fixtures, Mechatronics 34 (2016) 111–118.
[5] C.M.G. Bort, M. Leonesio, P. Bosetti, A model-based adaptive controller for chatter mitigation and productivity enhancement in CNC milling machines, Robot. Comput. Integr. Manuf. 40 (2016) 34–43.
[6] R.G. Landers, A.G. Ulsoy, Y.H. Ma, A comparison of model-based machining force control approaches, Int. J. Mach. Tools Manuf. 44 (7–8) (2014) 733–748.
[7] B. Denkena, T. Mörke, M. Krüger, J. Schmidt, H. Boujnah, J. Meyer, P. Gottwald, B. Spitschan, M. Winkens, Development and first applications of gentelligent components over their lifecycle, CIRP J. Manuf. Sci. Technol. 7 (2) (2014) 139–150.
[8] B. Denkena, K.M. Litwinski, D. Brouwer, H. Boujnah, Design and analysis of a prototypical sensory Z-slide for machine tools, Prod. Eng. Res. Dev. (WGP) 7 (1) (2013) 9–14.
[9] B. Denkena, K.M. Litwinski, H. Boujnah, Detection of tool deflection in milling by a sensory axis slide for machine tools, Mechatronics 34 (2016) 95–99.
[10] B. Denkena, D. Dahlmann, H. Boujnah, Tool deflection control by a sensory spindle slide for milling machine tools, in: CIRP Conference on Intelligent Computation, Manufacturing Engineering, Italy, 2016.
[11] R.C. Dorf, R.H. Bishop, Moderne Regelungssysteme, vol. 10, Pearson Studium - Elektrotechnik, 2007.

3.2.6 Tempering control of forging processes by integration of cavities

Bernd-Arno Behrens, Martin Bonhage, Irfan Yousaf Malik, Christian Bonk
Institute of Forming Technology and Machines, Leibniz Universität Hannover, Garbsen, Germany

3.2.6.1 Introduction

Powder metallurgy and its parts have gained significant importance in the metal processing industry in recent years. It is known for its high potential to produce near net-shape products. In view of a higher percentage of material utilization and comparably lower energy costs, powder metallurgy processes fulfill the requirements of modern manufacturing. The primary objective of this work is to develop a forging die which can "feel," "learn," and "control" autonomous reactions to process variations. Therefore, a hot forging die with integrated tempering channels (active) in a metal powder matrix as well as slots for passive tempering elements was developed. Moreover, one of the aims was to gain more knowledge about the tool condition by implementing newly developed fiber-optic temperature sensors. Fig. 3.2.6.1 shows the schematic of such a forging die.

Forging dies undergo high thermal and mechanical loads. High forging temperatures result in structural changes of the surface area which especially benefit mechanical wear mechanisms (abrasive wear) [1]. Furthermore, based on thermal expansion, the temperatures of a hot forging process affect the part accuracy, e.g. at process start, process variation, and process interruption (Fig. 3.2.6.2).

The thermal load of a tool is mainly influenced by the forging temperature, forming rate, amount of friction, cycle time, and heat flow caused by convection and radiation [2]. To counteract these thermal stresses and keep the basic die temperature steady,

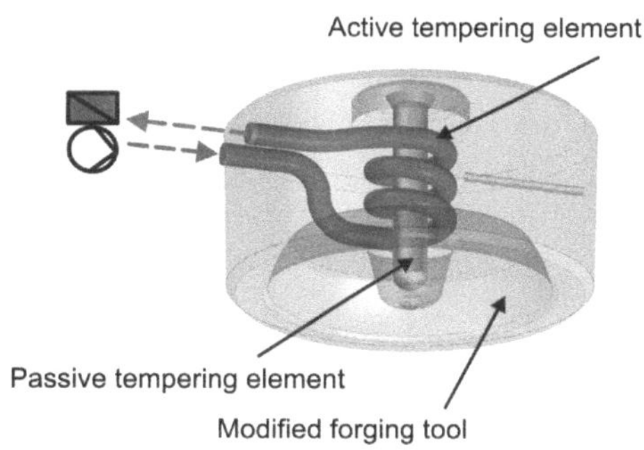

Fig. 3.2.6.1 Schematic of a forging die with integrated cooling channels [IFUM].

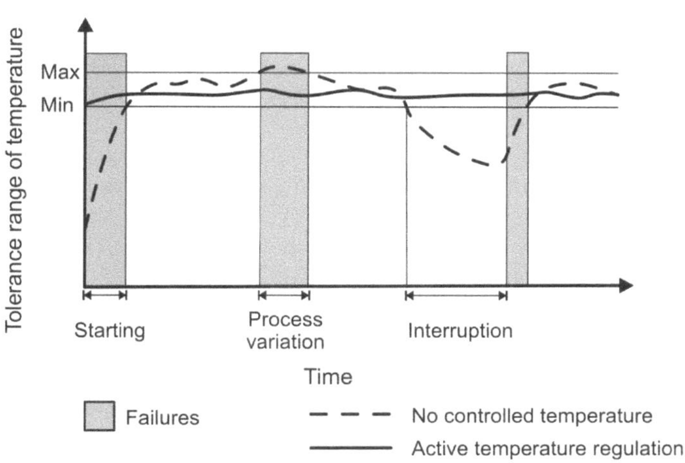

Fig. 3.2.6.2 Failures based on thermal variations of a forging process [2].

normally spray cooling systems are applied [3]. As the tools in hot forging processes undergo a combination of cyclic thermomechanical loads, tool wear as well as the service life of the tools is of particular importance in these processes [4]. Behrens et al. have presented a numeric estimation of abrasive tool wear as a major cause of tool failure in hot forging processes [3]. Cooling systems play a crucial role in tool life, process productivity, as well as product quality. Lubricants and a spray cooling system ensure a steady basic temperature [4]. A drawback of spray cooling systems is the risk of thermal shock-induced cracks as well as thermal softening, both resulting in an increased tool wear.

A supplementary cooling system based primarily on the concept of internal cooling channels which can influence the regions with high thermal stresses is the focal point of this work. The channels will accommodate the cooling medium whose temperature is to be regulated by means of an external temperature control system. Temperature recording will be realized by application of fiber-optic sensors, hence enabling continuous temperature measurement along the length of the fiber. Based on these measurements, the temperature of the cooling medium can be regulated as required. To obtain knowledge about the material flow of the workpiece material, the temperatures in the surface layer of the forging dies, as well as to carry out a realistic tool analysis already during the concept phase of the forging tools, computer-aided process design and FE investigations were used.

3.2.6.2 Integration of cooling channels in powder matrix

In order to influence the temperature of forging dies, normally spray cooling systems are used. During forging cycles, the cooling lubricant is applied to the outer surface resulting in a high temperature gradient between shell and ground temperature. Common temperature control routes can be complemented by the integration of internal channels. The advantage of these tempering channels is the ability to actively influence the basic temperature by heating and cooling. In a first attempt, tempering channels were integrated in a forging die by powder pressing and sintering.

Therefore, the melting behavior of liquefied copper and aluminum in a metal powder matrix (1.2999) was investigated. A selection of micrographs is shown in Fig. 3.2.6.3. The figures indicate that the inserted copper is thoroughly molten and has infiltrated the surrounding powder. The infiltration of the surrounding powder is enabled through capillary effects [5]. Furthermore, by means of copper more complex geometries can be molten out. A complete melting of a metric thread (M2) out of a steel powder is shown in Fig. 3.2.6.3 (right).

Contrary to copper, an infiltration with aluminum was not successful. During sintering, the density of the molten aluminum decreases up to 2.3 g/cm^3 at 1100°C [6] which leads to an increase in the inner pressure and finally causes the green compact to crack (Fig. 3.2.6.4).

Apparently, the volume increase of the molten aluminum is responsible. This situation is aggravated by the fact that the molten aluminum extracts iron atoms from the

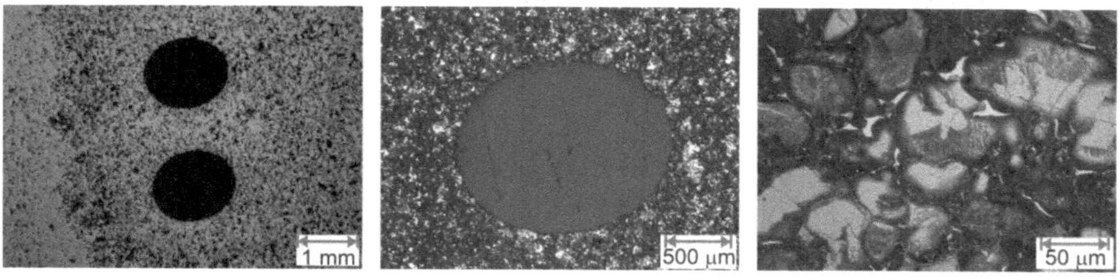

Fig. 3.2.6.3 Sintered part made of steel powder, infiltrated by integrated round copper rod (left, middle) and threaded rod (right) [IFUM].

Fig. 3.2.6.4 Broken specimen after infiltration with aluminum [IFUM].

powder matrix and causes the formation of solid intermetallic phases which might close the porosities and thus block the capillary effect as observed in Ref. [7].

Based on the knowledge for creating complex structures as shown in Fig. 3.2.6.3 (right), these geometries were, in general, identified to improve the overall heat transfer of a tempering system. From heat technique it is known that structured tubes can affect the Nusselt number [8]. The Nusselt number is a dimensionless number used to describe the convective heat transfer between a solid surface and a fluid flow. As shown in Eq. (3.2.6.1), the Nusselt number depends on the heat transfer coefficient α, the characteristic length L, and the thermal conductivity of the fluids:

$$Nu = \frac{\alpha \cdot L}{\lambda} \tag{3.2.6.1}$$

According to this, an increase in the Nusselt number, which occurs for structured tubes, is affected by an increasing heat transfer coefficient at same characteristic length. This can be explained by emergence of turbulent flows (Fig. 3.2.6.5) and thus by the modification of the heat transfer coefficient α.

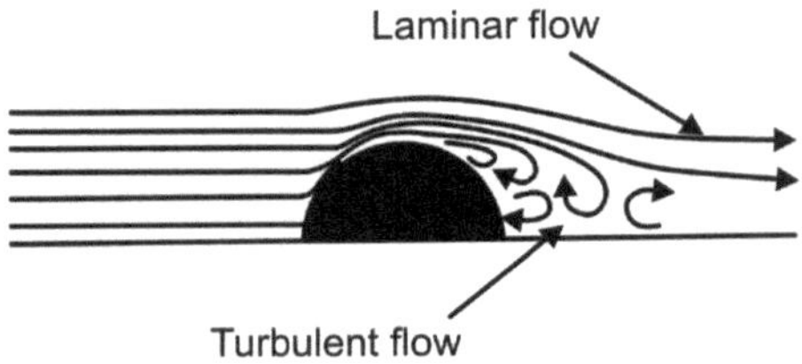

Fig. 3.2.6.5 Flow structure near the wall depending on wall geometry [8].

3.2.6.3 Pressing of metal powder with inserted foreign elements

The findings have shown that copper, as a low-melting metal, is suitable for the integration of tempering channels in a metal powder matrix. The application of tempering channels to hot forging dies enables controlling the basic temperature as well as influencing high thermally charged areas. The integration of copper in a steel matrix by pressing and sintering was identified as a suitable and time-efficient method [9]. To predict the location of the copper element after pressing, the interaction between steel powder matrix and copper element must be well known. Therefore, experimental studies of the forming behavior of copper elements in a powder matrix, depending on the location against the pressing force direction, were carried out. The examined positions were selected as vertical, horizontal, and helical to the force direction (Fig. 3.2.6.6).

The observed copper element had an initial diameter of Ø 8 mm and a length of 20 mm. Both of them were placed vertically and horizontally to the pressing direction. A compaction pressure of 600 MPa was applied and the powder matrix had a diameter of Ø 36 mm. A vertical position against the stamp results in an average increase in the diameter from 8 mm up to 10 mm (+25%) and an average decrease in length from 20 mm up to 15 mm (−25%) (upsetting). The resulting cavity volume remains the same (a). A horizontal location against the pressing force results in an elliptical average diameter of Ø 7.5 mm (−6%) and an average increase in width from 20 mm up to 22 mm (+10%). As with the vertical position, the resulting volume remains the same

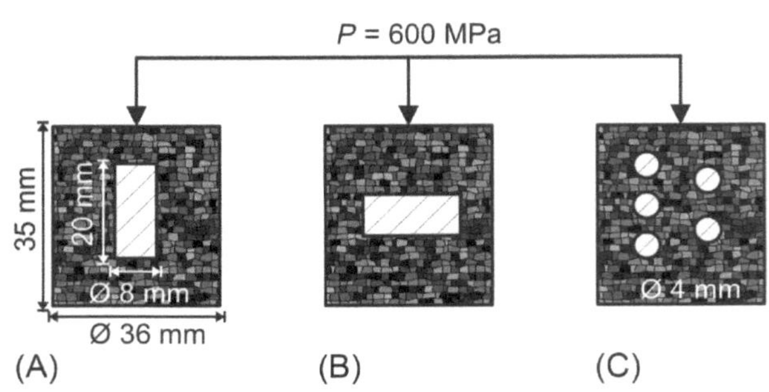

Fig. 3.2.6.6 Different orientations of copper elements against the pressing force direction [IFUM].

(b). The greater deformation of the vertically inserted copper element can be explained by higher local compression stress acting on the copper. The compaction of a spring in a steel powder matrix leads to both, an upsetting to an average elliptical wire diameter of Ø 7.5 mm and a reduction of the spring pitch by nearly 50% (c).

3.2.6.4 Influence of sintering on the copper infiltration

During the sintering of 1.2999 hot work tool steel powder (1250°C), the entire amount of copper liquefies and flows into the powder matrix through capillary forces in all cases. At lower temperatures the compound was not sustainable, due to extensive oxidation around the atomized powder (Fig. 3.2.6.7).

The sintering took place in a vacuum furnace for different holding times (30 and 60 min). As a result, longer holding times did not affect the infiltration depth of the steel powder matrix. At both holding times 100% of the negative cavity of the initial copper insert was generated. Nevertheless, regardless of component geometry (diameter, orientation) and sintering time, the distribution of copper into the metal powder matrix was not repeatable. On the contrary, the allocation of the molten copper seems randomly distributed. By means of metallographic preparation the infiltrated areas became visible. The effect has not yet been identified, but it might be influenced by the factors such as different particle size distribution, inhomogeneous density and heating, gravitation, or a combination of these.

Another drawback is the formation of cracks in areas with enhanced stress concentration (bottom of the foreign particle), as seen in Fig. 3.2.6.8. This observation is a major limitation for the integration of foreign elements into a metal powder matrix by cold pressing. The cyclic loads during a forging process will cause proceeding crack growth in these areas.

Through the liquefaction of copper, the steel matrix is modified by a copper phase around the cavity. To gain a better understanding of this effect, relating to mechanical strength, macro hardness measurements were carried out along the phases. In

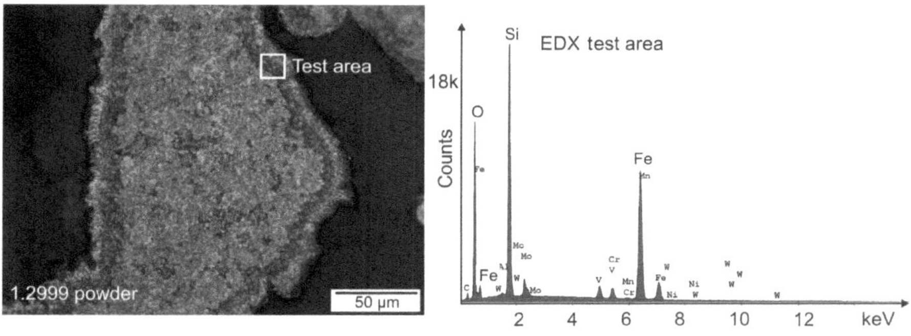

Fig. 3.2.6.7 Oxidation of water atomized powder tool steel (1.2999) during sintering [IFUM].

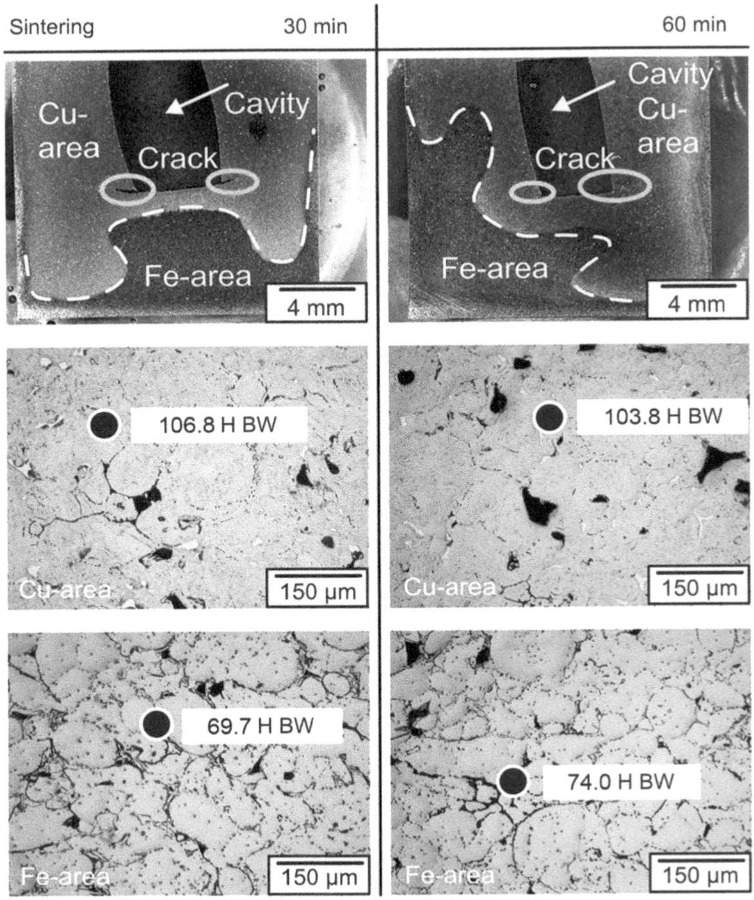

Fig. 3.2.6.8 Infiltration depth of molten copper in metal powder matrix (1.2999).

Fig. 3.2.6.8 the copper infiltration of vertically pressed samples is displayed. By etching the polished surfaces of the separated sintered parts, the copper-infiltrated area, marked with "Cu-area," becomes visible. It can be identified that the infiltrated areas differ from each other.

This observation is problematic since porosity and resulting hardness can be set in direct relation to the infiltrated areas. These areas have a significantly higher macro hardness of +40% and higher. As known from literature, especially small pores are closed by liquefied copper through capillary force, which is the main reason for the increasing macro hardness [10]. In infiltrated areas a Brinell hardness of nearly 100 H BW was measured, whereas noninfiltrated areas exhibit a hardness of only 70 H BW.

Given the irregular amount of pores in combination with crack formation due to stress shielding in a sintered forging tool, practical use does not appear realistic. Under cyclic loads, cracks will grow based on pores, especially from the highly stressed surface. Alternative solutions like hot pressing (HP) or selective laser melting (SLM) may cause a better matrix quality, but due to higher production times and costs these alternatives seem impractical as well. This might be a problem for further investigations since for reproducible components, as needed in the forging industry, heterogeneous component properties (hardness, density, alloy) are essential.

3.2.6.5 Identification of an alternative modular tool layout

As a consequence of the previous findings, a modular tool layout was identified as a promising approach having an active influence on the thermal behavior without the given drawbacks described earlier. By the integration of both, active and passive tempering inserts, an individual and efficient tempering profile will be achieved (Fig. 3.2.6.9).

The developed modular layout consists of three major components:

- An outer shell manufactured from conventional hot tool steel,
- An inner oil-cooled active insert manufactured from metal powder (SLM) and
- A passive insert consisting of high-strength copper alloy.

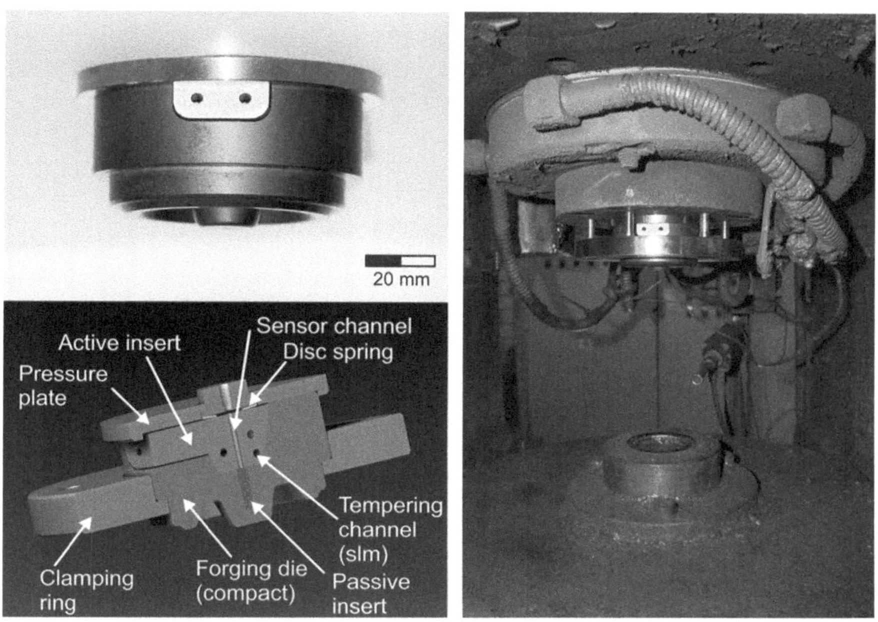

Fig. 3.2.6.9 Modular tool layout of a regulated forging tool [IFUM].

With this setup, already known surface coatings and spray cooling systems can be used and optimized. The heat transfer via passive insert will enable an individual inner tempering of forging dies in highly stressed areas. Furthermore, the complex inner active oil-cooled inserts, created by generative methods (SLM), will increase the dynamics of the passive cooling insert. The active insert is equipped with tempering channels and connected to the shell by pressure bonding (disc spring).

The advantage of the identified modular tool layout is the use of well-known and cost-efficient conventional tool steel in the highly stressed impact zone (shell). The complex active insert is designed to be reused, if the outer shell is damaged by wear which increases the overall efficiency as compared to tools completely made by SLM.

With the integration of sensory elements, a forging die becomes smarter in many ways. Especially by means of the detailed knowledge of heat distribution, present processes can be optimized and numerical simulation models can be validated. Additionally, effects like thermal softening and thermal shock, which are responsible for major wear mechanisms in hot forging applications, will be more comprehensible and can be better controlled.

In view of these facts, the local temperature is measured with a fiber-optic sensor. This sensor allows a high resolution temperature measurement with one measurement point every 1.25 mm along the entire length of the fiber. To implement a glass fiber sensor in an environment with high thermal loads without destroying the sensor, the primary coating of low-melting polyamide must be stripped beforehand (Fig. 3.2.6.10).

Without coating, the fiber loses its flexibility but gains more temperature resistance. To protect the stripped glass fiber, it is housed in a steel capillary tube. The capillary tube has an outer diameter of 2.5 mm and an inner diameter of 250 μm. The fiber used

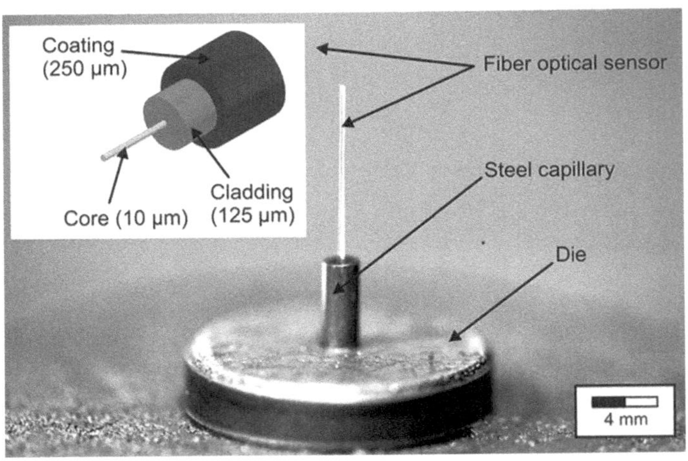

Fig. 3.2.6.10 Stripped fiber-optic sensor housed in a steel capillary and integrated in a forging die [IFUM].

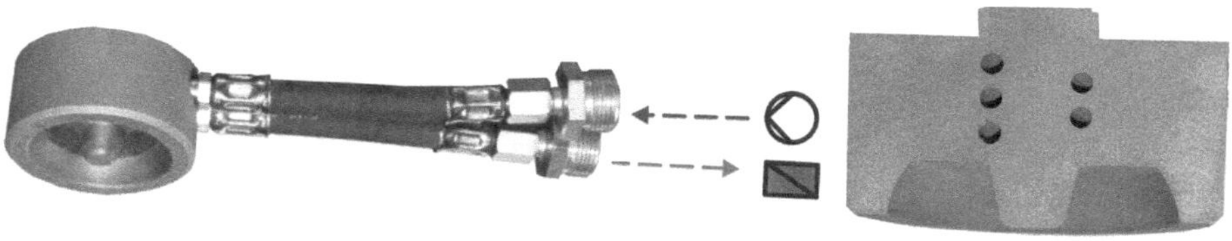

Fig. 3.2.6.11 Prototype (SLM) of forging die with inner cavities [IFUM].

has a diameter of 125 μm. With this configuration it was possible to implement a modified high-temperature glass fiber sensor in a forging die (Fig. 3.2.6.10).

The implementation of tempering channels has so far been demonstrated by means of a prototype. The prototype was completely created by SLM to evaluate the heating, cooling, and flow behavior by means of oil as a tempering medium (Fig. 3.2.6.11, left). Therefore the forging die was connected to a tempering device (HB Therm 5) which pumped tempered oil through the channels (diameter 4 mm) (Fig. 3.2.6.11, right). The used tempering device has a cooling capacity of 34 kW and a heating capacity of 16 kW which can preheat the die up to 250°C. For performing serial tests, the implementation of the presented modular tool layout (Fig. 3.2.6.9) in conjunction with the tempering device in an automated eccentric press is scheduled.

3.2.6.6 Background to the numerical investigation methods

The numerical investigation approach renders cost-intensive experiments redundant thus a finite element-based analysis is necessary to achieve optimum design, dimension, and location of cooling channels. As forging tools experience temperature variation in their surface layers due to contact with hot workpieces, thermomechanical numerical investigations are of great importance. Thermomechanical tool analysis not only allows for the calculation of the heat transfer between workpiece and tools, but also for determining the mechanical loads acting on the tools during the forging process. The goal of such investigations is an estimation and assessment of the loads acting on the tools during the forming process. Important process parameters for the assessment of thermal and mechanical loads are temperature distribution and stress state. A consideration of the von Mises stress ($\sigma_{v,\text{Mises}}$) and the maximum principal stress ($\sigma_{I,\max}$) is of particular importance for a detailed analysis of mechanical strengths. The von Mises stress is direction independent and indicates the beginning of plasticity. The first principal stress indicates the maximum tensile stress in the component. If the calculated first principal stress (maximum principal stress) reaches the ultimate tensile (normal) strength, the tool regions exposed to high tensile stresses have a high risk of failure due to crack initiation.

3.2.6.7 Initial investigation of the die behavior

Initially an insight into the die behavior without cooling channels was necessary. For this reason, a stress analysis was carried out for the forging die without channels in the FE program Simufact.forming [9].The FE model for this analysis consisted of an upper die, a lower die, and a billet. The upper die was assigned an elasto-plastic material behavior and the billet an elasto-viscoplastic material behavior. The lower die was modeled as a rigid body. The workpiece was defined to consist of steel 1.0503 (C45, AISI 1045) and for the tools the hot work tool steel 1.1545 (C105W1, AISI W1) was considered. To achieve a good accuracy in the FE results, the element size in the layer of the dies was set to 0.15 mm. The forging temperature of the raw part was set to 1200°C. As basic tool temperature, an operating temperature of 20°C was assumed and the combined coulomb-tresca model was used for describing the friction. Based on literature study, the friction coefficient was set to 0.1 and the friction shear factor to 0.3 [11]. For the modeling, rotational symmetry was assigned, but the model in Fig. 3.2.6.12 has been expanded to 180 degrees for purposes of clearer illustration.

The von Mises stress ($\sigma_{v.\mathrm{Mises}}$) was observed to have a local maximum value of 725 MPa (Fig. 3.2.6.13, left), whereas the first principal stress ($\sigma_{I,\max}$) decreased from a maximum of 330 to −280 MPa (Fig. 3.2.6.13, right). The yield stress of the die material 1.1545 is around 500 MPa and the ultimate tensile strength is 800 MPa at 700°C. As the observed maximum stress $\sigma_{v.\mathrm{Mises}}$ was higher than the material yield stress, micro plastifications of the surface could be expected. It was observed that high stresses exist only in the upper part of the mandrel and on the bottom plane surface of the die. The stresses in the other die areas were lower and plastifications did not occur. The critical stress value, the first principal stress, was observed to be below the mechanical limit of the material so that crack initiation was not expected. This shows that the geometrical dimensions of forging part and forming dies can be used

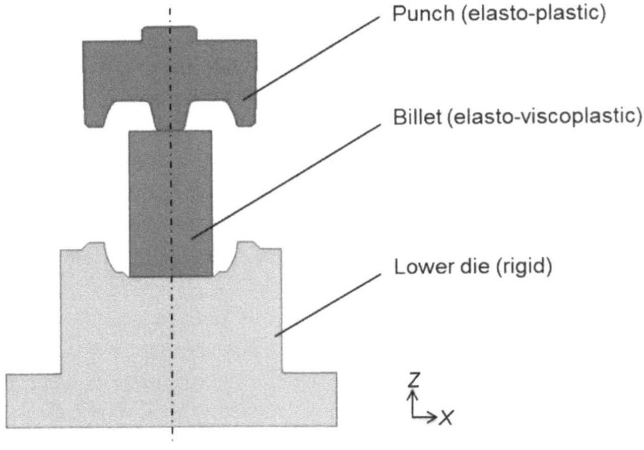

Fig. 3.2.6.12 Schematic model of the simulation process [9].

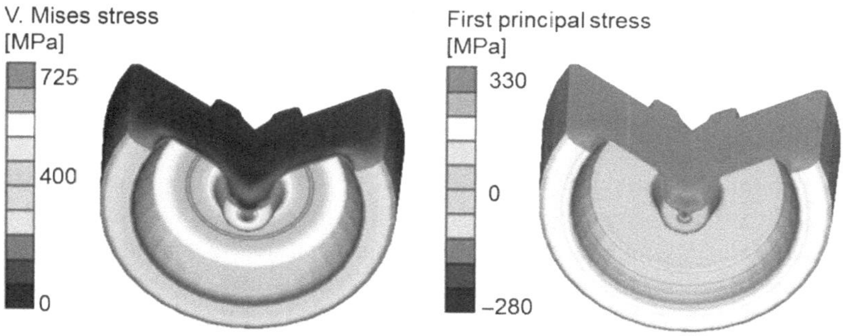

Fig. 3.2.6.13 von Mises stress distribution (left) and maximum principal stress (right) in the forging die [9].

without die failure. Moreover, the plastic strain of the billet was observed to be in a middle range of 1.5–2. Only in the area of the mandrel, a high plastic strain from 3 to 4.5 was observed (Fig. 3.2.6.14, left). The temperature distribution shows maximum values of about 670°C at the upper mandrel of the die by the end of the forming process (Fig. 3.2.6.14, right).

3.2.6.8 Evaluation of temperature field distributions in the forging die

In order to indicate the optimal positions of cooling channels as well as their dimensions, an investigation of the temperature field at the end of a forging step is also necessary. Thus considering the similar FE model approach as used for the initial stress analysis presented earlier, a comprehensive tool analysis was done using the FE program FORGE Transvalor which enables a cyclic repeat of the forging stages considering all phases in a forging process [12].

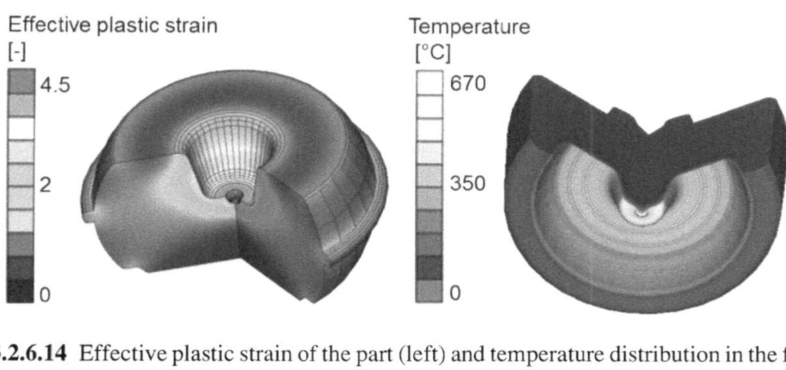

Fig. 3.2.6.14 Effective plastic strain of the part (left) and temperature distribution in the forging die (right) [9].

With an increasing number of forging cycles, a stationary temperature field developed in the forging die. This depends on the thermal material behavior, the heat transfer from workpiece to dies, the contact time between workpiece and dies, the transport time, and the cooling medium. To design cooling channels, the stationary temperature field has to be calculated. Therefore a transport time of 6 s and a cooling with water spray were applied. Heat transfer coefficient of 10,000 W/m^2 K was chosen between the dies and the water spray cooling. The cooling time was set to 2 s. For the calculation of a stationary temperature field a convergence criterion was needed which was set to 1°C temperature difference between two forging cycles. For the calculation, the absolute maximum and minimum temperatures in a forging cycle were used. A stationary temperature field occurred after 64 cycles. A maximum temperature of 706°C in the mandrel Radius 1 (R1) was observed at the end of the forging step (Fig. 3.2.6.15). Other die regions showed lower temperatures and the minimum temperature observed was 352°C.

The evaluation of the temperature evolution in the die region R1 showed a temperature of 680°C after the first forging step (Fig. 3.2.6.16). After the spray cooling phase the temperature decreased by 20°C. With an increasing number of cycles, the maximum temperature T_{max} increased to 706°C in the stationary case. In contrast, the minimum temperature T_{min} in R1 increased to 450°C. In the first cycle the temperature ranges around 660°C and in the stationary case it was measured to be 256°C. This shows that the temperature ranges vary for different die regions.

To investigate the temperature evolution depending on the depth in the FE calculation, the element nodal temperature was determined as a function of time. To gain knowledge about the temperature evolution in the die surface as well as in the inner die, the particles R2 and R3 on the radii (Fig. 3.2.6.17) and D1, D2, and D3 in the middle of the mandrel at a depth of 2 mm, 5 mm, and 11 mm were defined (Fig. 3.2.6.18).

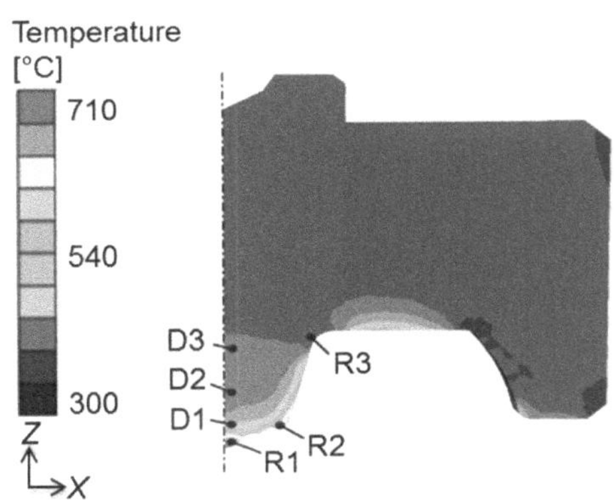

Fig. 3.2.6.15 Temperature field in the die in stationary case [12].

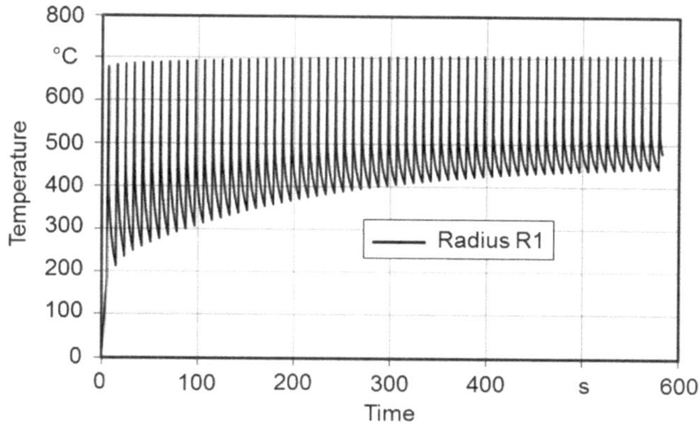

Fig. 3.2.6.16 Temperature-time evolution in radius R1 [12].

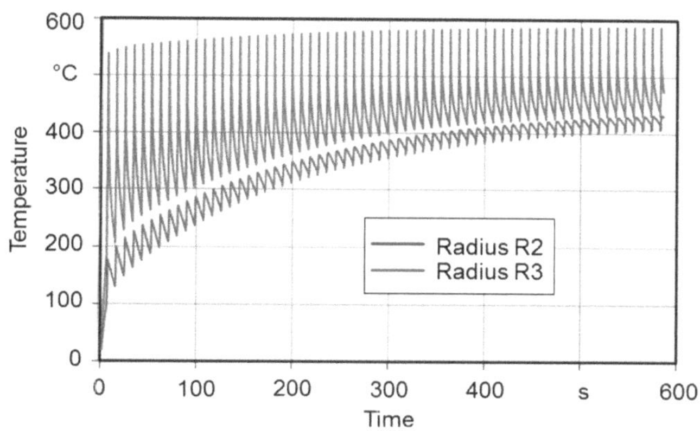

Fig. 3.2.6.17 Temperature-time evolution in radii R2 and R3 (left) [12].

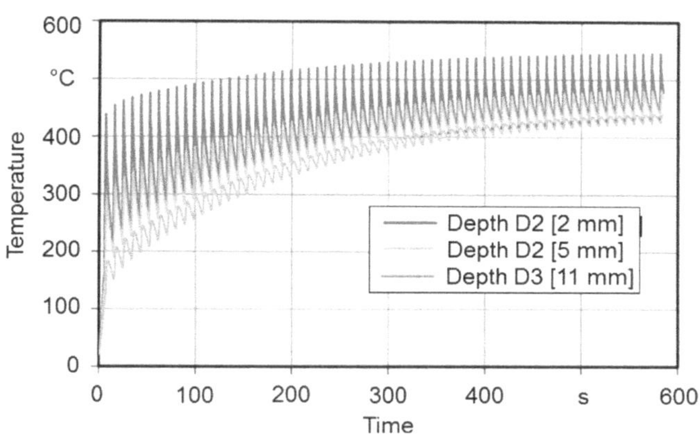

Fig. 3.2.6.18 Temperature-time evolution in depth points D1, D2, and D3 [12].

A comparison of the temperature-time evolution at R2 and R3 showed a great temperature difference between these two in different temperature ranges (Fig. 3.2.6.17). In the first cycle, the difference between T_{max} and T_{min} amounts to 520°C in R2 and 154°C in R3. The reason for the lower temperature and the decrease in the temperature range for the inner points D1–D3 was a temperature loss due to heat transfer. R2 is exposed to maximal temperatures of 590°C in the stationary case. In contrast, a lower maximum temperature in R3 was observed (425°C). The cyclic temperature range in the stationary case in R3 (21°C) was considerably lower than in R2 with 150°C. The reason for the lower temperature as well as the lower temperature range was the comparatively shorter contact time between workpiece and die. Due to the heat transfer of the die material the high temperatures in R1 and R2 lead to a heating of the whole mandrel region. In these regions, the points D1, D2, and D3 were exposed to cyclic temperature loads. The shapes of the temperature curves were very similar to each other. With higher depth the maximum temperatures as well as the temperature ranges in one forging cycle decreased (Fig. 3.2.6.18). At D1 the material was exposed to a temperature of 545°C, at point D2 to 477°C and at D3 to a temperature of 437°C. The temperature range in the stationary case decreased from 111°C at D1, to 36°C at D2 and to 9°C at point D3.

By means of the numerical result of the temperature fields in the dies, an optimized planning and dimensioning of cooling channels could be realized. It was concluded that the inner cooling of the dies needs to be carried out based on the stationary temperature field in the forging dies. To realize a higher cooling rate and to decrease the thermal shock load on the material, the cooling channels must be arranged in a die region with a high stationary temperature but low temperature range.

3.2.6.9 Integration of foreign elements into base powder

Once the temperature field distributions in the forging die were known, it was possible to integrate a cooling channel. Thus it was necessary to study the compaction behavior of powder and foreign element. The powder pressing process is highly nonlinear due to the large deformations and the contact as well as friction behavior. The use of an appropriate mathematical model is important for an accurate simulation of this process. One of the most commonly used models today is the Drucker-Prager Cap (DPC) model. It is evaluated using the continuum approach which has been adopted from soil mechanics. Cylindrical and Brazilian disc compression tests were performed for the characterization of the compaction behavior of metallic powders as well as for the parametrization of the DPC model. Furthermore, the compaction process for the integration of foreign functional elements (e.g., copper spiral) into the base powder was investigated. The interaction between foreign element and metal powder during the compaction as well as the deformation behavior of the foreign element was studied based on these investigations [13].

The computation tool Abaqus/Standard was used to carry out geometric nonlinear, dynamic implicit numerical simulations with a 1-second time period. The modified

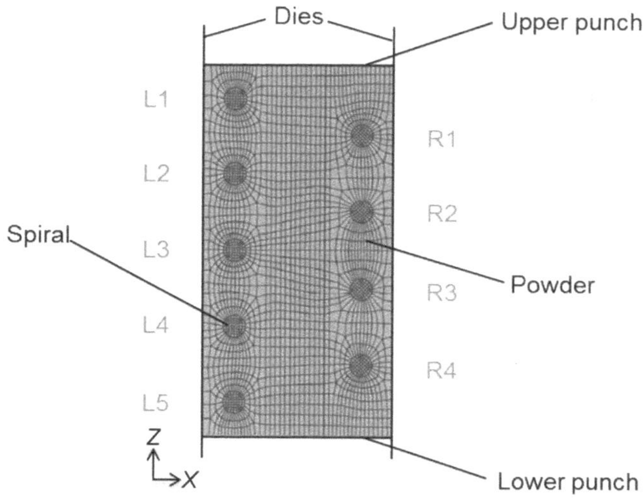

Fig. 3.2.6.19 FE model and mesh of the spiral-powder cylinder [13].

Drucker-Prager Cap model is used to simulate the process. The modified DPC is available as a built-in material plasticity model in Abaqus/Standard.

The FE model consists of a cylinder of powdered steel with copper spirals, a punch on the upper and lower surfaces of the cylinder, and a support die on each of the sides as shown in Fig. 3.2.6.19. Moreover, the numbering used for spiral circles of both, the left (L1–L5) and the right side (R1–R4) of the model, is also mentioned. The die and punches were modeled as analytical rigid bodies. The punches were allowed to move in vertical direction whereas all the degrees of freedom for the die were constrained. The cylinder consisted of the tool steel DIN 1.2343 (X37CrMoV5-1) and the spiral was assigned copper DIN 2.0240 (CuZn15). Surface-to-surface contact was defined and friction was described by using the coulomb model. A friction coefficient of 0.3 was used for the contact between die and powder steel as well as between the punches and the powder, whereas a friction coefficient of 0.5 was used for powder to copper spiral contact.

Fig. 3.2.6.20 provides contours of vertical displacement, von Mises stress, and relative density for the spiral-powder compact. The von Mises contours in this figure show an increase in the stress distribution for spiral circles as they are compressed toward the center of the cylinder. After a compaction of 12.5 mm on both ends of the cylinder, it was observed that the right spiral circles R2 and R3 experience the highest stresses, whereas R1 and R4 show high stress distribution on their top and bottom sides only. For the left side, L2 and L4 undergo the highest stresses, whereas the stress distribution in the spiral L3 (center circle on the left) is almost uniform, since in the vicinity of the center of the cylinder, the deformation is negligible, as can be seen in vertical displacement contours. Higher values of relative density for the powder material were observed above and below all the spiral circles. The lowest relative

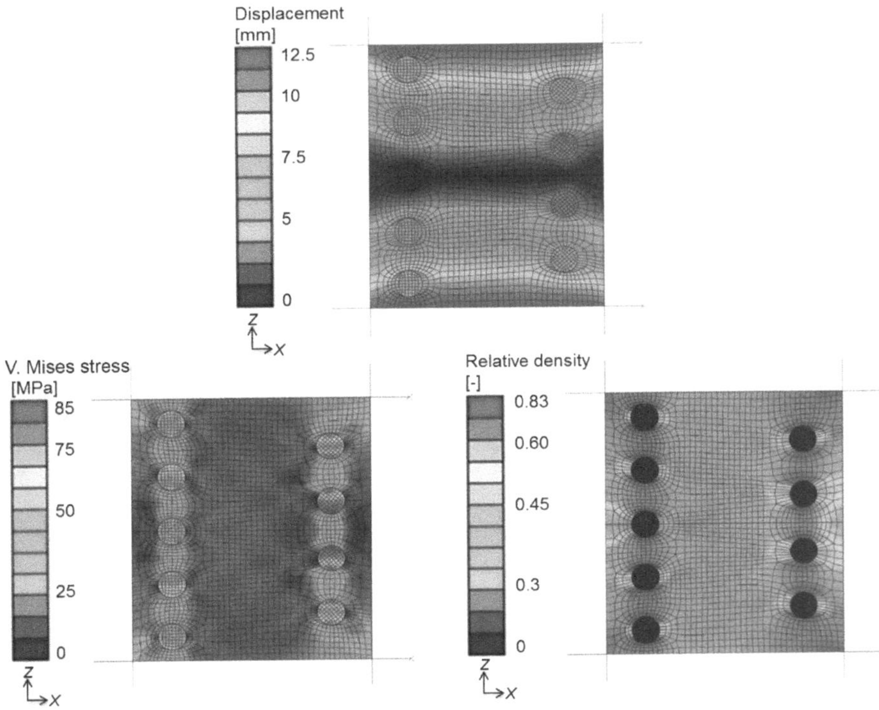

Fig. 3.2.6.20 Contours of vertical displacement, von Mises stress distribution, and relative density for the powder-copper spiral compact [13].

density can be seen along the center of the powdered cylinder, as this region is least affected during pressing and deformation.

3.2.6.10 Numerical investigation of different variations of the forging die with integrated cooling channels

The earlier presented numeric investigations provided a basis for the design and optimization of the size as well as the location of cooling channels. Moreover, the interaction and behavior of foreign elements with powder material during compaction was evaluated. The findings were applied to introduce cooling channels in the investigated channel die. The shape and size of the cooling system were optimized according to the numerical process simulation of thermal loads and mechanical stresses in collaboration with the Institute of Product Development (iPEG) of the Leibniz Universität Hannover (LUH) [14].

Based on the earlier presented investigations, two different design variations of forging dies with cooling channels were conceived. The new designs were numerically analyzed in order to study the mechanical strength of both, the die variations

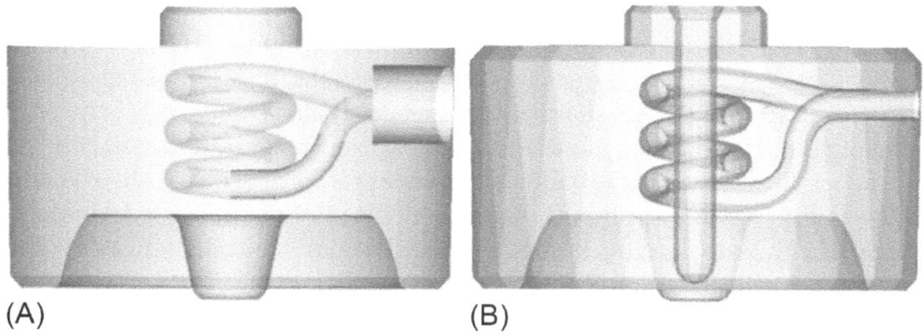

Fig. 3.2.6.21 Geometric variations of the forging die, model A (left) and model B (right) [15].

under operating conditions and their behavior under process boundary conditions [15]. Fig. 3.2.6.21 presents a geometric view of both variations. On the left (model A) is a die with integrated channels only, whereas the right-side figure shows the variation with an additional vertical bar slot at the center (model B) designed as a passive cooling insert with the aim to further extend the cooling effect to the mandrel by means of a high conductivity insert.

In order to determine the mechanical loads in the forging die as well as the workpiece, a mechanical tool analysis was performed. In addition to the elastic material behavior, the interactions between workpiece and tool were also taken into account. The FE model consisted of a lower die, a billet, and an upper die (with channels) as presented in Fig. 3.2.6.22.

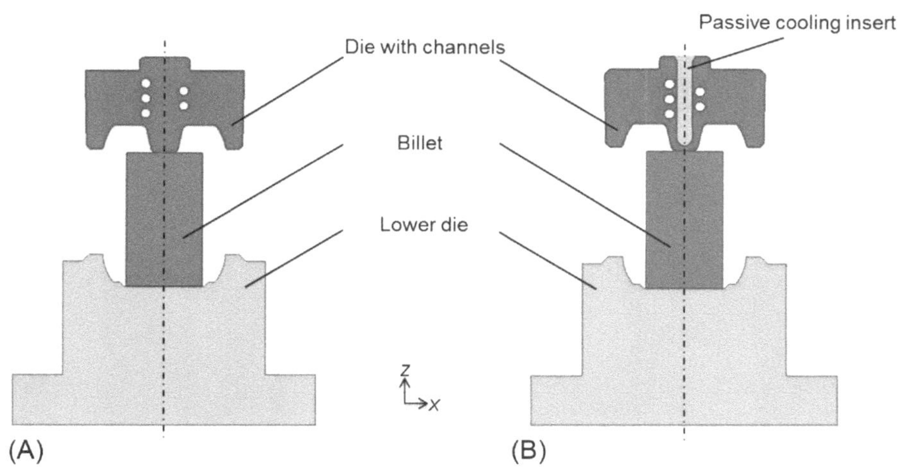

Fig. 3.2.6.22 Schematic model of the simulation process for both design variations: model A (left) and model B (right) [15].

The lower die was modeled as a rigid body. The upper die was modeled as an elasto-plastic material and assigned work tool steel 1.1545 (C105W1, AISI W1). The billet (workpiece) was assigned an elasto-viscoplastic material behavior with steel 1.0503 (C45, AISI 1045). Furthermore, model B also contained a bar slot (passive cooling insert) which was modeled as elasto-plastic material and assigned 1.1545 like the upper die. As used in the earlier simulations for forging dies without channels, 1200°C was assigned as the forging temperature of the raw part (billet). An initial temperature of 250°C was considered for the tool. The combined coulomb-tresca model was used to describe the friction with a friction coefficient of 0.1 and a friction shear factor of 0.3 based on work done by Jeong et al. [11]. A hydraulic press with a downward ($-Z$ direction) speed of 90 mm/s (0.09 m/s) was used to simulate the forging process.

von Mises stress and maximum principal stress for both design variations are shown in Figs. 3.2.6.23 and 3.2.6.24, respectively. It was observed that high values exist only in the vicinity of channel cavities and the curved inner walls of the die (regions a and b, respectively). The rest of the die exhibited $\sigma_{v.\mathrm{Mises}}$ values not higher

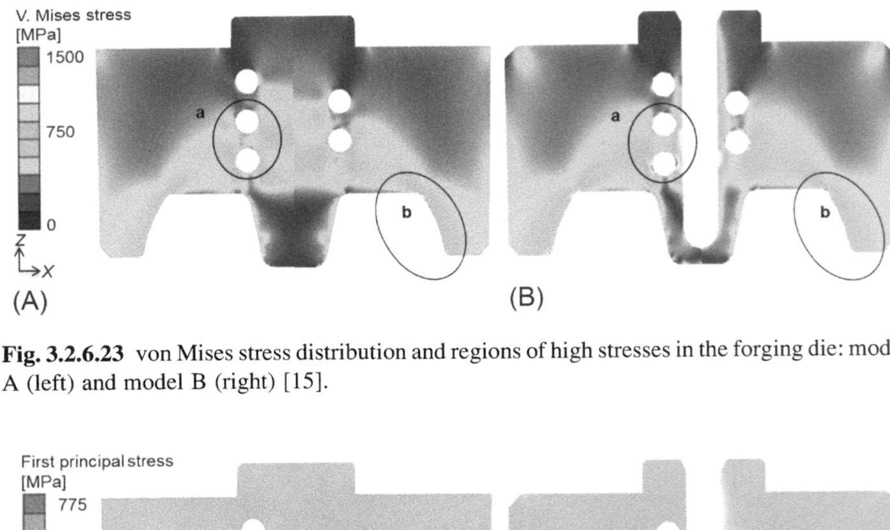

Fig. 3.2.6.23 von Mises stress distribution and regions of high stresses in the forging die: model A (left) and model B (right) [15].

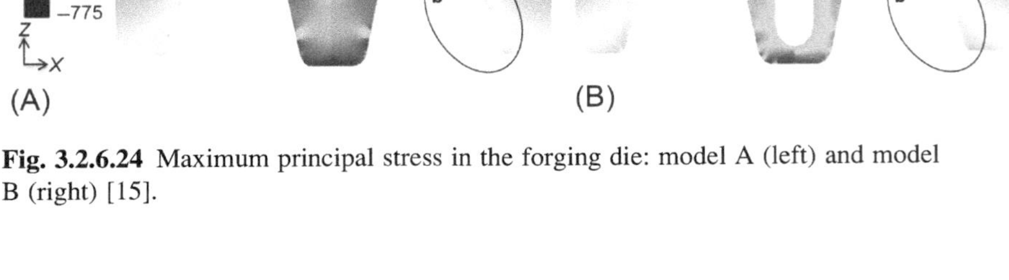

Fig. 3.2.6.24 Maximum principal stress in the forging die: model A (left) and model B (right) [15].

than 500 MPa. The $\sigma_{I,\max}$ was observed to decrease in the mandrel (−635 MPa for model A, whereas −775 MPa for model B) indicating a compression behavior contrary to the outer walls of the die and the channel cavities which experience tension. As mentioned earlier, the yield stress of the die material 1.1545 is around 500 MPa and the ultimate tensile strength is 800 MPa at 700°C. Both the investigated design variations exhibited higher local maximum values for $\sigma_{v.\text{Mises}}$ than the material yield stress, thus micro plasticity of the surface can be expected in the respective regions. Contrarily, stresses in the other die areas were observed to be lower than the yield stress, hence no large plasticization is visible. The critical stress value (maximum principal stress) was found to be below the ultimate tensile strength of the material throughout the process.

This investigation showed that the geometrical dimensions of forging dies can be used without die failure, but an extended use can cause cracks in the vicinity of cavities, particularly in the case of model B (with a vertical slot). Furthermore, it was observed that due to the presence of a vertical slot in the center of model B, it experiences higher stresses than model A, particularly around the channel cavities. It has been concluded that these issues can be resolved by varying the location of the cavities, e.g., at a greater distance from the slot.

3.2.6.11 Summary

In the present work a new forging tool layout is presented. An active influence on the temperature distribution was achieved by integrating internal channels in the forging die. The layout of tempering channels was validated by means of FEM simulations and optimized using detailed process data from a fiber-optic sensor. For hot forging applications it was identified that the channels manufactured by melting out copper elements are unsuitable. Here, an unpredictable distribution of molten copper phases, a highly porous metal matrix, crack formation, and deformation of the copper elements are the major drawbacks.

A modular tool layout consisting of active and passive tempering inserts was identified as a suitable alternative for hot forging applications. Therefore, active and passive inserts were integrated in a die of conventional hot work tool steel. As a result, common hot forging tool layout (material, coating, spray cooling) can be functionally extended by integrating process monitoring and temperature control.

References

[1] Z. Gronostajski, M. Kaszuba, M. Hawryluk, M. Zwierzchowski, A review of the degradation mechanisms of the hot forging tools, Arch. Civil Mech. Eng. 14 (2014) 528–539.

[2] B. Müssig, Temperierung von Schmiedewerkzeugen zur Erhöhung der Bauteilgenauigkeit, Dissertation Universität Hannover, 2002.

[3] B.A. Behrens, F. Schaefer, Prediction of wear in hot forging tools by means of finite-element-analysis, J. Mater. Process. Technol. 167 (2–3) (2005) 309–315.

[4] R. Seidel, Reduzierung der thermischen Schädigung von Schmiedewerkzeugen durch Kühlschmierung, Fortschritt-Berichte VDI, Reihe 2, Fertigungstechnik Nr. 399, 1996.
[5] H. Danninger, Homogenization and pore formation during sintering with transient liquid phase, Powder Metall. Int. 20 (1988) 21–25.
[6] F. Regelsberger, Chemische Technologie der Leichtmetalle und ihrer Legierungen, Springer-Verlag, Berlin, Heidelberg, 1926.
[7] B. Sundman, I. Ohnuma, N. Dupin, U. Kattner, S. Fries, An assessment of the entire Al-Fe system including D03 ordering, Acta Mater. 57 (10) (2009) 2896–2908.
[8] J. Mitrovic, Heat Exchanger and Condenser Tubes, Tube Types-Materials-Attributes-Machining, PUBLICO, 2004.
[9] B.A. Behrens, M. Bonhage, M. Kammler, A. Klassen, N. Vahed, Development of a powder metallurgical self-cooling forging die with inner cavities, Procedia Technol. 15 (2014) 456–464.
[10] T. Kodama, T. Ohmi, M. Iguchi, Microchanneling and lining layer formation by spontaneous infiltration in Fe-Cu system, J. Jpn. Soc. Exp. Mech. 10 (2010) 199–204.
[11] D.J. Jeong, D.J. Kim, J.H. Kim, B.M. Kim, T.A. Dean, Effects of surface treatments and lubricants for warm forging die life, J. Mater. Process. Technol. 113 (2001) 544–550.
[12] B.A. Behrens, A. Bouguecha, S.M. Bouguecha, A. Klassen, M. Bonhage, Conception of a hot forging die out of metal powder equipped with inner cooling channels, Powder Metall. 58 (3) (2015) 166–170, http://dx.doi.org/10.1179/0032589915Z.000000000239.
[13] B.A. Behrens, A. Bouguecha, M. Vucetic, M. Bonhage, A. Zaitsev, I.Y. Malik, Compaction of a copper spiral within a cylindrical die of steel powder and investigation of their deformation behavior, in: Proceedings of the World PM 2016 congress and exhibition, Hamburg, Germany, October 09th–13th, 2016.
[14] B.A. Behrens, R. Lachmayer, I.Y. Malik, M. Bonhage, B. Sauthoff, P. Gottwald, Sensorik in Schmiedewerkzeugen-Temperierung von Gesenken für die Warmmassivumformung steigert Präzision und reduziert Verschleiß, wt-online 10 (2015) 692–696.
[15] B.A. Behrens, A. Bouguecha, M. Vucetic, M. Bonhage, I.Y. Malik, Numerical investigation for the design of a hot forging die with integrated cooling channels, Procedia Technol. 26 (2016) 51–58.

3.3 Transfer of basic research into industrial applications

The aim of this chapter is to show which application areas of basic research topics lying outside the CRC 653 were identified. Several of these were pursued and enhanced toward industrially relevant topics. The vision of gentelligent technologies shows a high potential in the optimization of production processes as well as in process planning concerning the utilization of in-process data. Besides sensory components, the inherent connection of information and the component within the manufacturing process builds a substantial basis for this potential.

As an example, in Chapter 2.1.2 methods for the insertion of information to a workpiece and for reading the information from the workpiece are developed and investigated. A fast tool servo (FTS) was used to mill data into a workpiece's surface during a face milling process in the manufacturing stage of a component.

Further possible application areas are identified for actuated face milling using a FTS, besides the presented application. The investigations in Chapter 2.1.2 show that actuated face milling cannot only be used for information storage but also has the potential to influence the topography and, thus the optics and haptics of the workpiece. The reason for this is that apart from cutting defined micro patterns, the actuated face milling allows roughness and waviness to be generated in a defined way on a workpiece surface. This also leads to an influence on the subsurface of the workpiece, which depicts a further possible application area. Subsurface properties, such as residual stresses, have a positive effect on a workpiece's lifetime and may be specifically adjusted in order to influence the fatigue behavior. This shows high potential for industrial usage.

In order to allow the application of the presented prototypical system from Chapter 2.1.2 in an industrial environment, the tool system needs to be enhanced. Therefore research demand was identified to close the gap between basic research and a possible industrial application. The energy transfer of the presented FTS, in particular, presents a weakness concerning an industrial application. The energy transfer is realized by a slip ring, which requires a connection by cable. Therefore automated handling and changing of the tool are not possible. To counter this, an inductive, i.e., contactless energy transmission has to be realized. This challenge was approached in cooperation with an industrial partner.

The following chapter describes the extension of this earlier approach in detail. This includes the development of the FTS presented in Chapter 2.1.2 and the tool's transfer to a new application area in order to exploit the potential of the actuated face milling.

3.3.1 Defined manipulation of workpiece surface and subsurface properties by extending the field of application of a fast tool servo

Berend Denkena, Thilo Grove, Alexander Seibel, Christopher Schmidt
Institute of Production Engineering and Machine Tools, Leibniz Universität Hannover, Garbsen, Germany

3.3.1.1 Introduction

The following chapters present the results of the design, testing, and application of a new actuated tool system. The research is based on the joint advancement of an actuated tool holder (see Chapter 2.1.2 for reference) in cooperation with the Sauer GmbH, a producer of oscillated cutting tool and laser machining systems. The actuated tool holder is used to generate micro patterns on workpiece surfaces [1,2]. The major drawbacks hindering the industrial application of this tool system are the principle of energy transmission. The currently utilized slip ring system features a stationary ring connected by cable, which forbids automatic tool change. Further issues are a high weight, size, and slip ring wear [3,4]. The use of an inductive energy transmission allows for automatic tool change and handling within the machine tool. This research work is based on a new tool design realized in cooperation with the SAUER GmbH and then validated in order to enable application in an industrial environment.

In a first step, the requirements for the tool are specified. Next, the tool system's new design is presented and the tool is characterized. Once it is determined that the fast tool servo (FTS) meets the stated requirements it is applied to an actuated face milling process to investigate the impact on the workpiece's topography and subsurface. Following that, the tool is used in hammer peening in order to manipulate these characteristics selectively.

3.3.1.2 Actuated tool holder with inductive energy transmission

For the application of a piezo-actuated FTS in an industrial environment several requirements have to be met, which are not fulfilled by a present prototype [1,2]. In order to meet industrial needs, the requirements and applying standards are defined first.

Besides an inductive energy transmission, high application flexibility is targeted. For that matter, a variable use of tools and a wide range of rotational speed up to 18,000 min^{-1} have to be met. The weight and the number of mechanical parts prone to wear must be kept low. Further issues are the static and dynamic behavior, i.e., stiffness and natural frequencies. As the tool is designed for a milling process, it has to be suitable for relevant process forces and vibration loads. Next, the tool design is presented followed by the analysis of the static and dynamic behavior. The tool design that meets the defined requirements is shown in a sectional view in Fig. 3.3.1.1.

The main issue that prevented the industrial application of a prior prototype is the energy transmission using a slip ring. Consequently, an inductive energy transmission is realized. To guarantee a standardized automated handling and a variable use of tools, a HSK-63 tool holder and an ER-20 collet chuck according to DIN ISO 15488 are implemented. In order to reduce the amount of wear-sensitive parts, the mechanical transmission unit contains only few components. Parts that oscillate during operation are made of titanium in order to reduce the moving mass. The rotational speed ranges between 0 and 18,000 min^{-1}. The control of the piezo actuator is realized by a sine signal, which is defined by frequency f_s, voltage U, and offset voltage U_{basic}.

The energy transmission is based on a Villard circuit. This type of circuit is integrated to increase the electric potential in order to avoid negative voltages on the piezo actuator (Fig. 3.3.1.2). The circuit works as a voltage doubler circuit. The capacitor C_2 is charged to the voltage's peak value by the diode D_1 through the negative half wave of the alternating voltage, which leads to an increase of potential.

Next, the tool behavior is analyzed. This tool was designed for static and dynamic process forces that occur in milling of C45, including the structuring processes. In order to reach high dynamics of the actuation (natural frequency $f_0 = 4700$ Hz), the mass of moving parts is reduced to 320 g by using titanium for the vibrating body.

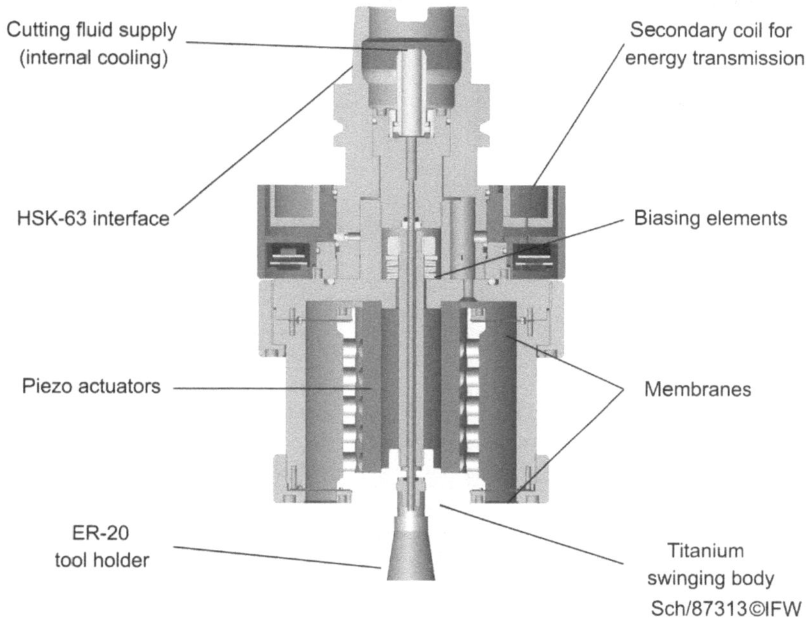

Fig. 3.3.1.1 Sectional view of the piezo-actuated tool holder with inductive energy transmission.

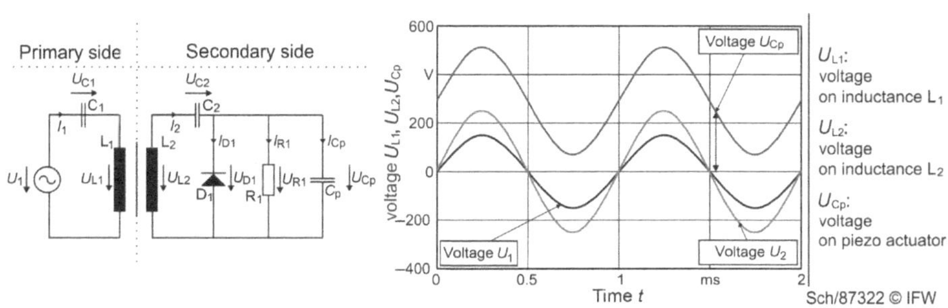

Fig. 3.3.1.2 Circuit diagram from input voltage U_1 to the actuator's voltage U_{cp}.

The tool's static and dynamic behavior are determined using a finite element model (FEM) and also validated experimentally. A laser vibrometer is used to identify the maximum oscillation amplitude of $a_s = 20$ μm for permanent operation, which can be increased to $a_s = 25$ μm in short-term operation. The oscillation frequency covers a range of $f_s = 500$ Hz to $f_s = 1500$ Hz. The balancing quality and the concentricity meet industrial requirements. The balancing quality of the designed tool meets the requirements of DIN ISO 1940 up to a rotational speed of 18,000 min^{-1}. The

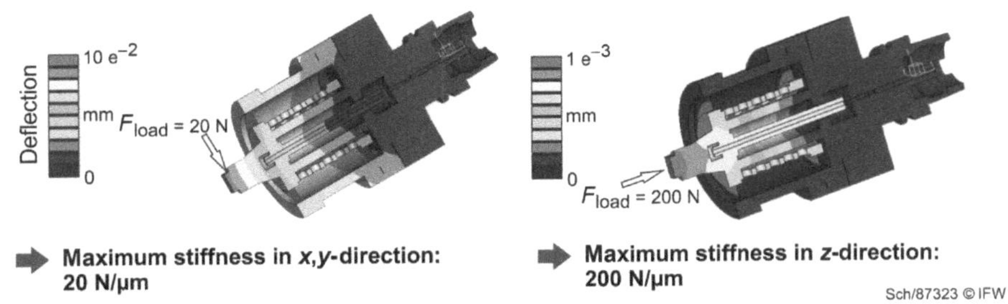

Fig. 3.3.1.3 Stiffness simulation of the system.

concentricity of the tool reaches 0.05 mm. The concentricity of the ER-20 collet chuck shows the same concentricity of 0.05 mm in its inside cone.

The system's stiffness is simulated using a FEM simulation and experimentally measured subsequently. The simulated stiffness in x- and y-direction is 20 N/µm whereas the experimental measurement yields 19 N/µm. The stiffness in x- and y-direction shows small differences to the simulated values shown in Fig. 3.3.1.3. The stiffness in z-direction reaches a value of 200 N/µm in the simulation and 160 N/µm in the measurement. This is a difference of 20% between simulation and measurement, which can be explained with the piezo actuator's stiffness depending on the input voltage. The stiffness values are measured for an unloaded actuator. If the actuator is loaded with an offset voltage, its stiffness as well as the complete system's stiffness consequently increase.

The system's response characteristics are determined by stimulation with a sine sweep, resulting in a natural frequency of $f_0 = 1700$ Hz for the complete system (Fig. 3.3.1.4A). Compared to this, a natural frequency of 2200 Hz can be measured when the piezo actuator is directly connected to the voltage source without inductive

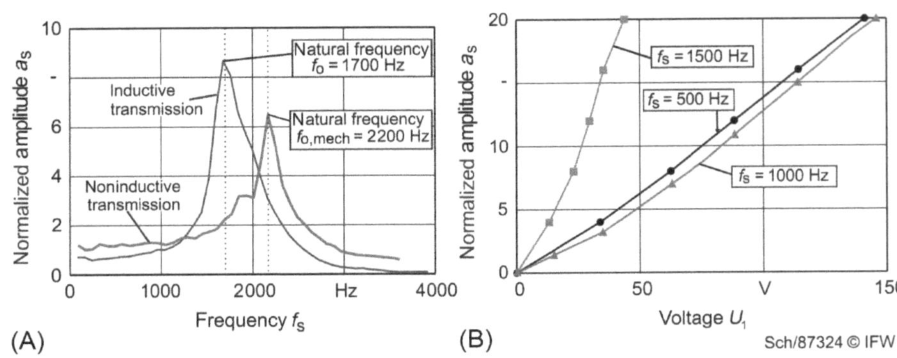

Fig. 3.3.1.4 (A) Frequency response with and without inductive energy transmission and (B) realizable amplitudes depending on the input voltage U_1.

energy transmission. Therefore it is to be noted that the inductive energy transmission has a high influence on the natural frequency f_0 of the complete system. The difference can be seen in Fig. 3.3.1.4A. The inductive energy transmission used here is designed for a frequency of 1600 Hz. Thus there is only a small deviation of $\Delta f = 100$ Hz toward the system's natural frequency of $f_0 = 1700$ Hz.

Additionally, the system's response behavior is investigated using the oscillation amplitude a_s as shown in Fig. 3.3.1.4B. The amplitude increases near the natural frequency f_0 at $f_s = 1500$ Hz. In contrast to this, there is only a small difference a_s between the amplitudes for operation frequencies f_s significantly below the natural frequency f_0 (here: $f_s = 500$ Hz; 1000 Hz).

Besides the static and dynamic behavior of the tool, the temperature generated during operation is measured. A rapid heating up to the critical piezo actuator's temperature of $T = 80°C$ occurs for an oscillation amplitude of $\Delta l = 20$ µm and a voltage of $U_1 = 146$ V on the primary side of the electrical circuit. Thus a permanent operation is limited to a duration of 80 s. In the case of further developments of the tool system, an improved heat coupling of the piezo actuator to the tool casing has to be guaranteed for better heat dissipation. Next, the new FTS is investigated toward its impact on the subsurface and surface quality of workpieces.

3.3.1.3 Impact of oscillated face milling on the subsurface and surface quality of the workpiece

In order to evaluate the influence of the oscillated face milling on the subsurface, the formation of residual stresses in this zone is investigated. An oscillated face milling with a double-edged face milling head is conducted on the four materials Al7075, C45, 42CrMo4, and TiAl6V4. In the experimental design, the oscillating frequency and the tooth feed are varied in two steps with a constant oscillating amplitude of $a_s = 20$ µm. Fig. 3.3.1.5 exemplarily shows the results of machining 42CrMo4.

As a reference, a conventional process with a feed of $f = 0.6$ mm without oscillation is carried out. The nonoscillating process generates tensile residual stresses. The principal residual stresses for this process reach $\sigma_1 = 346$ MPa and $\sigma_2 = 155$ MPa. Reducing the feed to $f = 0.4$ mm leads to residual stresses of $\sigma_1 = 191$ MPa and $\sigma_2 = 24$ MPa as shown in Fig. 3.3.1.5.

Next, oscillating tests are conducted. As a result, the surfaces' principal residual stresses are clearly reduced on both steel materials (C45, 42CrMo4) compared to the reference surface. Fig. 3.3.1.5 shows an example based on the principal residual stress measurement at 42CrMo4. The same behavior, only less intense, is shown for Al7075 and TiAl6V. The change due to the oscillation shows a frequency-dependent behavior, while no dependence on the feed per tooth f_z appears. The stronger influence toward compressive residual stresses depending on the frequency could be caused by the higher kinetic energy during the process. A significant change in the principal residual stresses' direction cannot be shown for all materials.

In addition to surface-near residual stresses, exemplary measurements of residual stress depth profiles are conducted on C45 as given in Fig. 3.3.1.6. The residual stress

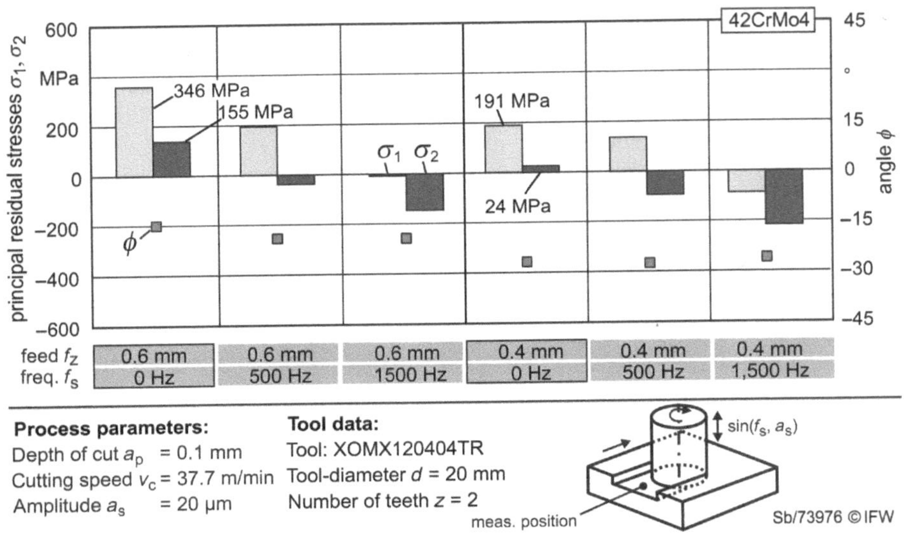

Fig. 3.3.1.5 Principal residual stresses depending on the oscillation frequency during face milling.

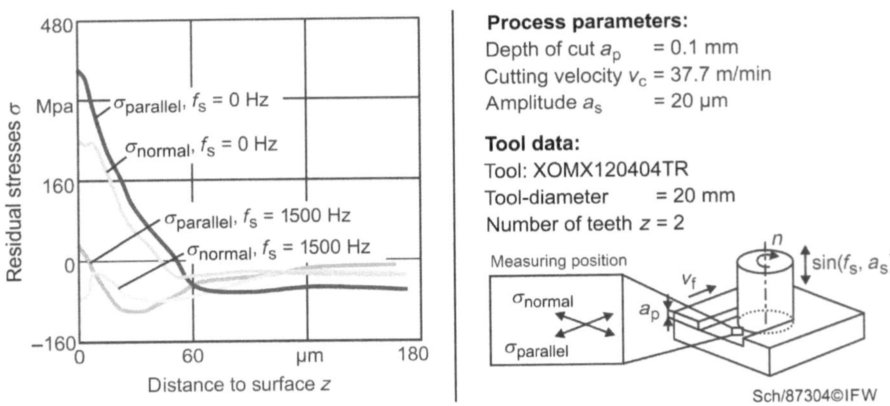

Fig. 3.3.1.6 Course of residual stresses after oscillated face milling of C45.

depth profiles of the three other workpiece materials show a similar course to Fig. 3.3.1.6. Here, residual stresses are investigated in a parallel and normal direction to feed. The results show an influence of the milling on the residual stresses up to a depth of about 60 µm. Directly at the surface, the residual stresses reach sum $\sigma_{\text{parallel}} = 380$ MPa parallel and $\sigma_{\text{parallel}} = 240$ MPa normal to the feed direction after conventional milling. Compared to this, the residual stresses after oscillated

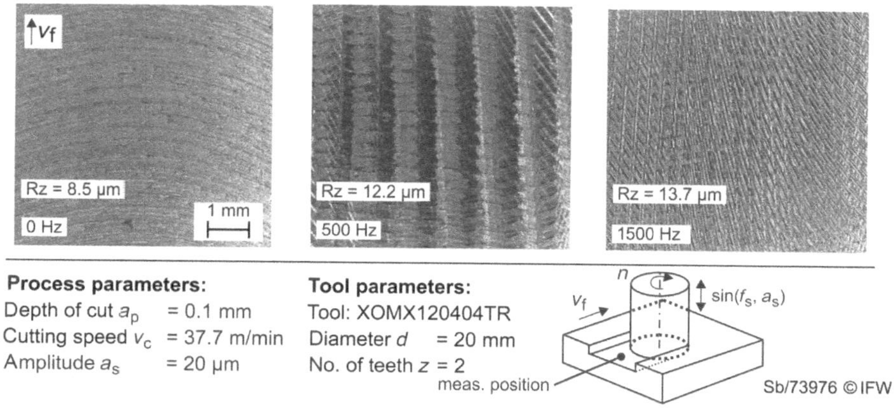

Fig. 3.3.1.7 Change of surface roughness due to increasing oscillation frequency.

machining with $f_s = 1500$ Hz reach $\sigma_{normal} = -78$ MPa parallel and $\sigma_{parallel} = 32$ MPa normal to the feed direction.

Besides the residual stresses, the generated surface roughness, after oscillated face milling is measured. The resulting surface roughness is influenced negatively with an increase of frequency by the oscillated face milling. In the process, the surface roughness Rz increases up to 62% compared to the reference surface, which is shown exemplarily in Fig. 3.3.1.7.

The conducted research work shows that, in general, the actuated face milling influences the residual stresses and the surface roughness of a workpiece.

3.3.1.4 Defined manipulation of the subsurface by hammer peening

It can be seen that the oscillating milling process influences the subsurface quality. Following this result, experiments are conducted in order to generate defined residual stresses in the workpiece's subsurface. For this investigation, tests on hammer peening are conducted with the oscillating tool. For this purpose, the FTS tool has to be changed concerning the tool holder.

The tool holder is equipped with a cemented carbide ball, which is moved over the workpiece surface in parallel paths while oscillating axially (see pictogram in Fig. 3.3.1.8). This leads to grooves, which are varied in their coverage ratio on three levels by changes of the feed velocity and the stepover distance on all four workpiece materials. The oscillation frequency is kept constant. Moreover, the impact energy is varied on three levels by changes of the distance of the sine wave's zero crossing to the workpiece surface. The results are given in Fig. 3.3.1.8.

The process is carried out on a face-milled surface. The hammer peening is performed on all four workpiece materials and the machining parameters are kept

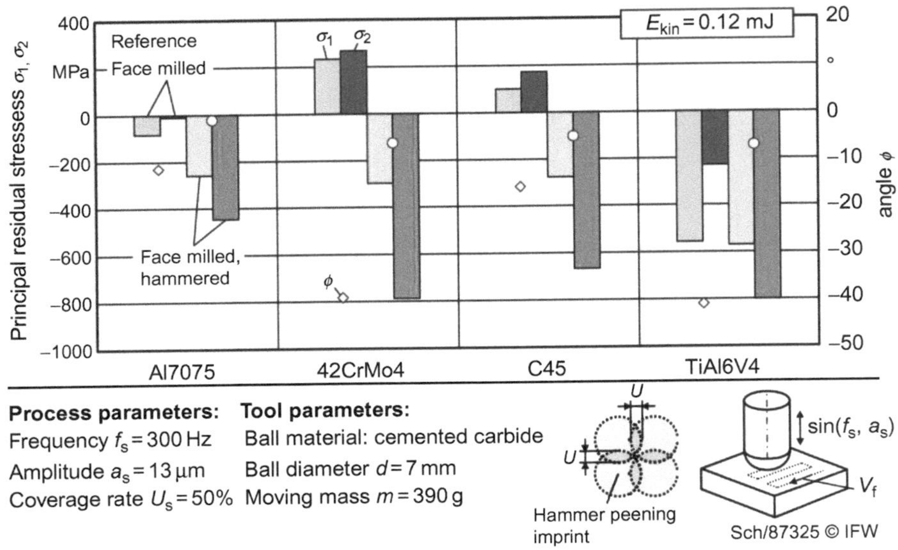

Fig. 3.3.1.8 Change of the principal residual stresses due to hammer peening.

constant for all materials. As a result of the material deformation, clear changes in the principal residual stresses toward pressure are identified.

Compared to the initial state, the change toward compressive residual stresses reaches differences of up to 1050 MPa. The forces of the hammer peening toward the surface reach a maximum of 160 N. Although the surface residual stresses are affected intensely with increasing coverage ratio, these changes are relatively low if they are related to the overall change compared to the initial state.

The surface roughness is also positively affected. The feed marks of the face milling are nearly leveled, as shown in Fig. 3.3.1.9 for Al7075. The Rz value shows a clear dependence on the coverage ratio, which is most distinct for the steel materials. Independently of the initial state, all materials clearly move toward a limit value. The kinetic energy only shows a linear improvement for the steel materials with a high initial roughness. The titanium and aluminum alloys show an improvement for the lowest energy. This improvement in roughness cannot be raised with increased energy and partly shows deteriorations.

3.3.1.5 Conclusions

This chapter investigated the influence of FTS-assisted face milling on the workpiece's topography and subsurface on different workpiece materials in cooperation with the industrial partner SAUER GmbH. The aim was to transfer basic research presented in Chapter 2.1.2 to industrial applications, a new tool concept with an

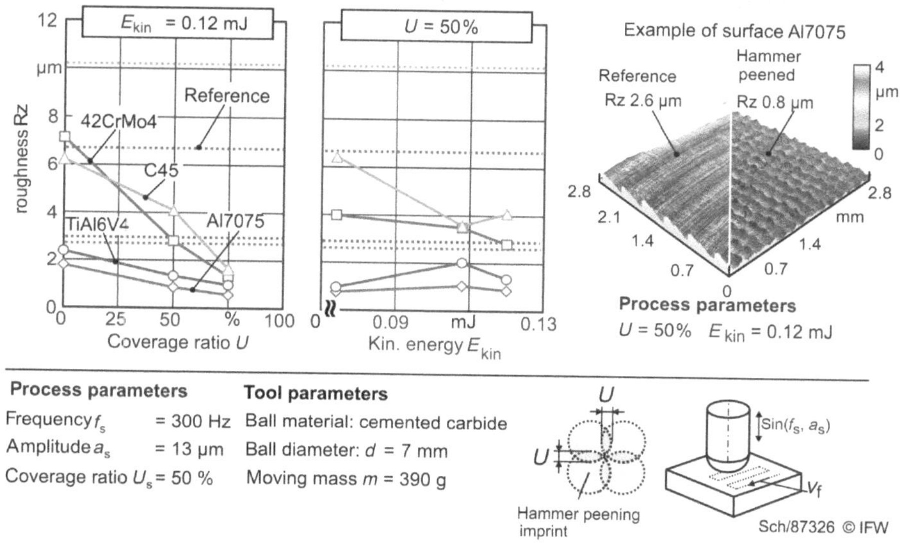

Fig. 3.3.1.9 Change of surface roughness due to increasing coverage ratio and kinetic energy.

inductive energy transmission being presented for this purpose. The new tool was characterized statically and dynamically in order to show its behavior. As this tool was used for oscillating face milling, it was shown that the chosen process parameters influence not only on the subsurface residual stresses of a workpiece but also its surface topography. The subsurface quality of a workpiece can be influenced by changing residual stresses by means of an oscillated milling process. Based on these findings, tests were conducted in order to selectively modify the surface roughness and the subsurface of a workpiece. For this, it was necessary to modify the FTS to enable a hammer peening process. It was shown that the process parameters can be used to influence the workpiece selectively. These results became possible due to the cooperation of academic and industrial partners and demonstrate the transfer of a basic research topic to industrial applications.

References

[1] B. Denkena, J. Köhler, A. Seibel, Experimental analysis of cutting forces in actuated face milling of micro patterns, Procedia Technol. 15 (2014) 424–430.

[2] J. Köhler, A. Seibel, FTS-based face milling of micro structures, Procedia CIRP 28 (2015) 58–63.

[3] B. Denkena, H. Henning, L.E. Lorenzen, Genetics and intelligence: new approaches in production engineering, Prod. Eng. Res. Dev. 4 (1) (2011) 65–73.

[4] B. Denkena, J. Köhler, T. Mörke, O. Gümmer, High performance cutting of micro patterns, Procedia CIRP 1 (2011) 144–149.

Applications in the product life cycle

4

Today, a steady tendency of transitioning from centralized management processes to decentralized models of collecting data, information processing, and decision-making is present. Additionally, the performance level and the autonomy of decentralized systems are growing continuously. Ultimately, such systems become active system components, capable of managing their manufacturing processes and of monitoring their operating conditions. Hence, it is possible to recognize major changes that affect many industries. Already the principles of modularity, the ability of transmitting and storing data, autonomy and flexibility, as well as the individualization of a product are at the forefront in the design of products. The feedback from the product user and the exchange of information between designer, manufacturer, and user play an important role. This exchange involves the systematic monitoring, collection, storage, and processing of data from production and usage and a consistent information support of the product throughout its life cycle. Thus the effective implementation of high potential ideas of "Industry 4.0" requires the development of new types of products or components, so-called gentelligent or smart components.

For this purpose, information and communication technologies which enable the acquisition, the inherent storage, and the exchange of information along the different phases of the product life cycle have been developed within the CRC 653. The presented approaches extend the definition of "Industry 4.0" by enabling the collection, storage, and processing of product data over the entire product life cycle, including not only production but also the usage and the development phase (Fig. 4.1). This represents the preconditions for the development of new product generations based on collected data from the production and usage of previous generations of the product. Here, a new engineering approach is required, referred to as *Technical Inheritance* (see Chapter 4.1), for the effective development of such systems. An appropriate way of developing technical systems and related technologies of data analysis and the analysis of the condition of a product are presented in Chapters 4.2 and 4.3. The corresponding processing systems provide data collection, analysis, and conditioning of the incoming data. Thereby, information on the current state of the products is obtained. Furthermore, based on this information, subsequent product generations are adapted to the specific loading during operation.

Cyber-Physical and Gentelligent Systems in Manufacturing and Life Cycle. http://dx.doi.org/10.1016/B978-0-12-811939-6.00004-2

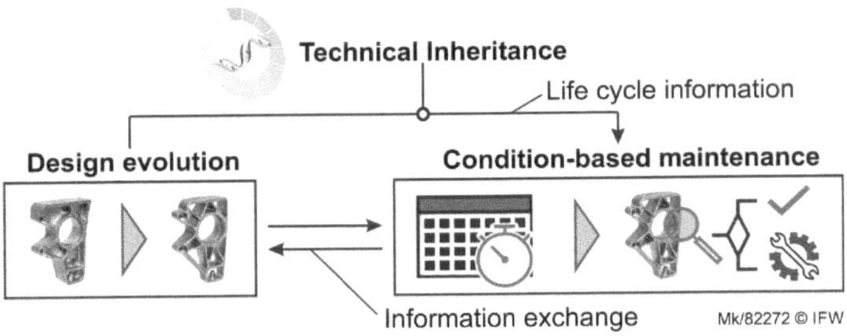

Fig. 4.1 Passing on life cycle information for an advanced design evolution and a condition-based maintenance.

See the following chapters for further information

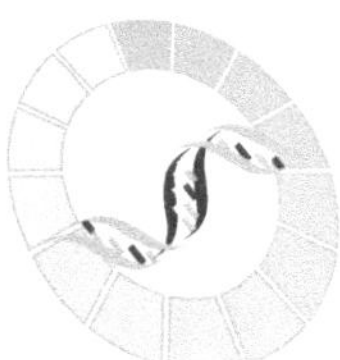

4.1 Paradigm of technical inheritance

*Iryna Mozgova**
*Institute of Product Development, Leibniz Universität Hannover, Hannover, Germany

4.1.1 Introduction

The approach of technical inheritance (TI) is based on the laws of biological and technical systems as well as the conduction of a comparative analysis of the development of technical systems. The CRC 653's research shows, that according to the biological principles of the transfer of hereditary information, the process of TI for smart products can be described on the basis of evolutionary mechanisms. TI is defined as a transfer of assembled and verified information from production and application to the next product generation [1,2]. The purpose of TI is the development of new generations of a product adapted to new or changed requirements based on information on the development, production, and exploitation of the existing generations of the product. Thus it is valid to speak about an evolution of technical products. Technical evolution is defined as a process of controlled, stepwise, and continuous change of technical systems, products, and processes as well as models with the aim to adapt to influences and requirements. Within the scope of paradigm of TI [1] the evolution of populations of technical products is observed. Every population consists of all generations of individuals of a technical system, product, and process as well as a model at the current time. Whereas, a generation is a group of individuals with the same level of development. The individual is the smallest considered technical system, product, process, or model in a population. A special role is given to tools, methods, and facilities of collection, storage, exchange, and analysis of data, as well as the algorithmic feedback of information collected by or stored in smart products to the production process and the development process. For this purpose the provided technologies are integrated in the closed loop of the phases of the design, production, usage, and return of information (Fig. 4.1.1).

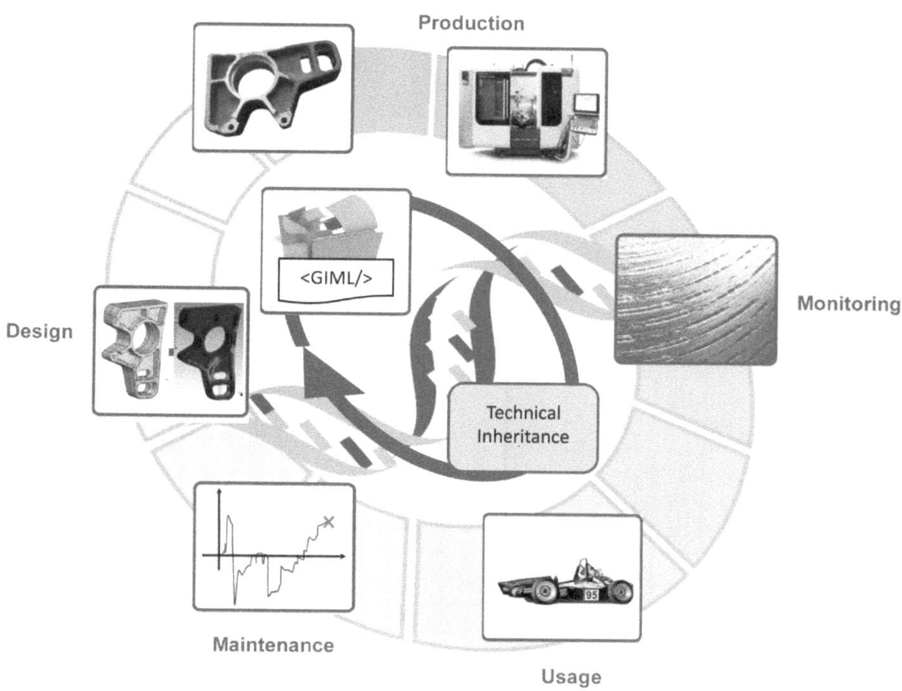

Fig. 4.1.1 Phases of technologies for the algorithmic feedback of information [2].

Each product generation generates its information during the phases of development, production, and usage. In order to provide efficient information management over the product life cycle, a unified GIML data exchange format has been developed by the CRC 653 [3].

The important aspect of the evolution of technical systems or products is an individual identification and the assessment of full information throughout the phases of development, production, and operation. For this, the concept of a product's genetic code (GC) is applied in TI. The genetic information summarizes data, which is necessary to identify or reproduce the product. It includes vital information such as geometric descriptions and information about materials. The genetic information can be stored as static, unchangeable data in the product and may have been inherited from previous product generations.

The GC of a product includes several parts and levels of parametrization. Its general scheme and characteristic information of a product are shown in Fig. 4.1.2.

The code elements, which describe the product in its design phase, are Alloy GC and Design GC. Design GC includes information on the parameterized skeleton, a fixed number of design areas and connection surfaces. These elements are the basic

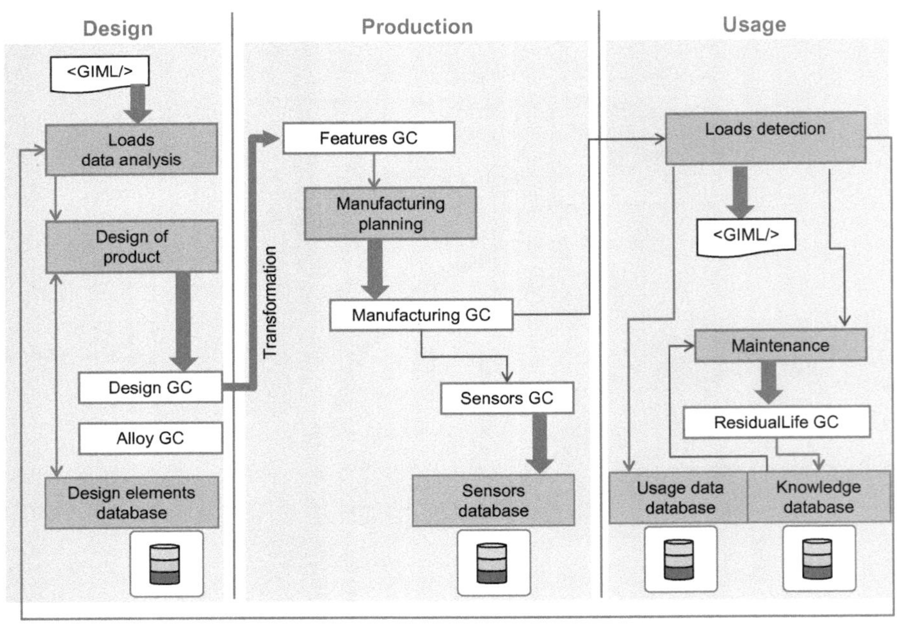

Fig. 4.1.2 Genetic code of a smart product.

components of the developed Generative Design Approach, which is described in detail in Chapter 4.1.1.

Separate design areas can be represented as objects of inheritance. Provided that the requirements corresponding to these areas as well as loads received during the life cycle will remain invariable, these are transferred to the next product generation. To manufacture a product, the Design GC has to be transformed into a Features GC. Information about the product manufacturing, used sensors, and the forecast residual life are stored in the Manufacturing GC, Sensors GC, and ResidualLife GC, respectively.

ResidualLife GC includes information about the load change number required during the product's life cycle, the load change number after the manufacturing process, the current load change number in the usage phase, and the estimated number of load changes as the result of a condition-based maintenance algorithm. The GC, unlike the full data about the product life cycle, is the short structured characteristic information obtained by means of specially developed data processing techniques. It is possible to uniquely identify and retrieve important information about the product's design, production, and operation based on the GC of the product. Data collected throughout its life cycle, for example, various loads during operation conditions, may be stored in the GIML format.

4.1.2 Methodologies of data analysis and information processing

The design optimization uses real life load information from the product's life cycle, provided by gentelligent or smart products (see Chapter 2). The subsequent design evolution elaborates different optimization strategies. These allow an adaptation of the shape of mechanical components. Here, the generative design approach in combination with a genetic algorithm for the objective represents the most suitable solution (see Chapter 4.2). This includes shape modifications in the topology as well as in the parameters of the products design, which also consider the manufacturing and functional restrictions at any time.

Taking into account the throughout the life cycle of the product received information about real loads, the products could be adapted to their actual environment. Thus the development of new generations of products is carried out. The evaluation of load data, detected during manufacturing and usage, may be used for the design and planning of maintenance processes and strategies. Due to the increasing productivity and cost pressure as well as complexity and susceptibility to failure, both resulting from a progressive technological development, maintenance becomes increasingly important for manufacturing companies and the achievement of their corporate objectives. For maintenance, three basic strategies may be discerned: preventive, reactive, and condition-based maintenance. In terms of modern condition-based maintenance, cost savings of up to 30% are expected. These approaches aim a safe consumption of the entire usage reserve. Unlike the mentioned approaches, a methodology for product status-driven maintenance was developed. The suggested methodology requires a continuous evaluation of maintenance-related data to determine the current product status. This also implies a reliable forecast of the residual life, which is a necessary premise for the optimal utilization of the product's usage reserve.

Previous approaches of condition-based maintenance (e.g., condition monitoring) target primarily abrasion as a wear out type and do not consider the material fatigue due to an insufficient cost-benefit ratio. Since fatigue-related failures usually occur spontaneously and unpredictably and thus have unplanned downtime consequence the relevance of fatigue is not negligible and is focused in the developed approach of product status-driven maintenance (see the Chapter 4.3).

4.1.3 Conclusion

Technical products equipped with sensing technologies can be enabled to inherently collect data about their production process and operation conditions. This offers new opportunities for the processes of product development, production, and usage and unfolds new opportunities to increase product individualization as well as the adaptation to user needs and operating conditions. Chapters 4.2 and 4.3 present advanced methods and tools, which allow the design and monitoring of gentelligent or smart products as well as the recognition of relevant information for development of a next generation of the product.

References

[1] R. Lachmayer, I. Mozgova, P. Gottwald, Formulation of paradigm of technical inheritance, in: Proceedings of the 20th International Conference on Engineering Design (ICED15), Milan, Italy, 2015, vol. 8, 2015, pp. 271–278.

[2] C. Demminger, I. Mozgova, M. Quirico, F. Uhlich, B. Denkena, R. Lachmayer, P. Nyhuis, The concept of technical inheritance in operation: analysis of the information flow in the life cycle of smart product, Procedia Technol. 26 (2016) 79–88.

[3] B. Denkena, R. Lachmayer, J. Ostermann, P. Nyhuis, Datenaustausch in den verschiedenen Phasen des Lebenszyklus von smarten Produkten, in: VDI IT & Production, 2016, Ausgabe 1–2, 2016.

4.2 Algorithmic design evolution based on product life cycle information

Roland Lachmayer, Iryna Mozgova, Bastian Sauthoff, Philipp Gottwald, Wieben Scheidel
Institute of Product Development, Leibniz Universität Hannover, Hannover, Germany

4.2.1 Introduction

The development and application of cyberphysical systems and smart products as well as the cross-linking of machine tools and products in context of "Industry 4.0" offers new opportunities for the design processes in product development [1–3]. The gentelligent technology provides a lot of possibilities for the developing of smart products or machines. With the combination of genetics and intelligence, gentelligent components represent one of the basics to create smart technical systems, but even more system of systems. These mechanical parts are enabled to sense life cycle data inherently [4]. By this innovative technology new possibilities are realized to get suitable information of technical systems of various disciplines in engineering design during the product life cycle with the focus on its application [5]. Life cycle information about technical systems could be applied in different ways. One practice represents the coupling of such information with the methodology of X-in-the-Loop [6]. Another example for using life cycle information is the linkage between collected data sets and product management systems [7]. Beyond life cycle information supports an integrated design engineering [8].

In this elaboration the focus is on an approach of an algorithmic data feedback from gentelligent components to influence the next-generation development. In contrast to the nature-inspired process model [9] as well as the autogenetic design theory [10] the investigations about the here so-called technical inheritance represents an intergenerational development process [3,11]. In context of the investigation the development process is analyzed and adapted. Therefore the included development methods and tools are elaborated, expanded, or redeveloped. Also different specification techniques have been evaluated and the most suitable technique, the Unified Modeling Language (UML) activity diagram, was applied for the process description. Moreover for the achievement of a goal-oriented product life cycle information feedback methods of data mining are investigated. The consideration of such methods directly in phases of product development is required to reduce the data volume and to extract the development-relevant information. In addition, different statistic operations are applied for the experience of gentelligent components during the usage phase. The theoretical results are transferred to the demonstrator of the CRC, a race car from the Formula Student of the Leibniz Universität Hannover. The collected and prepared life cycle information are used in the end to adapt the next product generation by a design evolution. As a part of this investigation different kind of optimization strategies are analyzed and applied. With the combination of process analysis, statistic operation with data mining methods, and the design evolution, the technical inheritance with an algorithmic data feedback of gentelligent life cycle information is realized to adapt smart components to their real environment.

4.2.2 Methodology

In this investigation an algorithmic data feedback of life cycle information into the phases of product development is in focus. This includes the assumption that smart products like the race car equipped with gentelligent technology produce and collect various data in the whole life cycle. In that way the intergenerational development process is modeled, which constitute the technical inheritance of intelligent and linked products in context of "Industry 4.0." The simplified approach of this process model is depicted in Fig. 4.2.1 [12,13].

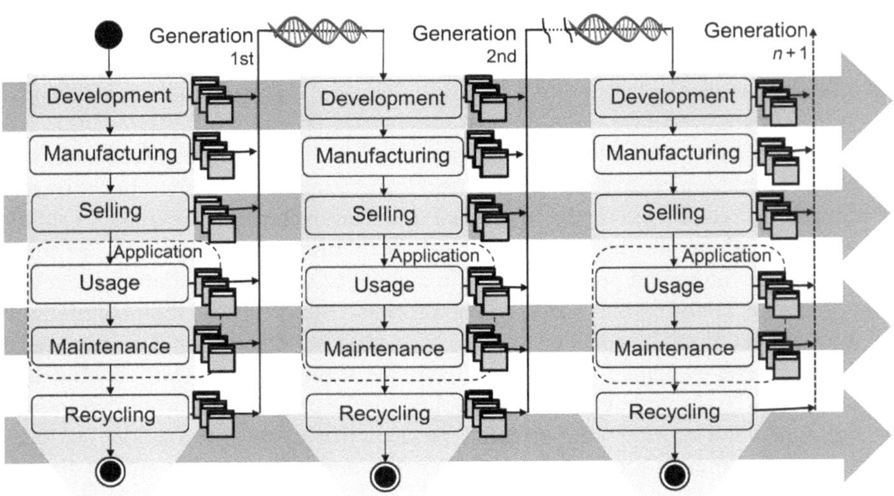

Fig. 4.2.1 Approach of intergenerational development process [12].

In the consideration of the development of this process a goal-oriented algorithmic data feedback is realized, which contains statistic methodologies and operations as well as a design evolution for product adaption in the development process. The process for an intergenerational development represents a model for the adaption of smart components or system of systems by the consideration of their life cycle experiences. Therefore the applied semantic in this context is defined as follows [14]:

Technical inheritance is a transfer of assembled and verified information from production and application to the next product generation.

The adaption of products by analyzing product life cycle while taking evolutionary mechanism into account is defined as *design evolution*.

An *algorithmic data feedback* means a goal-oriented monitoring of products implementing methods of data mining in order to extract life cycle data. These data are attributed to the development process of the next product generation.

4.2.2.1 Technical inheritance

The technical inheritance with the algorithmic data feedback regarding the implementation of the intergenerational development process is divided into the following three aspects. The three working packages are the "Design of Gentelligent Components," "Statistic Operator," and "Design Optimization." The approach of the technical inheritance for gentelligent components is depicted in Fig. 4.2.2 [15].

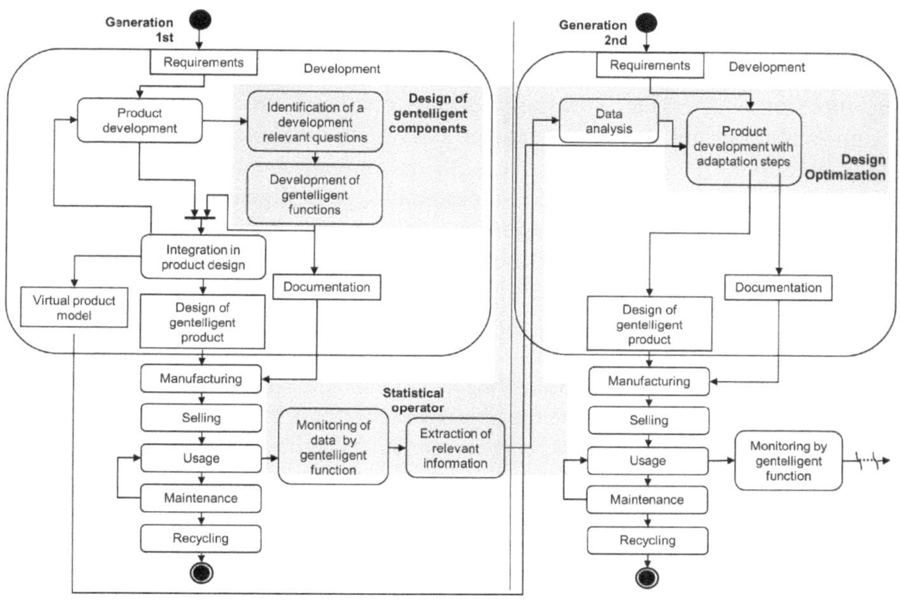

Fig. 4.2.2 Approach of technical inheritance.

In the working packages "Design of Gentelligent Components" methodologies for developing with the gentelligent technology as well as the specification of the process model are investigated. Moreover, approaches to transform the inherent gentelligent product data sets to useful life cycle information for the developer of next product generation were analyzed. Therefore the different life cycle information as well as the information flows is evaluated and specified. Thereby the goal-oriented data feedback forms the base for the intergenerational development process.

In the second working package "Statistic Operator" investigates various applications of data mining methods. The objectives are Identification and Classification of the development-relevant information in the huge life cycle data sets as well as the reduction of the data volume. Additionally bionic-based algorithms are elaborated and enabled for an optimal product-specific data analysis. Beyond the extracted life cycle information are linked with the product representing models of product development [16].

The collected, extracted, and transmitted life cycle data forms the base for the third working package, the "Design Optimization." Different kinds of optimization strategies like, for example, topology or parametric optimization are elaborated. In the context of optimization the generative design models are developed. This optimization method allows adaption of mechanical components in the concept as well as in the design phase of development process with the same product representing models. By the application of nature-based algorithms the product adaption to its real environment is realized [17].

4.2.2.2 Task

The three aspects of the working packages complete the investigations of the design evolution and the subsequent process of the technical inheritance for an intergenerational development process. With innovations about the development in the course of "Industry 4.0" like, for example, gentelligent technology, a lot of different aspects of a technical system could be monitored and measured. In consequence of the different possibilities, e.g., in telecommunication or automotive engineering, the objective for the data acquisition changes. By this way various applications, like the detection of failures or the reliability of products, according to Haken, are available [18].

The technical inheritance examines a goal-oriented data feedback into the phases of product development. In this case the objective represents the adaption of the components shape in consideration of its real environment. Thereby the following challenges are solved:

- By means of which methodology, an intergenerational development process could be specified, analyzed, and its information flow being visualized?
- Which algorithms support different approaches of data mining for, on the one hand, the reduction of the data volume and, on the other hand, an identification of product-specific development-relevant information?

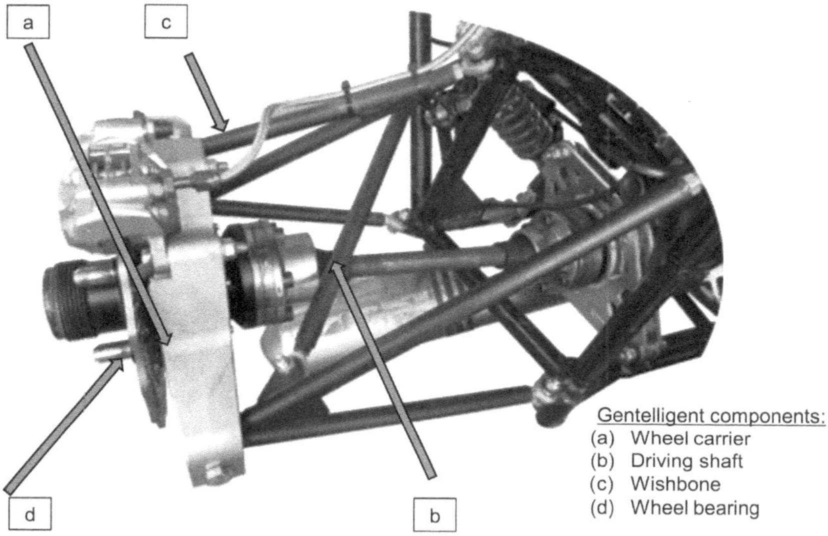

Fig. 4.2.3 Wheel suspension of a race car with its gentelligent components [12].

- In which ways life cycle information could be integrated into optimization strategies in order to adapt mechanical components optimally to its real environment?

These challenges and the affiliated solution approach are theoretically explained in the next chapters. Moreover, a practical example of the technical inheritance including the applied methods and tools is presented with a demonstrator of the wheel suspension of the Formula Student race car.

4.2.2.3 Demonstrator

The chosen wheel suspension is part of a race car from the Formula Student Team "HorsePower" of the Leibniz Universität Hannover. It was developed in the year 2009. The type of the suspension is a double wishbone. Here in this assembly various components are examined. The rear wheel suspension of the race is equipped with gentelligent technology, depicted in Fig. 4.2.3.

In this wheel suspension four components are enabled to collect life cycle data. This includes data sets from manufacturing process, experimental data from racing evaluations as well as from maintenance. With the application of this demonstrator the results of the analysis are exemplified and the technical inheritance validated.

4.2.3 Identification of life cycle information

First, the identification of life cycle information represents a main part of the "Design of Gentelligent Components." Therefore the information flow of the life cycle has to be modeled. For the realization different specification techniques could be applied. After defining a method and tool for the process of technical inheritance with its information flows the possible life cycle information are analyzed. After a general consideration of possible monitored data a transfer to the demonstrator and its development-relevant life cycle information is done.

4.2.3.1 Information flow in product life cycle

There exists a lot of descriptions and models of the point of view for the product development process, compared, for example, with [8,16]. The approach of technical inheritance of gentelligent components is used in the intergenerational development process. In this process different kinds of information can be detected in the whole product life cycle. Therefore the technical process of the life cycle of gentelligent products has to be specified. In consequence the specification technique of the activity diagram from the UML is chosen. By this selection the process of the technical inheritance could be analyzed as well as its information flows visualized.

4.2.3.1.1 Adaption of the notation for process specification

For a better comprehension of the linkages between different information flows as well as the relationships between the various steps in product life cycle the whole process is modeled with the specification technique of activity diagrams in UML. Moreover the notation is adapted for a simplified process and information analysis, depicted in Fig. 4.2.4

The first adaption of the notation concerns the objects in this specification technique. In this context the forms of the objects are designed differently for physical and digital objects. This is necessary in case of the digitization in context of "Industry 4.0." By the integration of intelligence in products, like the gentelligent components, as well as the transition to flexible production modes more and more digital aspects in the product life cycle are created. Moreover the linkages realized by arrowheads between activities and objects are adapted too. Thereby information flows of the manufacturing, the maintenance, and the application information and beyond the characterization of the algorithmic data feedback are changed. In dependency to the kind of information the arrowheads are adapted.

The new notation for activity diagrams supports the visualization and an analysis of the intergenerational development process to identify the product-specific life cycle information. In the next step a general life cycle is modeled with the applied notation and afterward transferred to the demonstrator.

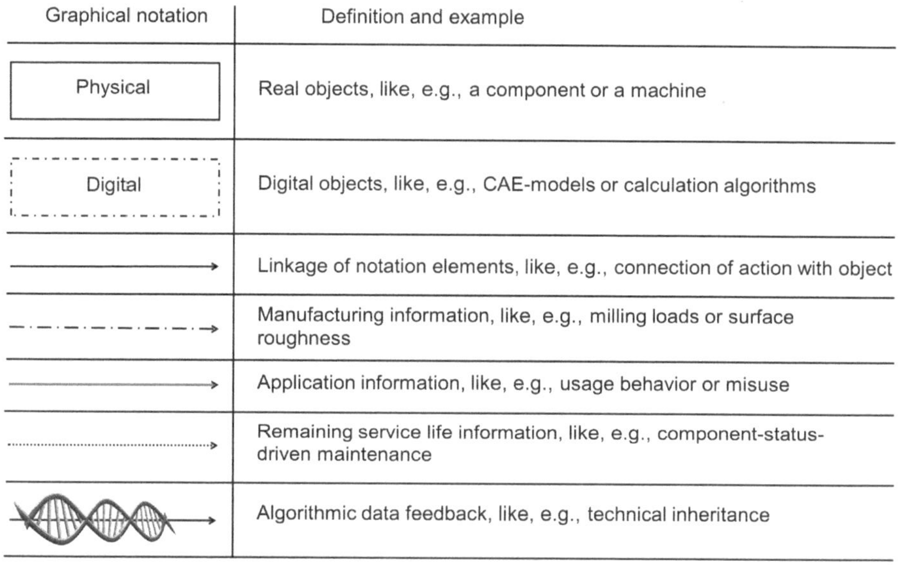

Fig. 4.2.4 Adaption of UML notations for illustration of technical inheritance [12].

4.2.3.1.2 Application of UML adaption

At the beginning a general product life cycle to visualize and analyze the inter-generational development process is modeled. In this approach the product life cycle is illustrated in a simplified form in case of a linearized process. For the explained activity, diagram of UML with the new notation is applied [14].

Based on this process model an elaboration of the product life cycle for the wheel suspension car is realized. Therefore the different phases are transferred for the situations of the race. The objective represented by the identification of the different information flows which could be detected for an implementation of the technical inheritance. This is necessary for a better understanding of the dependencies between the product-specific life cycle steps and their information content. A simplified excerpt of this model for the application phase for the race car is depicted in Fig. 4.2.5 [12].

In this excerpt the focus is placed on the wheel carrier from the wheel suspension. This component equipped with gentelligent technology, exactly with magnetic magnesium, represents the central component in this assembly. With this technology the loads during its application could be detected inherently. The possible life cycle information of these phases can be divided into three aspects. In this case the adaption notation shows the kind of information. For example, the

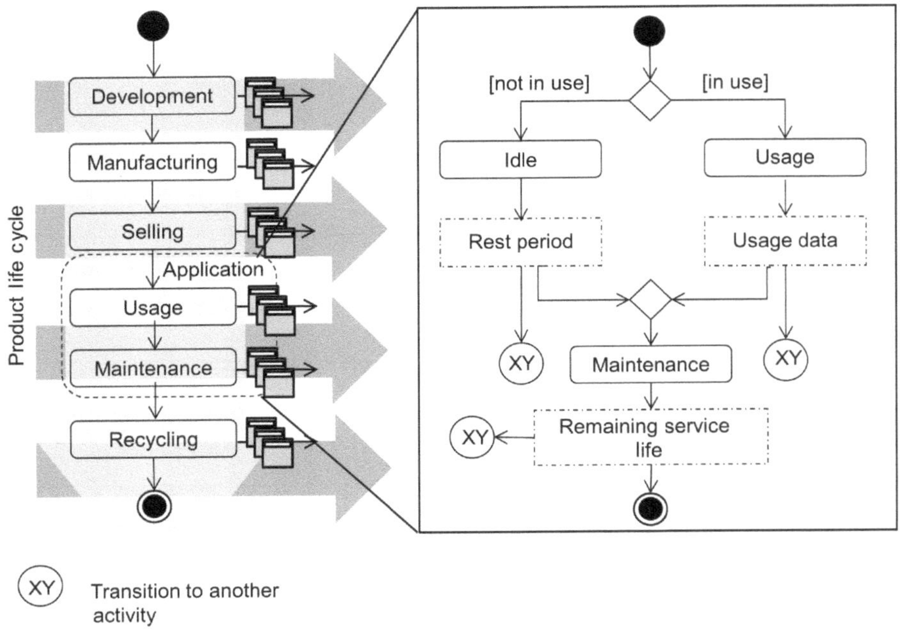

Fig. 4.2.5 Excerpt of the information flow at application phase in life cycle [12].

information about the usage, which are exemplary data sets about the rest period or the usage during the race, are differently marked as the possible data sets from the maintenance, especially the information about the detected remaining service life. All this information is stored in knowledge base which is linked to the product development.

With this applied notation the information flow for the implementation of the intergenerational development process is simplified presented considering the whole product-specific life cycle. This adapted methodology has two advantages. Different kinds of information as well as the dependencies between different steps of product life cycle can be visualized and analyzed. Moreover the information flow from various steps can be defined and identified to support the technical inheritance. With the identification of life cycle information the actual development could be supported, the service strategies could be influenced, and of course the information can be used for the development of the next product generation.

Nowadays the considerations of a product life cycle especially the view from product development is not described sequential [8]. But in this case the linearized view of different steps, sequences, and activities in the product life cycle allows a better comprehension of possible life cycle information as well as a simplified analysis of the process. Moreover, this method of process modeling demonstrates the potential of the identification of different kinds of life cycle information. This can be used to realize the technical inheritance with the gentelligent technology. In addition, the process of an intergenerational development regarding an integrated design approach is supported [3].

4.2.3.2 Analysis of life cycle information

After modeling the product specific life cycle of the race car with the focus on the wheel suspension, especially the wheel carrier, the possible life cycle information is analyzed. Physical parameters are measurable in the technical inheritance, defined as a technical-biological process [14], and the simultaneous integration of gentelligent technologies. In consequence by the analysis of possible product life cycle information, various types of them could be allocated. In every phase of the products life cycle, different physics characteristics could be identified, according to Roth [19].

In the methodology for the analysis of product life cycle information, four analysis steps for every phase are necessary. These can be worked out by the following questions for every object in the product-specific life cycle model [3]:

- Which physical measurements could be recorded or localized?
- What influence has the information to the physically or mentally of the humans and their environment?
- Which economic aspects could be identified?
- What general aspects are from special interest like, for example, standards or trends?

The results of this analysis are demonstrated in an excerpt for the usage data of the wheel suspension. Exemplarily different reasons of the four aspects are depicted in Fig. 4.2.6. In this case the four aspects are considered and some possible life cycle information of this phase could be identified. For example, the first aspect is information about the reliability for the wheel carrier. Moreover the second aspect shows, for example, that information about the loads could be associated to the safety of the product. From the economic point of view the energy costs for the race car can be reduced by an adaption of a lightweight designed component. In addition, information about the competitors could be giving a benefit for the development of the next product generation [12].

By this analysis the dependencies of different aspects become apparent, furthermore the illustration of the product life cycle in combination with the analysis of these four different aspects and possibilities of life cycle information allows an identification of development-relevant information.

4.2.3.3 Application race car

With the gentelligent technology the loads during the usage phase of the wheel suspension could be measured. These data sets could be used, on the one hand, for a proactive maintenance of the mechanical components. On the other hand,

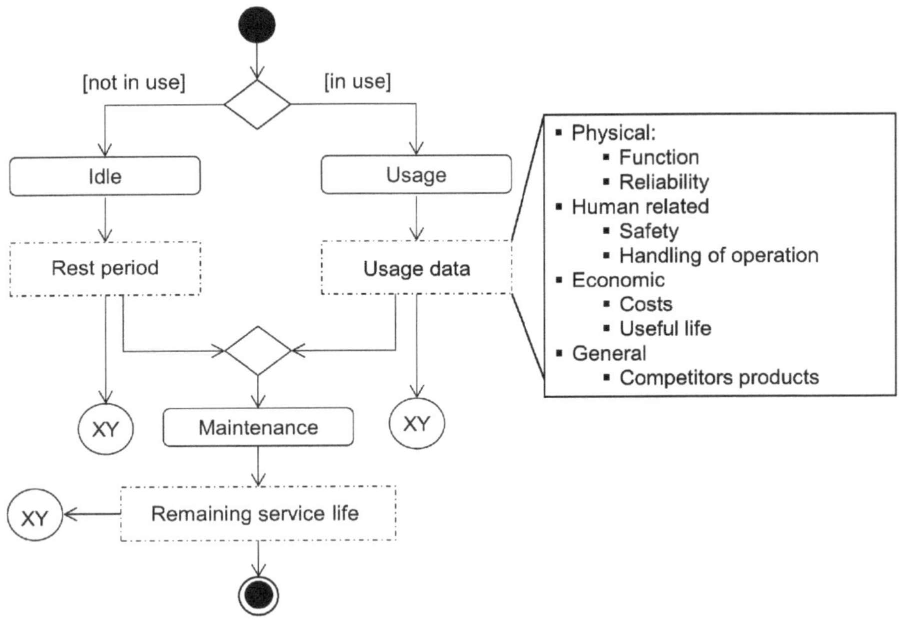

Fig. 4.2.6 Excerpt of the four possible information types in usage data [12].

this life cycle information represents a benefit for the design of the next product generation. Hence in this investigation the second aspect is focused to realize the technical inheritance. Therefore the race car has to be analyzed for the emergence of the applied loads.

The dynamics of the race car is divided into three aspects. The longitudinal dynamics results of the acceleration, braking, or sloping of the race car. Lateral dynamics are the consequence of cornering whereas the vertical dynamics are increased by the wheel load variation [20].

The dynamic reaction of the race car depends on the different driving situations during a race. Moreover every occurring situation generates another characteristic of loads. These driving situations could be divided according to [21] in stationary position as well as continuous, accelerated, retarded straight-line motion, and stationary or unsteady cornering.

After defining the loads during the usage phase as the information relevant for development, the transfer of the inherent data to the relevant positions can be realized by mechanical models in case of the application of gentelligent technologies. The measured data should be set to the identified driving situation by the application of data analysis methods.

4.2.4 Data analysis

With the gentelligent technology different loads during the product life cycle of the wheel carrier from the wheel suspension could be detected. On the one hand, the applied loads during the manufacturing of this component are appropriated to an optimized process planning [22]. On the other hand, collected data about the applied loads during its usage are transferred for a component status-driven maintenance [23]. In case of the technical inheritance the loads are linked with the product development. Therefore the detected life cycle information is attributed to the development of the next product generation. In this context the application data has to be assigned to the development-relevant information. In case of the wheel carrier these are defined as significant, maximum, and critical load cases [12].

4.2.4.1 Data processing

One of the encounters during the monitoring problems is the large amounts of heterogeneous diagnostic data such as signals, their parameters, indications of strength and strain gauges, etc.; as the construction of analytical or numerical models for the interpretation of diagnostic signals is a complex task, it is appropriate to apply a statistical analysis, allowing a high degree of confidence to carrying out structuring and interpretation of data. In developing of methods of the analysis the principles of horizontal scalability, fault tolerance and locality of data are used. Wherein the problem of storage and data management, taking into account the factor of unstructured data, the problem of obtaining structured information and save the data to the knowledge repository have to be solved.

The first step for the data analysis is the preparation of the data (see Fig. 4.2.7). It is important, that in spite of the fact that data sets can be very big, information, which is contained in them, has considerably smaller dimension. For example, while data collect every second or every minute, many parameters (e.g., temperature) remain stable over long time intervals. Thus it is appropriate to conduct a "smart" aggregation of data for modeling and optimization, which contain only the necessary information about the dynamic changes influencing overall performance of a method. Additional effective strategy of handling large data volumes is splitting data into segments and creating models for each segment separately, with further consolidation of results.

After the preprocessing is prepared, classification and analysis of data are carried out. In the beginning of the analysis the simplest algorithms are used. The part of data, which can be processed by means of simple fast algorithms and which are senseless, like noise or measurement errors, for processing with use of more difficult methods, is analyzed and excluded from the further processing. The remaining data is passed to the next stage with more sophisticated algorithms, and so on up the chain. Further analytical process is divided into two parts: the construction of the model and the use of

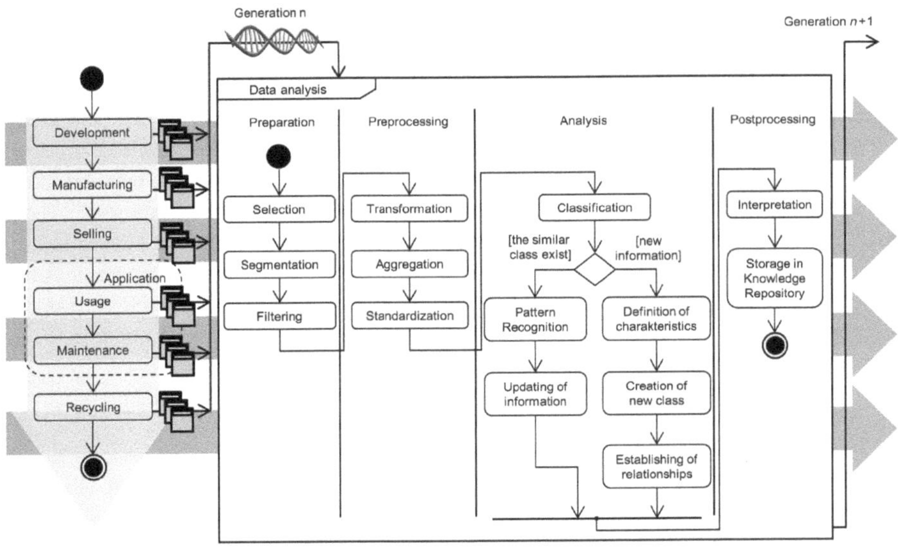

Fig. 4.2.7 Data processing.

the constructed model to new data. Depending on the applied algorithm data are cached and auxiliary parameters are calculated.

The developed analysis method for results of monitoring can be used for both set of measuring information: parameters and timing realization of signals. This method is based on a two-level clustering data: initial signals detected by individual measuring channels and "signal groups" formed on the basis of the primary clustering. The data processing involves a statistical analysis. The data sets of simulated and real force at the sensor points (see Fig. 4.2.9) are combined for the analysis. One reason is that for an equivalent informative over the loads during the product life cycle a huge number of data sets had been considered. In order to assign loads the critical certain driving situations, typical driving situations, and the forces and moments that occur are analyzed. Various track courses were selected for the preceding test situations which contain possibly many different driving situations. To create different possible loop trails, several narrow and longer right turns, left turns, and straight lines are combined.

The used analysis approach includes methods of the factorial analysis and creation of a phase portrait of a signal. The analysis of collected results for aforementioned driving situations shows that the phase portraits of signals reveal the typical situations which are define significant load cases were chosen in association with the vertical, lateral and longitudinal dynamics of a wheel suspension. In the case of uncertain signal processing the ways of structuring and grouping data are probabilistic cluster analysis or statistical signal processing. The EM-cluster analysis method and methods for the receiving of phase portraits of signals are applied. The time-dependent subjected load forces were developed, which represent realizations of a random load process in certain sensor points. Monitoring the force vectors indicates that at certain moments of time or quickly the length or direction, or all at once are changed. Comparison of typical situations the behavior of a racing car and using a combination of different statistical methods reveals the patterns of signals corresponding to the different situations where mentioned. This information about the loads is stored in knowledge repositories and is the basis for the adaption of the next generation.

4.2.4.2 Classification of detected life cycle information

At the beginning of the classification the inherent collected life cycle data has to be transferred to development relevant geometry positions of the monitored product. In this case there exist three important positions at the wheel carrier. These are depicted in Fig. 4.2.8 and represent the linkage to the wishbone and pushrod as well as the suspension which limits the dynamics of the whole assembly.

These three force application points (A, B, C) are determined for the next analysis. Also the application point represents the conditions for the adaption of the next product generation. This is necessary because a variation of this fixed position of linkages leads to other applied loads in use of the product and changes also the

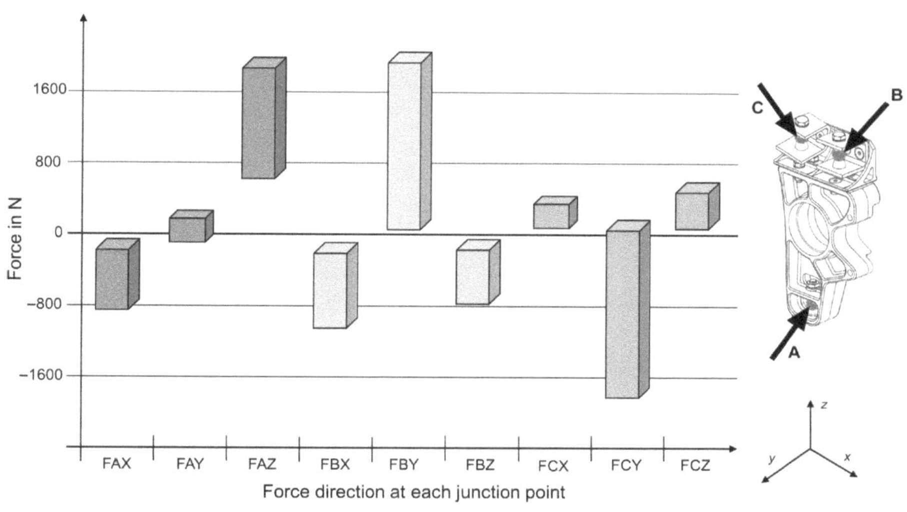

Fig. 4.2.8 Load case: acceleration from stationary position [12].

behavior of the race car. In the next step the collected and transferred data sets of the gentelligent component are investigated and linked with the three aspects of development-relevant information.

The assignment for the life cycle data is to reduce the data volume on the one side and to detect and collect the product-specific life cycle information on the other side. Therefore data mining methods are applied, especially in this case a genetic algorithm with the task of clustering, classification, deriving of rules, and pattern recognition. Based on this task for data analysis the genetic algorithm is enabled to investigate the collected data sets of the life cycle information from the wheel carrier [3].

With the data analysis the three aspects of development-relevant information could be detected. The first one, the significant loads, is defined as the most frequently occurring couple of loads which is set into dependency to the located driving situations. One of such driving situation, the acceleration from the stationary position of the race car and the associated cluster areas of the three directions of force to the described application points, is depicted in Fig. 4.2.8.

The second point of view represents the maximum load cases, which occurs to life cycle incidents, e.g., driving through a curb. The values of such unforeseen situations are from special interest, right for safety-relevant components. These load cases have also been considered in the design and dimensioning of the next generation in order to avoid hazards to the driver of the race car or the marshals.

The identified critical loads constitute the third development relevant part. These occurring load cases are directly related to the current shape of the monitored product. These are defined as such force combinations which generate stresses in the component with values in the area of short-term strains. Based on this definition the kinetics of the wheel suspension and the resulting possible load cases including their force combination are elaborated to identify such values and combination at the wheel carrier. This procedure is realized by thousands of static FEM analysis. The resultant two possible force combinations are depicted in Fig. 4.2.9 [3].

As an exception this life cycle information could only be used for plausibility studies or safety test of the current product and not for the development of the next product generation because the relationship of this kind of information and the shape of the actual product generation [3].

The results demonstrate different kind of information to add them to the development process. The first two life cycle information could be linked with the development of the next generation whereas the third one provides information for the current generation. In case of the algorithmic data feedback the significant and maximum load cases are focused. Normally decision-making mechanism has to be installed in which phase or step of the development the information has to be placed. In this case the analyzed load cases are directly linked with the design evolution which investigates und applies optimization strategies for the adaption of mechanical components.

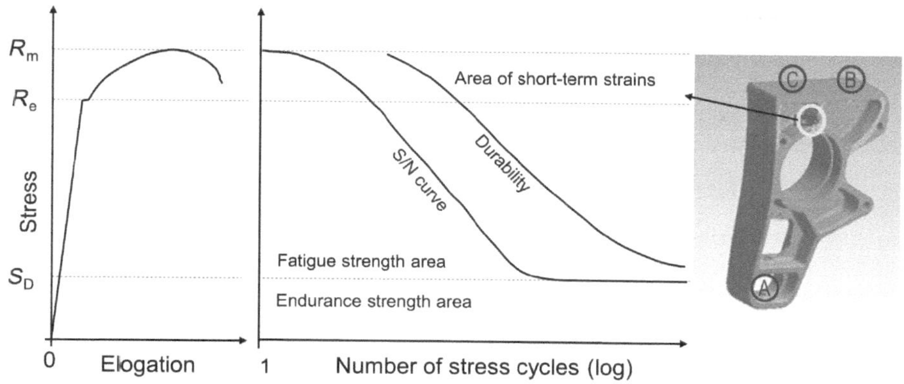

Fig. 4.2.9 Critical loads and its position at the wheel carrier [12].

4.2.5 Design evolution

In the context of technical inheritance the design evolution is an important element to link the life cycle information with product models to adapt the product, represented by a computer model, to the new resulting requirements. This aspect is focused to mechanical components and the loads applied during the life cycle. Hence there is the question how design representing computer models for synthesis and mechanical analysis can be developed in a way that they represent a resulting design solution space with a wide range of design solution options. If such a model can be established, the life cycle information can be used to find the best suited design option of the solution space for the current requirement situation. While feasible algorithms of the field of optimization, e.g., genetic algorithms, still exist, the development of such solution space representing design models is a major challenge. Thus the focus of the design evolution package is the development of such models.

The question of solution space representing computer models fits into the field of knowledge-based engineering (KBE) and structural optimization. Thus the common modeling techniques in this field are analyzed for the application in the context of life cycle information-based design evolution. While in the field of structural optimization mesh-based methods like topology and shape optimization are very common, in the field of KBE parametric-associative CAD models are the basis for more complex product models enriched with engineering knowledge. There are several advantages and disadvantages of the different methods, which are not discussed in detail in this chapter, but the main lacks of mesh-based model are options of engineering restrictions formulation, e.g., for manufacturing, as well as the further use of the models in the design process, e.g., for CAM, because of the mesh formulation and the limited methods for transformation into CAD. Parametric-associative models are well suited

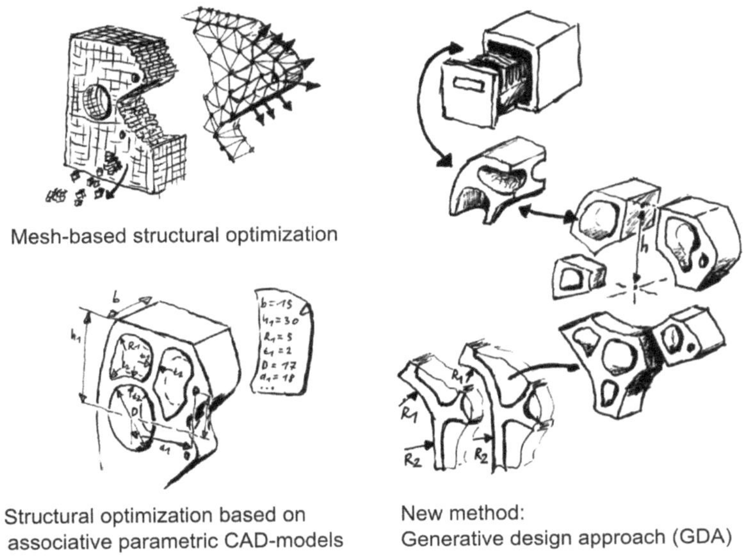

Fig. 4.2.10 Comparison of modeling approaches in the field of KBE and structural optimization.

for the representation of variable design solutions, but when the solution space represented by the model increases, the dependencies between parameters and constraints of the model become more complex. Instable geometry as well as a huge development effort is the implication of this issue. Thus a new modeling approach for parametric-associative design is developed, called Generative Design Approach (GDA), to organize all, geometric, technological and physical dependencies in a modular framework, which also supports conceptual design variations (Fig. 4.2.10).

The core idea of the GDA is the modularization of design by CAD partial models, called design elements. These represent geometric solutions of a local section of a part on a detailed level which is quite smaller than the size of the modeling part itself. Thus a part is represented by different assembled design elements in opposite to the common way to use one parametric part model (also which interlinkage to a higher level assembly file). All design elements are equipped with specified connection faces to guarantee closed surface transition after assembling. The global arrangement of the elements is organized by a parametric skeleton, which consists of position and characterization parameters for every connection face as well as global structure concept dependencies.

A variation of such a GDA model—in relation to the representation of a design solution space—can be executed on different levels: For global structural variation the skeleton parameters can be varied to shift the connection faces. Shape variation is done by contour parameter change of the connection faces. In addition, there is the option to replace an existing connection face by a different type stored in a connection face library. The same concept is also established for the design elements: In relation to the supported possible connection face combinations for all locations in the skeleton, where design elements can be placed, several parametric-associative design elements are stored in a design element library. Connectional variations are done by replacing an existing design element located in the model by a different one of the design element library. By this also topological changes are supported by GDA. Furthermore, size and inner contour of a design elements instance of the model can be modified by element parameter variations. In this situation design rules, which are implemented in the parametric structure of the element, control the consistence which the actual connection face design as well as the geometric stability conditions inside the element. These aspects are illustrated in Fig. 4.2.11.

By this parametric-associative design framework a wide range of design solutions—the solution space we talked about—can be supported. The modular structure introduces a quite organized dependencies network, where global and local design variations can be done with limited interdependencies. This approach of modeling fulfills the requirements for the earlier addressed design evolution loop for semiautomatic design adaption to new mechanical boundaries identified by life cycle data analysis. The framework for this design evolution cycle is discussed in the following section.

The GDA including the libraries is implemented in a CAD software environment by extensions implemented via available KBE functionalities and API-based scripts. By this the parameter sets of all levels—skeleton, connection faces, and design elements—are accessible and the change of geometry is organized by

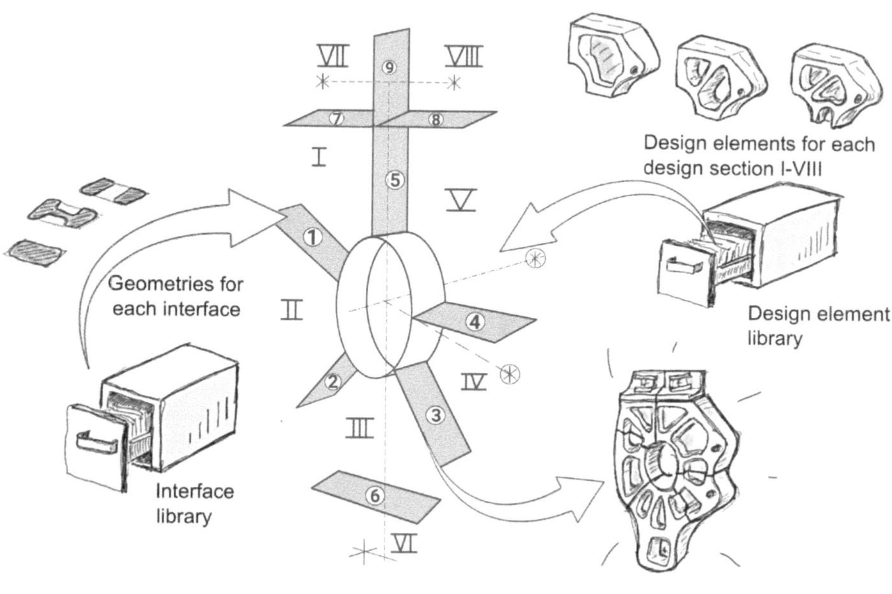

Fig. 4.2.11 Generative design approach.

the GDA functionalities. For the establishment of a design evolution cycle two further modules are required with the suited interfaces for data exchange: An analysis module for mechanical evaluation of a design solution generated by GDA as well as an optimization module to organize the parameter variation during the evolution. The design evolution process, illustrated in Fig. 4.2.11, is implemented containing such modules and interfaces: For mechanical analysis a FE software environment with scripting support on an object-oriented level is included in the loop. A design solution of a defined parameter set generated by GDA is transferred by two different interface formats to the FE software. The geometric information is exported to a (nonparametric) step file while the design element structure is preserved in this format. All additional information about FE boundaries (meshing, loads resulting from the life cycle information, supports, material data, etc.) are stored in a XML-based exchange format with a design element-oriented data structure. By scripts embedded in the FE software, these data files are imported, analyzed, and used for FE model generation (Fig. 4.2.12). After solving the resulting FE model the results are stored in an xml file and transferred to a genetic algorithm-based optimization toolbox, which is implemented as a stand-alone software module. By the results during the optimization process new parameter sets are generated and transferred by an interactive interface to the CAD environment to regenerate the GDA model. The loops are continued until the best fit is identified. The modular concept—of the GDA model by the library concept and of the software environments by defined exchange formats—can be adapted to different mechanical structure problems by extension of the library elements and the script functionalities.

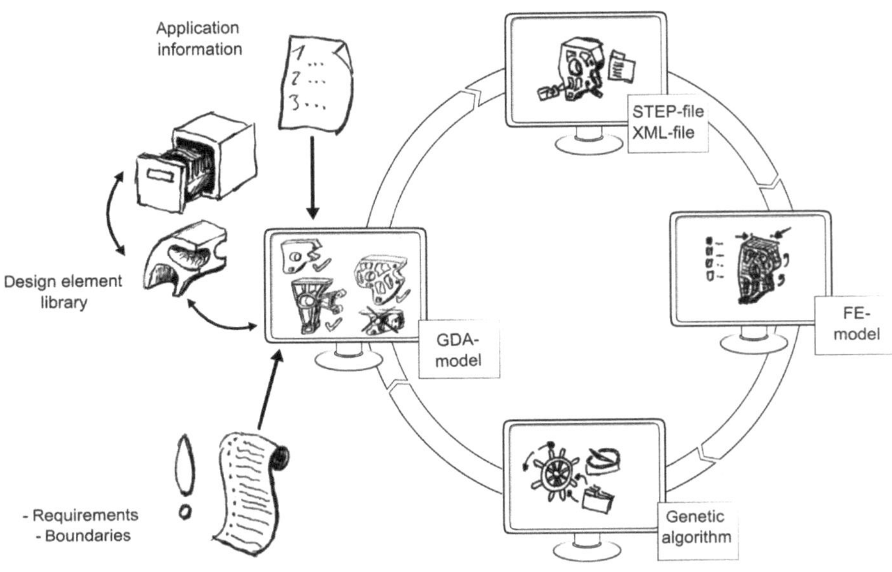

Fig. 4.2.12 Design evolution process.

4.2.6 Process model

With the three aspects the "Design of Gentelligent Components," "Statistic Operator," and "Design Optimization" the base for an algorithmic data feedback is realized. In consequence a process modell of intergenerational development process for developing products with gentelligent technologies within the technical inheritance is depicted in Fig. 4.2.13. This process is modeled similar to the structured process from VDI 2221 [24].

The first step deals with the identification of life cycle information. In this package the technical system is analyzed regarding its life cycle. In this content the different possible life cycle information as well as the behavior of the technical system can be worked out. Moreover by investigations of the product specifications in the development, process made assumptions could be identified to monitor these in the products life cycle. The monitored hypothesis forms the bases for the next development steps for an intergenerational development [3].

In the second step of this process model the implementation of a monitoring strategy is arranged. Therefore a design catalog as well as design rules for gentelligent technologies is developed. These tools support the integration of this new technology in smart and cyberphysical systems. After defining the technology for monitoring the life cycle information the position of the measurement has

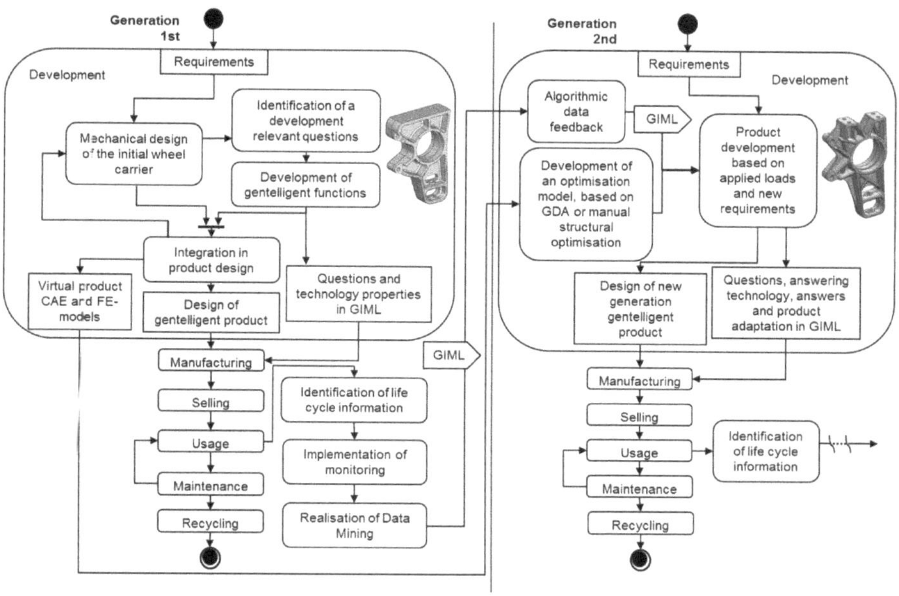

Fig. 4.2.13 Process model of intergenerational development process.

to be clarified. Therefore the kind of measuring is divided into two ways, direct and indirect. The direct form allows the data capture at the development relevant position of the component. In case of an indirect form, transformation models have to be developed. The monitoring strategy could be started with the realization of data analysis.

In the third step the methodologies of data mining are in focus. Here an integration of these methods in the phases of product development as well as the linkage of the results of the product representing models with the methods of data evaluation could be realized. First of all the clustering and also the deriving of rules for empty classes could be realized in this step. Afterward the data mining methodology can be adapted for product specific to the technical system.

The last step, described as "algorithmic data feedback" begins with the product-specific analysis of the technical system. With the results out of step one the typical life cycle situations could be identified. With this knowledge and the linkage with the product representing models, like dynamic simulations of a race car, the different driving situations for the wheel carrier could be defined. These are applied in the data mining method. In the following procedure the life cycle information has to be assigned to the phase of product development. Therefore decision-making mechanism could be developed. In context of the technical inheritance regarding the wheel carrier of the wheel suspension the life cycle information could be directly linked to the design evolution. This methodology presents approaches to adapt components in its shape to its real environment.

This process model supports the development with gentelligent technology as well as the realization of the technical inheritance in an integrated design environment. These expansions for the development phase are necessary for a goal-oriented algorithmic data feedback for a design evolution.

4.2.7 Conclusion

The technical inheritance with an algorithmic data feedback of life cycle information supports the intergenerational development process. The gentelligent technology forms the bases to collect, store, and transmit life cycle data inherently. With the three research packages "Design of Gentelligent Components," "Statistic Operator," and "Design Optimization" a process model to develop with the gentelligent technology for a technical inheritance is realized. The theoretical results are transferred and demonstrated by the wheel carrier, part of the wheel suspension of a race car.

In this conclusion the three challenges out of Section 4.2.2.2 should be considered again:

- By means of which methodology, an intergenerational development process could be specified, analyzed, and its information flow being visualized?

In this case an investigation of different specification techniques is realized. With the adaption of the activity diagrams a methodology is created to identify life cycle information as well as visualize the product-specific life cycle. Moreover by the differentiation of the kinds of life cycle information in the four aspects (physical, human oriented, economic, and trends) an approach to elaborate development-relevant information is implemented. So tools and methods to integrate gentelligent technology in smart or cyberphysical systems are developed to support the engineers in context of "Industry 4.0."

- Which algorithms support different approaches of data mining for, on the one hand, the reduction of the data volume, and on the other hand, an identification of product-specific development-relevant information?

By the transfer of nature-based algorithms into the techniques of data mining it is possible to generate product-specific data analyses. In addition, the realizable tasks like classification or deriving of rules provide itself advantages for a life cycle data reduction. Moreover the different classes and cluster could be adapted to product-specific situations. Also these identified solutions are linked to the monitored hypothesis and its verification or falsification. By this way the development-relevant information for an algorithmic data feedback could be detected in life cycle data sets.

- In which way life cycle information could be integrated into optimization strategies in order to adapt mechanical components optimally to its real environment?

The design evolution elaborates different optimization strategies. These allow an adaption of the shape of mechanical components. As a result the generative design approach in combination with a genetic algorithm for the objective was developed. This includes shape changes in the topology as well as in the parametric of the components design which also consider the manufacturing and functional restrictions at any time. With the integration of real loads the gentelligent components of the assembly could be adapted to its real environment.

In summary the process model of the intergenerational development process supports the developing with gentelligent technologies and the implementation of a technical inheritance. With specification techniques, methods of data mining and the design evolution smart or cyberphysical components could be monitored. With an algorithmic data feedback the component adaption to its real environment is realized.

In further works the technical inheritance will be applied for a resource-efficient product development. Moreover, engineers should be trained to use an algorithmic data feedback to optimize products with regard to their CO_2 consumption. Moreover the collected life cycle data could be analyzed for the trends of mass customization. With product specific and strictly personal life cycle information the product individualization could be aimed directly. In addition, decision-making mechanism should be developed to transmit the extracted life cycle information to the phases of product development where the information leads to a benefit.

References

[1] M. Abramovici, R. Stark (Eds.), Smart Product Engineering, Proceedings of the 23rd CIRP Design Conference, March 11th–13th, 2013, Bochum, Germany, Springer Verlag, Berlin, Heidelberg, 2013.
[2] T. Feld, M. Hoffemann, R. Schmidt, Industrie 4.0, Vom intelligenten Produkt zur intelligenten Produktion, Inf. Manage. Consult. 3 (2012) 38–42.
[3] P. Gottwald, Prozess einer generationsübergreifenden Produktentwicklung durch technische Vererbung, Dissertation Leibniz Universität Hannover, Berichte aus dem IPeG, PZH-Verlag, Garbsen, 2016.
[4] B. Denkena, H. Henning, L.E. Lorenzen, Genetics and intelligence: new approaches in production engineering, Prod. Eng. 4 (1) (2010) 65–73.
[5] B. Denkena, T. Mörke, M. Krüger, J. Schmidt, H. Boujnah, J. Meyer, P. Gottwald, B. Spitschan, M. Winkens, Development and first applications of gentelligent components over their life-cycle, CIRP J. Manuf. Sci. Technol. 7 (2) (2014) 139–150.
[6] T. Düser, X-in-the-Loop – ein durchgängiges Validierungsframework für die Fahrzeugentwicklung am Beispiel von Antriebstrangfunktionen und Fahrerassistenzsystemen, Dissertation Karlsruher Institut für Technologie, Forschungsberichte IPEK, Karlsruhe, 2010.
[7] M. Neubach, Wissensbasierte Rückführung von Produktnutzungsinformationen in die Produktentwicklung im Rahmen einer Product Lifecycle Management (PLM)-Lösung, Dissertation Ruhr-Universität Bochum, Bericht vom Lehrstuhl für Maschinenbauinformatik, Shaker Verlag, Aachen, 2010.
[8] S. Vajna, Integrated Design Engineering: Ein interdisziplinäres Modell für die ganzheitliche Produktentwicklung, Springer Verlag, Berlin, Heidelberg, 2014.
[9] M. Parvan, F. Miedl, U. Lindemann, Nature-Inspired Process Model for Concept Selection and Evaluation in Engineering Design, NordDesign, Aalborg, Denmark, 2012, pp. 486–493.
[10] S. Clement, Erweiterung und Verifikation der Autogenetischen Konstruktionstheorie mit Hilfe einer evolutionsbasierten und systematisch-oppertunistischen Vorgehensweise, Dissertation Otto-von-Guericke-Universität Magdeburg, Lehrstuhl für Maschinenbauinformatik, Magdeburg, 2006.
[11] R. Lachmayer, I. Mozgova, B. Sauthoff, P. Gottwald, Product evolution and optimization based on gentelligent components and product life cycle data, in: Proceedings of the 23rd CIRP Design Conference, March 11th–13th, 2013, Bochum, Germany, Springer Verlag, Berlin, Heidelberg, 2013, pp. 687–694.

[12] R. Lachmayer, P. Gottwald, Integrated development by the consideration of product experiences, in: A. Meyer, R. Schirmeyer, S. Vajna (Eds.), Proceedings of the 10th International Workshop on Integrated Design Engineering, University Magdeburg, Inst. of Machine Design, Magdeburg, 2014, pp. 73–82.
[13] R. Lachmayer, I. Mozgova, B. Sauthoff, P. Gottwald, Evolutionary approach for an optimized analysis of product life cycle data, in: Proc. 2nd International Conference on System-Integrated Intelligence: Challenges for Product and Production Engineering, SysInt 2014, July 2nd–4th, Bremen, Germany, 2014, pp. 359–368.
[14] R. Lachmayer, P. Gottwald, An Approach to Integrate Data Mining into the Development Process, Modelling and Management of Engineering Processes, Springer Verlag, Berlin, Heidelberg, 2014, pp. 175–185.
[15] R. Lachmayer, B. Sauthoff, P. Gottwald, Product optimization by analysis of gentelligent life cycle information, in: Proc. 1st Joint International Symposium on System-Integrated Intelligence: New Challenges for Product and Production Engineering, SysInt 2012, June 27th–29th, Hannover, Germany, 2012, pp. 28–31.
[16] R. Lachmayer, I. Mozgova, A. Deiters, B. Sauthoff, P. Gottwald, Design evolution by integration of life cycle information in knowledge repositories, consilience and innovation in design, in: 5th International Congress of International Association of Societies of Design Research, Tokyo, Japan, vol. 2, 2013, pp. 309–318.
[17] B. Sauthoff, R. Lachmayer, Optimierung von Bauteilen auf Basis von Produktnutzungsinformationen, in: Beitrag zum 24.DfX Symposium, Jesteburg, Germany, 2013, pp. 119–130.
[18] K.L. Haken, Grundlagen der Kraftfahrzeugtechnik – 2. Aktualisierte und erweiterte Auflage, Carl Hanser Verlag, München, 2012.
[19] K. Roth, Konstruieren mit Konstruktionskatalogen, Band I, Konstruktionslehre, Springer Verlag, Berlin, Heidelberg, 2000.
[20] P. Heinemann, Bewertung der Langzeitqualität am Beispiel von Fahrwerks-komponenten, Dissertation Technische Universität Braunschweig, Shaker Verlag, Aachen, 2012.
[21] F. Regh, Objektive Bestimmung der sicherheitsrelevanten Auswirkungen durch Fahrwerk Modifikationen, Dissertation Technische Universität Darmstadt, VDI-Verlag, Berlin, Heidelberg, 2012.
[22] B. Denkena, J. Schmidt, M. Krüger, Data mining approach for knowledge-based process planning, in: Proc. 2nd International Conference on System-Integrated Intelligence: Challenges for Product and Production Engineering, SysInt 2014, July 2nd–4th, Bremen, Germany, 2014, pp. 407–416.
[23] M. Winkens, J. Busch, P. Nyhuis, Method of component-status-driven maintenance for gentelligent components, in: Proceedings of the 19th International Conference JAABC—The Journal of American Academy of Business, Washington, DC, USA, 2013, pp. 111–117.
[24] VDI-Richtlinie 2221: Methodik zum Entwickeln und Konstruieren technischer Systeme und Produkte, Beuth, Berlin, 1993.

Further Reading

W. Engeln, Methoden der Produktentwicklung, Oldenbourg Industrieverlag, München, 2006.

4.3 Evaluating life cycle data to enable condition-based maintenance

Melissa Quirico, Maximilian Winkens, Peter Nyhuis
Institute of Production Systems and Logistics, Leibniz Universität Hannover, Garbsen, Germany

4.3.1 Maintenance in industrial productions

In industrial productions, enterprises are dependent on efficient processes for their economic success. Therefore when designing and optimizing high-performance processes, supportive tasks in addition to production processes need to be increasingly focused upon. Within this context, maintenance is a particularly key factor for the success of production enterprises [1]. Maintenance is defined as the combination of all technical, administrative, and management measures for maintaining and restoring the functionality of components or systems [2]. Selecting the relevant measures to be taken requires a positioning in the field of tension between ensuring a high availability of systems and attaining the lowest possible maintenance costs [3]. The time at which measures are conducted is determined by the maintenance strategy being pursued. With a reactive maintenance policy, maintenance measures are conducted only after damage has arisen. It is thus also referred to as damage-oriented maintenance. With preventative maintenance, measures are implemented already before damages arise in order to prevent damage. Maintenance is then usually conducted cyclically, thus causing planned, though occasionally unnecessary, standstills on functional systems. Condition-based maintenance is a special type of preventative maintenance. In addition to preventing damages, this policy aims to exhaust component utility as completely as possible [4], see Fig. 4.3.1. Estimates ascribe modern approaches to condition-based maintenance a cost savings of up to 30% [5].

Previous approaches to condition-based maintenance mainly concentrate on wear. This is, for example, detected by measuring vibrations, temperatures, oil quality, or electricity consumption [6]. However, taking material fatigue into account offers considerable additional potential with regard to preventing spontaneous failures of fatigue-endangered components, since this represents the most frequent cause of damage with cyclical loads [7]. Signs of fatigue though are generally not recognizable based on a component's appearance or operating behavior. Monitoring and determining the condition while operating is consequently difficult. A comprehensive analysis of the component load, which is considered the cause of material fatigue, usually only occurs during the component design stage when laboratory tests are conducted with strain gauges [8]. Determining the material fatigue and specific wear margin curves

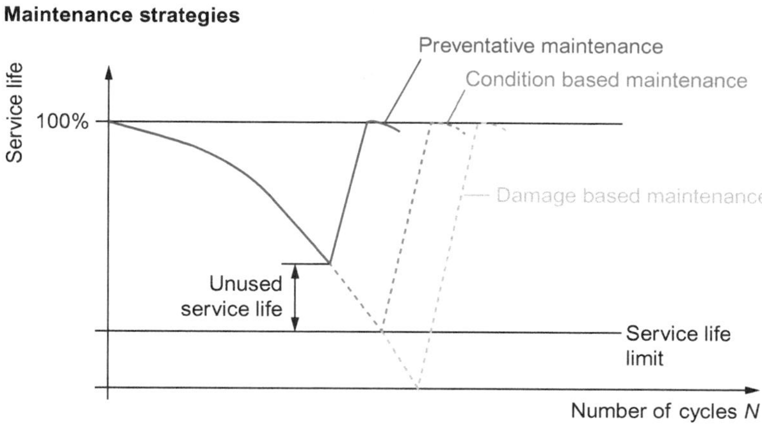

Fig. 4.3.1 Maintenance strategies [4].
Based on M. Strunz, Instandhaltung, Grundlagen Strategien Werkstätten, Springer, Berlin, Heidelberg, 2012.

while the component is in use has not previously been taken into account due to the insufficient cost-benefit ratio [4]. Signs of fatigue, however, may result in spontaneous, unplanned downtimes and perhaps even intensive reactive maintenance measures and should therefore be considered when planning maintenance [9]. Thus CRC 653 developed a method for component condition-based maintenance, which focuses on component fatigue as the cause of downtime. This approach makes it possible to determine the condition of a component based on life cycle data and thus facilitates efficiently maintaining components that are at risk for fatigue.

4.3.2 Life cycle data in Collaborative Research Centre 653 (Example of Application 5)

Gentelligent components enable the collection and storage of life cycle data, such as the load experienced while it is being manufactured and during its service life [10]. For example, sensory carriages and clamping systems on an intelligent tool machine compile and store production parameters and process forces during the manufacturing (see Chapter 3.2). If the component is subsequently used, new sensor technologies make it possible to continually compile the mechanical load during the component's service life (see Chapter 2). Subsequently, load data from both the manufacturing phase and service life phase are available via storage and information technologies and can be used as a basis for further analysis when designing the life cycle of high-performance components. Related to the Example of Application 5, the data of structurally identical components may be analyzed for design optimization (see Chapter 4.2) or condition-based maintenance. Following, we describe the approach developed for condition-based maintenance.

4.3.3 Condition-based maintenance of gentelligent components

New technologies developed within CRC 653 create the basis for an innovative approach to evaluating life cycle data for condition-based maintenance. Condition-based maintenance requires both a reliable evaluation of the component's state and an accurate prediction of its residual life [11]. This is dependent on the component's fatigue strength and the load experienced while in service [7].

The method developed in CRC 653 for condition-based maintenance centers on analyzing life cycle data and has three steps. First, the component fatigue strength is determined by means of a newly developed analysis of cutting forces while components are being manufactured [12]. Following that, the current state of the component is determined contingent on the loads during its service and the residual life is predicted [13,14]. The established condition is then used as a guide for selecting suitable maintenance measures [15]. Consequently, the basis for realizing this experience-based method is the collection of all the data into a comprehensive database [16]. Fig. 4.3.2 illustrates the procedure for condition-based maintenance with the aid of a control loop. Within the context of CRC 653, a racing car prototype, whose wheel carriers are manufactured from a sensitive material, served as a demonstrator.

In the following we will explain condition-based maintenance in more detail. In Section 4.3.3.1, we will describe the approach developed for determining the fatigue strength of components based on experience using load data from the manufacturing period. From there, in Section 4.3.3.2, we will clarify the residual life prediction in relation to the load data from the service life. Following that in Section 4.3.3.3, we will discuss selecting suitable maintenance measures.

4.3.3.1 Determining component fatigue strength based on experience

The residual life of a component is not determined just by the load while it is in use, but rather also from the load experienced during the manufacturing phase [3]. During machining, the manufacturing parameters, for example, influence the surface qualities considerably as well as the residual load and thus the strength of the manufactured component [7,17]. In order to take these influences into account, Winkens [12] developed a method for analyzing the cutting forces which allows the component fatigue

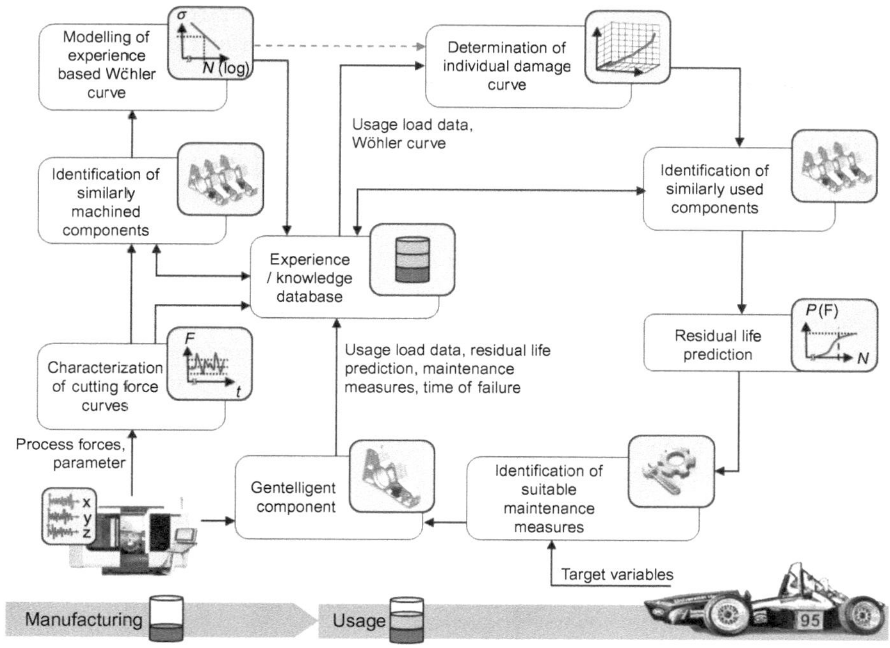

Fig. 4.3.2 Condition-based maintenance using life cycle data.

strength to be determined based on experience. This procedure is divided into four steps and described as follows. First the cutting force curves are processed and characterized, see Section 4.3.3.1.1. Based on this, components in the "experience database," which have comparable cutting force curves can be identified, see Section 4.3.3.1.2. These serve as the foundation for modeling the Wöhler curve based on experience and thus for describing the strength, see Section 4.3.3.1.3.

4.3.3.1.1 Characterizing cutting force curves

In order to account for manufacturing influences on the component fatigue strength, Winkens' approach [12] analyses the recorded signal characteristics of cutting forces. The measurement processes are dependent on the manufacturing methods of the implemented machinery as well as the tools used and component geometry [18]. Moreover the chosen cutting parameters have considerable influence, whereby the feed is particularly significant, causing reproducible influences. The rounding of the tool's cutting edge as well as the materials' wear and tear or inhomogeneity impacts the signal characteristics [17,19].

In addition to these influences, which are inherent in the processes, the signal characteristics can also be dependent on interferences in the measuring chain. However, in order to consider only process-dependent influences in the subsequent characterization, the signal characteristics are processed using a low-pass filter [20]. Afterward, the cutting force F_c of the individual cuts is identified in the signal characteristic. Since the last cut determines the component surface and thus has the greatest impact on the work product, we will focus solely on it during the rest of the analysis. With the start and end times derived from the cutting force curve for this last cut, the corresponding curves for the passive force F_p and the feed force F_f can be extracted from the respective signal characteristics. Based on the component geometry the signal characteristic of the last cut is subdivided into individual sections. By segmenting the cutting force curves, all of the geometric sections can be automatically identified in the cutting force curve. The geometric cut, critical to the component fatigue strength, is selected for further analysis. Fig. 4.3.3 depicts the division for selecting a segment based on the example of a standardized rotating bar bending fatigue test, in which the section particularly relevant to the component fatigue strength is the one with the smallest measurement.

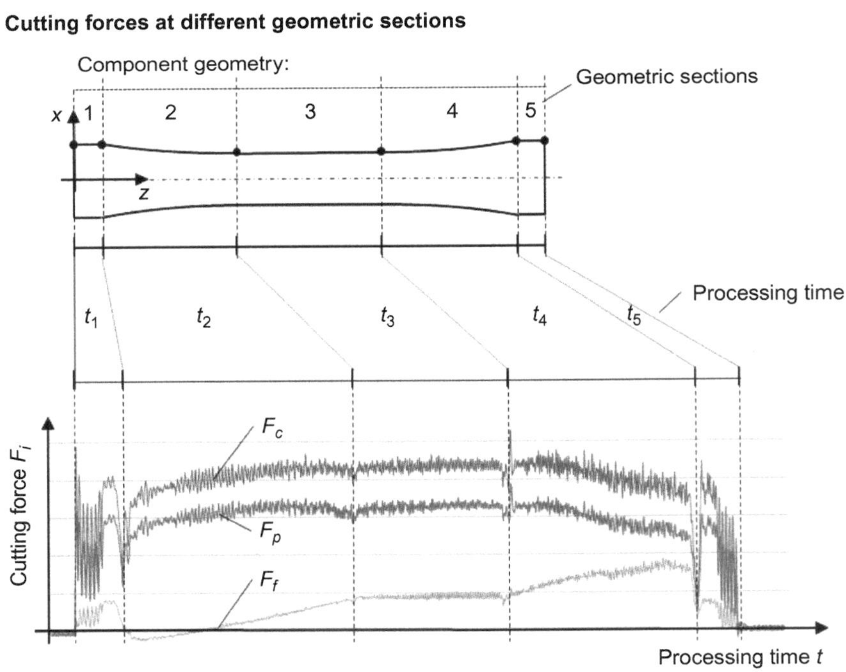

Fig. 4.3.3 Segmenting the cutting force curve according to component geometry [12].

Cutting force curves are unique for every component and will be characterized with regard to the degrees of force that occur and the changes in them. In order to accomplish this and to be able to take into consideration the relations between the individual force curves, a triparametric time-at-level count (TaL) is conducted. To do so, the curve segments being considered are first subdivided into 64 equidistant classes. This number of classes has frequently proven itself in the practice [21] and is thus also used in Winkens' approach. The width of the individual classes corresponds to the quotient of the maximum value of the respective cutting force and the number of classes, rounded off to a multiple of 0.5 N, see Eq. (4.3.1).

$$Cw_z = \left\lceil \frac{F_{z,max}}{64} \right\rceil_{0.5} \tag{4.3.1}$$

where Cw_z is the class width in N, $F_{z,max}$ is the maximum load, and z is the cutting force component, $z = \{c, p, f\}$.

For each of these classes the time-at-level count Δt_{TaL} is then determined. The three-dimensional class conditions $C^T = (c, p, f)$ are used to classify the curves. The force curves can thus be located in the same three-dimensional class a number of times. The individual time-at-level counts are subsequently summed for each class. The total time-at-level count per class represents the counter result and is stored in the experience database. On the left side of Fig. 4.3.4, curves for F_c, F_p, and F_f are exemplarily depicted with their classifications and on the right side, the counter result. In this example, the signal characteristic of the cutting force is subdivided into three classes (c_1 to c_3) and the curves for the passive and feed forces each into two (p_1, p_2 and f_1, f_2). The classes have widths Cw_c, Cw_p, and Cw_f.

The triparametric time-at-level count is conducted for the curve of the actual component as well as for the extracted curves of all n components in the experience database. So that the results can be compared, the same geometric segment is considered for all counts of the same class width Cb_z.

4.3.3.1.2 Identifying components with similar cutting force characteristics

According to Winkens [12] the next step in determining the fatigue strength based on experience involves identifying components in the database with similar cutting force curves. To assist with this, the degree of divergence is introduced here. The degree of divergence expresses the extent to which the experienced cutting force for component C_i, saved in the database deviates from the current component C_a. The degree of divergence is calculated by first organizing the results of the time-at-level count descending according to the total time-at-level count of the class and assigning these a rank in ascension, see Table 4.3.1.

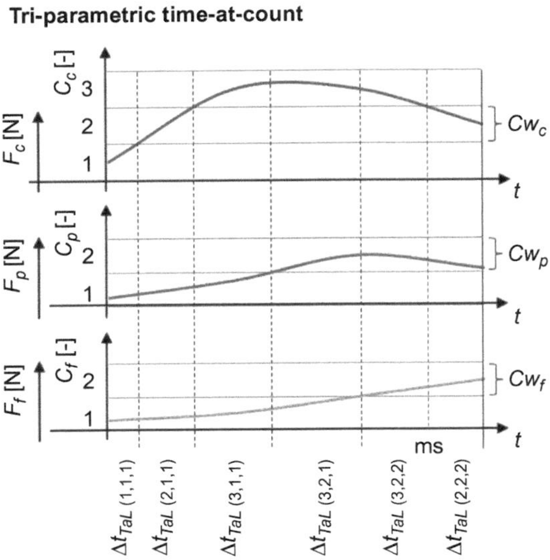

Counter results

C_c	C_p	C_f	$\Sigma\, \Delta t_{TaL}$
1	1	1	79
2	1	1	143
2	2	2	200
3	1	1	229
3	2	1	171
3	2	2	142

Legend:

C_c: cutting force class F_c
C_p: passive force class F_p
C_f: feed force class F_f
Cw_i: class width
t: processing time
Δt_{TaL}: time-at-level in ms

Fig. 4.3.4 Sample characterization of cutting force curves using time-at-level count [12].

Table 4.3.1 Sample ranking from time-at-level count [12]

Rank	$\sum \Delta t_{TaL}$	C_c	C_p	C_f
1	229	3	1	1
2	200	2	2	2
3	171	3	2	1
4	143	2	1	1
…	…	…	…	…

Following that, classes for the individual ranks are compared in pairs. The similarities for each rank r are calculated with the squared distance, see Eq. (4.3.2).

$$d_r^2(c,p,f) = (c_r[C_i] - c_r[C_a])^2 + (p_r[C_i] - p_r[C_a])^2 + (f_r[C_i] - f_r[C_a])^2 \tag{4.3.2}$$

where d^2 is the squared distance, and r is the considered ranks.

Subsequently, the degree of divergence D can be calculated for all n components in the database, according to Eq. (4.3.3). This corresponds to the weighted mean of the squared distances. The weighting factor per rank corresponds to the ratio of the respective time-at-level count to the total time, see Eq. (4.3.4).

$$D_{Ci} = \frac{\sum d_r^2(c,p,f) \cdot w_r}{\sum w_r} \tag{4.3.3}$$

where D is the degree of divergence and w_r is the weighting factor per rank r.

Whereby

$$w_r = \frac{\Delta t_{TaL,r}[C_a] + \Delta t_{TaL,r}[C_i]}{\sum (\Delta t_{TaL,r}[C_a] + \Delta t_{TaL,r}[C_i])} \tag{4.3.4}$$

Following that, components in the experience database with a small degree of divergence are identified since these have experienced similar cutting forces while being manufactured. We will refer to these later as experience components. To identify them, the n components in the database are arranged in ascending order according to their degree of divergence. The limit between experience components and those that are not experience components is set where the degree of divergence value climbs significantly. The first ten components in the degree of divergence rank are always declared experience components, in order to have a minimum number for modeling the Wöhler curve. Consequently, the search for the first significant increase begins with the 11th value. To do so a trend line is modeled for the first ten degree of divergence values. The value of the trend function at the eleventh position is compared with the eleventh degree of divergence. If the squared deviation of D_{C11} to $D_{C11,\ Trend}$ is greater than a tolerated deviation, there is a significant increase and the components after the eleventh cannot be used as experience components, see Fig. 4.3.5. If the difference is tolerable, the analysis continues iteratively: The trend line is always first modeled by the x previously verified components and the trend value for $(x+1)$ compared with component C_{x+1}, until a significant increase is determined.

4.3.3.1.3 Determining the experience-based Wöhler curve

The Wöhler curve is used to describe a component's resistance to vibrations [7]. Based on the data of the identified experience components it is modeled in accordance with the approach described by Winkens [12]. The modeling of the curve for the short-term strength, finite strength, and fatigue limit ranges occurs uniformly. This requires less precision in the short-term range. However, since the component should not be subject to any load in this range with typical operations, this assumption can be made in view of the effort entailed in the modeling.

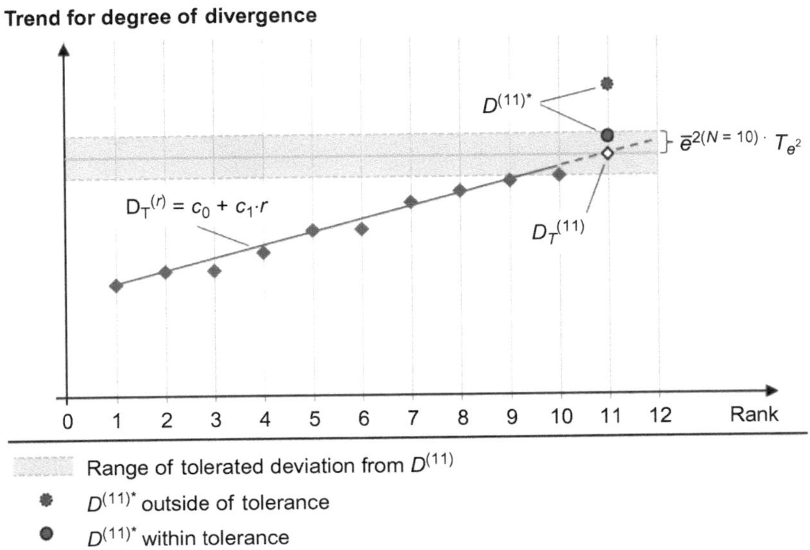

Fig. 4.3.5 Identifying the first significantly increased degree of divergence [12].

With a log-log plot, a linear trend can be assumed for the Wöhler curve in the finite strength range [22], which can be extended for the fatigue limit range. According to Corten-Dolan this Wöhler curve can be mathematically described with Eq. (4.3.5) [23]. The transition from the short-term strength range to the finite strength range is selected as the basepoint P_z for the modeling. For this the number of cycles is defined with $N_z = 10^4$ [24].

$$\log(\sigma_i) = -k \cdot \lg\left(\frac{N_i}{N_z}\right) + \lg(\sigma_z) \tag{4.3.5}$$

where σ_i is the load amplitude (variable), N_i is the number of cycles to failure (variable), N_z is the number of cycles at basepoint P_z, σ_z is the load amplitude at basepoint P_z, and k is the slope of the line.

In order to model the Wöhler curve the parameters m and σ_z are based on the data from the previously identified experience components. So that the Wöhler curve satisfies all of the experience components j, the following objective function applies:

$$g(\sigma_z, k) = \sum_{j=1}^{h}\left(1 - \frac{1}{\sigma_z} \cdot \sum_{i=1}^{m} \sigma_{ji} \cdot \left(\frac{n_{ji}}{N_z}\right)^k\right) \rightarrow min \tag{4.3.6}$$

where σ_{ji} is the load amplitude i of experience component j, and n_{ji} is the number of cycles i of experience component j.

This objective function for the determined parameters σ_z and k is solved with the gradient method [25], since it is frequently used due to the ease and speed with which it finds solutions [26]. After σ_z and k are determined using this method, they can be inserted into the linear equation for modeling the Wöhler curve. Experiments showed that the modeled Wöhler curve provides the best possible description of the vibration resistance based on the data from the identified experience components and thus the load capacity of the current component.

4.3.3.2 Predicting components' residual life

In order to predict the component's residual life, load data detected during the usage phase of the current component as well as all of the previously used and already failed components is analyzed. This data is stored in the experience database and is read from there as required. In its original version developed by van Thiel [13] the method described in the following sections evaluates uniaxial load data and has thus been extended by Quirico et al. [14] for the evaluation of multiaxial load data.

To determine the life span based on the component's loads, the loads experienced by the current component are compared with the load data from the experience database, in order to ascertain if they are typical. So that the condition of the component can be assessed and compared with other components, the loads have to be converted into key figures. Here, the linear accumulated damage is commonly used for quantification. However, this approach only considers the load level value [27]. Research has shown though that the sequence of the loads and time they arise impact the damage to a component [28]. The linear accumulated damage approach was thus extended in view of describing the condition. First, time-dependent loads are collectively depicted. Subsequently, an individual damage curve is calculated for the component, see Section 4.3.3.2.1. The determined theoretical damages form the basis for the later classification and comparability of the component damage curves from the experience database, see Section 4.3.3.2.2. The statistical prediction of the component's residual life then occurs based on this classification, see Section 4.3.3.2.3.

4.3.3.2.1 Determining damage curves for individual components

The load data from the component's service life that is detected and stored in the experience database includes information about the number, level, and time points of cyclical loads. For simplification, these are summarized into load collectives, using an extension of rainflow counting. In its original form rainflow counting only takes into account the load levels and frequency [29]. In order to also consider the sequence and time at which loads are experienced, the existing approach was extended to include analysis of individual cycles. The load levels are placed in a row in the sequence that

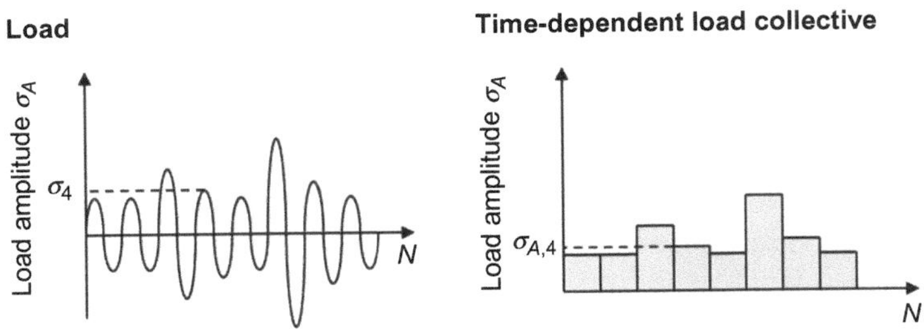

Fig. 4.3.6 Time-dependent load collectives [13].

they arise as load increments ($i = 1, \ldots, n$) in a time-dependent load collective, see Fig. 4.3.6. This is done for each spatial orientation ($j = x, y, z$) and is a simple technique for quantitatively depicting the experienced loads. Each load level corresponds to a load cycle.

This is accomplished by means of the experience-based Wöhler curve (see Section 4.2.2.2), since it takes into account the individual manufacturing conditions and provides the best possible description of the expected resistance.

Following Haibach [28], partial damages can be calculated for every load level in the time-dependent load collective, see Eq. (4.3.7).

$$d_{i,j} = \frac{n_j}{N_{j,i}} \tag{4.3.7}$$

where $n_{i,j}$ is the experienced load per load level and per spatial orientation, $N_{i,j}$ is the bearable load per load level and per spatial orientation, $d_{i,j}$ is the partial damages per load level and per spatial orientation, i is the load level ($i = 1, \ldots, n$), and j is the spatial orientation ($j = x, y, z$).

According to Eq. (4.3.7), the partial damages $d_{i,j}$ per load level i and per spatial orientation j can be calculated from the respective experienced load $n_{i,j}$ and the maximum bearable load $N_{i,j}$ (taken from the Wöhler curve). Subsequently, the partial damages calculated for each load level per different spatial orientation are used to determine a resulting damage vector. These damage vectors are then placed cumulatively in a row, so that a three-dimensional damage trend results, see Fig. 4.3.7. The established damage trend is specific to each component.

4.3.3.2.2 *Classifying damage trends and identifying comparable components*

The extent and frequency of the partial damages directly impact the total component damage. Moreover, the time at which an increased load arises and the sequence in which different higher loads occur influence a component's service life. In order to

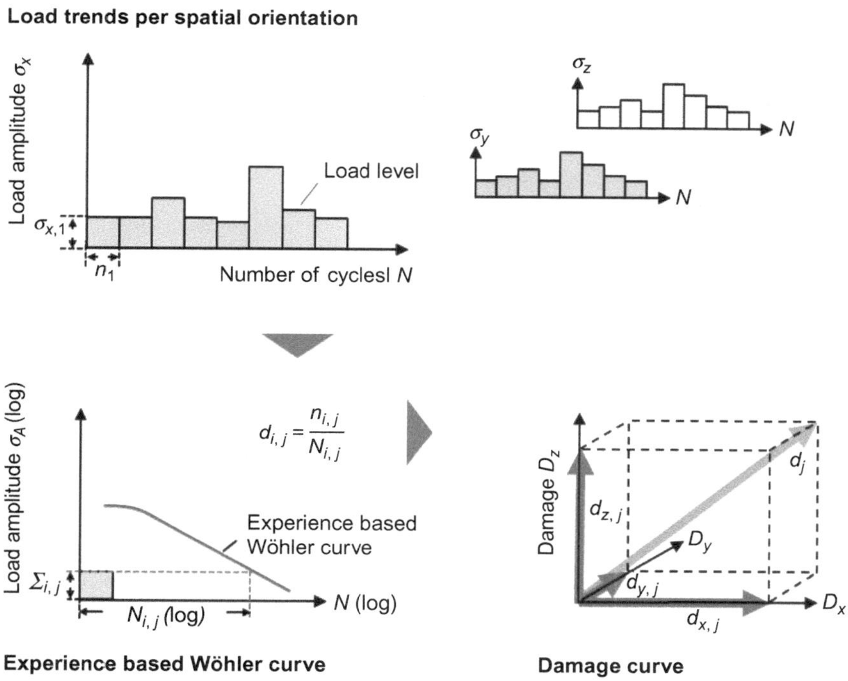

Fig. 4.3.7 Determining the component-specific damage trend [13,14].

be able to classify the component-specific damage trends, each of these influences has to be taken into account. Criteria used for classifying include the maximum damage per spatial orientation $D_{max,j}$, the mean damage per spatial orientation per range $D_{R,m,\,j}$, and the standard deviation of the mean damage per spatial orientation per range $\sigma_{R,j}$. For purposes of simplification, a number of load cycles can be combined into equidistant ranges. Thus for $j = x, y, z$ and $i = 1, \ldots, n$ nine criteria result for classifying multiaxial damage trends.

The maximum damage per spatial orientation is determined from the total partial damages per spatial orientation according to Eq. (4.3.8).

$$D_{max,j} = max\left\{d_{j,i}\right\} \tag{4.3.8}$$

where $D_{max,j}$ is the maximum damage per spatial orientation, $d_{j,i}$ is the partial damage per spatial orientation, j is the spatial orientation ($j = x, y, z$), and i is the load level ($i = 1, \ldots, n$).

In addition to the load peaks per spatial orientation, the mean partial damages per range are also considered classification features, see Eq. (4.3.9). Through the differentiation, not only are the extent and frequency of the loads and resulting partial damages taken into account, but also the sequence and times at which they arise. A similar course for the individual damage graphs for different components can consequently be derived based on this criterion. The number of ranges results from the desired number of cycles per range. The fewer cycles combined into a range, the more precise the classification.

$$D_{m,B,j} = \frac{\sum_a^b d_{i,j}}{b-a} \tag{4.3.9}$$

where $D_{m,B,j}$ is the mean partial damage per range per spatial orientation, $d_{i,j}$ is the partial damage per load level and spatial orientation, i is the load level ($i = 1, \ldots, n$), j is the spatial orientation ($j = x,y,z$), a is the lower interval limit of range, and b is the upper interval limit of range.

Through the mean partial damage per range per spatial orientation the component's damage trend is already very precisely classified; however, the standard deviation of the mean damage per range $\sigma_{B,j}$ is also determined as an additional classification feature. This describes the dispersion or variance of the partial damages per spatial orientation per range and thus indicates if components were subject to a constant or fluctuating load. It is calculated according to Eq. (4.3.10).

$$\sigma_{B,j} = \sqrt{\frac{1}{b-a} \cdot \sum_a^b \left(d_{j,i} - D_{m,B,j}\right)^2} \tag{4.3.10}$$

where $\sigma_{B,j}$ is the standard deviation per spatial orientation per range, $d_{i,j}$ is the partial damage per load level per spatial orientation, $D_{m,B,j}$ is the mean partial damage per range per spatial orientation, j is the spatial orientation ($j = x,y,z$), i is the load level ($i = 1, \ldots, n$), a is the lower interval limit of range, and b is the upper interval limit of range.

Based on the described criteria, the multidimensional damage trends for all n components stored in the experience database can be classified and compared. In the experience database, components with similar damage trends in comparison to the current component can be identified by entering a maximum tolerated percentage of deviation of the values per criterion. There are referred to as comparable components and form the basis for statistically predicting residual life.

4.3.3.2.3 Statistically predicting residual life

To predict the residual life, the damage trends of the identified comparable components are statistically analyzed. Up until the cycle number of the observed time point $N_{current}$, the damage trends of the comparable components are similar to the trend of the current component according to the applied criteria. In order

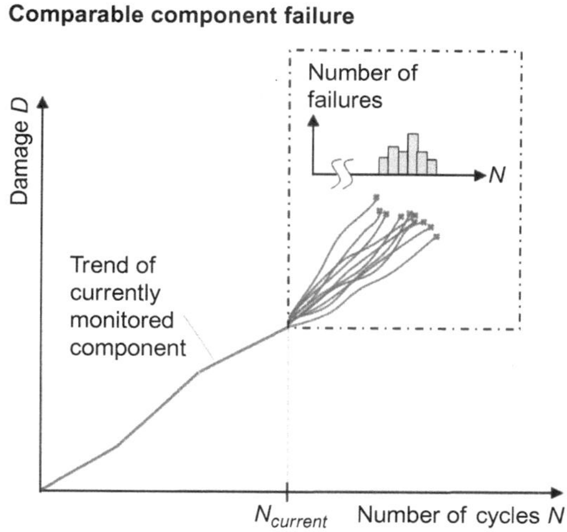

Fig. 4.3.8 Damage trends for identified comparable components [13].

to conduct a statistical analysis, the damage trends of the comparable components from $N_{current}$ on, stored in the experience database, are considered, see Fig. 4.3.8. Due to the close similarity, it is assumed that the loads will be comparable not only in the past but also in the future.

To predict the residual life, the times at which the comparable components fail are transferred to a histogram. The span can be subdivided into individual equidistant failure classes C. The number of failures per class corresponds to the frequency. The ratio of the frequency to the number of comparable components corresponds to the relative frequency of failures per class. From there, the relative cumulative frequency is calculated, which is taken into account as the empirical probability of a failure P. Consequently, the statistical analysis of the damage trends for the comparable components does not indicate an exact breakdown point, but rather marks the period where failure is probable. Based on the maintenance goals, the critical number of cycles until failure that corresponds to a definable probability of failure can be read from the empirical failure probability curve, see Fig. 4.3.9. This is then the recommended replacement time for the component.

In order to improve residual life predictions when there are a small number of comparable components, optional knowledge about the materials used can be taken into account in the earlier described derivation of the probability of a failure. For this extension, it is necessary to have an existing store of data garnered from experimental trials about materials' behaviors (e.g., rotating bar fatigue tests). Within CRC 653 trials such as these were conducted, for example, by

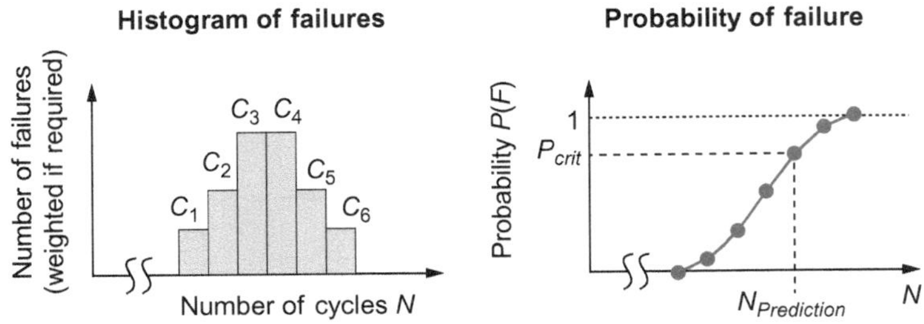

Fig. 4.3.9 Determining the probability of failure [13].

Demminger and Klose [30] in regards to the fatigue strength of magnetic magnesium alloys.

The range of the residual life with different load levels can generally be characterized by the mean and standard deviation. However, for predicting the residual life, the expected period during which failure is probable for each load level is particularly relevant. For this, the cracked surfaces of the samples are analyzed after experimental loads, in order to determine if a premature failure is expected due to the material, or if the sample's failure behavior can be considered representative. The results can then be used to derive representative periods during which failure is probable for each of the load levels.

To increase the quality of the former described residual life prediction [13,14], components with premature failures and components with representative failure can be considered on a differentiated basis. In this case, weighted frequencies are to be used in the statistical analysis for determining the probability of failure and not, as was previously described, absolute frequencies. The respective weighting is derived from the ratio of the number of premature failures to the total number of failures, see Eqs. (4.3.11), (4.3.12). Failures can be divided into early or representative based on the experimentally determined failure periods. Furthermore, after components have failed, they can be examined, allocated to one of the two categories based on the results and recorded accordingly in the experience database.

$$\omega_p = \frac{n_p}{n_f} \tag{4.3.11}$$

$$\omega_r = 1 - \frac{n_p}{n_f} \tag{4.3.12}$$

where ω_p is the weighting for premature failure, ω_r is the weighting for representative failure, n_p is the number of premature failures, and n_f is the total number of failures.

The approach described here for predicting components' residual life forms the basis for efficient maintenance planning with regards to the expected failure behavior of components at risk of fatigue. The high quality of predictions has been experientially proven by Winkens and Van Thiel, please refer to [12,13] for a detailed description of the validation.

4.3.3.3 Selecting suitable maintenance measures

In order to select suitable maintenance measures after the condition of the component has been evaluated, Winkens' [15] applied case-based reasoning [31]. Case-based reasoning postulates the selection of solutions, which have previously proven to be useful for a similar problem. If necessary, these solutions can be adapted to the current problem before being applied. Case-based reasoning is implemented using the component histories stored in the experience database. In addition to the load data, histories contain information about the maintenance measures conducted on all components. The maintenance measure is verified according to the control loop depicted in Fig. 4.3.10.

First, it is necessary to identify identical or similar cases in the experience database. To do so, the current component is compared with components stored in the database by means of various criteria. These include, for example, the number of hours it has already been in operation as well as the predicted service life. In addition to that, the experienced load is compared for all of the spatial orientations using the maximum load, the mean load, and the standard deviation. Based on the criteria, an overall degree of agreement is calculated for each of the identical or similar components. By extending the method [14], individual criteria can be weighted, allowing critical load directions to be given particular consideration.

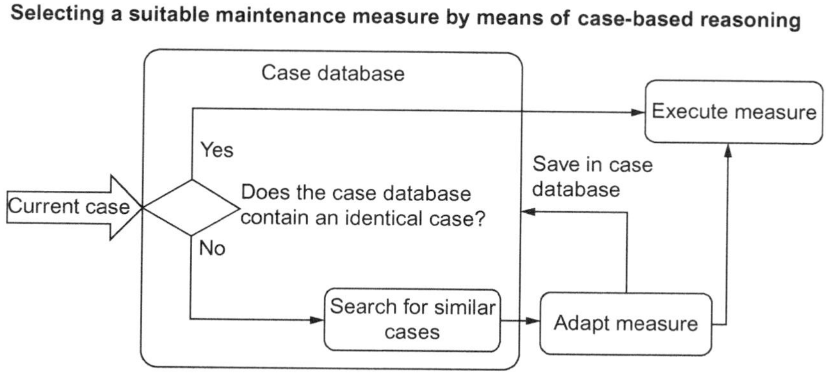

Fig. 4.3.10 Selecting suitable maintenance measures [15].

The result is a list of comparable cases from the experience database, sorted in descending order according to the degree of agreement. This provides support to maintenance staff, who in the end make the decision in accordance with knowledge gained from their own experience. Selected measures are implemented and then entered in the component history.

4.3.4 Conclusion

For production companies, efficient maintenance is increasingly critical for success [1]. Considering the positioning within the field of tension between high availability and low costs, condition-based maintenance, in particular, holds significant potential [4]. The growing permeation of production with information and communication technologies as well as the increasing availability of data allows a new approach to maintenance planning that allows the usage potential of components to get increasingly exhausted [9]. Within this context, a method for component condition-based maintenance was developed for maintaining components at risk of fatigue based on their condition. This method, which focuses on material fatigue as the cause of failure, is based on analyzing life cycle data from the manufacturing and service life phases. Using load data from the manufacturing, resistance can be determined based on experience [12]. While the component is in use, the residual life can be predicted based on these results and through statistical analyses of load data from the usage phase [13,14]. This makes it possible to efficiently plan maintenance times based on the component's condition. This experience-based method also provides support in selecting suitable maintenance measures [15]. This method thus makes it possible to effectively plan and implement condition-based maintenance.

References

[1] N.N., Instandhaltung – Zweitägiger Fachkongress bereitet Instandhalter auf die Herausforderungen von Industrie 4.0 vor, MaschinenMarkt 48 (2014) 16.

[2] DIN 31051:2012-09, Grundlagen der Instandhaltung, ICS-Notation, Beuth, Berlin, 2012.

[3] B. Bertsche, G. Lechner, Reliability in Automotive and Mechanical Engineering, Determination of Component and System Reliability, Springer, Berlin, Heidelberg, 2008.

[4] M. Strunz, Instandhaltung, Grundlagen Strategien Werkstätten, Springer, Berlin, Heidelberg, 2012.

[5] T. Bauernhansl, Die Vierte Industrielle Revolution – Der Weg in ein wertschaffendes Produktionsparadigma, in: T. Bauernhansl, M. ten Hompel, B. Vogel-Heuser (Eds.), Industrie 4.0 in Produktion, Automatisierung und Logistik, Anwendung Technologien Migration, Springer Fachmedien, Wiesbaden, 2014, pp. 5–35.

[6] G. Schuh, A. Kampker, B. Franzkoch, N. Wemhöner, Intelligent Maintenance, Potenziale zustandsorientierter Instandhaltung, Studie (WZL, IPT), Aachen, 2004.

[7] M. Oechsner, B. Pyttel, K.-H. Kloos, C. Berger, E1 Werkstoff und Bauteileigenschaften, in: K.-H. Grote, J. Feldhusen (Eds.), Dubbel, Taschenbuch für den Maschinenbau, 24. Auflage, Springer, Berlin, Heidelberg, 2014, pp. E2–E23.
[8] S. Zerbst, Global approach for early damage detection on rotor blades of wind energy converters, Dissertation Leibniz Universität Hannover, Mitteilungen des Instituts für Statik und Dynamik der Leibniz-Universität Hannover, 2011.
[9] M. Winkens, M. Schmidt, P. Nyhuis, Zustandsorientierte Instandhaltung - Potentiale bei der Lebensdauerprognose von Bauteilen im Sonderforschungsbereich 653, Werkstatttechnik Online 105 (3) (2015) 104–108.
[10] B. Denkena, H. Henning, L.-E. Lorenzen, Genetics and intelligence, new approaches in production engineering, Prod. Eng. 4 (1) (2010) 65–73.
[11] M. Schenk (Ed.), Instandhaltung technischer Systeme – Methoden und Werkzeuge zur Gewährleistung eines sicheren und wirtschaftlichen Anlagenbetriebs, Springer, Berlin, Heidelberg, 2010.
[12] M. Winkens, Methodik zur Prognose der Bauteillebensdauer unter Berücksichtigung von Fertigungseinflüssen, Dissertation Leibniz Universität Hannover, Berichte aus dem IFA, PZH-Verlag, Garbsen, 2017.
[13] B. van Thiel, Entwicklung einer Methodik zur Zustandsüberwachung von Bauteilen aus sensitiven Werkstoffen, Dissertation Leibniz Universität Hannover, Berichte aus dem IFA, PZH-Verlag, Garbsen, 2013.
[14] M. Quirico, M. Winkens, P. Nyhuis, Auswertung mehrachsiger Belastungszustände zur Lebensdauerprognose von gentelligenten Bauteilen, ZWF Zeitschrift für wirtschaftlichen Fabrikbetrieb 3 (2016) 109–112.
[15] M. Winkens, P. Nyhuis, Determining a suitable maintenance measure for gentelligent components using case-based reasoning, Int. J. Mech. Aerosp. Indust. Mechatron. Eng. 9 (1) (2015) 257–260.
[16] B. van Thiel, P. Nyhuis, Database for a gentelligent maintenance planning, in: Proceedings of the IJIE 2010, 15th Annual International Conference on Industrial Engineering—Theory, Applications & Practice, October 17th–20th, Mexico City, 2010.
[17] B. Breidenstein, Oberflächen und Randzonen hoch belasteter Bauteile, Dissertation Leibniz Universität Hannover, Berichte aus dem IFW, PZH-Verlag, Garbsen, 2011.
[18] E. Haibach, Betriebsfeste Bauteile, Ermittlung und Nachweis der Betriebsfestigkeit, konstruktive und unternehmerische Gesichtspunkte, Springer, Berlin, Heidelberg, 1992.
[19] W. Degner, H. Lutze, E. Smejkal, Theorie, Berechnung, Richtwerte, 17. aktualisierte Auflage, Carl Hanser Verlag, München, 2015.
[20] J. Köhler, Berechnung der Zerspankräfte bei variierenden Spanungsquerschnittsformen, Dissertation Leibniz Universität Hannover, Berichte aus dem IFW, PZH-Verlag, Garbsen, 2010.
[21] A. Westermann-Freidrich, H. Zenner, Zählverfahren zur Bildung von Kollektiven aus Zeitfunktionen, Vergleich der verschiedenen Verfahren und Bespiele, FVA-Merkblatt 0/14, Forschungsvereinigung Antriebstechnik, Frankfurt am Main, 1988.
[22] G. Schott, Werkstoffermüdung, Verhalten metallischer Werkstoffe unter wechselnden mechanischen und thermischen Beanspruchungen, 3. überarbeitete Auflage, VEB Deutscher Verlag für Grundstoffindustrie, Leipzig, 1977.
[23] V. Läpple, Einführung in die Festigkeitslehre, Lehr- und Übungsbuch mit 49 Tabellen, 2. Auflage, Vieweg+Teubner, Wiesbaden, 2008.

[24] DIN 50100:1978-02, Werkstoffprüfung, Dauerschwingversuch, Begriffe, Zeichen, Durchführung, Auswertung, Beuth, Berlin, 1978.
[25] S.N. Sivanandam, S.N. Deepa, Introduction to Genetic Algorithms, Springer, Berlin, Heidelberg, 2008.
[26] P. Mendes, K. Douglas, Non-linear optimization of biochemical pathways. Applications to metabolic engineering and parameter estimation, Bioinformatics 10 (14) (1998) 869–883.
[27] FVA-Richtlinie, Zählverfahren zur Bildung von Kollektiven und Matrizen aus Zeitfunktionen, Forschungsvereinigung Antriebstechnik, Frankfurt am Main, 2011.
[28] E. Haibach, Betriebsfestigkeit, Verfahren und Daten zur Bauteilberechnung, 3. Auflage, Springer, Berlin, Heidelberg, 2006.
[29] DIN 45667, Classification methods for evaluation of random vibrations, DIN German Institute for Standardization, 1969.
[30] C. Demminger, C. Klose, H.J. Maier, Mechanical Properties and fatigue strength of extruded cobalt-containing magnetic magnesium alloys, in: N.S. Kiran et al., (Eds.), Magnesium Technology, Proceedings of a Symposium Sponsored by Magnesium Committee of the Light Metals Division of The Minerals, Metals & Materials Society (TMS), 146th Annual Meeting & Exhibition, February 26th–March 2nd, San Diego, California, USA, 2017.
[31] B. Freudenthaler, Case-based Reasoning (CBR): Grundlagen und ausgewählte Anwendungsgebiete des fallbasierten Schließens, VDM Verlag Dr, Müller, Saarbrücken, 2008.

Further Reading

C. Demminger, C. Klose, P. Taptimthon, H.J. Maier, Material-inherent data storage using magnetic magnesium-cobalt alloys, in: Proc. 2nd International Conference on System-Integrated Intelligence: Challenges for Product and Production Engineering, SysInt 2014, July 2nd–4th, Bremen, Germany, 2014, pp. 188–197.
H.-C. Möhring, K. Litwinski, O. Gümmer, Process monitoring with sensory machine tool components, CIRP Ann. Manuf. Technol. 59 (1) (2010) 383–386.
I. Mozgova, C. Demminger, M. Quirico, F. Uhlich, B. Denkena, R. Lachmayer, H.-J. Maier, P. Nyhuis, The concept of technical inheritance in operation, analysis of the information flow in the lifecycle of smart products, in: Proc. 3rd International Conference on System-Integrated Intelligence: New Challenges for Product and Production Engineering, SysInt 2016, June 13th–15th, Paderborn, Germany, 2016, pp. 79–88.

Examples of application

5

5.1 Storing data within materials and its potential during production and product life cycle

Benjamin Spitschan
Institute of Information Processing, Leibniz Universität Hannover, Hannover, Germany

A smart production demands for smart components, which are capable of identifying themselves at any time. Furthermore, relevant information on this particular component must be available not only during manufacturing but also at any time during the life cycle of the product. The CRC 653 meets these requirements with its vision of merging a physical component and corresponding data. Relevant information on a component such as its manufacturer and its purpose, the date of its next overhaul, or details about unusual occurrences during its production or application will always be available. To make this vision come true, advanced technologies have been developed within the CRC 653, enabling data storage within a component. These technologies not only make use of magnetic effects but also exploit geometrical properties as well as properties of the material's microstructure to realize read only and random access memories.

Both the optimization of the production process itself and the process planning show huge potentials for utilizing in-process data. In a manufacturing environment, it is often necessary to identify individual products and components during their journey through the different stages of the production chain. Unique identification of individual components is a second major example for the use of inherent information for production and during a component's life cycle.

For instance, the production of a particular gearwheel requires it to pass through several manufacturing stations. Following an initial forming step, the semifinished part is subject to a machining process such as turning and thus gets its final geometry. After each production step, the component is deposited in a warehouse. Subsequently, it acquires additional properties and material characteristics in a finishing step by being hardened and tempered. Additionally, coating is applied in order to improve physical and chemical properties of the surface. Further subsequent steps of a component's life cycle follow, such as quality

Cyber-Physical and Gentelligent Systems in Manufacturing and Life Cycle. http://dx.doi.org/10.1016/B978-0-12-811939-6.00005-4

control or logistics, e.g., warehousing and dispatch to the customer, and long-term monitoring during maintenance (also see Chapter 4). Each of the manufacturing stations needs to operate with a set of parameters. The way when and how often these manufacturing parameters need to be exchanged, however, differs strongly depending on the batch sizes. Traditional high-volume mass production has long setup times and therefore seldom requires reconfigurations; a continuous closed-loop optimization of process parameters, though, is often performed (also see Chapter 5.2). Smaller batches usually travel through the manufacturing process closely together, e.g., stacked in containers. The appropriate parameter set is commonly attached to these containers, either physically carrying all necessary information on paper sheets (kanban cards) or electronically by using an identifier on the container such as the batch number that allows to retrieve the corresponding parameter record from a database. Nowadays, production scenarios implementing the lot-size one paradigm gain increasing importance. In these environments, each component exhibits different properties such as geometry, surface quality, material characteristics, etc. This implies that at every station the parameter set of the machine must be adapted before each single component is processed, increasing the data handling requirements. To facilitate this permanent switching of parameter sets, they need to be closely linked to that very corresponding component. Automatic technologies to establish and to read out identifiers from physical objects are generally known as *Auto-ID* (see Fig. 5.1.1). Auto-ID, also called Automatic Identification and Data Capture (AIDC), is the automatic identification of physical objects, the collection of accompanying data, and the ingress into computer systems. One primary application is the self-identification of components, i.e., the emission of an identifier such as an identification number upon request. This close tie between a workpiece

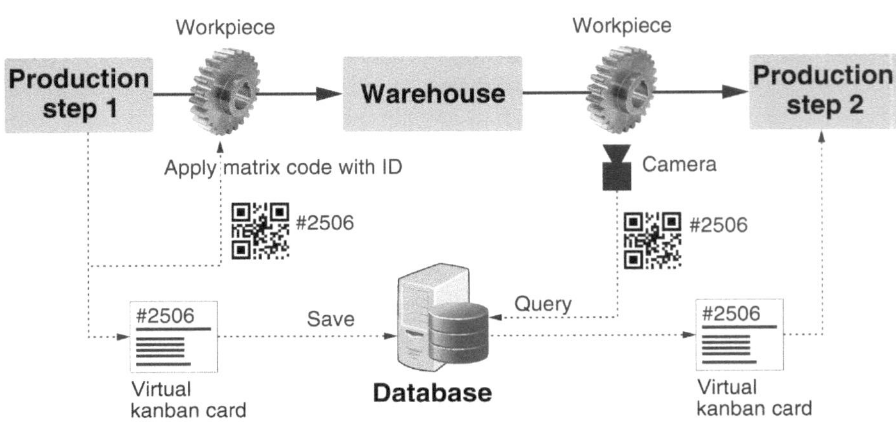

Fig. 5.1.1 Example of an Auto-ID workflow using matrix codes.

and its associated information enables the component itself to control production when it passes through the production chain.

Previous approaches use identifiers such as two-dimensional matrix codes or RFID (radio frequency identification) tags that are attached to the component. Especially in case of high volumes and low piece costs, these solutions may turn out to be too costly, too complex, too time consuming, or simply too large in dimension in many instances. Moreover, the inclusion of an additional station performing the attachment or application of these conventional identifiers may be expensive in terms of capital investment, cycle time, or floor space. In Fig. 5.1.1, a workflow is depicted in which matrix codes such as QR or DataMatrix codes serve as identifiers for Auto-ID use cases. A fully intrinsic solution is shown in Chapter 2.1.2. It allows for "engraving" data into the surface during a machining process in the manufacturing stage of a component directly using the machine tool. To accommodate for transmission of electrical energy, a slip ring is used. During operation, piezo actuators displace the tool axially at high frequencies and result in the cutting inserts being offset according to the excitation signal. No additional stations are required for writing data. For reading engraved information, off-the-shelf image processing algorithms can be used if appropriate data preprocessing, for instance, rearrangement of information bits into matrix codes, channel coding, and modulation, is carried out prior to engraving. Suitable algorithms for decoding matrix codes can be found frequently in specialized contemporary Auto-ID readers or even in software-based solutions that are available on multipurpose handheld devices such as generic end-user smartphones.

A fingerprinting method using the surface topography of a component for passive identifiers is presented in Chapter 2.1.1. In a markerless Auto-ID system, no additional tags or markers need to be physically attached or applied to the component. In particular in conjunction with high-volume components such as simple screws, etc., individual tracing of single parts will become feasible since any additional marker that would increase the costs significantly is obsolete. A part, e.g., a gearwheel, is first manufactured in a machining process, i.e., turning. Directly after this manufacturing step, the gearwheel's surface is scanned at a defined location and a "fingerprint" is extracted subsequently. The relationship between the physical surface and a newly allocated serial identification number is now perpetually established. At any later point in time, the component's surface is scanned again, and its ID number is obtained by extracting the fingerprint from the surface and matching it with the database (Fig. 5.1.2).

Depending on the application, specialized storage techniques are required. A novel method for applying robust, invisible identifiers onto components is presented in Chapter 2.2.3. It allows for placing inherent component markers by modification of the local microstructure. By using fiber lasers, a spatially confined heat treatment is carried out which allows well-defined adjustments of graded properties of the microstructure. Exploiting the depth and the strength of the microstructural change as an additional dimension, the achieved data density

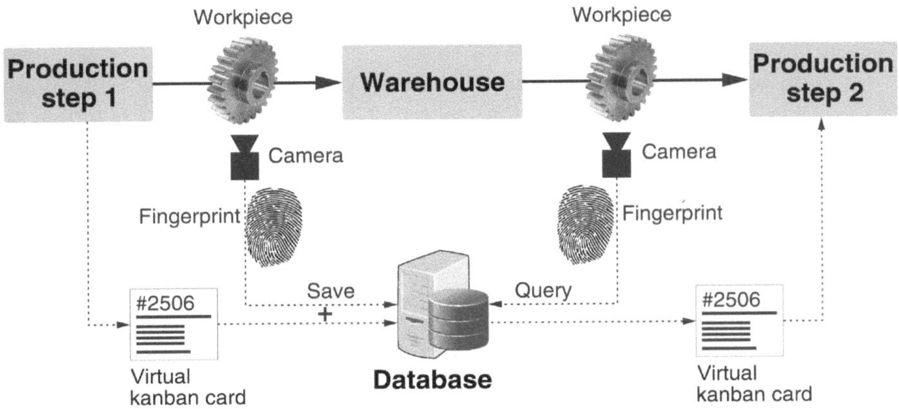

Fig. 5.1.2 Example of a markerless Auto-ID workflow using intrinsic surface fingerprints.

can be significantly increased compared to binary two-dimensional approaches. Thus, tamperproof, robust product markers in the subsurface can be created. Those markers are to a large extent insusceptible to external influences, both mechanical and thermal. The presented technology even works underneath coating and lacquering layers. When needed at a later time, the information carried by microstructural changes can be read out again with high-resolution eddy current sensors. For ease of operation and increased speed, a miniaturized sensor array was developed, facilitating fast and secure read out.

A system based upon magnetic storage is shown in Chapter 2.3.3. Inherent magnetic data recording that is carried out directly on the surface of technical components is highly promising, especially when the components are employed in harsh environments. Compared to other Auto-ID technologies, a significant advantage of magnetic storage is that it is rewritable and can be updated during the component's life cycle. Also, magnetically stored data are resistant to weathering. High temperatures and high magnetic field intensities, however, are able to degrade or even delete magnetically stored data. This limitation can be mitigated using a medium with higher coercivity which can withstand strong external magnetic fields and high temperatures. As a consequence of higher coercivity, a higher recording field is required to magnetize the medium. A heat-assisted magnetic recording (HAMR) system was developed as a way to temporarily reduce coercivity of a recording medium during writing in order to realize the required field intensities. The HAMR method shows high potential for further increasing the achieved data density and enables rewritable component-inherent data storage in "Industry 4.0" scenarios for the first time.

5.2 Optimization of production processes with sensitive machines and workpieces

Berend Denkena, Florian Uhlich, Haythem Boujnah
Institute of Production Engineering and Machine Tools, Leibniz Universität Hannover, Garbsen, Germany

Increasing competitive pressure and rising demands on small-lot production put high requirements on today's manufacturing technologies. On the one hand this requires the flexibility of a workshop production, whereas on the other hand the efficiency of a high-volume production is needed to meet the customers' pricing requirements. Digitization of the production meets this challenge and depicts a vision of smart factories and smart products. In contrast to the current situation, parts and machines will be independent entities that are capable to sense their own condition. Within this vision both, production machines and workpieces, communicate and share information, perceptions, and knowledge among each other. Thus, production processes can be monitored with detailed information measured by the machine tool or the workpiece itself and a comprehensive picture of every manufacturing process arises.

Key elements to turn this vision into reality are the development and application of advanced sensor and communication technologies. Their realization and integration enables implementation of sensory components directly into the flux of force and closely located to the process. Therefore within the CRC 653, machine tools are enabled to sense vibrations and forces, which occur during machining processes. In milling, clamping systems (see Chapter 3.2.3) and slides carrying the work spindle are most suitable for such applications (see Chapter 3.2.1). Within the research of the CRC 653, the load detection is realized by strain measurements. Machine tool components usually have a high stiffness in order to achieve the requirements for high machining accuracy and process stability. Therefore the strains in the structure are low and can hence hardly be measured. The development of micro strain gauges (see Chapters 2.4.1, 2.4.2, and 2.4.3) which are applied, for instance, in small notches of the slide represents a promising approach to improve sensitivity as well as the sensor integration into the machine tool component. Equipped with several strain sensors, the sensory slide is able to detect process forces. The provided process signals can be used in many applications, such as the optimization of machine components and represent a fundamental information source for process and condition monitoring systems. They form a basis for the following prototypical application, which has been put into practice by the CRC 653 (see Fig. 5.2.1, left and center).

With respect to milling processes, the forces detected by the sensory slide allow for the identification of the static tool stiffness and enable the online in-process detection of the tool deflection. A measurement of the tool stiffness is executed in the machine

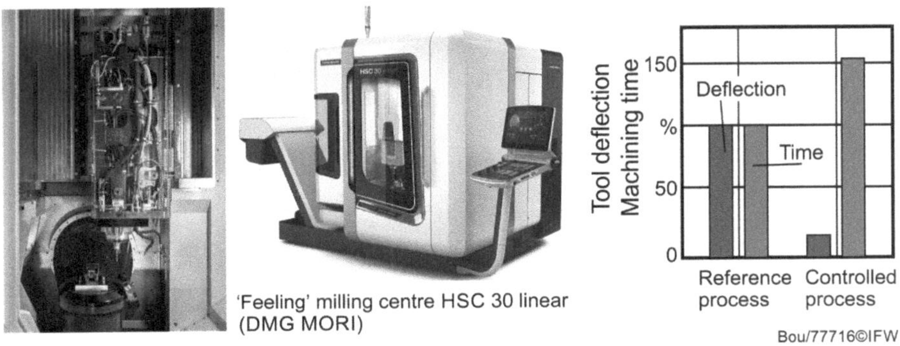

Fig. 5.2.1 Integration of advanced microsensors for the measurement of process forces. Left: Sensory spindle slide. Center: "Feeling" machine tool. Right: Reduction of deflection due to controlled process.

by a controlled soft collision of tool and workpiece. With the known set distance and the measured variation of the collision force, the tool stiffness is calculated and used for deflection estimation. Based on the communication between monitoring system and machine control, tool deflection is regulated online in the milling center by adjusting axis feed or axis position. The approach shows considerable improvement regarding surface accuracies (see Fig. 5.2.1, right).

Furthermore, in analogy to a human worker's ability to "feel" and "react," "feeling" machines are able to observe their state of processing themselves. As they are enabled to report important incidents, appropriate actions can be taken if necessary. By storing past "experiences" and making them accessible for analyses, the manufacturing system is equipped with a memory. This memory is accessed and evaluated by appropriate algorithms. By comparing past process parameters and results, production knowledge arises and can be used for subsequent processes. Before manufacturing data, for instance, process forces and quality information, can be analyzed they have to be put in an applicable context. On this account, the so-called *Virtual Planner* has been developed. It uses analytical forecast models as well as simulations that allow for predictions of the workpiece characteristics based on raw data. By using data mining approaches, empirical models are built up that model causal relationships and correlations. The learned relations support the decision-making for process design by recommending appropriate process parameters. As soon as new data are available, the models are refined before the consequent part is manufactured. If necessary, the *Virtual Planner* can interfere running processes to adjust parameters (see Chapter 3.1.3).

Based on the same process signals, teachless online process monitoring and control systems are presented in Chapter 3.2.4. The main goal is to increase process security of production facilities for single-item production. This, for example, allows for the production of individualized products resulting from the adaptive technology (see Chapter 4.2). The presented approaches offer robust and reliable monitoring for a production starting with the first manufactured item. To substitute ramp-up processes that were necessary so far, a reference for parameterization of monitoring limits is now simulated during the process planning phase. Relevant information from process planning is used to simulate monitoring limits. Hence, teaching of the monitoring system during ramp-up is completely substituted by a simulation during process planning. Thus, an NC-based simulation is not only used to validate and optimize the planned process, but it also provides process conditions to calibrate the monitoring limits (see Fig. 5.2.2).

Another example for an application of a sensitive tool for an optimization of a production process is a "feeling" forging die. It has been developed allowing for a bulk metal forming process to be monitored (see Fig. 5.2.3). Based on integrated cooling channels, the process can be controlled (see Chapter 3.2.6).

A regulated temperature control for dies permits an effective prevention of defects as well as a reduction of process interruptions by means of temperature stabilization. A reduction of thermally induced wear leads to a favorable decrease of the tool costs. Such a die temperature control is of particular importance for processes like precision forging in which geometrical integrity of produced parts is the main concern.

Fig. 5.2.2 Implemented monitoring system. External screen visualizes machine signals and confidence limits for online process fault detection.

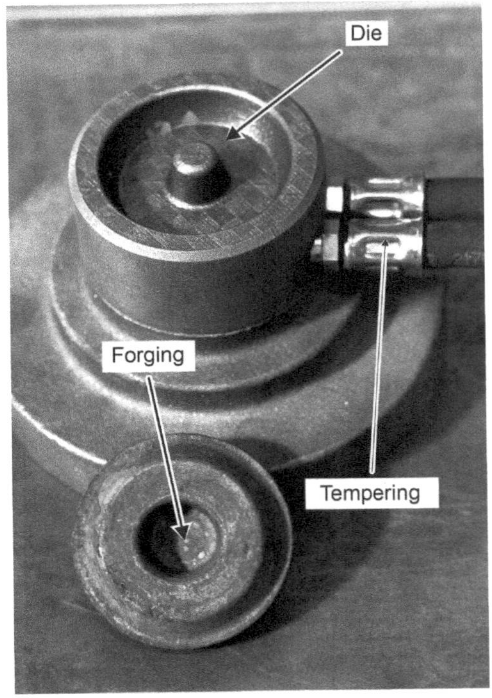

Fig. 5.2.3 Prototype of a "feeling" forging die. Cavities within the die enable controlled tempering.

5.3 Benefits of cyberphysical systems in work and process planning

Florian Winter, Quang Huy Dao†, Philipp von Witzendorff‡*
*Institute of Production Engineering and Machine Tools, Leibniz Universität Hannover, Garbsen, Germany, †Institute of Microwave and Wireless Systems, Leibniz Universität Hannover, Hannover, Germany, ‡Laser Zentrum Hannover e.V., Hannover, Germany

High productivity and product quality are essential to succeed with international competition. Companies are especially challenged by the increasing trend of individualized manufacturing and a greater importance of flexibility. In single-item and small-lot production, this tendency leads to a vast variety of products and a large number of possible processes and process chains. Particularly in the mentioned case of single-item and small-lot production, decisions concerning production are often based on unstructured practical knowledge. In order to find the optimal solution, this knowledge can be extended by an effective data analysis and process monitoring. The CRC 653's vision of a cross-linked production pursues the controlling and adjustment of the manufacturing process based on real-time data communicated by machine and workpiece. This information is processed by a so-called *Virtual Planner*. Therefore a reliable data transmission and an exact capturing of process data are required. To clarify the role of communication between workpieces and the *Virtual Planner* an analogy can be used. The driver of a car can choose between different routes to reach his destination. Indeed, all routes are limited by a certain capacity and as a consequence some dynamic effects like traffic jams can occur. According to the preferences (e.g., fastest way, shortest way, most cost-effective route), the selection of the "best" route is a decision based on the dynamic effects and their interdependencies. In order to optimize the flow of cars through a city, the vision of autonomous driving has been developed. All cars communicate information such as their current position, desired destination, or weather effects for the purpose of finding the best collaborative solution for the whole traffic. Because interruptions like closed roads within the traffic can occur, the system needs to be dynamic in terms of route planning, which means that the route of one single car may change while it is already on its way. To evaluate the current traffic, information on every car needs to be collected and communicated. The same effects and aims exist within a modern job shop manufacturing system. Several jobs that differ from each other have to be carried out by means of a limited amount of manufacturing resources, e.g., machines. Due to these limited resources, queues and interruptions in the manufacturing system may occur, comparable to traffic jams.

However, which elements are missing for implementing a modern job shop manufacturing system comparable to modern car navigation systems? One has to have the possibility to incorporate specific data, store them on the workpiece, and communicate it if required. Moreover, communication and evaluation should occur in real time. Based on these data, the workpiece is able to choose the optimal process chain and optimal process parameters autonomously as it is demonstrated in Chapter 3.1.1. For these purposes, some of the major aims are the planning, benchmarking, and controlling of jobs within such a manufacturing system.

First of all, the workpiece needs detailed information about all possible sequences of processing steps. An algorithm which calculates all possible combinations of processing steps has been developed. Design features of the workpieces can be processed by different combinations of machines and tools. This information is stored within an ontology which is able to draw conclusions, e.g., to extend the amount of processing possibilities. The resulting process chains are called nonlinear process plans resembling a tree structure. An adaptive process planning is enabled by this nonlinear process plan. The plan is generated automatically and contains all alternative routes of the manufacturing system. It creates the possibility to react very fast in case of unavailability or disturbance of a machine. This feature is very similar to car navigation systems and is comparable to find an alternative routing in case of roadblocks. Fig. 5.3.1 depicts that process steps 2 and 3 have alternative process steps $2'$ and $3'$. The preferred process step 2 is undisturbed and available, therefore the workpiece chooses this way. Process step 3 is unavailable. Therefore the workpiece automatically uses the alternative process step $3'$.

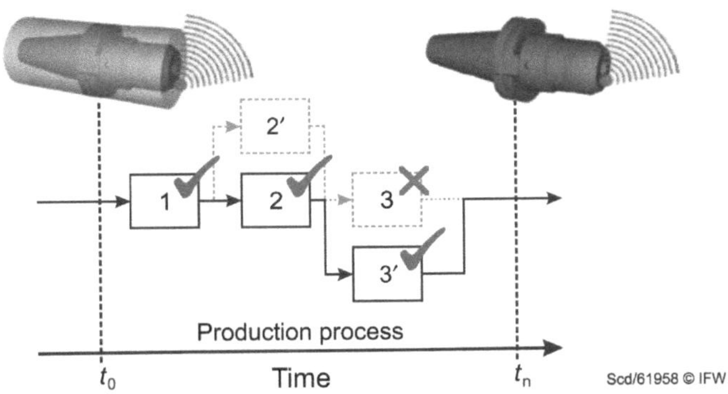

Fig. 5.3.1 Adaptive process planning based on current state information.

Beyond the opportunity of determining all possible sequences of processing steps and reacting in case of inaccessible routes in the manufacturing system, adaptive process planning enables even more benefits. One of these benefits is the definition of the "best" accessible route and which criterion is important for that definition. Equally to a car navigation system which can define the fastest, shortest, or most cost-effective way, the adaptive process planning system has the ability to define specific preferences. According to the characteristics of the production order, the preferences—time, cost, quality—are set and the optimal process parameters are figured out by an Analytical Hierarchic Process (AHP) based on their importance.

Information on the actual state of the workpiece are recorded and stored directly on the workpiece during the manufacturing process. As a next step the information needs to be communicated by a wireless system. One of the major challenges of communication modules is size reduction. These units have to be miniaturized in order to be integrated in small workpieces, e.g., screws (see Fig. 5.3.2). Furthermore, a self-sufficient energy supply and a high resistance against several external influences like high temperatures, dust, and fluids are required. Moreover, the transmission range of the communication module is limited in order to prevent overlays and interferences inside a factory. Based on these requirements, a self-sufficient wireless sensor node is developed (see Chapter 2.5.1). It provides production data and information on process temperature. Furthermore, the amount of information can be extended in order to capture data about the acceleration and the position of the workpiece. An exact definition of the workpiece's position is important in the manufacturing process (e.g., small inaccuracies can lead to an unacceptable quality). In terms of power supply a batteryless optoelectronic solution with a high efficient solar cell is used as part of the energy harvesting concept. Thanks to the supercapacitor acting as an energy storage and the use of low-power components the communication module is able to process and transmit data for several minutes in a complete darkness condition. Gathered information by sensor systems can be stored in integrated memory and communicated

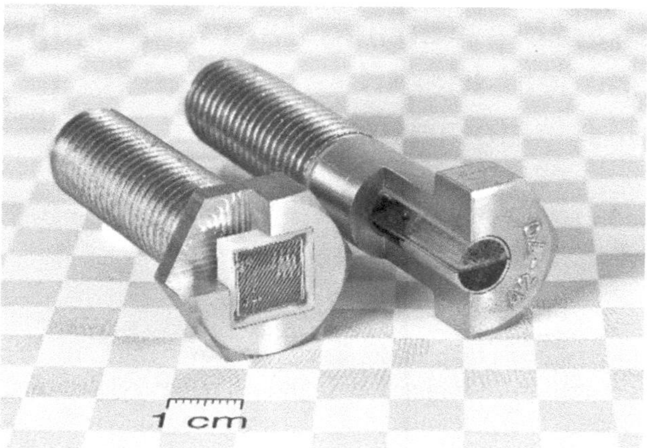

Fig. 5.3.2 Radio frequency communication systems integrated into screws; left: new design concept; right: old design.

via the transmitter unit consisting of an optically transparent radio frequency circuit and an antenna. A further special feature of this module is the highly integrated multilayer composition of electrical and optical components combined with a three-dimensional molded interconnect device (3D-MID).

Incidentally, the conjunction between operation scheduling and manufacturing control is visible. The manufacturing control is monitoring the real-time information of the workpiece and based on these data the operation scheduling adjusts the process parameters depending on the chosen preferences, which is processed by the developed *Virtual Planner*. Furthermore, the *Virtual Planner* utilizes self-learning models to describe the process parameters, demonstrated in Chapter 3.1.3. In the future, more sensors and more data about workpiece state, process state, and machine state are expected. The greater amount of data causes challenges concerning the analysis of data and the necessity of defining the relevant data for the individual propose. An economical handling and the realization of capable transmission are key aspects concerning the management of manufacturing systems in the future.

Referring to the example, the user of a car navigation system is interested in information about the chosen route and therefore the data are filtered. Based on the specific information, a prediction of the journey time with current status information is created. In case the user wants to know the expected journey time at a specific date in the future, the system is using experience-based data. The adaptive process planning works comparably. In order to find similar process steps, a clustering method is integrated and a structured data bank is provided. They guarantee the repatriation of experience-based and application-specific knowledge into the adaptive process planning.

5.4 Inseparably merging data and component for component identification and for protection against plagiarism

Benjamin Spitschan
Institute of Information Processing, Leibniz Universität Hannover, Hannover, Germany

Plagiarism of mechanical components has grown to significant amounts, causing immense losses especially in innovative, high-technology related industry branches. Falsified components also bear the risk of safety and liability issues. Plagiarism protection is therefore becoming increasingly important. This involves both plagiarism *prevention*, i.e., constructive means to impede counterfeiting, and plagiarism *detection*, i.e., the ex-postanalysis of a given component to determine whether it is original or not (see Fig. 5.4.1).

In Chapter 2.1.1 a method for extracting intrinsic passive identifiers from surface topographies of machined components is demonstrated. This patented system [1] allows for unambiguous identification of single individual components with a sufficient degree of surface roughness. Surface topographies of a component manufactured in a machining process, i.e., turning, milling, grinding, etc., exhibit two common features. On the one hand, the deterministic fraction accounts for the ideal nominal geometry plus manufacturing tolerances as well as waviness and cutting grooves. On the other hand, the stochastic fraction is caused by random micro-fluctuations of the local height that lead to deviations from the smooth surface.

Though usually undesirable in manufacturing, acceptable levels of roughness can be utilized for a "fingerprinting" scheme which provides true physical security. Distinct unique features in the stochastic fraction are then used to build a set of individual,

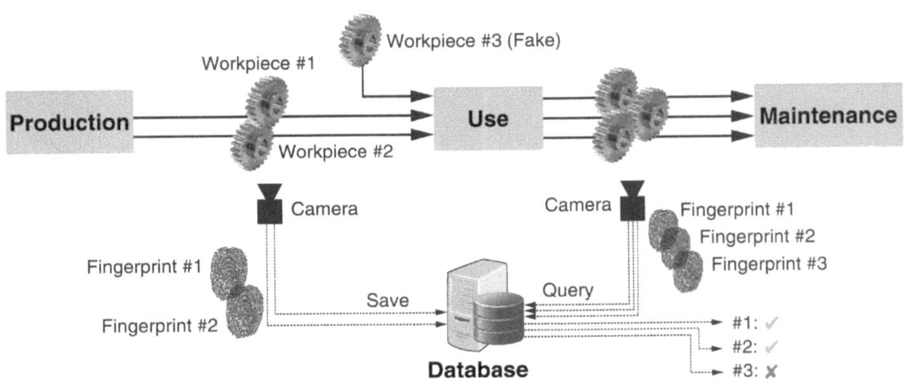

Fig. 5.4.1 Detection of a fake part during maintenance.

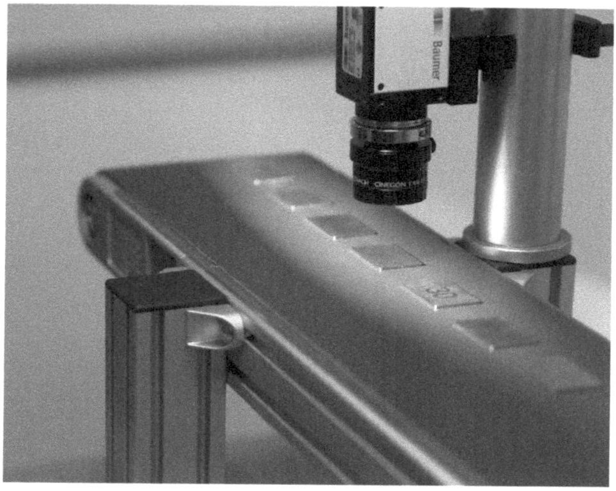

Fig. 5.4.2 Demonstration prototype implementing plagiarism detection of ground workpieces.

characteristic markers of a particular surface patch. Such a data set is saved as a record in a database and can be retrieved at any later point of time, alongside with any piece of information that is linked to that particular component, such as manufacturing parameter sets, data sheets, CAD drawings, process parameters, maintenance instructions, etc. Since it only exploits random information that was *passively* generated during machining, no additional station or modification of the production process is required.

A prototype for ground surfaces was realized (see Fig. 5.4.2), employing an industrial off-the-shelf camera for surface imaging and an embedded computing system for fingerprint analysis, data management, and database connectivity

It is able to detect fakes among 500 ground samples. For real-world applications in a production environment, however, the number of workpieces is not limited. Experiments with 200,000 simulated sample surfaces were successfully carried out. The processing time to determine whether a particular workpiece is original or not is less than 500 ms. The false-positive rate, i.e., the probability that a fake component is erroneously detected as original, was determined to be less than 10^{-20}.

Reference

[1] Patent WO2013024161 A1, Method, device and computer program for recognizing three-dimensional objects, 2013.

Index

Note: Page numbers followed by *f* indicate figures, and *t* indicate tables.

D

H

I

K

L

M

N

O

P

T

FSC
www.fsc.org
MIX
Papier | Fördert
gute Waldnutzung
FSC® C083411

Kolibri 360 GmbH
Ferdinand-Jühlke-Straße 7
99095 Erfurt, Deutschland
produktsicherheit@kolibri360.de